Advances in Intelligent Systems and Computing

Volume 230

Series Editor

Janusz Kacprzyk, Warsaw, Poland

For further volumes:
http://www.springer.com/series/11156

Józef Korbicz · Marek Kowal
Editors

Intelligent Systems in Technical and Medical Diagnostics

Springer

Editors
Józef Korbicz
Institute of Control and Computation Engineering
University of Zielona Góra
Zielona Góra
Poland

Marek Kowal
Institute of Control and Computation Engineering
University of Zielona Góra
Zielona Góra
Poland

ISSN 2194-5357 ISSN 2194-5365 (electronic)
ISBN 978-3-642-39880-3 ISBN 978-3-642-39881-0 (eBook)
DOI 10.1007/978-3-642-39881-0
Springer Heidelberg New York Dordrecht London

Library of Congress Control Number: 2013943919

Printed on acid-free paper

Springer is part of Springer Science+Business Media (www.springer.com)

Preface

For many years technical and medical diagnostics has been the area of intensive scientific research. It covers well-established topics as well as emerging developments in control engineering, artificial intelligence, applied mathematics, pattern recognition and statistics. At the same time, a growing number of applications of different fault diagnosis methods, especially in electrical, mechanical, chemical and medical engineering, is being observed. The rapidly increasing complexity of automation in industry as well as the continuing need to ensure reliability and safety at the highest level require ongoing research and development of innovative fault diagnosis approaches. Furthermore, special attention is paid to fault-tolerant and self-reconfiguring control systems, which are crucial wherever maintenance or repair cannot be realized immediately.

This book contains selected papers presented at the 11th International Conference on *Diagnostics of Processes and Systems, DPS 2013*, held in Łagów Lubuski, Poland, on 8–11 September 2013. The conference was organized by the Institute of Control and Computation Engineering of the University of Zielona Góra in cooperation with the Warsaw and Gdańsk Universities of Technology. The previous *DPS* conferences took place in Podkowa Leśna (1996), Łagów Lubuski (1997, 2001, 2013), Jurata (1998), Kazimierz Dolny (1999), Władysławowo (2003), Rajgród (2005), Słubice (2007), Gdańsk (2009) and Zamość (2011), and attracted a large number of authors as well as internationally recognized plenary speakers: Ronald J. Patton, Paul Frank, Tadeusz Kaczorek, Dominique Sauter, Ryszard Tadeusiewicz, Czesław Cempel, Jakob Stoustrup, Michèle Basseville, Jan Maciejowski, Eduardo F. Camacho, Antoni Nowakowski, Jan Lunze, Christopher Edwards, Louise Travé-Massuyès, Piotr Tatjewski, Steven Ding and Ewaryst Rafajłowicz.

In general, conference topics correspond to the subject area of the IFAC symposiums on *Fault Detection, Supervision and Safety for Technical Processes, SAFEPROCESS*, extended to the broad area of medical diagnostics. The aim of this collective work is to show the bridge between technical and medical diagnostics based on artificial intelligent methods and techniques. The book is divided into four parts:

I. Soft Computing in Technical Diagnostics,
II. Medical Diagnostics and Biometrics,
III. Robotics and Computer Vision,
IV. Various Problems of Technical Diagnostics.

I wish to thank all participants, plenary speakers and paper reviewers for their scientific and personal contribution to the conference. My particular appreciation goes to the authors of the papers published in this collective book.

Acknowledgment

I am grateful to the *DPS 2013* Organising Committee Chair, Marek Kowal, for his and his co-workers' enormous effort put into making the conference such a successful scientific event. I would also like to acknowledge the technical and administrative support of Paweł Kasza, Ewa Lehmann and Agnieszka Rożewska.

Zielona Góra, June 2013

Józef Korbicz
DPS 2013 Programme Committee Chair

Organisation

DPS 2013 was organised by the Lubuskie Scientific Society in Zielona Góra and the Institute of Control and Computation Engineering of the University of Zielona Góra, Poland.

DPS Chairs

General Chair

Józef Korbicz	University of Zielona Góra, Poland

Co-Chairs

Jan Maciej Kościelny	Warsaw University of Technology, Poland
Zdzisław Kowalczuk	Gdańsk University of Technology, Poland

DPS Programme Committee

Christophe Aubrun, France
Stanisław Bańka, Poland
Andrzej Bartoszewicz, Poland
Wojciech Batko, Poland
Piotr Bielawski, Poland
Liliana Byczkowska-Lipińska, Poland
Jõao Calado, Portugal
Alessandro Casavola, Italy
Paolo Castaldi, Italy
Czesław Cempel, Poland
Wojciech Cholewa, Poland
Vincent Cocquempot, France
Steven Ding, Germany
Stefan Domek, Poland
Christopher Edwards, UK
Krzysztof Fujarewicz, Poland
Jerzy Głuch, Poland
Ming Hou, UK
Krzysztof Janiszowski, Poland
Ireneusz Jóźwiak, Poland
Jacek Kabziński, Poland
Jacek Kluska, Poland
Krzysztof Kozłowski, Poland
Kazimierz Kosmowski, Poland
Dušan Krokavec, Slovakia
Andrzej Królikowski, Poland
Piotr Kulczycki, Poland
Krzysztof Latawiec, Poland

Krzysztof Lewenstein, Poland
Antoni Ligęza, Poland
Jan Lunze, Germany
Krzysztof Malinowski, Poland
Letitia Mirea, Romania
Wojciech Mitkowski, Poland
Wojciech Moczulski, Poland
Henrik Niemann, Denmark
Krzysztof Patan, Poland
Ronald Patton, UK
Andrzej Pieczyński, Poland
Vicenç Puig, Spain
Joseba Quevedo, Spain
Ewaryst Rafajłowicz, Poland
Leszek Rutkowski, Poland
Ryszard Rojek, Poland
José Sá da Costa, Portugal
Dominique Sauter, France
Miroslav Šimandl, Czech
Silvio Simani, Italy
Piotr Skrzypczyński, Poland
Jakob Stoustrup, Denmark
Piotr Szczepaniak, Poland
Roman Śmierzchalski, Poland
Mirosław Świercz, Poland
Andrzej Świerniak, Poland
Ryszard Tadeusiewicz, Poland
Piotr Tatjewski, Poland
Didier Theilliol, France
Leszek Trybus, Poland
Dariusz Uciński, Poland
Wiesław Wajs, Poland
Marcin Witczak, Poland
Marian Wysocki, Poland

DPS Organising Committee

Marek Kowal, Chair
Krzysztof Patan, Vice-Chair
Marcin Witczak
Andrzej Czajkowski
Marcel Luzar
Ewa Lehmann
Agnieszka Rożewska
Paweł Kasza

Sponsors and Patrons

Polish Ministry of Science and Higher Education
Polish Academy of Sciences, Committee on Automatic Control and Robotics
Lubuskie Scientific Society in Zielona Góra
University of Zielona Góra, Institute of Control and Computation Engineering

Contents

Part I
Soft Computing in Technical Diagnostics

Bridges between Diagnosis Theories from Control and AI Perspectives

L. Travé-Massuyès[1,2]

[1] CNRS, LAAS, 7, avenue du Colonel Roche, F-31400 Toulouse, France
[2] Univ de Toulouse, LAAS, F-31400 Toulouse, France
louise@laas.fr
http://www.laas.fr

Abstract. Diagnosis is the process of identifying or determining the nature and root cause of a failure, problem, or disease from the symptoms arising from selected measurements, checks or tests. The different facets of the diagnosis problem and the wide spectrum of classes of systems make this problem interesting to several communities and call for bridging theories. This paper presents diagnosis theories proposed by the Control and the AI communities and exemplifies how they can be synergically integrated to provide better diagnostic solutions and to interactively contribute in fault management architectures.

Keywords: Model-based diagnosis, data-based diagnosis, Bridge diagnosis track, abstractions, learning diagnostic models, fault management.

1 Introduction

The goal of diagnosis is to identify the possible causes explaining observed symptoms. A set of concomitant tasks contribute to this goal and the following three tasks are commonly identified:

- *fault detection*, which aims at discriminating normal system states from abnormal ones, i.e. states which result from the presence of a fault,
- *fault isolation*, also called *fault localization*, whose goal is to point at the faulty components of the system,
- *fault identification*, whose output is the type of fault and possibly the model of the system under this fault.

In front of the diversity of systems and different views of the above problems, several scientific communities have addressed these tasks and contribute with a large spectrum of methods. The Signal Processing, Control and Artificial Intelligence (AI) communities are on the front.

Diagnosis works from the signals that permit efficient fault detection towards the upper levels of supervision that call for qualitative interpretations. Signal processing provided specific contributions in the form of statistic algorithms for detecting changes in signals, hence detecting faults. This track has been surveyed

J. Korbicz and M. Kowal (eds.), *Intelligent Systems in Technical and Medical Diagnostics*, Advances in Intelligent Systems and Computing 230,
DOI: 10.1007/978-3-642-39881-0_1,

in several reference books and papers [7,8,9,33,34] and remains out of the scope of this paper.

Interfaces between continuous signals and their abstract interpretations, in symbolic or event-based form, implement the qualitative interpretations of the signals that are required for supervision. To do that, discrete formalisms borrowed from Artificial Intelligence find a natural link with continuous models from the Control community. These two communities have their own model-based diagnosis track :

- the FDI (Fault Detection and Isolation) track, whose foundations are based on engineering disciplines, such as control theory and statistical decision making,
- the DX (Diagnosis) track, whose foundations are derived from the fields of logic, combinatorial optimization, search algorithms and complexity analysis.

In the last decade, there has been a growing number of researchers in both communities, who tried to understand and incorporate approaches from the FDI and DX fields to build better, more robust and effective diagnostic systems. In this paper, the concepts and results of the FDI and DX tracks are put in correspondence and the lessons learned from this comparative analysis are pointed out.

Data-based diagnosis approaches based on machine learning techniques are present in both the Control and AI communities and complement model-based approaches to provide solutions to a variety of diagnostic problems where difficulty arises from the scarce nature of the instrumentation or, conversely, from the massive amounts of data to be interpreted for the emergence of hidden knowledge. Interesting bridges can be foreseen between model-based and data-based approaches and these are illustrated in this paper with the problem of learning the models that support model-based diagnosis reasoning.

Other bridges can be found when considering that diagnosis is not a goal per se but a component in fault management architectures. It takes part in the solutions produced for tasks such as design, failure-mode-and-effects analysis, sensor placement, on-board recovery, condition monitoring, maintenance, repair and therapy planning, prognosis. The contribution of diagnosis in such architectures means close links with decision tasks such as control and planning and calls for innovative integrations.

In this paper, different facets of diagnosis investigated in the Control or the AI fields are discussed. While [101,102,103] provide three interesting surveys of the different approaches that exist in these fields, this paper aims at reporting the works that integrate approaches of both sides, hence creating "bridges".

The paper is organized as follows. After the introduction section, section 2 first presents a brief overview of the approaches proposed by the model-based diagnosis communities, FDI and DX, in subsections 2.1 and 2.2, respectively. Although quite commonplace, this overview is necessary because it provides the basic concepts and principles that form the foundations of any diagnosis method. It is followed by subsection 2.3 that compares the concepts and techniques used by the FDI and DX communities and presents the lessons learned from this comparative analysis. Section 3 is concerned with the trends that integrate and take advantage of techniques from both sides, in particular causal model-based diagnosis in subsection 3.1 and diagnosis of hybrid systems in subsection 3.2. Section 4 then raises the problem of obtaining the models supporting diagnosis reasoning and discusses bridges that can contribute to learning them in an automated manner. Section 5 widens the scope of diagnosis and is concerned with diagnosis as a component of fault management architectures, discussing several links with control and planning. Finally, section 6 concludes the paper.

2 DX and FDI Model-Based Diagnosis Bridge

The FDI and DX streams both approach the diagnosis problem from a *system* point of view, hence resulting in large overlaps, including the name of the tracks: *Model-Based Diagnosis* (MBD).

The diagnosis principles are the same, although each community has developed its own concepts and methods, guided by different modelling paradigms. FDI relies on analytical models, linear algebra, and non linear system theory whereas DX takes its bases in logic formalisms. In the 2000s, catalyzed by the BRIDGE group *"Bridging AI and Control Engineering model-based diagnosis approaches "* [114] within the Network of Excellence MONET II [115] and its French counterpart, the IMALAIA group *"Intégration de Méthodes Alliant Automatique and IA"* supported by GDR MACS [116], GDR I3 [117], as well as AFIA [118], there were more and more researchers who tried to understand and synergistically integrate methods from the two tracks to propose more efficient diagnostic solutions. This collaboration resulted in the organization of several events:

- a BRIDGE Workshop in 2001 in the framework of DX'01, 12th International workshop on Principles of Diagnosis, Sansicario, Via Lattea, Italy, 5-9 Mars 2001 [119].
- the co-location of the two main events of the FDI and the DX communities, namely the Symposium IFAC Safeprocess 2003 and the International Workshop "Principles of Diagnosis" DX 2003, in Washington DC (USA) in June 2003 with a BRIDGE Workshop in the form of a join day.

This events were followed by the publication of a special issue of the IEEE SMC Transactions, Part B, on the topic *Diagnosis of Complex Systems: Bridging the methodologies of the FDI and DX Communities* in 2004 by [19].

The *Bridge track* was launched and is still active today. Lets's mention the two invited sessions "*AI methods for Model-based Diagnosis*" and *"Bridge between Control Theory and AI methods for Model-based Diagnosis"*, recently organized in the framework of the 7th IFAC Symposium on Fault Detection, Supervision and Safety of Technical Processes Safeprocess'09, Barcelona, Spain, 30 July-3 August 2009 [120].

The next subsections first summarize the foundations of the FDI and DX methods, then proceed to a comparative analysis that allows us to draw some practical assessments in the form of lessons learned.

2.1 Brief Overview of FDI Approaches

The detection and diagnosis methods of the FDI communauty rely on behavioral models that establish the constraints between system inputs and outputs, i.e. the set of measurable variables Z, as well as the internal states, i.e. the set of unknown variables X. The variables $z \in Z$ et the variables $x \in X$ are functions of time. The typical model may be formulated in the temporal domain, then known as a *state-space model* of the form:

$$\begin{aligned} BM &: dx/dt = f(x(t), u(t), \theta) \\ OM &: y(t) = g(x(t), u(t), \theta). \end{aligned} \tag{1}$$

where $x(t) \in \Re^{n_x}$ is the state vector, $u(t) \in \Re^{n_u}$ is the input vector and $y(t) \in \Re^{n_p}$ is the output vector. BM is the behavioral model and OM is the observation model. The whole model is noted $SM(z,x)$. The equations of $SM(z,x)$ may be associated to components but this information is not represented explicitly. The models can also be formulated in the frequency domain (transfer functions in the linear case).

Models are used in three families of methods:

- the methods based on *parameter estimation* that focus on the value of parameters as representing physical features of the system
- the methods based on *state estimation* that rely on the estimation of unknown variables
- the methods based on the *parity space* that rely on the elimination of unknown variables

The books [39], [20], [31], [77] provide excellent surveys, which cite the original papers that the reader is encouraged to consult. The equivalence between observers, parity and paramater estimation has been proved in the linear case [78].

The concept central to FDI methods is the concept of *residual* and one of the main problems is to *generate residuals.* Let's consider the model $SM(z,x)$ of a system in the form (1). $SM(z,x)$ is said to be consistent with an observed trajectory z, or simply *consistent with measurements* z, if there exists a trajectory of x such that the equations of $SM(z,x)$ are satisfied.

Definition 1 (Residual generator for $SM(z,x)$). *A system that takes as input a sub-set of measured variables $\tilde{Z} \subseteq Z$ and generates as output a scalar r,is a residual generator for the model $SM(z,x)$ if for all z consistent with $SM(z,x)$, we have $\lim_{t\to\infty} r(t) = 0$.*

When the system model is consistent with measurements, the residuals tend to zero as t tends to infinity, otherwise some residuals may be different from zero. The residuals are often optimized to be robust to disturbancies [82] and to take into account uncertainties [2]. The evaluation of residuals and assigning them a boolean value (0 or non 0) generally calls for statistical decision techniques [31].

The methods based on parameter estimation are used for linear as well as non linear systems [80]. Fault detection is achieved by comparing the estimated parameter values to their nominal values. With these methods, fault detection, isolation, and identification are achieved at once, provided that model parameters can be put in correspondence with physical parameters. They rely on the identifiability of the system (see [41] and recently [47]) that casted the identifiability problem in the set-membership framework).

The methods based on state estimation take the form of *observers* or optimized *filters*, such that the Kalman filter, which provide an estimation of the state of the system. Specific filter architectures are often designed to structure the residuals, i.e. to provide different residuals for different faults, and isolate the faults [35]. Numerous diagnosis solutions rely on state estimation, particularly for hybrid systems. These works are surveyed in section 3.2. In this case, the continuous state is augmented by a discrete state that corresponds to the operation mode (normal or faulty) of the system components.

The methods based on the parity space [23] generate residuals from relations that are inferred from the system model. Theses relations, called *Analytical Redundancy Relations*(ARR), are determined off-line.

Definition 2 (ARR for $M(z,x)$). *An equation of the form $r(z, \dot{z}, \ddot{z}, \ldots) = 0$ is an ARR for the model $SM(z,x)$ if for all z consistent with $SM(z,x)$, the equation is satisfied.*

ARRs are constraints that only involve measured input and output variables and their derivatives. For linear systems, ARRS are obtained by eliminating unknown state variables by a linear projection on a particular space, so called the *parity pace*, [23]. An extension to non linear systems is proposed in [92]. On the other hand, the structural approach [93,5] is an interesting alternative because it allows one to obtain, for linear or non linear systems, the just overdeterminated sets of equations from which ARRs can be derived.

Every ARR can be put in the form $r(t) = 0$, where $r(t)$ is the *residual.* If the behavior of the system satisfies the model constraints, then the residuals are zero because the ARRs are satisfied. Otherwise, some of them may be different from zero as the corresponding ARRs are violated. A *theoretical signature* $FS_j = [s_{1j}, s_{2j}, \ldots, s_{nj}]$ given by the Boolean evaluation (0 or not 0) of each residual is associated to each fault F_j. The *signature matrix* is then defined s follows.

Definition 3 (Signature Matrix). *Given a set of n ARRs, the signature matrix associated to a set of n_f faults $F = [F_1, F_2, \ldots, F_{n_f}]$ is the matrix that crosses ARRs as rows and faults as columns and whole columns are given by the theoretical signatures of the faults.*

Diagnosis is achieved by comparing the observed signature, i.e. the Boolean residual values obtained from the actual measurements, and the theoretical signatures of the n_f faults.

2.2 Brief Overview of the DX Logical Diagnosis Theory

In the model-based logical diagnosis theory of DX as proposed by [83,54], the description of the system is driven by components and relies, in its original version, on first order logic. A system is given by a tuple $(SD, COMPS, OBS)$ where:

- SD is the *system description* in the form of a set of first order logic formulas with equality,
- $COMPS$ represents the *set of components* of the system given by a finite set of constants,
- OBS is a set of first order formulas, which represent the *observations.*

SD uses the specific predicate AB, meaning *abnormal.* Applied to a component c of $COMPS$, $\neg AB(c)$ means that C is normal and $AB(c)$ that c is faulty. The model of an adder would be given by :

$$ADD(x) \wedge \neg AB(x) \Rightarrow Output(x) := Input1(x) + Input2(x)$$

Definition 4 (Diagnosis). *A diagnosis for the system* $(SD, COMPS, OBS)$ *is a set* $\Delta \subseteq COMPS$ *such that* $SD \cup OBS \cup \{AB(C) \mid C \in \Delta\} \cup \{\neg AB(C) \mid C \in COMPS - \Delta\}$ *is satisfiable.*

The above definition means that the assumption stating that the components of Δ are faulty and all the others are normal is consistent with the observations and the system description. A diagnosis hence consists in the assignment of a mode, normal or faulty[1], to each component of the system, which is consistent with the model and the observations.

Definition 5 (Minimal diagnosis). *A* minimal diagnosis *is a diagnosis* Δ *such that* $\forall \Delta' \subset \Delta$, Δ' *is not a diagnosis.*

Following the parsimony principle, preference is always given to minimal diagnoses. Minimal diagnoses are also interesting because in several cases, they characterize the whole set of diagnoses, i.e. all the supersets of a minimal diagnosis are diagnoses [54].

To obtain the set of diagnoses, it is usual to proceed in the following steps:

1. The set of conflicts are first determined. An *R-conflict*[2] is a set $s \subseteq COMPS$ such that the assumption that all the components of s are normal is not consistent with SD and OBS. A *minimal conflict* is a conflict set that does not contain any other conflict.
2. The set of diagnoses is generated from the set of R-conflicts. [83] proved that minimal diagnoses are given by the *hitting sets*[3] of the set of minimal R-conflicts. An algorithm based on the construction of a tree, known as the HS-tree, was originaly proposed by [83].

[1] This framework has been later extended to fault models [54].

[2] An R-conflict is a conflict in the sense of Reiter [83]. This concept has been later generalized [54].

[3] The hitting sets of a collection of sets are given by the sets that intersect every set of the collection.

2.3 Lessons Learned from Comparing of the FDI and DX Approaches

This section summarizes the results issued from the comparative analysis of the DX approach and the parity space FDI approach performed by [24]. It then discusses the lessons learned.

It should be noticed that the modeling paradigm of FDI does not make explicit use of the concept of component. The behavioral model 1 describes the system as a whole. On the contrary, the DX approach models every component independently in a generic way, and specifies the topology of the system. Another important difference is that the assumption of correct behavior is represented explicitly in SD thanks to the predicate AB. If $\mathcal{F}$ is a formula describing the normal behavior of a component, SM only contains $\mathcal{F}$ whereas SD contains the formula $\neg AB \Rightarrow \mathcal{F}$.

The comparison of the two approaches is only possible if the models on both sides represent the same system and the observations/measurements capture the same reality. This is formalized by the *System Representation Equivalence* (SRE) property, which means that SM is obtained from SD by setting to *false* all the occurrences of the predicate AB. It is also assumed that the same observation language is used, i.e. OBS is a conjunction of equality relations, which assign a value to every measured variable. In addition, the faults relate to the same entities that are assumed to be components, without loss of generality.

The comparison is based on the concept of signature matrix FS, as defined in definition 3, which crosses ARRs in rows and components in columns (since faults are univocally associated to components), and relies on the concept of *ARR Support*:

Definition 6 (ARR Support). *The* support *of an ARR ARR_i is the set of components (columns of the signature matrix FS) whose corresponding matrix cell on the ARR_i line is non zero.*

Let us also introduce the two following properties, which refer to *detectability* indicated by a "d" and to *isolability* indicated by an "i" :

Property 1 (ARR-d-completness). A set E of ARRs is said to be *d-complet* if:

- E is finite;
- $\forall OBS$, if $SM \cup OBS \models \perp$, then $\exists ARR_i \in E$ such that $\{ARR_i\} \cup OBS \models \perp$.

Property 2 (ARR-i-completness). A set E of ARRs is said to be *i-complet* if:

- E is finite;
- $\forall C$, set of components such that $C \subseteq COMPS$, and $\forall OBS$, if $SM(C) \cup OBS \models \perp$, then $\exists ARR_i \in E$ such that the support of ARR_i is included in C and $\{ARR_i\} \cup OBS \models \perp$.

ARR-d-completeness and ARR-i-completeness express the theoretical capability of a set of ARRs to be sensitive, hence to detect, any inconsistency between the model SM and observations OBS They are hence key to the comparison, which includes the following results :

Proposition 1. *Assuming the SRE property and that OBS is the set of observations for the system given by SM (or SD), then :*

1. *If ARR_i is violated by OBS, then the support of ARR_i is an R-conflict;*
2. *If E is a d-complete set of ARRs, and if C is an R-conflict for $(SD, COMPS, OBS)$, then there exists $ARR_i \in E$ that is violated by OBS;*
3. *If E is an i-complete set of ARRs, then given an R-conflict C for $(SD, COMPS, OBS)$, there exists $ARR_i \in E$ that is violated by OBS and whose support is included in C.*

The result 1. is intuitive and can be explained by the fact that the inconsistencies between the model and observations are captured by R-conflicts in the DX approach and by ARRs violated by OBS in the FDI approach. Consequently, the support of an ARR can be defined as a *potential R-conflict.* This concept is called *possible conflict* by [81].

The results 2. and 3. are existence and completeness results, the first one being related to *fault detectability* and the second to *fault isolability.*

Let's now analyze the results from a more practical point of view, and point at the differences of the two approaches and how there advantages can be integrated.

Lesson One: Redundant ARRs – An important result about redundant ARRs comes from i-completeness. In FDI, it is generally accepted that if ARR_j is obtained from a linear combination of two other ARRs, ARR_{i_1} and ARR_{i_2}, then it is redundant (unless some considerations about noises and sensitivity to faults come into play). But the property of i-completeness states that not only the analytical expression of the ARR but also its support must be taken into account to conclude about its redundancy.

Proposition 2. *A given ARR_j is redundant with respect to a set of ARR_is, $i \in I, j \notin I, iff \exists I \subseteq I$ such that 1) $\forall OBS$, if all ARR_is, $i \in I$, are satisfied by OBS, then ARR_j is satisfied by OBS 2) the support of ARR_j contains the support of each ARR_i.*

Lesson Two: Exoneration Assumptions – The exoneration assumptions, *ARR-exoneration* and *component-exoneration*, used by DX and FDI, respectively, are different and cannot be expressed in the other formalism of the streams.

Definition 7 (ARR-exoneration). *Given OBS, any component in the support of an ARR satisfied by OBS is exonerated, i.e. considered as normal.*

Definition 8 (Component-exoneration). *Given OBS and $c \in COMP$, if $SM(c) \cup OBS$ is consistent, then c is exonerated, i.e. considered as normal.*

Importantly, the FDI approach generally uses the ARR-exoneration assumption but it is not stated explicitly. On the other hand, the DX approach is generally used without exoneration assumption and it is represented explicitly when it does[4]. This is just what guarantees the logical correctness of the DX diagnoses with DX methods.

It has been shown that under the same assumptions, in particular in the case of no exoneration, the diagnoses that are obtained by the DX and the FDI approach are the same.

Theorem 1. *Under the i-completeness and no exoneration assumptions, the diagnoses obtained by the FDI approach are identical to the (non empty) diagnoses obtained by the DX approach.*

Lesson Three: Single and Multiple Faults – In the FDI approach, because the fault signatures are determined off-line for every fault, the number of considered faults is generally limited. Most of the time, only single faults are considered. On the contrary, the DX approach naturally deals with multiple faults. A consequence is that the number of diagnoses is exponential and this is why it is common to introduce preference criteria, like fault probabilities, to order the diagnoses. Several methods have been proposed to search for the preferred diagnoses first (see for instance [111,89]).

Lesson Four: Off-line versus On-line – In the FDI approach, ARRs are determined off-line and only a simple consistency check is evaluated on-line. This may be very quite relevant for real time applications with hard temporal constraints. Inversely, in the DX approach, the whole diagnosis process is on-line, one advantage being that only the models need to be update if any change. The two approaches have been integrated to obtain the advantages of both: some DX works have used the idea of the FDI community to construct ARRs off-line [49,64,108,81]. Other works, presented in section 3, take advantage of an explicit representation of causal influences underlying the system model or deal with hybrid phenomena that call for continuous (as used in FDI) and discrete formalisms (as used in DX).

3 Bridging the FDI and DX Approaches

This section is concerned with two tracks that are quite representative of the proposed integrations of the FDI and DX model-based theories.

---

[4] If a component c is exonerated, its model is written as $COMP(c) \wedge \neg AB(c) \iff SM(c)$, where the simple logical implication has been replaced by a double implication.

3.1 Causal Model-Based Diagnosis

In the 90s, the synergies between the AI communities of qualitative reasoning (QR) [26,98,109] and model-based diagnosis concretized in a series of studies proposing the use of *causal models* for diagnostic reasoning. Diagnosis methods based on causal models are a perfect example of an integration of the FDI approach used for the fault detection phase and the DX approach used for the fault isolation phase.

Causal models are based on a graph structure expressing explicitly the dependencies between variables by edges called *influences*. They can then exhibit explanations for the values (normal or abnormal) of variables. This structure is called a *Causal Graph*. The principle of exploiting the dependencies between variables is the basis of many diagnostic algorithms. While the standard algorithms of logical diagnosis, such as GDE [28] use the technique known as "dependence recording" (implemented with an ATMS for example) that records dependencies during the inferences, the dependencies understood by causal models are exhibited from the outset. These are obtained either directly from expert knowledge as in [61,38], or by techniques of causal ordering of the QR community as in [97] or from bond-graphs models as in [71,32].

Causal diagnosis uses the causal graph as an abstract parameterized representation of the system in which every influence is associated to a physical component of $COMP$. Early work was limited to labeling causal influences by the signs giving the direction of change of the influenced variable in relation to the cause variable, thus obtaining a *Signed Oriented Graph*. Subsequently, the labeling of influences became more sophisticated and local quantitative models as used in FDI were attached.

Locally, FDI standard techniques for evaluating the residuals are used for fault detection. Fault detection is an (online) procedure that determines, at each time t and for each measured variable y, the consistency of the measured value $\bar{y}_t$ and the predicted value $\hat{y}_t$ obtained with the – causal – behavioral model of the system. It hence evaluates the residual:

$$r_t^y = \bar{y}_t - \hat{y}_t$$

Globally, diagnostic reasoning along the DX approach is underpinned by the causal structure labelled by an abstraction of the local models, such as delay times as in [71,99]. The detection of a variable as being abnormal is interpreted as the violation of one of the influences involved in the estimation of the variable, i.e. one of the upstream influences in the causal graph. Each influence being associated with a component, this set of influences can be identified as an R-conflict as defined in section 2.2 (or the support of the ARR that could be built from the local models by performing the variable eliminations indicated by the causal chains).

In [99], the influences have a *delay* attribute corresponding to the time required for a perturbation on the source variable to propagate to the effect variable. This information is used to generate R-conflicts whose elements are labeled by a temporal label indicating the latest time at which the fault occurred on this component. Diagnoses are obtained from the R-conflicts by an incremental algorithm that generates the hitting sets while managing the temporal labels.

3.2 Model-Based Diagnosis of Hybrid Systems

Hybrid systems [113,66,42,65] address the need to capture phenomema at different time scales in the same model. They are more particularly devoted to represent systems that have double continuous and discrete dynamics. A hybrid system is modeled by a state transition system whose states represent its modes of operation (normal or faulty) for which the continuous dynamics are specified. The modeling of such systems and the associated diagnostic algorithms make use of continuous and discrete formalisms, so that hybrid systems are a perfect field for the integration of methods from the FDI and DX communities. FDI methods, in the form of state estimators, parameter estimators or sets of ARRs are indeed coupled to search methods used in AI to deal with the combinatorial explosion of the number of trajectories to be tracked or to discrete event based methods present in both communities.

In the DX community, researchers first tried to abstract the continuous dynamics in qualitative terms. The Livingstone diagnostic engine [79,110], from NASA, that flew on board the DS-1 probe, was among the first to be qualified as hybrid. This engine was rooted in the DX approach, with a model written in propositional logic, and behavioral equations reflecting the continuous dynamics in the form of logical relationships (qualitative constraints). The qualitative abstraction required *monitors* between the sensors and the model to interpret continuous signals in terms of discrete modalities. This approach suffered from the two drawbacks: the thresholds are extremely difficult to determine, and detection sensitivity is quite poor. Subsequent research thus replaced the qualitative abstractions by classical differential equations, resulting in real hybrid models interlinking continuous and discrete states.

A hybrid model may, as proposed in [12], be given as a tuple:

$$S = (\zeta, Q, \Sigma, T, C, (q_0, \zeta_0)) \tag{2}$$

where:

- ζ is the set of continuous variables including state variables, input/output variables, and possibly noise, which are functions of time t. Some are measured, others do not.
- Q is the set of discrete system states. Each state $q_i \in Q$ represents a mode of operation of the system.
- Σ is the set of events that correspond to discrete control inputs, autonomous mode changes and fault occurrences. Events corresponding to autonomous mode changes are issued upon guards that depend on continuous variables.
- $T \subseteq Q \times \Sigma \rightarrow Q$ is the partial transition function. A transition $t(q_i, \sigma_{ij}, q_j)$ may be guarded by a condition given as a set of equations $\mathcal{G}(t(q_i, \sigma_{ij}, q_j)) = g_{ij}(x, \theta_g) = 0$, θ_g being a constant parameter vector. Then σ_{ij} results from

the state $x(t)$ hitting the guard g_{ij} at some time instant t^* and is not observable. A reset map $\mathcal{R}_{ij}$, possibly equal to the identity, is specified.
- $C = \bigcup_i C_i$ is the set of system constraints linking continuous variables. C_i denotes the set of constraints associated to the mode q_i, which are given in the state-space by the following continuous time state-evolution and output equations:

$$\begin{cases} \dot{x}(t) = f_i(x(t), u(t), \epsilon(t)) \\ y(t) = g_i(x(t), u(t), \epsilon(t)) \end{cases} \tag{3}$$

where $u \in \mathbb{R}^{n_u}$, $x \in \mathbb{R}^{n_x}$, $y \in \mathbb{R}^{n_y}$ are the input, output, state vectors, respectively, and $\epsilon \in \mathbb{R}^{n_\epsilon}$ denotes some noise vector. The variables gathered in these vectors belong to ζ.
- $(\zeta_0, q_0) \in \zeta \times Q$ is the initial condition of the hybrid system.

In the hybrid state (ζ, Q), only the discrete state Q is representative of the operation mode of the system and it provides the diagnosis. Because the evolution of discrete states is closely tied to the evolution of the continuous states, the diagnosis problem is often brought back to the problem of estimating the full hybrid state.

In theory, hybrid estimation must consider all the possible mode sequences with the associated continuous evolutions, which is exponentially complex. Consequently, many sub-optimal methods have been proposed, of which we can distinguish the following three classes:

- methods known as *multiple-model estimation*, rather rooted in the field of control [1,21,62]
- *particle filtering* methods [36,105,74], which are present both in FDI and DX
- methods that address hybrid aspects in a *dedicated* manner [43,15,73,17], taking advantage of FDI and DX approaches.

Multiple-model estimation methods, inspired by the IMM algorithm of [21], are formulated in a probabilistic framework. Estimation *hypotheses*, in the form of sequences of modes and their associated continuous evolution, are tracked over a limited time window. Continuous estimates are merged according to a likelihood measure to provide a *belief state* in the form of a probability distribution over states at the current time. The likelihood measure is provided by every continuous filter and expresses the degree of consistency between this hypothesis and observations but it can also take into account the information carried by the probabilities of transitions, if any. This family of methods includes [107], which suggests combining a process of continuous state estimation given by a Kalman filter (KF) with an estimation process of the discrete state based on a POMDP (Partially Observed Markov Decision Process).

[43] also unifies continuous state observers with a belief calculated from Hidden Markov Models (HMM). The approach is quite similar to the one of [107] in that it uses a set of KFs, but adopts an aggressive strategy for adjusting the number of tracked trajectories to the computing resources available. This strategy uses an A^* algorithm, which selects the most likely state trajectory

branches and takes into account unknown fault modes. It is implemented in the hME engine by [44].

Besides, it was shown by [17] that taking into account the uncertainty as bounded rather than probabilistic uncertainty allows one to merge the hypotheses with the same discrete state estimate by computing the hull of continuous estimates. This fusion, possible possible in a set-membership framework, controls very effectively the number of hypotheses followed and may avoid truncating/approximating the set of hypotheses.

Particle filtering methods are based on a Bayesian update of prior beliefs. They proceed by sampling the initial probability distribution. With enough samples, they approach the optimal Bayesian estimate. We refer the reader to [95] for a tutorial on particle filtering methods.

These methods, which accept probability distributions of any kind, are very popular. However, it should be noted that they are difficult to apply to diagnosis problem since the probabilities of faults are generally very low compared to those of nominal states. The number of particles needed to track the faulty states is very high. Different strategies were tested to provide solutions to this problem [106,96,75].

As we have seen, the combinatorial explosion inherent to hybrid estimation leads to solutions providing an approximated belief state. Consequently, all previous methods suffer from the problem of *state loss*, which may be critical for a diagnosis application. Indeed, the set of modes estimated over time, i.e. the diagnoses, takes the form of a set Γ of trajectories given by a decision process that abandons the trajectories whose belief is below a certain threshold.

Unfortunately, Γ is reused in the calculation of future estimates and it may happen that the actual trajectory of the system is no longer in Γ, thus producing an erroneous diagnosis. Several studies have addressed this problem. [57] proposes a solution based on a *a posteriori* progressive exploration of the space guided by the faults *rank*. However, this requires considering a wide time window backwards from the current time, which may even reach the initial time. In this case, the current state mustbe re-predicted from all the system's history.

The diagnostic engine KOALA [16] proposes a variant of the previous algorithm with a *revision* procedure that produces diagnoses from the current trajectory by performing partial changes (minimal in the best case). There have been other attempts to remedy the loss of solutions. In [60] [59], the authors reduce the set of tracked trajectories by aggregating those using similar assumptions. In addition, they develop a smoothing strategy that propagates the current beliefs back in time to improve the prediction of system states.

More recently, an alternative approach, called rME, to hybrid state estimation has been proposed. rME focuses on mode estimation, the actual diagnosis, and ignores continuous state estimation [10]. rME is based on the parity space approach as presented in section 2.1 for processing continuous data. A set of ARRs is determined for each mode of the hybrid system from the associated continuous model. They are all put together in a vector. When evaluated with the measurements of a given mode, this vector defines its *mode signature*. The mode signature changes resulting from the transition from one mode to another are then abstracted as a set of events, called *signature events*, by the abstraction function f_{Sig}:

$$\begin{aligned} f_{Sig} : Q \times T(Q, \Sigma) &\longrightarrow \Sigma^{Sig} \\ (q_i, q_j) &\longmapsto \delta_{ij} \end{aligned} \quad (4)$$

The event δ_{ij} is observable and noted Ro_{ij} if the mode signature of the source mode q_i is different from the mode signature of the destination mode q_j ($Sig(q_i) \neq Sig(q_j)$). δ_{ij} is unobservable and noted Ruo_{ij} otherwise[5]. Hence Σ^{Sig} is partitionned in a set of observable signature-events Σ_o^{Sig} and a set of unobservable signature-events Σ_{uo}^{Sig}.

The set $\Sigma^{Sig} = \Sigma_o^{Sig} \cup \Sigma_{uo}^{Sig}$ is used to enrich the underlying DES $M = (Q, \Sigma, T, q_0)$, producing an automaton that captures both the abstraction of the continuous dynamics and the discrete dynamics.

The classical diagnoser approach [90] for DES permits to track the current mode of the system based on observable events received, i.e. to provide the diagnosis. Lately, the mode estimation rME approach was combined with the state estimation approach hME [45] in [86], leading to mutual enrichment. On the one hand, hME brings an estimate of the continuous state that is not provided by rME. On the other hand, it was shown that among all modes corresponding to the current hypotheses of hME, rME can identify those that are consistent with observations, providing a reduced set of possible modes to hME and significantly reducing the number of hypotheses to be tracked by the hybrid estimator.

4 Bridges for Learning Diagnosis Models

The main weakness of MBD diagnosis approaches, either from FDI or from DX, is to require well-formalized models that one must generate from the available knowledge about the system. On the other hand, a particularity of nowadays systems is to generate big amounts of data that are stored. Data-driven diagnosis approaches take direct advantage of these data and use a whole spectrum of machine learning methods to assess the state of a system, i.e. the diagnosis.

But another way to go is to use machine learning to generate the models automatically, hence bridging with MBD diagnosis methods. Whereas regression analysis is well-known for learning continuous models, there is still an open field for learning more abstract models, in particular discrete-event models. This is a promising research track that can provide methods benefiting from all the available information in the form of data and knowledge.

To exemplify this track, let us consider the problem of generating discrete-event models automatically from raw data in the form of time series. Some works have been done to generate automata or Markov chain models from classification results, for example [51]. In this paper, we illustrate the generation of chronicle models, which are rather used in the AI community [94].

[5] By construction, mode signatures cannot change while being in the same mode.

Chronicles are high level temporal patterns that are able to capture the behavior of dynamic processes at an abstracted level based on events. They can be used as signatures of specific situations and associated to decision rules specifying which actions must be taken in such and such situation.

In the approach proposed by [3], time series data can be abstracted in terms of sequences of temporally constrained *events*. Events are a symbolic representation of time-stamped specific singularities of the signal. Temporal patterns take the form of *chronicles* [29]. They are very useful to characterize dynamic situations manifesting as specific signal patterns in many application domains and come together with powerful event recognition engines that are able to recognize chronicles on the fly given a sequence of events as input [6].

Chronicle Concept. Chronicles are a rich formalism allowing one to describe the observable patterns corresponding to behaviors one wants to detect. With chronicles one may want to track dynamic situations like the ones below:

- event A is immediately followed by event B,
- event A is followed by events B and C after less than 10 time units
- event B occurs after n occurrences of event A.

A chronicle is the description of a temporal pattern in terms of dated events and time relations i.e. time constraints between event occurrence dates. Time constraints are durations expressed as bounds on the difference between two time points or event occurrence dates. A chronicle model c is then a pair $(\mathcal{S}, \mathcal{T})$ where $\mathcal{S}$ is a set of events and $\mathcal{T}$ a set of constraints between their occurrence dates. Figure 1 is an example of chronicle in which $\mathcal{S} = \{e_1, e_2, e_3\}$ and $\mathcal{T} = \{\{t_1 < t_3$ with $3 \leq t_3 - t_1 \leq 6\}, \{t_2 < t_3$ with $3 \leq t_3 - t_2 \leq 9\}\}$.

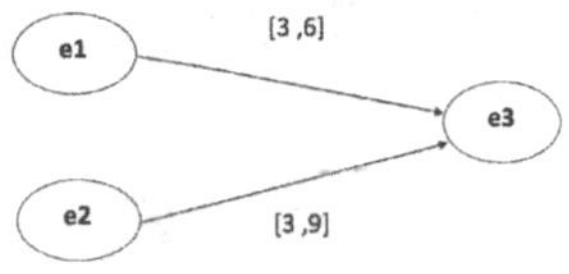

Fig. 1. Example of chronicle

Let us consider a set of scenarios labelled with their corresponding situation in the form of a set of time series, one for each relevant signal. The goal is to learn one characterizing chronicle for each situation. This chronicle is thus to be learned from all the scenarios labelled by this situation. The method includes two steps:

– Step1: every scenario data is clustered into a set of classes and the conditions that underly the transitions between classes are identified. The transitions between classes are then assimilated to events. The output of this step is a set of dated event sequences representing an abstraction of each scenario data.

- Step 2: symbolic learning methods of the same type as [25] are applied to the set of sequences obtained at step 1 to extract the chronicle that characterizes the most precisely all the sequences in the set.

Step 1 is illustrated by figure 2 in which the fuzzy clustering method LAMDA [4,50] has been used. The three sequences with same label have been given to LAMDA concatenated, hence the repetition of the same clustering pattern in three classes appearing on the top. The bottom bar diagrams provide a characterization of the three classes. Figure 3 illustrate that an event is generated at the transition of the classes.

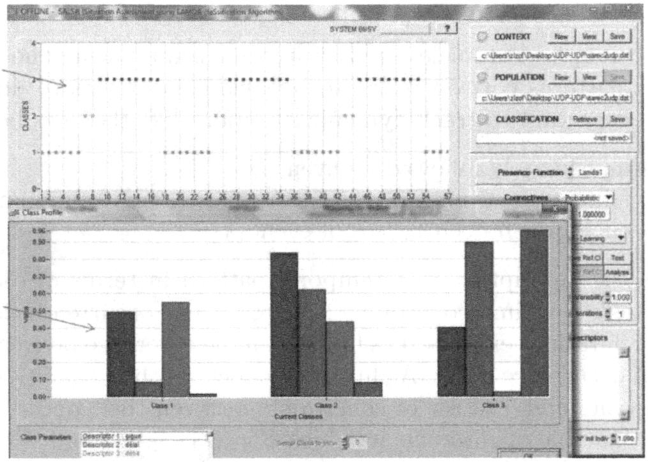

Fig. 2. Clustering step with LAMDA

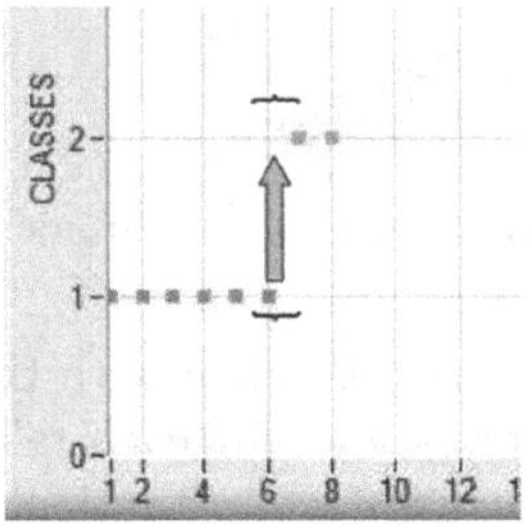

Fig. 3. Generation of events

Step 2 is illustrated by figures 4 and 5, which provide an example of three event sequences obtained from three time series corresponding to the same situation and the chronicle that characterizes the three sequences, respectively.

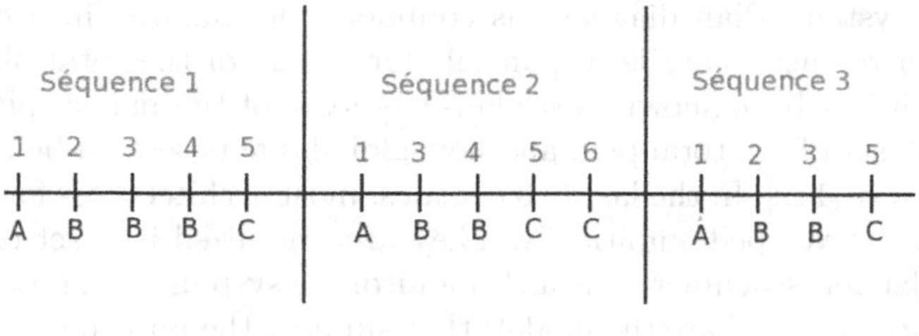

Fig. 4. Example of three sequences with same label

```
message A
message B
message C

chronicle Chronicle1
{
        event(A,t1)
        event(B,t2)
        event(C,t3)

        t2-t1 in [2.0,2.0]
        t3-t1 in [4.0,4.0]
        t3-t2 in [2.0,3.0]
}
```

Fig. 5. Chronicle characterizing the three sequences

5 Diagnosis in Fault Management Architectures

In the last few years, there has been a significant trend for accounting for fault analysis and diagnosis in the solutions produced for autonomous architectures on one side and maintenance programs on the other side. Diagnosis is thus seen as a piece of the puzzle that constitutes the whole fault management architecture. This section does not aim at detailing the literature existing in this areas but rather at discussing the openings that exist and the challenges for future research.

For autonomous systems, diagnosis must cooperate with control and planning so as to jointly solve the state tracking/decision problem. The problem of fault tolerant control has been intensively studied by our community [20]. Tools for deciding about system reconfigurability have been proposed as well as solutions to reconfigure control laws on line in order to guarantee the stability of the system and to keep the state of the system close to the desired state even when a fault occurs.

Hence nowadays challenges are rather put on bridging diagnosis with high level planning [18]. Planning generates a control program, also known as a *plan*, which describes the sequence of actions necessary to achieve some pre-defined goals. As proposed by [48], diagnosis may as well be applied to plans for determining which action(s) failed and what went wrong in terms of the internal states of the system. Plan diagnosis is commonly formulated in a distributed framework where each *agent* is responsible for a part of the total plan [68]. In this context, it has been shown how plan-diagnoses of the partial plans are related to diagnoses of the total plan and how global diagnoses can be obtained in a distributed way [112]. In the last two decades, most architectures for autonomy are ad-hoc and developed empirically. They are organized in a set of modules, typically a planner/scheduler, a health monitoring system (diagnoser), and an executive which do not share the models that support the implemented methods [72,46]. This obviously limits the articulation of diagnosis and the other functional modules, in particular the planner, and some work should be done towards unifying the underlying models.

Active diagnosis is another aspect of the picture in which diagnosis and decision technologies must cooperate. Active diagnosis relies on applying specific inputs to the system so that additional symptoms that permit refining the diagnosis are exhibited. As proposed in [22,27,13], starting with an ambiguous diagnosis state, the diagnoser can be coupled with a conditional planner to provide a plan of actions that work towards refining the diagnosis. Optimal input design, implemented on-line, can be seen as a continuous control version of active planning [67]. However, when some discrete actions are possible like changing operation mode, active diagnosis calls for planning algorithms rather rooted in the AI field [40].

In domains in which faults are accounted for through maintenance programs, diagnosis appears as an ingredient for achieving condition monitoring and allows one to make adjusted predictions about the remaining useful life of the system and its components. Prognostic is often the result of some statistics drawn from a set of scenarios implementing different stress conditions [100,91]. But for a specific component in operation, its is possible to assess its current healh status, resulting from the specific stress it has undergone, and this is just the aim of diagnosis. Diagnosis thus allows one to perform adaptive prognosis [85,84,70,58]. Diagnosis determines the state of the system (and each of its components) as being consistent with the current observations, then prognosis can determine the state that will be reached in the future and that is consistent with the current diagnosis and with the ageing model of the system. Hence the remaining useful life can be evaluated for every component in adaptation with the health status reported by diagnosis. In this area, the tools that are generally in use are statistics, machine learning and estimation methods. Coupling diagnosis and prognosis is still young and more work has to be done to better understand and benefit from this link.

6 Conclusion

In this paper, different facets of diagnosis investigated in the Control and the AI fields are discussed. Diagnosis benefits from a wide spectrum of methods and each of them brings in different potentialities. Having the whole picture helps understand the different variants of the diagnosis problem in relation with the different application domains. This is also essential to propose suitable and relevant solutions to real diagnosis problems.

This paper does not survey the different diagnosis approaches (see [101,102,103] for a survey of this kind) but it surveys the *bridging* works that integrate approaches from the Control and the AI fields.

The model-based FDI and DX tracks are given special attention because they can be put in correspondence and the lessons learned from this comparative analysis are very informative in showing how the two approaches deal with different aspects of the diagnosis problem. In section 4, learning the models that support diagnosis reasoning is shown to be a rich field for bridging theories. Finally, diagnosis is discussed in relation with theories that participate to provide global solutions to fault management problems.

It is shown that diagnosis benefits from a wide spectrum of methods and that each of them has its own specificity, to be associated with the requirements of the application domains. It is also shown that diagnosis is a critical function in fault management architectures and that maximal benefit can be achieved when it is formally integrated with other tasks.

Throughout the paper several openings central to the development of the diagnosis field are pointed out, and that call for integrative solutions.

Acknowledgements. I do thank all the members of the *Diagnosis and Supervisory Control* (DISCO) team of *LAAS-CNRS*, University of Toulouse, France, for their support and interesting comments. This paper gives a picture of many of the DISCO research topics, and of the multidisciplinary approach adopted by this team to address diagnosis.

References

1. Ackerson, G., Fu, K.: On state estimation in switching environments. IEEE Transactions on Automatic Control 15, 10–17 (1970)
2. Adrot, O., Maquin, D., Ragot, J.: Fault detection with model parameter structured uncertainties. In: Proceedings of the European Control Conference, ECC 1999, Karlsruhe, vol. 99 (1999)
3. Aguilar-Castro, J., Subias, A., Travé-Massuyès, L., Zouaoui, K.: Situation assessment in autonomous systems. In: Proceeding of the 4th Global Information Infrastructure and Networking Symposium, GIIS 2012, Choroni, Venezuela, pp. 1–6 (2012)
4. Aguilar-Martin, J., López de Mántaras, R.: The process of classification and learning the meaning of linguistic descriptors of concepts. Approximate Reasoning in Decision Analysis, 165–175 (1982)
5. Armengol, J., Bregon, A., Escobet, T., Gelso, E., Krysander, M., Nyberg, M., Olive, X., Pulido, B., Travé-Massuyès, L.: Minimal Structurally Overdetermined sets for residual generation: A comparison of alternative approaches. In: Proceedings of the 7th IFAC Symposium on Fault Detection, Supervision and Safety of Technical Processes Safeprocess 2009, Barcelona, Spain (2009)
6. Artikis, A., Skarlatidis, A., Portet, F., Paliouras, G.: Logic-based event recognition. Knowledge Engineering Review 27(4), 469–506 (2012)

7. Basseville, M.: Detecting changes in signals and systems: a survey. Automatica 24(3), 309–326 (1988)
8. Basseville, M., Nikiforov, I.: Detection of abrupt changes: theory and application. Citeseer (1993)
9. Basseville, M., Mevel, L., Goursat, M.: Statistical model-based damage detection and localization: subspace-based residuals and damage-to-noise sensitivity ratios. Journal of Sound and Vibration 275(3-5), 769–794 (2004)
10. Bayoudh, M., Travé-Massuyès, L., Olive, X.: Hybrid systems diagnosis by coupling continuous and discrete event techniques. In: Proceedings of the IFAC World Congress, Seoul, Korea, pp. 7265–7270 (2008)
11. Bayoudh, M., Travé-Massuyès, L., Olive, X.: Towards Active Diagnosis of Hybrid Systems. In: Proceedings of the 19th International Workshop on Principles of Diagnosis, DX 2008, Blue Mountains, Australia, pp. 231–237 (2008)
12. Bayoudh, M., Travé-Massuyès, L., Olive, X.: Coupling continuous and discrete event system techniques for hybrid system diagnosability analysis. In: Proceedings of the 18th European Conference on Artificial Intelligence, ECAI 2008, July 21-25, pp. 219–223. IOS Press, Patras (2008)
13. Bayoudh, M., Travé-Massuyès, L., Olive, X.: Active diagnosis of hybrid systems guided by diagnosability properties. In: Proceeding of the 7th IFAC Symposium on Fault Detection, Supervision and Safety of Technical Processes, Safeprocess 2009, Barcelona, Spain, pp. 1498–1503 (2009)
14. Bayoudh, M., Travé-Massuyès, L.: Diagnosability analysis of hybrid systems cast in a discrete-event framework. Journal of Discrete Event Dynamic Systems, JDEDS (2012), doi:10.1007/s10626-012-0153-z
15. Benazera, E., Travé-Massuyès, L., Dague, P.: State tracking of uncertain hybrid concurrent systems. In: Proceedings of the 13th International Workshop on Principles of Diagnosis, DX 2002, Semmering, Austria, pp. 106–114 (2002)
16. Benazera, E., Travé-Massuyès, L.: The Consistency approach to the on-line prediction of hybrid system configurations. In: Proceedings of the International IFAC Conference on Analysis and Design of Hybrid Systems, ADHS 2003, St. Malo, Brittany, France, June 16-18, pp. 241–247. Elsevier Science (2003)
17. Benazera, E., Travé-Massuyès, L.: Set-theoretic estimation of hybrid system configurations. IEEE Transactions on Systems, Man, and Cybernetics. Part B, Cybernetics: A Publication of the IEEE Systems, Man, and Cybernetics Society 39(6), 1277–1291 (2009)
18. Benazera, E.: On the Articulation of Planning and Diagnosis. Contract report: ROSACE, RTRA Project, LAAS internal report N 09536, 29 p. (2009)
19. Biswas, G., Cordier, M., Lunze, J., Travé-Massuyès, L., Staroswiecki, M.: Diagnosis of complex systems: Bridging the methodologies of the FDI and DX communities. IEEE Transactions on Systems, Man, and Cybernetics, Part B 34(5), 2159–2162 (2004)
20. Blanke, M., Kinnaert, M., Lunze, J., Staroswiecki, M.: Diagnosis and fault-tolerant control. Springer (2003)
21. Blom, H., Bar-Shalom, Y.: The interacting multiple model algorithm for systems with markovian switching coefficients. IEEE Transactions on Automatic Control 33, 780–783 (1988)
22. Chanthery, E., Pencole, Y., Bussac, N.: An AO*-like algorithm implementation for active diagnosis. In: Proceedings of the International Symposium on Artificial Intelligence, Robotics and Automation in Space, i-SAIRAS 2010, Sapporo, Japan, pp. 378–385 (2010)

23. Chow, E., Willsky, A.: Analytical redundancy and the design of robust failure detection systems. IEEE Transactions on automatic control 29(7), 603–614 (1984)
24. Cordier, M., Dague, P., Lévy, F., Montmain, J., Staroswiecki, M., Travé-Massuyès, L.: Conflicts versus analytical redundancy relations: a comparative analysis of the model-based diagnosis approach from the artificial intelligence and automatic control perspectives. IEEE Transactions on Systems, Man, and Cybernetics, Part B 34(5), 2163–2177 (2004)
25. Cram, D., Mathern, B., Mille, A.: A complete chronicle discovery approach: application to activity analysis. Expert Systems 29(4), 321–346 (2012)
26. Dague, P., Travé-Massuyès, L.: Raisonnement causal en physique qualitative. Intellectica 38, 247–290 (2004)
27. Daigle, M., Koutsoukos, X., Biswas, G.: Improving diagnosability of hybrid systems through active diagnosis. In: Proceedings of the 7th IFAC Symposium on Fault Detection, Supervision and Safety of Technical Processes, Safeprocess 2009, Barcelona, Spain (2009)
28. De Kleer, J., Williams, B.: Diagnosing multiple faults. Artificial Intelligence 32(1), 97–130 (1987)
29. Dousson, C., Gaborit, P., Ghallab, M.: Situation recognition: representation and algorithms. In: Proceedings of the International Joint Conference on Artificial Intelligence, IJCAI 1993, Chambéry, France, pp. 166–172 (1993)
30. Dousson, C., Duong, T.V.: Discovering chronicles with numerical time constraints from alarm togs for monitoring dynamic systems. In: Proceedings of the Int. Joint Conf. on Artificial Intelligence, pp. 620–626. Lawrence Erlbaum Associates Ltd. (1999)
31. Dubuisson, B.: Automatique et statistiques pour le diagnostic. Hermes Science Europe Ltd. (2001)
32. Feenstra, P., Mosterman, P., Biswas, G., Breedveld, P.: Bond graph modeling procedures for fault detection and isolation of complex flow processes. Simulation Series 33(1), 77–84 (2001)
33. Fillatre, L., Nikiforov, I.: Non-Bayesian detection and detectability of anomalies from a few noisy tomographic projections. IEEE Transactions on Signal Processing 55(2), 401–413 (2007)
34. Fouladirad, M., Freitag, L., Nikiforov, I.: Optimal fault detection with nuisance parameters and a general covariance matrix. International Journal of Adaptive Control and Signal Processing 22(5), 431–439 (2008)
35. Frank, P.: On-line fault detection in uncertain nonlinear systems using diagnostic observers: a survey. International Journal of Systems Science 25(12), 2129–2154 (1994)
36. de Freitas, N.: Rao-blackwellised particle filtering for fault diagnosis. In: Proceedings of the IEEE Aerospace Conference 2002, vol. 4, pp. 1767–1772 (2002)
37. Fukunaga, K.: Introduction to statistical pattern recognition. Academic Press, New York (1972)
38. Gentil, S., Montmain, J., Combastel, C.: Combining FDI and AI approaches within causal-model-based diagnosis. IEEE Transactions on Systems, Man, and Cybernetics, Part B: Cybernetics 34(5), 2207–2221 (2004)
39. Gertler, J.: Fault Detection and Diagnosis in Engineering Systems. Marcel Deker (1998)
40. Ghallab, M., Nau, D., Traverso, P.: Automated Planning: theory and practice. Morgan Kaufmann (2004)
41. Grewal, M., Glover, K.: Identifiability of linear and nonlinear dynamical systems. IEEE Transactions on Automatic Control 21(6), 833–837 (1976)

42. Henzinger, T.: The theory of hybrid automata. In: Proceedings of the 11th Annual IEEE Symposium on Logic in Computer Science, LICS 1996, New Brunswick, New Jersey, pp. 278–292 (1996)
43. Hofbaur, M.W., Williams, B.C.: Mode estimation of probabilistic hybrid systems. In: Tomlin, C.J., Greenstreet, M.R. (eds.) HSCC 2002. LNCS, vol. 2289, pp. 253–266. Springer, Heidelberg (2002)
44. Hofbaur, M.W., Williams, B.C.: Hybrid diagnosis with unknown behavioral modes. In: Proceedings of the Thirteenth Internatinal Workshop on Principles of Diagnosis, DX 2002, Semmering, Austria, pp. 97–105 (2002)
45. Hofbaur, M.W., Williams, B.C.: Hybrid estimation of complex systems. IEEE Transactions on Systems, Man, and Cybernetics - Part B: Cybernetics 34(5), 2178–2191 (2004)
46. Ingrand, F., Chatila, R., Alami, R.: An architecture for dependable autonomous robots. In: Proceedings of the 8th IEEE International Conference on Emerging Technologies and Factory Automation, vol. 2, pp. 657–658 (2001)
47. Jauberthie, C., Verdière, N., Travé-Massuyès, L.: Set-membership identifiability: Definitions and analysis. In: Proceedings of the IFAC World Congress, Milan, Italy, pp. 12024–12029 (2011)
48. de Jonge, F., Roos, N., Witteveen, C.: Diagnosis of multi-agent plan execution. In: Fischer, K., Timm, I.J., André, E., Zhong, N. (eds.) MATES 2006. LNCS (LNAI), vol. 4196, pp. 86–97. Springer, Heidelberg (2006)
49. Katsillis, G., Chantler, M.: Can dependency-based diagnosis cope with simultaneous equations. In: Proceedings of the 8th International Workshop on Principles of Diagnosis, DX 1997, pp. 51–59 (1997)
50. Kempowsky, T., Aguilar-Martin, J., Subias, A., Le Lann, M.V.: Classification tool based on interactivity between expertise and self-learning techniques. In: Proceedings of the International Symposium on Fault Detection, Supervision and Safety of Technical Processes, Safeprocess 2003, Washington DC, USA (2003)
51. Kempowsky, T., Subias, A., Aguilar-Martin, J., Travé-Massuyès, L.: A discrete event model for situation awareness purposes. In: Proceedings of the International Symposium on Fault Detection, Supervision and Safety of Technical Processes, Safeprocess 2006, Beijing, China, pp. 1288–1293 (2006)
52. Kempowsky, T., Subias, A., Aguilar-Martin, J.: Process Situation Assessment: From a fuzzy partition to a finite state machine. Eng. App. of Artificial Intelligence 19, 461–477 (2006)
53. Kieffer, M., Walter, E.: Guaranteed estimation of the parameters of nonlinear continuous time models: Contributions of interval analysis. Int. Journal of Adaptive Cont. and Signal Processing 25(3), 191–207 (2011)
54. Kleer, J., Mackworth, A., Reiter, R.: Characterizing diagnoses and systems. Artificial Intelligence 56(2-3), 197–222 (1992)
55. Krysander, M., Aslund, J., Nyberg, M.: An efficient algorithm for finding minimal overconstrained subsystems for model-based diagnosis. IEEE Transactions on Systems, Man and Cybernetics, Part A: Systems and Humans 38(1), 197–206 (2008)
56. Kuhn, L., Price, B., Do, M., Liu, J., Zhou, R.: Pervasive diagnosis: The Integration of Diagnostic Goals into Production Plans. In: Proceedings of the 23rd AAAI Conference on Artificial Intelligence, Chicago, USA, pp. 1306–1312 (2008)
57. Kurien, J., Nayak, P.P.: Back to the future for consistency-based trajectory tracking. In: Proceedings of the National Conference on Artificial Intelligence, pp. 370–377. AAAI Press, MIT Press, Menlo Park, CA (2000)

58. Hedjazi, L., Le Lann, M.V., Kempowsky, T., Aguilar-Martin, J., Dalenc, G.F., Despenes, S.L.: From chemical process diagnosis to cancer prognosis: an integrated approach for diagnosis and sensor/marker selection. In: Proceedings of the European Symposium on Computer-Aided Process Engineering, ESCAPE 21, Chalkidiki, Greece, pp. 1510–1514 (2011)
59. Lerner, U., Moses, B., Scott, M., McIlraith, S., Koller, D.: Monitoring a complex physical system using a hybrid dynamic bayes net. In: Proceedings of the Eighteenth Conference on Uncertainty in Artificial Intelligence, UAI 2002, pp. 301–310. Morgan Kaufmann Publishers Inc. (2002)
60. Lerner, U., Parr, R., Koller, D., Biswas, G.: Bayesian fault detection and diagnosis in dynamic systems. In: Proceedings of the National Conference on Artificial Intelligence, AAAI 2000, Menlo Park, CA, pp. 531–537 (2000)
61. Leyval, L., Gentil, S., Feray-Beaumont, S.: Model-based causal reasoning for process supervision. Automatica 30(8), 1295–1306 (1994)
62. Li, X., Bar-Shalom, Y.: Multiple-model estimation with variable structure. IEEE Transactions on Automatic Control 41, 478–493 (1996)
63. Liu, F., Qiu, D.: Safe diagnosability of stochastic discrete-event systems. IEEE Transactions on Automatic Control 53(5), 1291–1296 (2008)
64. Loiez, E., Taillibert, P.: Polynomial temporal band sequences for analog diagnosis. In: Proceedings of the Fifteenth International Joint Conference on Artificial Intelligence, IJCAI 1997, Nagoya, Japan, August 23-29, pp. 474–479 (1997)
65. Lunze, J., Lamnabhi-Lagarrigue, F. (eds.): Handbook of hybrid systems control: theory, tools, applications. Cambridge University Press (2009)
66. McIlraith, S.A., Biswas, G., Clancy, D., Gupta, V.: Hybrid systems diagnosis. In: Lynch, N.A., Krogh, B.H. (eds.) HSCC 2000. LNCS, vol. 1790, pp. 282–295. Springer, Heidelberg (2000)
67. Mehra, R.: Optimal input signals for parameter estimation in dynamic systems–survey and new results. IEEE Transactions on Automatic Control 19(6), 753–768 (1974)
68. Micalizio, R., Torasso, P.: Plan diagnosis and agent diagnosis in multi-agent systems. In: Proceedings of the 10th Congress of the Italian Association for Artificial Intelligence AI*IA 2007: Artificial Intelligence and Human-Oriented Computing, Rome, Italy, Springer Berlin Heidelberg, pp. 434–446. Springer, Heidelberg (2007)
69. Milanese, M., Vicino, A.: Optimal estimation theory for dynamic systems with set membership uncertainty: an overview. Automatica 27(6), 997–1009 (1991)
70. Mohanty, S., Chattopadhyay, A., Peralta, P.: Adaptive residual useful life estimation of a structural hotspot. Journal of Intelligent Material Systems and Structures 21(3), 321–335 (2010)
71. Mosterman, P., Biswas, G.: Diagnosis of continuous valued systems in transient operating regions. IEEE Transactions on Systems, Man, and Cybernetics, Part A 29(6), 554–565 (1999)
72. Muscettola, N., Dorais, G.A., Fry, C., Levinson, R., Plaunt, C.: Idea: Planning at the core of autonomous reactive agents. In: Proceedings of the 3rd International NASA Workshop on Planning and Scheduling for Space (2002)
73. Narasimhan, S., Biswas, G.: An approach to model-based diagnosis of hybrid systems. In: Tomlin, C.J., Greenstreet, M.R. (eds.) HSCC 2002. LNCS, vol. 2289, pp. 308–322. Springer, Heidelberg (2002)
74. Narasimhan, S., Dearden, R., Benazera, E.: Combining particle filters and consistency based approaches for monitoring and diagnosis of stochastic hybrid systems. In: Proceedings of the 15th International Workshop on Principles of Diagnosis (DX 2004), pp. 123–128 (2004)

75. Narasimhan, S., Dearden, R., Benazera, E.: Combining particle filters and consistency-based approaches for monitoring and diagnosis of stochastic hybrid systems. In: Proceedings of the15th International Workshop on Principles of Diagnosis (DX 2004). Citeseer, Carcassonne (2004)
76. Omlin, C.W., Thornber, K.K., Giles, L.C.: Fuzzy Finite-State Automata Can be Deterministically Encoded into Recurrent Neural Networks. IEEE Trans. on Fuzzy Systems 6(1), 8–79 (1998)
77. Patton, R.J., Frank, P.M., Clarke, R.N.: Fault diagnosis in dynamic systems: theory and application. Prentice-Hall, Inc. (1989)
78. Patton, R.J., Chen, J.: A re-examination of the relationship between parity space and observer-based approaches in fault diagnosis. European Journal of Diagnosis and Safety in Automation 1(2), 183–200 (1991)
79. Pell, B., Bernard, D., Chien, S., Gat, E., Muscettola, N., Nayak, P., Wagner, M., Williams, B.: An autonomous spacecraft agent prototype. Autonomous Robots 5(1), 29–52 (1998)
80. Pouliezos, A., Stavrakakis, G., Lefas, C.: Fault detection using parameter estimation. Quality and Reliability Engineering International 5, 283–290 (1985)
81. Pulido, B., Gonzalez, C.: Possible conflicts: A compilation technique for consistency-based diagnosis. IEEE Transactions on Systems, Man, and Cybernetics – Part B: Cybernetics 34(5), 2192–2206 (2004)
82. Qiu, Z., Gertler, J.: Robust FDI and H_{inf} optimization. In: Proceedings of the 32nd IEEE Conference on Control and Decision, CDC 1993, San Antonio, Texas, pp. 247–252 (1993)
83. Reiter, R.: A theory of diagnosis from first principles. Artificial Intelligence 32(1), 57–95 (1987)
84. Ribot, P., Pencolé, Y., Combacau, M.: Prognostics for the maintenance of distributed systems. In: Proceedings of the International Conference on Prognostics and Health Management (PHM 2008), Denver, USA, pp. 6–10 (2008)
85. Ribot, P., Pencolé, Y., Combacau, M.: Diagnosis and prognosis for the maintenance of complex systems. In: Proceedings of the IEEE International Conference on Systems, Man and Cybernetics SMC 2009, San Antonio, Texas, USA, pp. 4146–4151 (2009)
86. Rienmuller, T., Bayoudh, M., Hofbaur, M., Travé-Massuyès, L.: Hybrid estimation through synergic mode-set focusing. In: Proceedings of the 7th IFAC Symposium on Fault Detection, Supervision and Safety of Technical Processes, Safeprocess 2009, Barcelona, Spain (2009)
87. Rienmuller, T., Hofbaur, M., Bayoudh, M., Travé-Massuyès, L.: Mode set focused hybrid estimation. International Journal of Applied Mathematics and Computer Science (AMCS) 23(1), 131–144 (2013)
88. Russel, S., Norvig, P.: Artificial Intelligence, A modern Approach, 2nd edn. Prentice Hall Series in Artificial Intelligence (2003)
89. Sachenbacher, M., Williams, B.: Diagnosis as semiring-based constraint optimization. In: Proceedings of the European Conference on Artificial Intelligence, ECAI 2004, vol. 16, pp. 873–879 (2004)
90. Sampath, M., Sengupta, R., Lafortune, S., Sinnamohideen, K., Teneketzis, D.: Diagnosability of discrete-event systems. IEEE Transactions on Automatic Control 40(9), 1555–1575 (1995)
91. Saxena, A., Celaya, J., Saha, B., Saha, S., Goebel, K.: Metrics for offline evaluation of prognostic performance. International Journal of Prognostics and Health Management 1(1), 20 (2010)

92. Staroswiecki, M., Comtet-Varga, G.: Analytical redundancy relations for fault detection and isolation in algebraic dynamic systems. Automatica 37(5), 687–699 (2001)
93. Staroswiecki, M., Declerck, P.: Analytical redundancy in non linear interconnected systems by means of structural analysis. In: Proceedings of the IFAC Symposium on Advanced Information Processing in Automatic Control, pp. 51–55 (1989)
94. Subias, A., Exposito, E., Chassot, C., Travé-Massuyès, L., Drira, K.: Self-adapting strategies guided by diagnosis and situation assessment in collaborative communicating systems. In: Proceedings of the International Workshop on Principles of Diagnosis (DX 2010), Portland, USA, pp. 329–336 (2010)
95. Arulampalam, M.S., Maskell, S., Gordon, N., Clapp, T.: A tutorial on particle filters for online nonlinear/non-gaussian bayesian tracking. IEEE Transactions on Signal Processing 50(2), 174–188 (2002)
96. Thrun, S., Langford, J., Verma, V.: Risk sensitive particle filters. In: Advances in Neural Information Processing Systems 2, pp. 961–968. MIT Press (2002)
97. Travé-Massuyès, L., Pons, R., Tornil, S., Escobet, T.: The CA-En Diagnosis System and its Automatic Modelling Method. Computación y Sistemas 5(2), 128–143 (2001)
98. Travé-Massuyès, L., Dague, P.: Modèles et raisonnements qualitatifs, Lavoisier (2003)
99. Travé-Massuyès, L., Calderon-Espinoza, G.: Timed fault diagnosis. In: Proceedings of the European Control Conference, ECC 2007, Kos, Greece (2007)
100. Vachtsevanos, G., Lewis, F., Roemer, M., Hess, A., Wu, B.: Intelligent fault diagnosis and prognosis for engineering systems. Wiley Online Library (2006)
101. Venkatasubramanian, V., Rengaswamy, R., Yin, K., Kavuri, S.N.: A review of process fault detection and diagnosis Part I: Quantitative model-based methods. Journal of Computers and Chemical Engineering 27(3), 293–311 (2003)
102. Venkatasubramanian, V., Rengaswamy, R., Kavuri, S.N.: A review of process fault detection and diagnosis Part II: Qualitative models and search strategies. Journal of Computers and Chemical Engineering 27(3), 313–326 (2003)
103. Venkatasubramanian, V., Rengaswamy, R., Kavuri, S.N., Yin, K.: A review of process fault detection and diagnosis Part III: Process history based methods. Journal of Computers and Chemical Engineering 27(3), 327–346 (2003)
104. Vento, J., Puig, V., Sarrate, R., Travé-Massuyés, L.: Fault Detection and Isolation of Hybrid Systems using Diagnosers that Reason on Components. In: Proceedings of the IFAC Symposium on Fault Detection, Supervision and Safety of Technical Processes, Safeprocess 2012 8(1), Mexico City, Mexico, pp. 1250–1255 (2012)
105. Verma, V., Gordon, G., Simmons, R., Thrun, S.: Real-time fault diagnosis. IEEE Robotics and Automation Magazine 11(2), 56–66 (2004)
106. Verma, V., Thrun, S., Simmons, R.: Variable resolution particle filter. In: Proceedings of the International Joint Conference on Artificial Intelligence (IJCAI 2003), pp. 976–981 (2003)
107. Washington, R.: On-board real-time state and fault identification for rovers. In: Proceedings of the IEEE International Conference on Robotic and Automation, ICRA 2000, San Francisco, CA, USA (2000)
108. Washio, T., Motoda, H., Niwa, Y.: Discovering admissible model equations from observed data based on scale-types and identity constraints. In: Proceedings of the 16th International Joint Conference on Artificial Intelligence, IJCAI 1999, vol. 16, pp. 772–779. Lawrence Erlbaum, Mahwah (1999)
109. Weld, D., De Kleer, J.: Readings in qualitative reasoning about physical systems. Morgan Kaufmann Publishers Inc., San Francisco (1989)

110. Williams, B., Nayak, P.: A model-based approach to reactive self-configuring systems. In: Proceedings of the National Conference on Artificial Intelligence, AAAI 1996, Portland, Oregon, pp. 971–978 (1996)
111. Williams, B., Ragno, R.: Conflict-directed A* and its role in model-based embedded systems. Journal of Discrete Applied Mathematics (2003)
112. Witteveen, C., Roos, N., van der Krogt, R., de Weerdt, M.: Diagnosis of single and multi-agent plans. In: Proceedings of the 4th International Joint Conference on Autonomous Agents and Multiagent Systems, pp. 805–812. ACM (2005)
113. Zaytoon, J. (ed.): Systèmes dynamiques hybrides. Hermès Science Publication, Paris (2001)
114. BRIDGE group: Bridging AI and Control Engineering model-based diagnosis approaches, `http://monet.aber.ac.uk:8080/monet/monetinfo/monetbridge.htm`
115. Network of Excellence MONET II, `http://monet.aber.ac.uk:8080/monet/index.html`
116. GDR MACS: Groupement de Recherche Modélisation, Analyse et Conduite des Systémes Dynamiques, `http://www.univ-valenciennes.fr/GDR-MACS/`
117. GDR I3: Groupement de Recherche Information, Interaction, Intelligence, `http://www.irit.fr/GDR-I3/`
118. AFIA: Association Française d'Intelligence Atificielle, `http://www.afia.asso.fr/`
119. BRIDGE Workshop (2001), `http://www.di.unito.it/~dx01`
120. 7th IFAC Symposium on Fault Detection, Supervision and Safety of Technical Processes Safeprocess 2009 (2009), `http://safeprocess09.upc.es/`

Diagnosis of Actuator Parameter Faults in Wind Turbines Using a Takagi-Sugeno Sliding Mode Observer

Sören Georg and Horst Schulte

HTW Berlin, Department of Engineering I, Control Engineering
Wilhelminenhofstr. 75A, 12459 Berlin, Germany
{soeren.georg,horst.schulte}@htw-berlin.de
www.rt.htw-berlin.de

Abstract. A Takagi-Sugeno sliding mode observer (TS SMO) is used to detect and reconstruct parameter faults in the actuator dynamics of wind turbines. The actuator dynamics of the pitch system and of the generator are modeled as first or second order delay models. By choosing appropriate actuator fault matrices for the TS SMO design, parameter changes or faults in the actuator dynamics can be calculated from the reconstructed additive fault signal. Example simulations are performed using both a reduced-order model and the aero-elastic wind turbine simulation FAST by NREL. The presented fault reconstruction method appears promising to be used for fast fault detection schemes. The accuracy of the reconstruction, however, is still in need of improvement and the method lacks robustness with respect to model uncertainties, since no satisfactory reconstruction can be achieved in FAST simulations.

Keywords: Fault Diagnosis, Wind Turbines, Actuator Dynamics, Sliding Mode Observer, Takagi-Sugeno Models.

1 Introduction

Fault diagnosis and fault-tolerant control for wind turbines are ever more important research topics. Especially for offshore wind turbines, reliable fault diagnosis algorithms are key to developing more intelligent control concepts that allow to keep the turbine running in the presence of non-critical faults.

In this paper, a Takagi-Sugeno sliding mode observer (TS SMO) is used to detect and partially reconstruct parameter faults in the actuator dynamics of wind turbines. The actuator faults are reflected in the delay time constant of a first oder model or in the characteristic frequency and the damping of a second order delay model. The parameter variations due to faults are formulated in the framework of an observer-based fault detection scheme as disturbance terms [1].

Other widely used techniques for fault detection and isolation (FDI) are parameter estimation schemes, notably least squares algorithms [2]. In [3], for example, a block least squares technique is used for fault diagnosis for a wind turbine FDI benchmark model [4].

J. Korbicz and M. Kowal (eds.), *Intelligent Systems in Technical and Medical Diagnostics*, Advances in Intelligent Systems and Computing 230,
DOI: 10.1007/978-3-642-39881-0_2, © Springer-Verlag Berlin Heidelberg 2014

In contrast to FDI using online system identification techniques, in this paper the parameters are not estimated in a statistical sense but reconstructed using a continuous approximation of the discontinuous switching term of the TS sliding mode observer. This enables faster detection of parameter changes compared to the recursive least squares (RLS) estimator. However, the accuracy of the reconstruction is still in need of improvement.

A comparison in [2] has shown that in the case of abrupt faults, state reconstruction reacts faster to a fault occurrence than parameter estimation. Another disadvantage of parameter estimation in the context of FDI is the need to ensure a sufficient excitation level during the estimation process. If the level of excitation is not high enough, problems may arise, since the estimator becomes more and more sensitive and small disturbances may lead to an unstable estimation of parameter faults.

This paper is organised as follows. In section 2, a brief overview of the TS sliding mode observer concept is given. In section 3, the TS SMO is used for the reconstruction of parameter faults in the pitch angle and generator torque actuators in a dynamic wind turbine simulation. The methods to reconstruct the parameter faults are described separately for the case of first and second order delay models for the actuator dynamics.

2 Takagi-Sugeno Sliding Mode Observer

In this work, the Takagi-Sugeno sliding mode observer concept from [5] is used, which is a nonlinear extension of the sliding mode observer concept by Edwards and Spurgeon [6], [7]. A detailed introduction to Takagi-Sugeno models [8] can be found in [9].

The TS SMO design is based on a nonlinear MIMO system subject to unmodeled dynamics $\boldsymbol{\xi}$ and actuator faults $\mathbf{f}_a$ in TS structure:

$$\dot{\mathbf{x}} = \sum_{i=1}^{N_r} h_i(\mathbf{z}) \left(\mathbf{A}_i\mathbf{x} + \mathbf{B}_i\mathbf{u} + \mathbf{D}_i\boldsymbol{\xi} + \mathbf{E}_i\mathbf{f}_a \right) , \qquad \mathbf{y} = \mathbf{C}\mathbf{x} , \tag{1}$$

where h_i denote the TS membership functions, for which the convexity condition $\sum_{i=1}^{N_r} h_i(\mathbf{z}) = 1$ is fulfilled. $\mathbf{z}$ denotes the vector of premise variables, which can consist of system states and inputs as well as of external variables. Sensor faults are not considered here.

Transformed form of the TS Sliding Mode Observer. The sliding mode observer concept by Edwards and Spurgeon [6] makes use of a series of linear coordinate transformations, whereby the underlying system is separated into the measurable system states $\mathbf{y}$ and non-measurable system states $\mathbf{x}_1$. The transformed state vector is given by $\left(\mathbf{x}_1^T \;\; \mathbf{y}^T\right)^T$. This separation is achieved by introducing a transformation $\mathbf{T}_c = \left[\mathbf{N}_\mathbf{C}^T \;\; \mathbf{C}\right]^T$, where $\mathbf{N}_\mathbf{C}$ denotes the null-space of $\mathbf{C}$. Applying the series of coordinate transformations[1] $\mathbf{T}_i = \bar{\mathbf{T}}_{L,i}\,\tilde{\mathbf{T}}_{DE,i}\,\mathbf{T}_c$, the transformed TS system is obtained as

$$\dot{\mathbf{x}}_1 = \sum_{i=1}^{N_r} h_i(\mathbf{z})\left(\mathcal{A}_{11,i}\,\mathbf{x}_1 + \mathcal{A}_{12,i}\,\mathbf{y} + \mathcal{B}_{1,i}\,\mathbf{u}\right) \tag{2}$$

$$\dot{\mathbf{y}} = \sum_{i=1}^{N_r} h_i(\mathbf{z})\left(\mathcal{A}_{21,i}\,\mathbf{x}_1 + \mathcal{A}_{22,i}\,\mathbf{y} + \mathcal{B}_{2,i}\,\mathbf{u} + \mathcal{D}_{2,i}\,\boldsymbol{\xi} + \mathcal{E}_{2,i}\,\mathbf{f}_a\right), \tag{3}$$

where the transformed system matrices are given by

$$\mathcal{A}_i = \mathbf{T}_i\mathbf{A}_i\mathbf{T}_i^{-1} = \begin{bmatrix}\mathcal{A}_{11,i} & \mathcal{A}_{12,i}\\ \mathcal{A}_{21,i} & \mathcal{A}_{22,i}\end{bmatrix}, \mathcal{B}_i = \mathbf{T}_i\,\mathbf{B}_i = \begin{bmatrix}\mathcal{B}_{1,i}^T & \mathcal{B}_{2,i}^T\end{bmatrix}^T,$$

$$\mathcal{D}_i = \mathbf{T}_i\,\mathbf{D}_i = \begin{bmatrix}\mathbf{0}^T & \mathcal{D}_{2,i}^T\end{bmatrix}^T, \mathcal{E}_i = \mathbf{T}_i\,\mathbf{E}_i = \begin{bmatrix}\mathbf{0}^T & \mathcal{E}_{2,i}^T\end{bmatrix}^T.$$

For a stable observer to exist, the following three existence conditions have to be fulfilled [7], [5]:

Condition 1. The uncertainties and actuator faults are unknown but bounded: $\|\boldsymbol{\xi}^T(t)\ \mathbf{f}_a^T(t)\|^T \le \Xi$. Furthermore, individual bounds exist: $\|\boldsymbol{\xi}(t)\|^T \le \Xi_\xi$ and $\mathbf{f}_a(t) \le \Xi_{f_a}$. Moreover, the system states and inputs are assumed to be bounded.

Condition 2. Let q be defined as the number of columns of the combined matrices $[\mathbf{D}_i\ \mathbf{E}_i]$. Then, the condition $q = \mathrm{rank}(\mathbf{C}[\mathbf{D}_i\ \mathbf{E}_i]) = \mathrm{rank}[\mathbf{D}_i\ \mathbf{E}_i]$ must be fulfilled. Furthermore, it must hold that $p > q$, where p is the number of measurable system states.

Condition 3. All invariant zeros of $(\mathbf{A}_i, [\mathbf{D}_i\ \mathbf{E}_i], \mathbf{C})$ must lie in $\mathbb{C}_-$.
The TS sliding mode observer in transformed form is given by

$$\dot{\hat{\mathbf{x}}}_1 = \sum_{i=1}^{N_r} h_i(\mathbf{z})\left(\mathcal{A}_{11,i}\,\hat{\mathbf{x}}_1 + \mathcal{A}_{12,i}\,\hat{\mathbf{y}} + \mathcal{B}_{1,i}\,\mathbf{u} - \mathcal{A}_{12,i}\,\mathbf{e}_y\right) \tag{4}$$

$$\dot{\hat{\mathbf{y}}} = \sum_{i=1}^{N_r} h_i(\mathbf{z})\left(\mathcal{A}_{21,i}\,\hat{\mathbf{x}}_1 + \mathcal{A}_{22,i}\,\hat{\mathbf{y}} + \mathcal{B}_{2,i}\,\mathbf{u} - (\mathcal{A}_{22,i} - \mathcal{A}_{22}^s)\,\mathbf{e}_y + \boldsymbol{\nu}\right), \tag{5}$$

where $\mathbf{e}_y = \hat{\mathbf{y}} - \mathbf{y}$ denotes the output error and $\mathcal{A}_{22}^s$ is a stable design matrix. In this work, the following weighted discontinuous term, introduced in [10], is used:

$$\boldsymbol{\nu} = \begin{cases} -\rho\,\dfrac{\mathbf{P}_2\,\mathbf{W}\,\mathbf{e}_y}{\|\mathbf{P}_2\,\mathbf{W}\,\mathbf{e}_y\|} & , \quad (\mathbf{e}_y \ne \mathbf{0}) \\ \mathbf{0} & , \quad (\mathbf{e}_y = \mathbf{0}) \end{cases} \tag{6}$$

[1] See [7] and [5] for a description of the transformation matrices. The MATLAB® commands from [7] can be used to obtain the transformations in practical applications.

where $\boldsymbol{\rho} = \text{diag}(\rho_1 \cdots \rho_p)$ is a diagonal, positive definite gain matrix. The weighting matrix $\mathbf{W}$ is a positive definite, diagonal matrix consisting of the reciprocal values of the estimated maximum absolute values of the output vector components: $\mathbf{W} = \text{diag}(W_1 \cdots W_p) = \text{diag}\left(\frac{1}{|y_{\max,1}|} \cdots \frac{1}{|y_{\max,p}|}\right)$. $\mathbf{P}_2$ is the unique symmetric positive definite (s.p.d) solution of the Lyapunov equation $\mathbf{P}_2 \mathcal{A}_{22}^s + \mathcal{A}_{22}^{s\,T} \mathbf{P}_2 = -\mathbf{Q}_2$, where $\mathbf{Q}_2$ is a symmetric positive definite design matrix. Once the sliding surface $\mathcal{S} = \{\mathbf{e}(t) \in \mathbb{R}^n : \mathbf{e}_y = \mathbf{0}\}$ is reached, the observer tries to maintain the sliding motion on $\mathcal{S}$.
In addition to the fault and uncertainty bounds in existence condition 1, the error vector $\mathbf{e}_1 = \hat{\mathbf{x}}_1 - \mathbf{x}_1$ of the unmeasurable states is assumed to be bounded: $\|\mathbf{e}_1\| < \Gamma$. Investigating the stability of the error system $\mathbf{e}_1$ of the unmeasurable states leads to an LMI (linear matrix inequality) design condition, which was presented in [5] and is omitted here for the sake of brevity.

Existence Condition for Sliding Motion. In [10], it was shown, that the TS SMO using the modified switching term (6) is able to establish an ideal sliding motion in finite time if the following condition is fulfilled:

$$\mathbf{e}_y^T \left[(\boldsymbol{\rho} - \mathcal{K}_{\max} \mathbf{I}_{p\times p}) (\mathbf{P}_2 \mathbf{W})^2\right] \mathbf{e}_y > \mathbf{0}, \tag{7}$$

where $\mathcal{K}_{\max}$ is given by $\mathcal{K}_{\max} := \|\mathcal{A}_{21}\|_{\max} \Gamma + \|[\mathcal{D}_2\, \mathcal{E}_2]\|_{\max} \Xi$, in a case without sensor faults.
The weighted switching term (6) was introduced in [10] to achieve good simultaneous reconstruction of fault signals in systems were the orders of magnitude of the outputs differ significantly. In the case of wind turbines, this is the case for the two control signals, the pitch angle and the generator torque, which are also included as system states and outputs in the model equations.

Fault Reconstruction. For simulations, the discontinuous switching term (6) is replaced by a continuous approximation: $\boldsymbol{\nu}_{\text{eq}} = -\boldsymbol{\rho} \frac{\mathbf{P}_2 \mathbf{W} \mathbf{e}_y}{\|\mathbf{P}_2 \mathbf{W} \mathbf{e}_y\| + \delta}$, with a small positive constant δ. The equivalent output injection signal $\boldsymbol{\nu}_{\text{eq}}$ describes the average behaviour of the discontinuous term (6). If no sensor faults are present, unmodeled dynamics ($\boldsymbol{\xi}$) and actuator faults ($\mathbf{f}_a$) can be reconstructed using the relation $\left(\hat{\boldsymbol{\xi}}^T\ \hat{\mathbf{f}}_a^T\right)^T = \left(\sum_{i=1}^{N_r} h_i(\mathbf{z}) [\mathcal{D}_{2,i}\, \mathcal{E}_{2,i}]\right)^+ \boldsymbol{\nu}_{\text{eq}}$ [7], [5].

3 Actuator Parameter Faults in Wind Turbines

In this section, the TS SMO is used to reconstruct parameter faults in the pitch and generator dynamics of wind turbines. As a nominal simulation model, the reduced-order model from [11] with four degrees of freedom (torsion-flexible drivetrain, fore-aft tower top deflection, flap-wise blade tip deflection) is used with the parameters of the NREL 5 MW reference turbine [12]. For comparison, two faults are also simulated using the aero-elastic code FAST by NREL [13]. The underlying observer model for the TS SMO contains only drivetrain degrees of freedom and first or second order delay models for the pitch and generator dynamics. The derivation of the TS model is described in [10] and is omitted here. The observer state vector is $\mathbf{x} = (\theta_s \quad \omega_r \quad \omega_g \quad \beta \quad T_g)^T$, where θ_s denotes the shaft torsion angle, ω_r and ω_g the rotor and generator angular velocities, β the pitch angle and T_g the generator torque. The input vector is $\mathbf{u} = (\beta_d \quad T_{g,d})^T$, with the demanded pitch angle and generator torque. All states except for the torsional angle θ_s are measurable: $\mathbf{y} = (\omega_r \quad \omega_g \quad \beta \quad T_g)^T$. Furthermore, the wind speed v, which is part of the TS premise variable vector $\mathbf{z}$, is assumed to be measurable for the TS SMO design. By making this assumption, it is easier to validate the stability of the observer than for a case with unmeasurable premise variables [14]. However, in first tests it could be verified that when using a wind speed estimate obtained from a TS observer [15], the simulation results hardly differ from those presented in this work.

The following parameters were used for the TS SMO design: $\mathcal{A}_{22}^s = -10 \cdot \mathbf{I}_{4\times 4}$, $\mathbf{Q}_2 = \mathbf{I}_{4\times 4}$, $\delta = 0.05$. To obtain the weighting matrix $\mathbf{W}$, the maximum absolute values of the individual states were estimated as: $\omega_{r,\max} = \omega_{g,\max} = 1.6\,\frac{\text{rad}}{\text{s}}$, $\beta_{\max} = 1.5708\,\text{rad}$, $T_{g,\max} = 4.2 \cdot 10^6\,\text{Nm}$. The elements of the sliding mode gain matrix were chosen as: $\rho_{\omega_r} = 200$, $\rho_{\omega_g} = 200$, $\rho_\beta = 200$, $\rho_{T_g} = 6 \cdot 10^8$. No $\mathbf{D}_i$-matrices for unmodeled dynamics were included for the observer design, as this yielded no improved fault reconstruction. The choice of the actuator fault matrices $\mathbf{E}_i$ is described below for the respective examples.

Altered Actuator Dynamics for First Order Delay Models. The pitch and generator dynamics are modeled as first order delay models: $\dot{\beta} = -\frac{1}{\tau}\beta + \frac{1}{\tau}\beta_d$, $\dot{T}_g = -\frac{1}{\tau_g}T_g + \frac{1}{\tau_g}T_{g,d}$, where the delay time constants in the fault free case are chosen as $\tau = 0.1$ s and $\tau_g = 0.02$ s.
Alterations in the actuator dynamics of pitch angle and generator torque can be examined by means of changed time delay constants τ and τ_g. Consider a change of the pitch delay time constant: $\tau \to \tilde{\tau}$ and its reciprocal value: $\tilde{a} := \frac{1}{\tilde{\tau}}$, which can be written as a sum of the nominal value $a = \frac{1}{\tau}$ and an offset value: $\tilde{a} = a + \Delta a$. The altered dynamics is then given by

$$\dot{\beta} = -\tilde{a}\,\beta + \tilde{a}\,\beta_d = -a\,\beta + a\,\beta_d - \Delta a\,\beta + \Delta a\,\beta_d\,. \tag{8}$$

Analogously, altered generator torque dynamics can be examined using the reciprocal value a_g of the delay time constant τ_g:

$$\dot{T}_g = -a_g\,T_g + a_g\,T_{g,d} - \Delta a_g\,T_g + \Delta a_g\,T_{g,d}\,. \tag{9}$$

Both offset values are time-dependent: $\Delta a = \Delta a\,(t)$, $\Delta a_g = \Delta a_g\,(t)$. Defining the offset matrices $\Delta\mathbf{A}$ and $\Delta\mathbf{B}$ as

$$\Delta\mathbf{A} = \begin{bmatrix} \mathbf{0}_{n-2\times n-2} & \mathbf{0}_{n-2\times 1} & \mathbf{0}_{n-2\times 1} \\ \mathbf{0}_{1\times n-2} & -\Delta a & 0 \\ \mathbf{0}_{1\times n-2} & 0 & -\Delta a_g \end{bmatrix}, \qquad \Delta\mathbf{B} = \begin{bmatrix} \mathbf{0}_{n-2\times 1} & \mathbf{0}_{n-2\times 1} \\ \Delta a & 0 \\ 0 & \Delta a_g \end{bmatrix}, \tag{10}$$

where n denotes the number of states of the observer system, the observer system dynamics (without feedback terms) with altered actuator dynamics can be written as

$$\dot{\mathbf{x}} = \sum_{i=1}^{N_r} h_i(\mathbf{z})\left(\mathbf{A}_i\mathbf{x} + \mathbf{D}_i\,\boldsymbol{\xi}\right) + \mathbf{B}\mathbf{u} + \Delta\mathbf{A}\,\mathbf{x} + \Delta\mathbf{B}\,\mathbf{u}\,. \tag{11}$$

In order to reconstruct the altered dynamics parameters Δa and Δa_g, a fault matrix $\mathbf{E}$ and an actuator fault vector $\mathbf{f}_a$ must be found such that

$$\begin{aligned}\mathbf{E}\,\mathbf{f}_a &= \Delta\mathbf{A}\,\mathbf{x} + \Delta\mathbf{B}\,\mathbf{u} \\ &= \begin{pmatrix}0 & \cdots & 0 & \Delta a\,(\beta_d - \beta) & \Delta a_g\,(T_{g,d} - T_g)\end{pmatrix}^T .\end{aligned} \tag{12}$$

Choosing

$$\mathbf{E} = \begin{bmatrix}0 \cdots 1\,0 \\ 0 \cdots 0\,1\end{bmatrix}^T \quad \text{and} \quad \mathbf{f}_a = \begin{pmatrix}\Delta a\,(\beta_d - \beta) \\ \Delta a_g\,(T_{g,d} - T_g)\end{pmatrix}, \tag{13}$$

condition (12) is fulfilled. Furthermore, in the case when no unmodeled dynamics are considered ($\mathbf{D} = \mathbf{0}$), this choice of $\mathbf{E}$ guarantees that existence condition 2 for the TS sliding mode observer is fulfilled: Since ω_r, ω_g, β and T_g are always measurable, the number of measurable states p is greater than q, the number of columns of $\mathbf{E}$. Also, $q = \operatorname{rank}(\mathbf{E})$.
The reconstructed actuator fault vector is thus given by

$$\hat{\mathbf{f}}_a = \begin{pmatrix}\Delta\hat{a}\,(\beta_d - \beta) \\ \Delta\hat{a}_g\,(T_{g,d} - T_g)\end{pmatrix}, \tag{14}$$

and is obtained using the equivalent output injection signal $\boldsymbol{\nu}_{\text{eq}}$ via the relation $\hat{\mathbf{f}}_a = \boldsymbol{\mathcal{E}}_2^+\,\boldsymbol{\nu}_{\text{eq}}$ (since all $\mathbf{E}_i$ are equal), such that the individual reconstructed altered dynamics parameters can be obtained as

$$\Delta\hat{a} = \frac{\hat{f}_{a,1}}{\beta_d - \beta} \quad (\beta_d \neq \beta) \qquad \Delta\hat{a}_g = \frac{\hat{f}_{a,2}}{T_{g,d} - T_g} \quad (T_{g,d} \neq T_g)\,. \tag{15}$$

In order to avoid large peaks in the reconstruction of τ and τ_g in case the differences between the demanded and actual actuator signals are close to zero, $\Delta\hat{a}$ and $\Delta\hat{a}_g$ are only calculated using the equations in (15) if $|\beta_d - \beta| > \varepsilon_{\beta,\beta_d}$ and $|T_{g,d} - T_g| > \varepsilon_{T_g,T_{g,d}}$, with small positive constants $\varepsilon_{\beta,\beta_d}$ and $\varepsilon_{T_g,T_{g,d}}$. Otherwise, $\Delta\hat{a}$ and $\Delta\hat{a}_g$ are fixed at their respective values of the previous timestep. If $\hat{f}_{a,1}$ assumes 0 for more than one timestep, there is no pitch dynamics fault present. In this case, the fault parameter $\Delta\hat{a}_1$ is set to zero. Equally, if $\hat{f}_{a,2}$ is 0 for more than one timestep, $\Delta\hat{a}_g$ is set to zero. In practice, these conditions are checked by defining small constants ε_1 and ε_g and checking whether

$\left|\hat{f}_{a,1}\right| < \varepsilon_1$ and $\left|\hat{f}_{a,g}\right| < \varepsilon_g$. The faulty delay time constants can be calculated as $\tilde{\tau} = \frac{1}{\tilde{a}} = \frac{1}{a+\Delta\hat{a}}$ and $\tilde{\tau}_g = \frac{1}{\tilde{a}_g} = \frac{1}{a_g+\Delta\hat{a}_g}$.

As an example for altered pitch actuator dynamics, a simulation of the nominal model was performed with turbulent wind input and a pitch actuator parameter fault active between 50 s and 150 s (step of τ from 0.1 s to 0.3 s). The reconstructed delay time constant $\hat{\tau}$ is depicted in the upper part of Fig. 1.

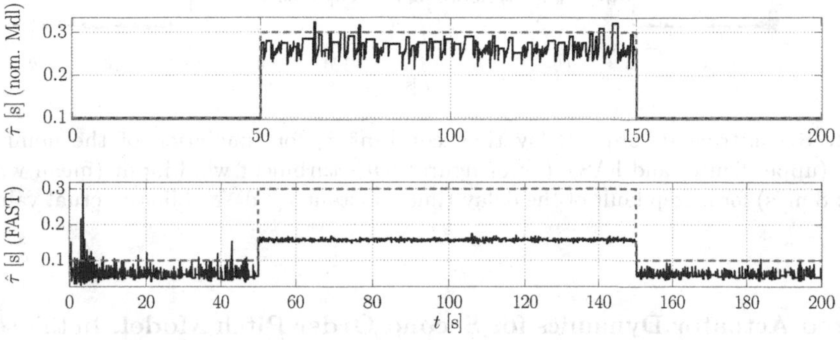

Fig. 1. Reconstructed pitch dynamics constant $\hat{\tau}$ for simulation of the nominal model (upper figure) and FAST (lower figure) with turbulent wind input (mean wind speed: 18 m/s) for a fault of the delay time constant τ (dashed lines: actual values of τ)

The magnitude of the faulty delay time constant (0.3 s) in Fig. 1 is reconstructed to a degree of about 80 %. However, due to the relatively large choice of $\varepsilon_{\beta,\beta_d} = 0.005$ rad, $\Delta\hat{a}$ in (15) is calculated only at certain instances, such that the time-series of $\hat{\tau}$ is discretized. Choosing a smaller value of $\varepsilon_{\beta,\beta_d}$, on the other hand, would lead to more marked spikes in the reconstruction of $\hat{\tau}$. ε_1 was set to $\varepsilon_1 = 10^{-6}\,\frac{\text{rad}}{\text{s}}$. The same fault was simulated with FAST. The constant $\varepsilon_{\beta,\beta_d}$ was set to 0.002 rad in this case. The reconstructed delay time constant $\hat{\tau}$ is depicted in the lower part of Fig. 1. In this case, the faulty value of 0.3 is reconstructed only to a degree of about 25 %.

The quality of the fault reconstruction in both cases, especially when using FAST, might not be high enough for fault-tolerant control purposes. However, a fast fault detection would be possible by defining appropriate thresholds. Using the parameter $\varepsilon_{\beta,\beta_d}$ as above, the reaction time to the fault is about 0.1 s.

A torque actuator dynamics fault was also simulated in separate simulations with a mean wind speed of 8 m/s. A step fault of the delay time constant τ_g (step from 0.02 s to 0.1 s) was active between 40 s and 160 s. The reconstructed delay time constant is depicted in Fig. 2. The constant $\varepsilon_{T_g,T_{g,d}}$ was set to 10000 Nm, which is still a relatively small value given that the order of magnitude of the generator torque is in the region of 10^6 Nm. ε_g was set to $\varepsilon_g = 10^{-4}\,\frac{\text{Nm}}{\text{s}}$.

As in the case of the pitch actuator fault, the reconstruction is better when using the nominal model than when using FAST. The fault reaction times are 0.1 s (nom. model) and 0.2 s (FAST), respectively.

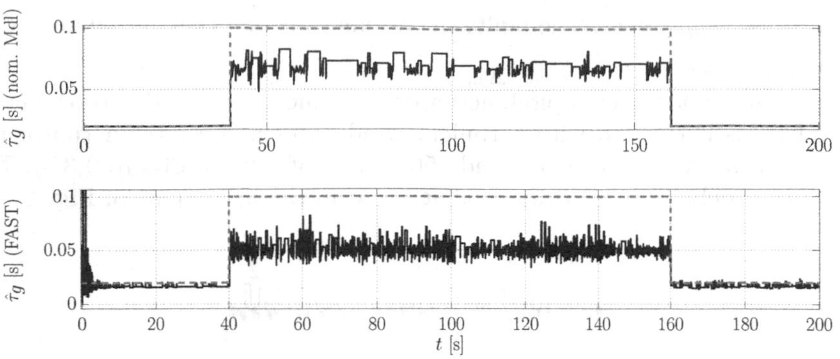

Fig. 2. Reconstructed torque delay time constant $\hat{\tau}_g$ for simulation of the nominal model (upper figure) and FAST (lower figure) with turbulent wind input (mean wind speed: 8 m/s) for a step fault of the delay time constant τ_g (dashed lines: actual values of τ_g)

Altered Actuator Dynamics for Second Order Pitch Model. In this section, it is investigated whether altered pitch dynamics can also be reconstructed for a second order delay model. In this case, the pitch dynamics is given by

$$\begin{pmatrix}\dot{\beta}\\ \ddot{\beta}\end{pmatrix} = \begin{bmatrix}0 & 1\\ -\omega_n^2 & -2\,\zeta\,\omega_n\end{bmatrix}\begin{pmatrix}\beta\\ \dot{\beta}\end{pmatrix} + \begin{pmatrix}0\\ \omega_n^2\end{pmatrix}\beta_d\,, \tag{16}$$

where ω_n and ζ are the characteristic frequency and the damping of the second order delay model, respectively. The (measurable) pitch rate $\dot{\beta}$ is then included in the state vector and in the output vector of both the simulation model and of the TS SMO.

Due to the nonlinear terms ω_n^2 and $-2\,\zeta\,\omega_n$, the modeling of the altered actuator dynamics for the second order pitch model is not as straightforward as for the first order delay model. As a first step, it is advantageous to substitute the nonlinear terms by linear parameters (a_1 and a_2), such that an alteration in actuator dynamics (parameters ω_n and ζ) can then be formulated as an additive offset on the substituted parameters:

$$a_1 := \omega_n^2\,, \qquad a_2 := 2\,\zeta\,\omega_n \tag{17}$$

$$a_1 \;\rightarrow\; \tilde{a}_1 = a_1 + \Delta a_1\,, \qquad a_2 \;\rightarrow\; \tilde{a}_2 = a_2 + \Delta a_2 \tag{18}$$

The altered pitch dynamics is then given by

$$\begin{pmatrix}\dot{\beta}\\ \ddot{\beta}\end{pmatrix} = \begin{bmatrix}0 & 1\\ -a_1 & -a_2\end{bmatrix}\begin{pmatrix}\beta\\ \dot{\beta}\end{pmatrix} + \begin{pmatrix}0\\ a_1\end{pmatrix}\beta_d + \begin{bmatrix}0 & 1\\ -\Delta a_1 & -\Delta a_2\end{bmatrix}\begin{pmatrix}\beta\\ \dot{\beta}\end{pmatrix} + \begin{pmatrix}0\\ \Delta a_1\end{pmatrix}\beta_d$$

Analogously to the case of first order pitch dynamics, offset matrices $\Delta\mathbf{A}$ and $\Delta\mathbf{B}$ can be defined as

$$\Delta\mathbf{A} = \begin{bmatrix} \mathbf{0}_{n-2\times n-3} & & \mathbf{0}_{n-2\times 3} & \\ \mathbf{0}_{1\times n-3} & -\Delta a_1 & -\Delta a_2 & 0 \\ \mathbf{0}_{1\times n-3} & 0 & 0 & -\Delta a_g \end{bmatrix} \qquad \Delta\mathbf{B} = \begin{bmatrix} \mathbf{0}_{n-2\times 1} & \mathbf{0}_{n-2\times 1} \\ \Delta a_1 & 0 \\ 0 & \Delta a_g \end{bmatrix},$$

such that the dynamics of the observer system (without feedback terms) with altered actuator dynamics can be written as

$$\dot{\mathbf{x}} = \sum_{i=1}^{N_r} h_i(\mathbf{z})\left(\mathbf{A}_i\mathbf{x} + \mathbf{D}_i\boldsymbol{\xi}\right) + \mathbf{B}\mathbf{u} + \Delta\mathbf{A}\,\mathbf{x} + \Delta\mathbf{B}\,\mathbf{u}\,.$$

The actuator dynamics fault term $(\Delta\mathbf{A}\,\mathbf{x} + \Delta\mathbf{B}\,\mathbf{u})$ is given by

$$\Delta\mathbf{A}\,\mathbf{x} + \Delta\mathbf{B}\,\mathbf{u} = \Big(0 \quad \cdots \quad 0 \quad \Delta a_1\,(\beta_d - \beta) - \Delta a_2\,\dot{\beta} \quad \Delta a_g\,(T_{g,d} - T_g)\Big)^T,$$

and can be written as $\Delta\mathbf{A}\,\mathbf{x} + \Delta\mathbf{B}\,\mathbf{u} = \mathbf{E}\,\mathbf{f}_a$ by choosing

$$\mathbf{E} = \begin{bmatrix} 0 \cdots 1\,0 \\ 0 \cdots 0\,1 \end{bmatrix}^T \quad \text{and} \quad \mathbf{f}_a = \begin{pmatrix} \Delta a_1\,(\beta_d - \beta) - \Delta a_2\,\dot{\beta} \\ \Delta a_g\,(T_{g,d} - T_g) \end{pmatrix}. \tag{19}$$

The torque actuator fault Δa_g can be reconstructed as described in the previous section. For the pitch actuator faults Δa_1 and Δa_2, however, there is now only one equation available: $f_{a,1} = \Delta a_1\,(\beta_d - \beta) - \Delta a_2\,\dot{\beta}$. As this equation is underdetermined, Δa_1 and Δa_2 cannot be directly reconstructed from it.

A possible way to proceed is to exploit the fact that the pitch rate $\dot{\beta}$ oscillates around zero in full-load operation of the wind turbine, so that there are occurrences when $\dot{\beta} = 0$ and thus $f_{a,1} = \Delta a_1\,(\beta_d - \beta)$. Consequently, at these single-point occurrences, $\Delta\hat{a}_1$ can be reconstructed as

$$\Delta\hat{a}_1 = \frac{\hat{f}_{a,1}}{\beta_d - \beta} \qquad \left(\dot{\beta} = 0,\quad \beta_d \neq \beta\right). \tag{20}$$

Then, keeping track of the reconstructed values of $\Delta\hat{a}_1$, for example by calculating a running average $\Delta\hat{a}_{1,\text{mean}}$, the second dynamics fault parameter Δa_2 can be reconstructed (when $\dot{\beta} \neq 0$) as

$$\Delta\hat{a}_2 = \frac{1}{\dot{\beta}}\left(\Delta\hat{a}_{1,\text{mean}}\,(\beta_d - \beta) - \hat{f}_{a,1}\right) \qquad \left(\dot{\beta} \neq 0\right). \tag{21}$$

In order to exclude large peaks in the reconstruction of $\Delta\hat{a}_2$ when $\dot{\beta}$ is close to zero, a threshold $\varepsilon_{\dot{\beta}} > 0$ can be defined and equation (21) is only calculated when $|\dot{\beta}| > \varepsilon_{\dot{\beta}}$. Otherwise, $\Delta\hat{a}_2$ is fixed to its value of the previous timestep.

Similarly to the fault detection algorithm for first order pitch dynamics, if $\hat{f}_{a,1}$ assumes 0 for more than one timestep, the fault parameter $\Delta\hat{a}_1$ is set to zero.

To check whether $\hat{f}_{a,1} \approx 0$ and $\dot{\beta} \approx 0$ is best done by specifying small constants ε_1 and ε_2 close to zero and testing whether $\left|\hat{f}_{a,1}\right| < \varepsilon_1$ and $|\dot{\beta}| < \varepsilon_2$.

From the definitions in (17) and (18), the reconstructed parameters $\hat{\omega}_n$ and $\hat{\zeta}$ can be calculated from the reconstructed values of $\Delta\hat{a}_{1,\text{mean}}$ and $\Delta\hat{a}_2$ as

$$\hat{\omega}_n = \sqrt{\Delta\hat{a}_{1,\text{mean}} + \omega_n^2}\,, \qquad \hat{\zeta} = \frac{1}{\hat{\omega}_n}\left(\frac{\Delta\hat{a}_2}{2} + \zeta\,\omega_n\right). \tag{22}$$

Example for Second Order Pitch Parameter Fault. As an example for a fault in the pitch actuator dynamics using a second order pitch model, a fault from the FDI benchmark model [4] is used. The default parameters for the (hydraulic) pitch model are: $\omega_n = 11.11\,\frac{\text{rad}}{\text{s}}$ and $\zeta = 0.6$. One fault scenario described in [4] is a pressure drop in the hydraulic pitch system, with the following faulty parameters: $\tilde{\omega}_n = 5.73\,\frac{\text{rad}}{\text{s}}$, $\tilde{\zeta} = 0.45$. A fault in the torque actuator dynamics is not considered here, such that $\mathbf{E}$ in (19) consists only of one column and $\mathbf{f}_a$ is reduced to a scalar $f_a = f_{a,1} = \Delta a_1\,(\beta_d - \beta) - \Delta a_2\,\dot{\beta}$.

The small threshold values were set to $\varepsilon_{\beta,\beta_d} = 0.0005$ rad, $\varepsilon_{\dot{\beta}} = 0.04\,\frac{\text{rad}}{\text{s}}$, $\varepsilon_1 = 10^{-6}\,\frac{\text{rad}}{\text{s}^2}$ and $\varepsilon_2 = 10^{-4}\,\frac{\text{rad}}{\text{s}}$. A simulation was performed using the nominal model, with the pitch actuator fault active between 50 s and 150 s.

The left upper part of Fig. 3 shows the pointwise reconstruction of the fault parameter $\Delta\hat{a}_1$, at occurrences when $\dot{\beta} \approx 0$. At these instances, sharp negative peaks are visible in the figure with values around -80 $\frac{\text{rad}}{\text{s}^2}$. The true value for Δa_1 is $\tilde{a}_1 - a_1 = \tilde{\omega}_n^2 - \omega_n^2 = 5.73^2\,\frac{\text{rad}}{\text{s}^2} - 11.11^2\,\frac{\text{rad}}{\text{s}^2} = -90.6\,\frac{\text{rad}}{\text{s}^2}$.

From the running average of the reconstructed values of $\Delta\hat{a}_1$, $\Delta\hat{a}_2$ is calculated at instances when $|\dot{\beta}| > \varepsilon_{\dot{\beta}}$. The time-series of $\Delta\hat{a}_2$ is displayed in the upper right

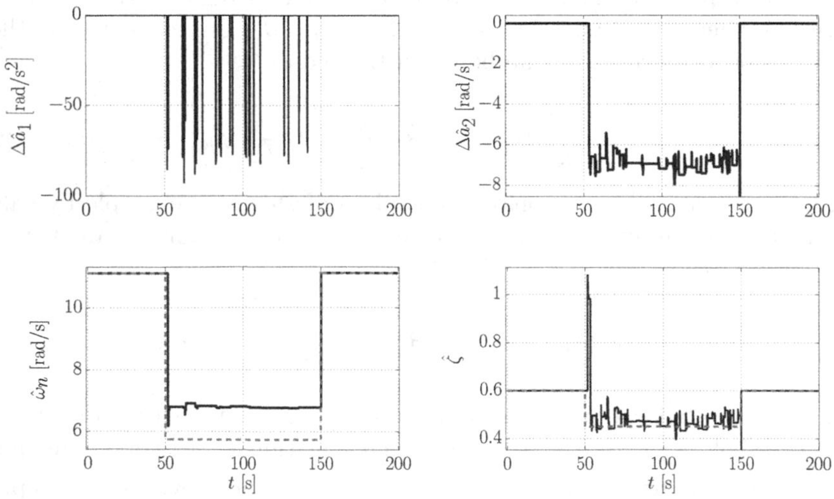

Fig. 3. Upper figures: reconstructed fault parameters $\Delta\hat{a}_1$ and $\Delta\hat{a}_2$ for a fault in a second order delay pitch model in a simulation of the nominal model (turbulent wind input; mean wind speed: 18 m/s). Lower figures: reconstructed frequency and damping parameters $\hat{\omega}_n$ and $\hat{\zeta}$ (dashed lines: actual values of ω_n and ζ)

part of Fig. 3. When the fault is active, the values of $\Delta\hat{a}_2$ oscillate around $-7\,\frac{\mathrm{rad}}{\mathrm{s}}$, compared to the true value of
$\Delta a_2 = 2\left(\hat{\zeta}\,\hat{\omega}_n - \zeta\,\omega_n\right) = 2\,\left(0.45 \cdot 5.73\,\frac{\mathrm{rad}}{\mathrm{s}} - 0.6 \cdot 11.11\,\frac{\mathrm{rad}}{\mathrm{s}}\right) = -8.175\,\frac{\mathrm{rad}}{\mathrm{s}}$.

Using the equations in (22), the reconstructed dynamics parameters $\hat{\omega}_n$ and $\hat{\zeta}$ are calculated and displayed in the lower part of Fig. 3. The faulty frequency parameter $\tilde{\omega}_n$ is approximated to a degree of about 80 %, with an average reconstructed value of $\hat{\tilde{\omega}}_n \approx 6.8\,\frac{\mathrm{rad}}{\mathrm{s}}$. The fault reaction time (about 1.6 s) is larger than for the first order delay models. This is due to the fact that (20) is calculated only when $\dot{\beta} \approx 0$, which may not be the case at the onset of a fault. An initial erroneous peak occurs in the reconstruction of the faulty damping parameter $\hat{\tilde{\zeta}}$. The constant $\varepsilon_{\dot{\beta}} = 0.04\,\frac{\mathrm{rad}}{\mathrm{s}}$ was chosen relatively large to avoid an even more noisy reconstruction due to the division by $\dot{\beta}$ in the calculation method for $\Delta\hat{a}_2$ in (21).

As a general rule, both for the first and for the second order delay models, the reconstruction of the dynamics parameters is strongly influenced by the choice of the ε-constants. Increasing the constants $\varepsilon_{\beta,\beta_d}$, $\varepsilon_{T_g\,T_{g,d}}$ and $\varepsilon_{\dot{\beta}}$ leads to a less noisy, but also more discretized and thus slower reconstruction.

4 Conclusion

In this paper, a Takagi-Sugeno sliding mode observer (TS SMO) was used to reconstruct actuator parameter faults in a dynamic wind turbine simulation. For the TS SMO design, appropriate actuator fault matrices were chosen such that the faulty dynamics parameters could be calculated from the reconstructed additive fault signals. For the first order delay models, this procedure is pretty straightforward. For the second order delay model, however, the parameters can only be reconstructed at defined instances, such that a two-step procedure had to be applied. The fault reconstruction is better when using the reduced-order model than when using FAST, which indicates that the fault reconstruction procedure is not very robust with respect to model uncertainties. Even when using the reduced-order model, the quality of the fault reconstruction might not be high enough to be used in fault-tolerant control applications. However, due to the fast fault reaction times, the presented method may be well applicable for fast fault detection schemes.

Acknowledgements. This research project is funded by the German Federal Ministry of Education and Research under grant no. 17N1411.

References

1. Chen, J., Patton, R.J.: Robust Model-Based Fault Diagnosis for Dynamic Systems. Kluwer Academic Publishers (1999)
2. Isermann, R.: Fault-Diagnosis Systems. Springer, Heidelberg (2006)

3. Rotondo, D., Nejjari, F., Puig, V., Blesa, J.: Fault Tolerant Control of the Wind Turbine Benchmark using Virtual Sensors/Actuators. In: 8th IFAC Symposium on Fault Detection, Supervision and Safety of Technical Processes, Mexico City, Mexico, pp. 114–119 (2012)
4. Odgaard, P.F., Stoustrup, J., Kinnaert, M.: Fault Tolerant Control of Wind Turbines - a benchmark model. In: 7th IFAC Symposium on Fault Detection, Supervision and Safety of Technical Processes, Barcelona, Spain, pp. 155–160 (2009)
5. Gerland, P., Gross, D., Schulte, H., Kroll, A.: Design of Sliding Mode Observers for TS Fuzzy Systems with Application to Disturbance and Actuator Fault Estimation. In: 49th IEEE Decision and Control Conf., Atlanta, USA, pp. 4373–4378 (2010)
6. Edwards, C., Spurgeon, S.K.: On the Development of Discontinuous Observers. International Journal of Control 59(5), 1211–1229 (1994)
7. Edwards, C., Spurgeon, S.K.: Sliding Mode Control: Theory and Applications. Taylor & Francis, Boca Raton (1998)
8. Takagi, T., Sugeno, M.: Fuzzy Identification of Systems and Its Applications to Modeling and Control. IEEE Transactions on Systems, Man, and Cybernetics 15(1), 116–132 (1985)
9. Tanaka, K., Wang, H.O.: Fuzzy Control Systems Design and Analysis: A Linear Matrix Inequality Approach. John Wiley & Sons, Inc. (2001)
10. Georg, S., Schulte, H.: Takagi-Sugeno Sliding Mode Observer with a Weighted Switching Action and Application to Fault Diagnosis for Wind Turbines. In: 11th International Conference on Diagnostics of Processes and Systems, Łagów Lubuski, Poland (accepted for publication, 2013)
11. Bianchi, F.D., De Battista, H., Mantz, R.J.: Wind Turbine Control Systems: Principles, Modelling and Gain Scheduling Design. Springer-Verlag London Limited (2007)
12. Jonkman, J., Butterfield, S., Musial, W., Scott, G.: Definition of a 5-MW Reference Wind Turbine for Offshore System Development. NREL TP-500-38060. National Renewable Energy Laboratory, Golden, Colorado (2009)
13. Jonkman, J.M., Buhl Jr., M.L.: FAST User's Guide. NREL TP-500-38230. National Renewable Energy Laboratory, Golden, Colorado (2005)
14. Lendek, Z., Guerra, T.M., Babuška, R., De Schutter, B.: Stability Analysis and Nonlinear Observer Design Using Takagi-Sugeno Fuzzy Models. Springer, Heidelberg (2010)
15. Georg, S., Mueller, M., Schulte, H.: Wind Turbine Model and Observer in Takagi-Sugeno Model Structure. In: Conference 'The Science of Making Torque from Wind', European Academy of Wind Energy (EAWE), Oldenburg, Germany (2012)

Takagi-Sugeno Sliding Mode Observer with a Weighted Switching Action and Application to Fault Diagnosis for Wind Turbines

Sören Georg and Horst Schulte

HTW Berlin, Department of Engineering I, Control Engineering,
Wilhelminenhofstr. 75A, 12459 Berlin, Germany
{soeren.georg,horst.schulte}@htw-berlin.de
www.rt.htw-berlin.de

Abstract. Takagi-Sugeno Sliding Mode Observers can be effectively employed for fault diagnosis in nonlinear systems. In this paper, the discontinuous observer term responsible for establishing the sliding motion is modified, such that the individual system output errors can be weighted using appropriate gains. This approach opens up new design possibilities for systems where the orders of magnitude of the individual outputs differ significantly. A formal proof for the existence of the sliding motion when using the modified Takagi-Sugeno Sliding Mode Observer (TS SMO) is given by showing both that the reachability condition can be fulfilled and that the sliding surface is reached in finite time. As an example, the modified TS SMO is used to reconstruct actuator offset faults for wind turbines within a simulation using the aero-elastic code FAST.

Keywords: Takagi-Sugeno, Sliding Mode Observer, Fault Diagnosis, Actuator Faults, Wind Turbines.

1 Introduction

Sliding mode observers have been shown to be appropriate means for fault detection and isolation, notably in [2] and [3], where actuator and sensor faults are directly reconstructed using the equivalent output injection signal. Compared to other observer-based fault detection methods that rely on evaluating residual signals, direct fault reconstruction can be advantageous because it provides a direct estimate of the fault magnitude [3].

The sliding mode observer concept from Edwards and Spurgeon [1], [2] is valid for linear systems. For nonlinear systems, an extension to this concept was presented in [4], where the observer is implemented within a Takagi-Sugeno (TS) model structure to account for system nonlinearities. This Takagi-Sugeno sliding mode observer (TS SMO) was successfully used in [5] to detect and reconstruct sensor faults in a dynamic pitch system model of a wind turbine. When applying fault diagnosis not only to submodels of wind turbines but to a complete wind turbine model, one is faced with the fact that the two main actuators, the pitch angle and the generator torque, are of significantly different orders of magnitude.

J. Korbicz and M. Kowal (eds.), *Intelligent Systems in Technical and Medical Diagnostics*, Advances in Intelligent Systems and Computing 230,
DOI: 10.1007/978-3-642-39881-0_3, © Springer-Verlag Berlin Heidelberg 2014

In tests, actuator faults for both the pitch angle and the generator torque could not be simultaneously reconstructed using the TS SMO from [4], where only one scalar gain factor is available to tune the equivalent output injection signal, and thus implicitly the maximum height of the fault reconstruction.

In this paper, a modification of the TS SMO from [4] is presented, which provides enhanced design possibilities for systems where the orders of magnitude of the individual outputs differ significantly. This is achieved by modifying the discontinuous observer term with a weighting matrix for the individual output errors and by using a gain matrix instead of a scalar gain factor. As an example, the modified TS SMO is used for actuator fault diagnosis for wind turbines within the aero-elastic simulation FAST [6] by NREL.

This paper is organized as follows. In section 2, a brief overview of the TS sliding mode observer is given. Section 3 introduces the idea to modify the TS SMO using a weighted switching term. In section 3.1, it is shown that the η-reachability condition can be fulfilled for the modified TS SMO and that the sliding surface is reached in finite time. In section 4, actuator fault diagnosis for wind turbines is studied in simulations using FAST.

2 Takagi-Sugeno Sliding Mode Observer

In this section, the TS SMO is briefly presented together with necessary existence conditions. In this paper, the TS SMO from [4] is used, which is a nonlinear extension of the sliding mode observer by Edwards and Spurgeon [1], [2]. For a detailed introduction to Takagi-Sugeno models [8], see for example [9].

Consider a nonlinear MIMO system subject to unmodeled dynamics $\boldsymbol{\xi}$, actuator faults $\mathbf{f}_a$, and sensor faults $\mathbf{f}_s$ in TS structure:

$$\dot{\mathbf{x}} = \sum_{i=1}^{N_r} h_i(\mathbf{z})\left(\mathbf{A}_i\mathbf{x} + \mathbf{B}_i\mathbf{u} + \mathbf{D}_i\boldsymbol{\xi} + \mathbf{E}_i\mathbf{f}_a\right), \qquad \mathbf{y} = \mathbf{C}\mathbf{x} + \mathbf{f}_s, \tag{1}$$

where N_r denotes the number of linear submodels and the TS membership functions h_i fulfill the convexity condition $\sum_{i=1}^{N_r} h_i(\mathbf{z}) = 1$. The vector of premise variables $\mathbf{z}$ can consist of system states and inputs, and of external variables.

Transformed Form of the TS Sliding Mode Observer. The TS SMO is designed in a transformed system, which is separated into the measurable system states $\mathbf{y}$ and non-measurable system states $\mathbf{x}_1$. The transformed state vector is $\left(\mathbf{x}_1^T \;\; \mathbf{y}^T\right)^T$. This separation is achieved by introducing a transformation $\mathbf{T}_c = \left[\mathbf{N}_\mathbf{C}^T \;\; \mathbf{C}\right]^T$, where $\mathbf{N}_\mathbf{C}$ denotes the null-space of $\mathbf{C}$. Applying a series of linear coordinate transformations[1] $\mathbf{T}_i = \bar{\mathbf{T}}_{L,i}\,\tilde{\mathbf{T}}_{DE,i}\,\mathbf{T}_c$ to the linear submodels, the transformed TS system is obtained as

[1] See [2] and [4] for a description of the transformation matrices. The MATLAB® commands from [2] can be used to obtain the transformations in practical applications.

$$\dot{\mathbf{x}}_1 = \sum_{i=1}^{N_r} h_i(\mathbf{z}) \left(\boldsymbol{\mathcal{A}}_{11,i}\,\mathbf{x}_1 + \boldsymbol{\mathcal{A}}_{12,i}\,\mathbf{y} + \boldsymbol{\mathcal{B}}_{1,i}\,\mathbf{u} \right), \tag{2}$$

$$\dot{\mathbf{y}} = \sum_{i=1}^{N_r} h_i(\mathbf{z}) \left(\boldsymbol{\mathcal{A}}_{21,i}\,\mathbf{x}_1 + \boldsymbol{\mathcal{A}}_{22,i}\,\mathbf{y} + \boldsymbol{\mathcal{B}}_{2,i}\,\mathbf{u} + \boldsymbol{\mathcal{D}}_{2,i}\,\boldsymbol{\xi} + \boldsymbol{\mathcal{E}}_{2,i}\,\mathbf{f}_a \right), \tag{3}$$

with the transformed system matrices

$\boldsymbol{\mathcal{A}}_i = \mathbf{T}_i \mathbf{A}_i \mathbf{T}_i^{-1} = \begin{bmatrix} \boldsymbol{\mathcal{A}}_{11,i} & \boldsymbol{\mathcal{A}}_{12,i} \\ \boldsymbol{\mathcal{A}}_{21,i} & \boldsymbol{\mathcal{A}}_{22,i} \end{bmatrix}$, $\boldsymbol{\mathcal{B}}_i = \mathbf{T}_i \mathbf{B}_i = \begin{bmatrix} \boldsymbol{\mathcal{B}}_{1,i}^T & \boldsymbol{\mathcal{B}}_{2,i}^T \end{bmatrix}^T$,

$\boldsymbol{\mathcal{D}}_i = \mathbf{T}_i \mathbf{D}_i = \begin{bmatrix} \mathbf{0}^T & \boldsymbol{\mathcal{D}}_{2,i}^T \end{bmatrix}^T$, $\boldsymbol{\mathcal{E}}_i = \mathbf{T}_i \mathbf{E}_i = \begin{bmatrix} \mathbf{0}^T & \boldsymbol{\mathcal{E}}_{2,i}^T \end{bmatrix}^T$.

For a stable observer in the transformed system to exist, three conditions have to be fulfilled:

Condition 1. The uncertainties and actuator faults are unknown but bounded: $\|\boldsymbol{\xi}^T(t)\; \mathbf{f}_a^T(t)\|^T \leq \Xi$. Furthermore, individual bounds exist: $\|\boldsymbol{\xi}(t)\|^T \leq \Xi_\xi$ and $\mathbf{f}_a(t) \leq \Xi_{f_a}$. The sensor faults and their derivatives are assumed to be bounded, too: $\|\mathbf{f}_s\| \leq \Psi$, $\|\dot{\mathbf{f}}_s\| \leq \Psi_d$. Moreover, the system states and inputs are assumed to be bounded.

Condition 2. Let q be defined as the number of columns of $[\mathbf{D}_i\; \mathbf{E}_i]$. Then, the condition $q = \operatorname{rank}(\mathbf{C}[\mathbf{D}_i\; \mathbf{E}_i]) = \operatorname{rank}[\mathbf{D}_i\; \mathbf{E}_i]$ must be fulfilled. Furthermore, it must hold that $p > q$, where p is the number of measurable system states.

Condition 3. All invariant zeros of $(\mathbf{A}_i, [\mathbf{D}_i\; \mathbf{E}_i], \mathbf{C})$ must lie in $\mathbb{C}_-$.

The TS sliding mode observer in transformed form is given by

$$\dot{\hat{\mathbf{x}}}_1 = \sum_{i=1}^{N_r} h_i(\mathbf{z}) \left(\boldsymbol{\mathcal{A}}_{11,i}\,\hat{\mathbf{x}}_1 + \boldsymbol{\mathcal{A}}_{12,i}\,\hat{\mathbf{y}} + \boldsymbol{\mathcal{B}}_{1,i}\,\mathbf{u} - \boldsymbol{\mathcal{A}}_{12,i}\,\tilde{\mathbf{e}}_y \right), \tag{4}$$

$$\dot{\hat{\mathbf{y}}} = \sum_{i=1}^{N_r} h_i(\mathbf{z}) \left(\boldsymbol{\mathcal{A}}_{21,i}\,\hat{\mathbf{x}}_1 + \boldsymbol{\mathcal{A}}_{22,i}\,\hat{\mathbf{y}} + \boldsymbol{\mathcal{B}}_{2,i}\,\mathbf{u} - (\boldsymbol{\mathcal{A}}_{22,i} - \boldsymbol{\mathcal{A}}_{22}^s)\,\tilde{\mathbf{e}}_y + \boldsymbol{\nu} \right), \tag{5}$$

where $\tilde{\mathbf{e}}_y := \hat{\mathbf{y}} - (\mathbf{y} + \mathbf{f}_s)$ denotes the output error including sensor faults and $\boldsymbol{\mathcal{A}}_{22}^s$ is a stable design matrix. In (4)-(5), the premise variables $\mathbf{z}$ are assumed to be measurable. For a TS SMO with unmeasurable premise variables, see [4]. The discontinuous switching term necessary to maintain a sliding motion is given by

$$\boldsymbol{\nu} = \begin{cases} -\rho \dfrac{\mathbf{P}_2\,\tilde{\mathbf{e}}_y}{\|\mathbf{P}_2\,\tilde{\mathbf{e}}_y\|} & , \quad \tilde{\mathbf{e}}_y \neq \mathbf{0} \\ \mathbf{0} & , \quad \tilde{\mathbf{e}}_y = \mathbf{0} \end{cases} \tag{6}$$

where $\rho > 0$ is a scalar gain factor. $\mathbf{P}_2$ is the unique symmetric positive definite (s.p.d) solution of the Lyapunov equation

$$\mathbf{P}_2\,\boldsymbol{\mathcal{A}}_{22}^s + \boldsymbol{\mathcal{A}}_{22}^{s\;T}\,\mathbf{P}_2 = -\mathbf{Q}_2\,, \tag{7}$$

where $\mathbf{Q}_2$ is a symmetric positive definite design matrix. Once the sliding surface $\mathcal{S} = \{(\mathbf{e}_1, \tilde{\mathbf{e}}_y) : \tilde{\mathbf{e}}_y = \mathbf{0}\}$ is reached, the observer tries to maintain the sliding motion on $\mathcal{S}$. The error vector $\mathbf{e}_1 = \hat{\mathbf{x}}_1 - \mathbf{x}_1$ of the transformed unmeasurable states is assumed to be bounded: $\|\mathbf{e}_1\| < \Gamma$. Investigating the stability of the error system $\mathbf{e}_1$ leads to an LMI (linear matrix inequality) design condition, which was presented in [4] and is omitted here for the sake of brevity.

Fault Reconstruction. For simulations, the switching term (6) is replaced by a continuous approximation: $\boldsymbol{\nu}_{\text{eq}} = -\rho \frac{\mathbf{P}_2 \tilde{\mathbf{e}}_y}{\|\mathbf{P}_2 \tilde{\mathbf{e}}_y\| + \delta}$, with a small positive constant δ. The equivalent output injection signal $\boldsymbol{\nu}_{\text{eq}}$ describes the average behavior of the discontinuous term (6). If no sensor faults are present, unmodeled dynamics ($\boldsymbol{\xi}$) and actuator faults ($\mathbf{f}_a$) can be reconstructed using $\boldsymbol{\nu}_{\text{eq}}$ [2], [4]:

$$\left(\hat{\boldsymbol{\xi}}^T \; \hat{\mathbf{f}}_a^T\right)^T = \left(\sum_{i=1}^{N_r} h_i(\mathbf{z}) \left[\mathcal{D}_{2,i} \; \mathcal{E}_{2,i}\right]\right)^+ \boldsymbol{\nu}_{\text{eq}} , \tag{8}$$

where $(\cdot)^+$ denotes the left pseudo-inverse.

3 TS SMO with a Weighted Switching Term

In [4] it was shown that for a system structure (2)-(3), a stable TS sliding mode observer (4)-(5) exists with the discontinuous term (6), where a constant scalar ρ is used as a gain factor.

This structure for the discontinuous term imposes a limitation for fault reconstruction for systems where the orders of magnitude of the outputs differ significantly. This can be understood by inspecting the switching term (6). Due to the normalization of the vector $\mathbf{P}_2 \tilde{\mathbf{e}}_y / \|\mathbf{P}_2 \tilde{\mathbf{e}}_y\|$, the gain ρ determines the maximum magnitude that can be applied by $\boldsymbol{\nu}$. If the orders of magnitude of some components of the output vector $\mathbf{y}$ significantly outweigh those of the other components, this will also be reflected in the components of the error vector $\tilde{\mathbf{e}}_y$. Unless the matrix $\mathbf{P}_2$ introduces a significant weighting on the different components of $\tilde{\mathbf{e}}_y$, merely those components with large orders of magnitude will contribute to the normalized vector $\mathbf{P}_2 \tilde{\mathbf{e}}_y / \|\mathbf{P}_2 \tilde{\mathbf{e}}_y\|$ and, consequently, to $\boldsymbol{\nu}$. Since $\boldsymbol{\nu}_{\text{eq}}$ is used for fault reconstruction in (8), the choice of ρ implicitly determines the maximum magnitude for fault reconstruction. In order to reconstruct faults also in those channels with very large orders of magnitude, ρ would have to be chosen very large. This, however, can lead to an erroneous observer behavior, where no meaningful reconstruction of any fault is possible.

For the case of wind turbines, the two main control inputs (pitch angle and generator torque) are of significantly different orders of magnitude. The pitch angle is in the range between 0 and 1.57 rad, whereas the generator torque for multi-megawatt wind turbines can be in the order of 10^4 Nm for turbines with gearbox or even 10^6 Nm for turbines without gearbox. Both control inputs are also modeled as system states and outputs in the model equations (section 4).

To overcome the limitation imposed by the structure of the switching term (6), two modifications are applied in this work for the calculation of $\boldsymbol{\nu}$:

1. Modification. The output error vector is modified with a diagonal, positive definite weighting matrix for the individual output components: $\tilde{\mathbf{e}}_y \rightarrow \mathbf{W}\,\tilde{\mathbf{e}}_y$, where the weighting matrix is given by the reciprocal values of the estimated maximum absolute values of the output vector components:
$\mathbf{W} = \mathrm{diag}\,(W_1 \cdots W_p) = \mathrm{diag}\,(1/\,|y_{\mathrm{max,1}}| \cdots 1/\,|y_{\mathrm{max,p}}|)$.
Choosing this weighting matrix, the components of the modified output error vector $\mathbf{W}\,\tilde{\mathbf{e}}_y$ are of comparable order of magnitude.

2. Modification. The scalar sliding mode gain ρ is replaced by a diagonal, positive definite gain matrix: $\boldsymbol{\rho} = \mathrm{diag}\,(\rho_1 \cdots \rho_p)$, in order to be able to set individual gains for the reconstruction of the individual faults. The standard relation (6) for the discontinuous term $\boldsymbol{\nu}$ is thus replaced by

$$\boldsymbol{\nu} = \begin{cases} -\boldsymbol{\rho}\,\frac{\mathbf{P}_2\,\mathbf{W}\,\tilde{\mathbf{e}}_y}{\|\mathbf{P}_2\,\mathbf{W}\,\tilde{\mathbf{e}}_y\|} & , \quad (\tilde{\mathbf{e}}_y \neq \mathbf{0}) \\ \mathbf{0} & , \quad (\tilde{\mathbf{e}}_y = \mathbf{0}) \end{cases} \tag{9}$$

Again, for simulations, a continuous approximation of (9) is used:
$\boldsymbol{\nu}_{\mathrm{eq}} = -\boldsymbol{\rho}\,\frac{\mathbf{P}_2\,\mathbf{W}\tilde{\mathbf{e}}_y}{\|\mathbf{P}_2\,\mathbf{W}\tilde{\mathbf{e}}_y\|+\delta}$. Unmodeled dynamics ($\boldsymbol{\xi}$) and actuator faults ($\mathbf{f}_a$) can then be reconstructed using relation (8). The particular structure in (9) was chosen as it proved successful for fault reconstruction in simulations studies.

3.1 Proof of an Ideal Sliding Motion

In this section, a formal proof is given that a sliding motion can be established using a TS SMO with the modified switching term (9) and that the sliding surface is reached in finite time. The proof follows along the lines of those given in [1], [2] for the sliding mode observer and in [4] for the TS SMO with the standard discontinuous term (6).

Proposition. For the TS structure (2)-(3) and the TS SMO (4)-(5) with the modified discontinuous term (9), an ideal sliding motion takes place on
$\mathcal{S} = \{(\mathbf{e}_1\,,\,\tilde{\mathbf{e}}_y) : \tilde{\mathbf{e}}_y = 0\}$ and the sliding surface is reached in finite time.
Necessary conditions are: $\boldsymbol{\mathcal{A}}_{22}^{s}$, $\mathbf{P}_2$, $\mathbf{W}$, and $\boldsymbol{\rho}$ must be diagonal matrices. The conditions that $\boldsymbol{\mathcal{A}}_{22}^{s}$ and $\mathbf{P}_2$ must be diagonal is not an overly restrictive constraint in practice, since a natural choice for $\boldsymbol{\mathcal{A}}_{22}^{s}$ is a diagonal matrix with the desired poles for the dynamics of $\mathbf{e}_1$ [4]. Choosing also $\mathbf{Q}_2$ as diagonal in the Lyapunov equation (7) for $\boldsymbol{\mathcal{A}}_{22}^{s}$ yields a diagonal $\mathbf{P}_2$.

Proof 1 (Reachability). It must be shown that a Lyapunov function $V_s(\tilde{\mathbf{e}}_y)$ exists that fulfills an η-reachability condition. Let $\tilde{\mathbf{P}}_2$ be defined as $\tilde{\mathbf{P}}_2 := \mathbf{P}_2\mathbf{W}$ and a candidate Lyapunov function as $V_s = \tilde{\mathbf{e}}_y^T \tilde{\mathbf{P}}_2 \tilde{\mathbf{e}}_y$. Since $\mathbf{W}$ and $\mathbf{P}_2$ are positive definite, $\tilde{\mathbf{P}}_2$ is also positive definite. It directly follows that $V_s > 0 \quad (\tilde{\mathbf{e}}_y \neq 0)$. For the proof of $\dot{V}_s < 0$, the following relation will first be shown:

$$2\,\tilde{\mathbf{e}}_y^T\,\tilde{\mathbf{P}}_2\,\boldsymbol{\mathcal{A}}_{22}^s\,\tilde{\mathbf{e}}_y \;=\; -\,\tilde{\mathbf{e}}_y^T\,\mathbf{Q}_2\,\mathbf{W}\,\tilde{\mathbf{e}}_y \;\leq\; 0\,. \tag{10}$$

The second inequality is obvious, as $\mathbf{Q}_2$ and $\mathbf{W}$ are positive definite.

Proof of the first equality making use of the Lyapunov equation (7):

$$\begin{aligned}
2\,\tilde{\mathbf{e}}_y^T\,\tilde{\mathbf{P}}_2\,\boldsymbol{\mathcal{A}}_{22}^s\,\tilde{\mathbf{e}}_y &= \tilde{\mathbf{e}}_y^T\,\tilde{\mathbf{P}}_2\,\boldsymbol{\mathcal{A}}_{22}^s\,\tilde{\mathbf{e}}_y + \tilde{\mathbf{e}}_y^T\left(\tilde{\mathbf{P}}_2\,\boldsymbol{\mathcal{A}}_{22}^s\right)^T\tilde{\mathbf{e}}_y \\
&= \tilde{\mathbf{e}}_y^T\,\mathbf{P}_2\,\mathbf{W}\,\boldsymbol{\mathcal{A}}_{22}^s\,\tilde{\mathbf{e}}_y + \tilde{\mathbf{e}}_y^T\,(\boldsymbol{\mathcal{A}}_{22}^s)^T\,\mathbf{W}\,\mathbf{P}_2\,\tilde{\mathbf{e}}_y \\
&= \tilde{\mathbf{e}}_y^T\,\underbrace{\left(\mathbf{P}_2\,\boldsymbol{\mathcal{A}}_{22}^s + (\boldsymbol{\mathcal{A}}_{22}^s)^T\,\mathbf{P}_2\right)}_{=-\mathbf{Q}_2}\,\mathbf{W}\,\tilde{\mathbf{e}}_y \\
&= -\tilde{\mathbf{e}}_y^T\,\mathbf{Q}_2\,\mathbf{W}\,\tilde{\mathbf{e}}_y \leq 0\,.
\end{aligned}$$

Here it was exploited that $\mathbf{W}^T = \mathbf{W}$, $\mathbf{W}\,\mathbf{P}_2 = \mathbf{P}_2\,\mathbf{W}$ and $\mathbf{W}\,\boldsymbol{\mathcal{A}}_{22}^s = \boldsymbol{\mathcal{A}}_{22}^s\,\mathbf{W}$, since $\mathbf{W}$, $\mathbf{P}_2$, and $\boldsymbol{\mathcal{A}}_{22}^s$ are diagonal matrices. For V_s to be a Lyapunov function, the relation $\dot{V}_s < 0$ must hold. From the definition of V_s it follows:

$\dot{V}_s = 2\,\tilde{\mathbf{e}}_y^T\,\tilde{\mathbf{P}}_2\,\dot{\tilde{\mathbf{e}}}_y$. From equations (3), (5) and $\tilde{\mathbf{e}}_y = \hat{\mathbf{y}} - \mathbf{y} - \mathbf{f}_s$, the derivative of the output error is obtained as

$$\dot{\tilde{\mathbf{e}}}_y = \sum_{i=1}^{N_r} h_i(\mathbf{z}) \left(\boldsymbol{\mathcal{A}}_{21,i}\,\mathbf{e}_1 + \boldsymbol{\mathcal{A}}_{22,i}\,\mathbf{f}_S + \boldsymbol{\mathcal{A}}_{22}^s\,\tilde{\mathbf{e}}_y + \boldsymbol{\nu} - [\boldsymbol{\mathcal{D}}_{2,i}\,\boldsymbol{\mathcal{E}}_{2,i}]\begin{pmatrix}\boldsymbol{\xi}\\ \mathbf{f}_a\end{pmatrix} - \dot{\mathbf{f}}_s\right) \tag{11}$$

Inserting (11) into $\dot{V}_s = 2\,\tilde{\mathbf{e}}_y^T\,\tilde{\mathbf{P}}_2\,\dot{\tilde{\mathbf{e}}}_y$ and using relation (10) it thus follows:

$$\dot{V}_s \leq \sum_{i=1}^{N_r} h_i(\mathbf{z}) \left(2\,\tilde{\mathbf{e}}_y^T\tilde{\mathbf{P}}_2\left[\boldsymbol{\mathcal{A}}_{21,i}\,\mathbf{e}_1 + \boldsymbol{\mathcal{A}}_{22,i}\,\mathbf{f}_S + \boldsymbol{\nu} - [\boldsymbol{\mathcal{D}}_{2,i}\,\boldsymbol{\mathcal{E}}_{2,i}]\begin{pmatrix}\boldsymbol{\xi}\\ \mathbf{f}_a\end{pmatrix} - \dot{\mathbf{f}}_s\right]\right).$$

With condition 1 for the TS SMO, the following relations are readily obtained:

$$2\,\tilde{\mathbf{e}}_y^T\,\tilde{\mathbf{P}}_2\,\boldsymbol{\mathcal{A}}_{21,i}\,\mathbf{e}_1 \leq 2\,\|\boldsymbol{\mathcal{A}}_{21,i}\|\,\Gamma\,\left\|\tilde{\mathbf{P}}_2\,\tilde{\mathbf{e}}_y\right\|, \tag{12}$$

$$2\,\tilde{\mathbf{e}}_y^T\,\tilde{\mathbf{P}}_2\,\boldsymbol{\mathcal{A}}_{22,i}\,\mathbf{f}_s \leq 2\,\|\boldsymbol{\mathcal{A}}_{22,i}\|\,\Psi\,\left\|\tilde{\mathbf{P}}_2\,\tilde{\mathbf{e}}_y\right\|, \tag{13}$$

$$-2\,\tilde{\mathbf{e}}_y^T\,\tilde{\mathbf{P}}_2\,[\boldsymbol{\mathcal{D}}_{2,i}\,\boldsymbol{\mathcal{E}}_{2,i}]\left(\boldsymbol{\xi}^T\;\mathbf{f}_a^T\right)^T \leq 2\,\|[\boldsymbol{\mathcal{D}}_{2,i}\,\boldsymbol{\mathcal{E}}_{2,i}]\|\,\Xi\,\left\|\tilde{\mathbf{P}}_2\,\tilde{\mathbf{e}}_y\right\|, \tag{14}$$

$$-2\,\tilde{\mathbf{e}}_y^T\,\tilde{\mathbf{P}}_2\,\dot{\mathbf{f}}_s \leq 2\,\Psi_d\,\left\|\tilde{\mathbf{P}}_2\,\tilde{\mathbf{e}}_y\right\|, \tag{15}$$

where any appropriate matrix norm can be used. In this work, the spectral norm is used. Applying relations (12) - (15), the following estimate for $\dot{V}_s$ holds:

$$\dot{V}_s \leq \sum_{i=1}^{N_r} h_i(\mathbf{z}) \left(2 \left\| \tilde{\mathbf{P}}_2 \tilde{\mathbf{e}}_y \right\| \mathcal{K}_i + 2 \tilde{\mathbf{e}}_y^T \tilde{\mathbf{P}}_2 \boldsymbol{\nu} \right), \quad (16)$$

with $\mathcal{K}_i := \|\boldsymbol{\mathcal{A}}_{21,i}\| \, \Gamma + \|\boldsymbol{\mathcal{A}}_{22,i}\| \, \Psi + \|[\boldsymbol{\mathcal{D}}_{2,i} \, \boldsymbol{\mathcal{E}}_{2,i}]\| \, \Xi + \Psi_d$. Due to the convexity condition of the TS membership functions ($\sum_{i=1}^{N_r} h_i(\mathbf{z}) = 1$) it holds that $\sum_{i=1}^{N_r} h_i(\mathbf{z}) \, \mathcal{K}_i \leq \mathcal{K}_{\max} := \|\boldsymbol{\mathcal{A}}_{21}\|_{\max} \, \Gamma + \|\boldsymbol{\mathcal{A}}_{22}\|_{\max} \, \Psi + \|[\boldsymbol{\mathcal{D}}_2 \, \boldsymbol{\mathcal{E}}_2]\|_{\max} \, \Xi + \Psi_d$.
It thus follows from (16): $\dot{V}_s \leq 2 \left\| \tilde{\mathbf{P}}_2 \tilde{\mathbf{e}}_y \right\| \mathcal{K}_{\max} + 2 \tilde{\mathbf{e}}_y^T \tilde{\mathbf{P}}_2 \boldsymbol{\nu}$.

Let $\eta > 0$ be a small scalar. If

$$2 \tilde{\mathbf{e}}_y^T \tilde{\mathbf{P}}_2 \boldsymbol{\nu} = -2 \left\| \tilde{\mathbf{P}}_2 \tilde{\mathbf{e}}_y \right\| \eta - 2 \left\| \tilde{\mathbf{P}}_2 \tilde{\mathbf{e}}_y \right\| \mathcal{K}_{\max} , \quad (17)$$

the following η-reachability condition is fulfilled:

$$\dot{V}_s \leq -2 \eta \left\| \tilde{\mathbf{P}}_2 \tilde{\mathbf{e}}_y \right\| . \quad (18)$$

A condition for the diagonal matrix $\boldsymbol{\rho}$ must now be determined such that both equation (17) and thereby relation (18) hold. Inserting the definition for the modified discontinuous term (9) in equation (17) yields

$$\begin{aligned}
& -2 \tilde{\mathbf{e}}_y^T \tilde{\mathbf{P}}_2 \boldsymbol{\rho} \frac{\tilde{\mathbf{P}}_2 \tilde{\mathbf{e}}_y}{\left\| \tilde{\mathbf{P}}_2 \tilde{\mathbf{e}}_y \right\|} \stackrel{(17)}{=} -2 \left\| \tilde{\mathbf{P}}_2 \tilde{\mathbf{e}}_y \right\| \eta - 2 \left\| \tilde{\mathbf{P}}_2 \tilde{\mathbf{e}}_y \right\| \mathcal{K}_{\max} \\
\Leftrightarrow \quad & \tilde{\mathbf{e}}_y^T \tilde{\mathbf{P}}_2 \boldsymbol{\rho} \tilde{\mathbf{P}}_2 \tilde{\mathbf{e}}_y = (\eta + \mathcal{K}_{\max}) \left\| \tilde{\mathbf{P}}_2 \tilde{\mathbf{e}}_y \right\|^2 \\
\Leftrightarrow \quad & \tilde{\mathbf{e}}_y^T \tilde{\mathbf{P}}_2 \boldsymbol{\rho} \tilde{\mathbf{P}}_2 \tilde{\mathbf{e}}_y = \tilde{\mathbf{e}}_y^T \tilde{\mathbf{P}}_2 \, (\eta + \mathcal{K}_{\max}) \, \tilde{\mathbf{P}}_2 \tilde{\mathbf{e}}_y \\
\Leftrightarrow \quad & \tilde{\mathbf{e}}_y^T \tilde{\mathbf{P}}_2 \boldsymbol{\rho} \tilde{\mathbf{P}}_2 \tilde{\mathbf{e}}_y > \tilde{\mathbf{e}}_y^T \tilde{\mathbf{P}}_2 \mathcal{K}_{\max} \tilde{\mathbf{P}}_2 \tilde{\mathbf{e}}_y \qquad \text{(since } \tilde{\mathbf{P}}_2 \text{ pos. def.)}, \\
\Leftrightarrow \quad & \tilde{\mathbf{e}}_y^T \left[(\boldsymbol{\rho} - \mathcal{K}_{\max} \mathbf{I}_{p \times p}) \, \tilde{\mathbf{P}}_2^2 \right] \tilde{\mathbf{e}}_y > \mathbf{0} \qquad \text{(since } \tilde{\mathbf{P}}_2 \text{ and } \boldsymbol{\rho} \text{ are diagonal)} . \quad (19)
\end{aligned}$$

Since $\tilde{\mathbf{P}}_2$ is positive definite, a sufficient, but not necessary, condition for (19) to be fulfilled is that the diagonal matrix $(\boldsymbol{\rho} - \mathcal{K}_{\max} \mathbf{I}_{p \times p})$ is positive definite, which is fulfilled if and only if for every element of $\boldsymbol{\rho}$ it holds that

$$\rho_i > \mathcal{K}_{\max} \qquad (i \in \{1, \ldots, p\}) . \quad (20)$$

Then, (17) holds and thus the reachability condition (18) is fulfilled.

Proof 2 (Sliding Surface Reached in Finite Time). Arguing as in [2], it is now shown that the sliding surface $\mathcal{S}$ is reached in finite time. Using the condition

$$\left\|\tilde{\mathbf{P}}_2\,\tilde{\mathbf{e}}_y\right\|^2 = \left(\tilde{\mathbf{P}}_2^{\frac{1}{2}}\,\tilde{\mathbf{P}}_2^{\frac{1}{2}}\,\tilde{\mathbf{e}}_y\right)^T\left(\tilde{\mathbf{P}}_2^{\frac{1}{2}}\,\tilde{\mathbf{P}}_2^{\frac{1}{2}}\,\tilde{\mathbf{e}}_y\right) = \left(\tilde{\mathbf{e}}_y^T\,\tilde{\mathbf{P}}_2^{\frac{1}{2}}\right)\tilde{\mathbf{P}}_2\underbrace{\left(\tilde{\mathbf{P}}_2^{\frac{1}{2}}\,\tilde{\mathbf{e}}_y\right)}_{=:\,\tilde{\mathbf{e}}_{yP}} = \tilde{\mathbf{e}}_{yP}^T\,\tilde{\mathbf{P}}_2\,\tilde{\mathbf{e}}_{yP}$$

it holds that $\|\tilde{\mathbf{e}}_{yP}\|^2 = \tilde{\mathbf{e}}_y^T\,\tilde{\mathbf{P}}_2\,\tilde{\mathbf{e}}_y = V_s$. For any symmetric matrix $\mathbf{A}$, the Rayleigh principle is fulfilled: $\lambda_{\min}(\mathbf{A})\,\|\mathbf{x}\|^2 \le \mathbf{x}^T\,\mathbf{A}\,\mathbf{x} \le \lambda_{\max}(\mathbf{A})\,\|\mathbf{x}\|^2$, where $\lambda_{\min}$ and $\lambda_{\max}$ denote the minimum and maximum eigenvalues of $\mathbf{A}$. Applying the Rayleigh principle to $\tilde{\mathbf{e}}_{yP}$ and $\tilde{\mathbf{P}}_2$ yields:
$\lambda_{\min}\left(\tilde{\mathbf{P}}_2\right)\|\tilde{\mathbf{e}}_{yP}\|^2 \le \tilde{\mathbf{e}}_{yP}^T\,\tilde{\mathbf{P}}_2\,\tilde{\mathbf{e}}_{yP} \le \lambda_{\max}\left(\tilde{\mathbf{P}}_2\right)\|\tilde{\mathbf{e}}_{yP}\|^2$ and thus:

$$\lambda_{\min}\left(\tilde{\mathbf{P}}_2\right)V_s \le \left\|\tilde{\mathbf{P}}_2\,\tilde{\mathbf{e}}_y\right\|^2 \le \lambda_{\max}\left(\tilde{\mathbf{P}}_2\right)V_s\,. \tag{21}$$

Using relation (21), the reachability condition (18) can be written independently of the output error $\tilde{\mathbf{e}}_y$:

$$\dot{V}_s \le -2\,\eta\left\|\tilde{\mathbf{P}}_2\,\tilde{\mathbf{e}}_y\right\| \le -2\,\eta\sqrt{\lambda_{\min}\left(\tilde{\mathbf{P}}_2\right)}\sqrt{V_s}\,. \tag{22}$$

Applying the chain rule for differentiation, the following equality holds: $\frac{d}{dt}\sqrt{V_s} = \frac{1}{2\sqrt{V_s}}\dot{V}_s$. Defining t_r as the time when the sliding surface is reached and making use of $V_s(t_r) = \tilde{\mathbf{e}}_y^T(t_r)\,\tilde{\mathbf{P}}_2\,\tilde{\mathbf{e}}_y(t_r) = 0$, the reachability condition (22) can be integrated and rearranged to obtain an estimate for t_r:

$$\frac{1}{2\sqrt{V_s}}\dot{V}_s \le -\eta\sqrt{\lambda_{\min}\left(\tilde{\mathbf{P}}_2\right)} \;\Rightarrow\; \int_0^{t_r}\frac{d}{dt}\left(\sqrt{V_s}\right)dt \le -\eta\sqrt{\lambda_{\min}\left(\tilde{\mathbf{P}}_2\right)}\int_0^{t_r}dt$$

$$\Leftrightarrow\; \underbrace{\sqrt{V_s}(t_r)}_{=0} - \sqrt{V_s}(0) \le -\eta\sqrt{\lambda_{\min}\left(\tilde{\mathbf{P}}_2\right)}\,t_r \;\Leftrightarrow\; t_r \le \frac{1}{\eta}\,\frac{\sqrt{V_s(0)}}{\sqrt{\lambda_{\min}(\mathbf{W}\,\mathbf{P})}}\,.$$

The sliding surface $\mathcal{S}$ is thus reached in finite time. This completes the proof of the proposition on page 45.

4 Actuator Fault Diagnosis for Wind Turbines

As an application example, the TS SMO with the modified switching term (9) is used in this section to detect and reconstruct actuator offset faults in a dynamic wind turbine simulation. As a simulation model, the NREL 5 MW reference turbine [7] is used and the simulation is carried out using the aero-elastic code FAST [6]. For the observer, a reduced-order model similar to the one described in [10], is used. Here, however, only the drivetrain degrees of freedom and first-order delay models for the pitch angle and the generator torque are included. The observer model equations are:

$$\dot{\theta}_s = \omega_r - \omega_g \tag{23}$$

$$\dot{\omega}_r = \frac{1}{J_r}\left(-k_s\,\theta_s - d_S\,(\omega_r - \omega_g) + T_a\right) \tag{24}$$

$$\dot{\omega}_g = \frac{1}{J_g}\left(k_s\,\theta_s + d_S\,(\omega_r - \omega_g) - T_g\right) \tag{25}$$

$$\dot{\beta} = -\frac{1}{\tau}\,\beta + \frac{1}{\tau}\,\beta_d \tag{26}$$

$$\dot{T}_g = -\frac{1}{\tau_g}\,T_g + \frac{1}{\tau_g}\,T_{g,d}\,, \tag{27}$$

where θ_s is the shaft torsion angle, ω_r and ω_g are the rotor and generator angular velocities, β is the pitch angle and T_g the generator torque. β_d and $T_{g,d}$ denote the demanded pitch angle and generator torque, respectively. k_S and d_S are the torsional stiffness and damping constants, J_r and J_g the rotor and generator inertias. The nonlinear aerodynamic rotor torque is given by [11]

$T_a = \frac{1}{2}\,\rho\,\pi\,R^3\,v^2\,C_Q\,(\lambda, \beta)$, where R denotes the rotor radius, ρ the air density, v the wind speed and $C_Q(\lambda, \beta)$ the rotor torque coefficient in dependence on the tip speed ratio ($\lambda = R\frac{\omega_r}{v}$) and the pitch angle. $C_Q(\lambda, \beta)$ can be calculated from look-up tables or analytical functions [12].

The following model parameters were used, which correspond to the parameters of the NREL 5 MW reference turbine, adapted for the reduced-order model (23)-(27): $J_r = 38759227\,\mathrm{kg\,m^2}$, $J_g = 5025347\,\mathrm{kg\,m^2}$, $k_S = 867637000\,\mathrm{Nm}$, $d_S = 6215000\,\mathrm{Nm\,s}$, $\tau = 0.1\,\mathrm{s}$, $\tau_g = 0.02\,\mathrm{s}$, $\rho = 1.225\,\frac{\mathrm{kg}}{\mathrm{m^3}}$, $R = 63\,\mathrm{m}$.

Introducing the observer state vector $\mathbf{x} = (\theta_s \quad \omega_r \quad \omega_g \quad \beta \quad T_g)^T$ and the input vector $\mathbf{u} = (\beta_d \quad T_{g,d})^T$, the observer model can be derived in TS structure using the sector nonlinearity approach [9]: $\dot{\mathbf{x}} = \mathbf{A}\,\mathbf{x} + \sum_{i=1}^{2} h_i\,(\mathbf{z})\,\tilde{\mathbf{A}}_i\,\mathbf{x} + \mathbf{B}\,\mathbf{u}$, where $\mathbf{A}$ and $\mathbf{B}$ are the matrices of the linear system part of (23)-(27). The membership functions are given by $h_1\,(\mathbf{z}) = \frac{f(\mathbf{z})-\underline{f}}{\overline{f}-\underline{f}}$ and $h_2\,(\mathbf{z}) = \frac{\overline{f}-f(\mathbf{z})}{\overline{f}-\underline{f}}$, with the nonlinear function $f\,(\mathbf{z}) := \frac{T_a(\mathbf{z})}{\omega_r} \quad (\omega_r > 0)$. The TS premise variable vector is given by $\mathbf{z} = (\omega_r \quad \beta \quad v)^T$. $\underline{f}$ and $\overline{f}$ denote the minimum and maximum values of $f\,(\mathbf{z})$ and were estimated here as $\underline{f} \approx 7.4 \cdot 10^{-6}\,\mathrm{Nm\,s}$ and $\overline{f} \approx 335.6\,\mathrm{Nm\,s}$. It holds that $h_1\,(\mathbf{z})\,\overline{f} + h_2\,(\mathbf{z})\,\underline{f} = f\,(\mathbf{z})$. The TS matrices $\tilde{\mathbf{A}}_i$ are given by

$$\tilde{\mathbf{A}}_1 = \begin{bmatrix} & \mathbf{0}_{1\times 5} & \\ 0 & \overline{f} & \mathbf{0}_{1\times 3} \\ & \mathbf{0}_{3\times 5} & \end{bmatrix},\ \tilde{\mathbf{A}}_2 = \begin{bmatrix} & \mathbf{0}_{1\times 5} & \\ 0 & \underline{f} & \mathbf{0}_{1\times 3} \\ & \mathbf{0}_{3\times 5} & \end{bmatrix},\ \text{so that } \sum_{i=1}^{2} h_i\,\tilde{\mathbf{A}}_i\,\mathbf{x} = \begin{pmatrix} 0 \\ T_a \\ \mathbf{0}_{3\times 1} \end{pmatrix}.$$

All states except for the torsional angle θ_s are measurable:

$\mathbf{y} = (\omega_r \quad \omega_g \quad \beta \quad T_g)^T$. Furthermore, the wind speed v, which is part of the TS premise variable vector $\mathbf{z}$, is assumed to be measurable. By making this assumption, it is easier to validate the stability of the TS SMO than for a case with unmeasurable premise variables. However, in first tests it could be verified that when using an estimated wind speed obtained from a TS observer [14], the simulation results hardly differ from those presented in this work.

Actuator offset faults can be modeled by adding an offset to the input signal: $\tilde{\mathbf{u}} = \mathbf{u} + \mathbf{f}_a$. All actuator fault matrices $\mathbf{E}_i$ in equation (1) are thus equal to the input matrix $\mathbf{B}$. No $\mathbf{D}_i$-matrices for unmodeled dynamics were included for the observer design, since this yielded no advantages in the fault reconstruction. The actuator faults are thus reconstructed from the equivalent output injection signal using the relation $\hat{\mathbf{f}}_a = \boldsymbol{\mathcal{E}}_2^+ \boldsymbol{\nu}_{\text{eq}}$. The following parameters were used for the TS SMO design: $\boldsymbol{\mathcal{A}}_{22}^s = -10 \cdot \mathbf{I}_{4\times4}$, $\mathbf{Q}_2 = \mathbf{I}_{4\times4}$, $\delta = 0.05$. To obtain the weighting matrix $\mathbf{W}$, the maximum absolute values of the individual outputs were estimated as: $\omega_{r,\text{max}} = \omega_{g,\text{max}} = 1.6\,\frac{\text{rad}}{\text{s}}$, $\beta_{\text{max}} = 1.5708\,\text{rad}$, $T_{g,\text{max}} = 4.2 \cdot 10^6\,\text{Nm}$. The elements of the sliding mode gain matrix were chosen as: $\rho_{\omega_r} = 200$, $\rho_{\omega_g} = 200$, $\rho_\beta = 200$, $\rho_{T_g} = 1.1 \cdot 10^9$.

The NREL 5 MW reference turbine was simulated in FAST using turbulent wind input with a mean wind speed of 18 m/s. For both actuators, constant offset faults were simulated. For the pitch angle, an offset of 0.1 rad was active between 30 s and 70 s. For the generator torque, an offset of 1000 Nm (which corresponds to 97000 Nm in the observer model, because the gearbox ratio of $n_g = 97$ in the FAST model is not included in the reduced order model) was active between 40 s and 80 s. This torque offset fault is taken from a wind turbine FDI benchmark model [13]. The simulation results for rotor speed, pitch angle, generator torque and the reconstructed faults are depicted in Fig. 1.

It can be seen from Fig. 1 that both faults are reasonably well reconstructed apart from short peaks that occur when the faults are switched on/off. In both cases, the true fault values are not fully reached but reconstructed to a degree of approximately 93 %. An erroneous transient oscillation occurs in the reconstructed generator torque fault signal after the onset of the pitch angle fault. When the pitch fault is switched on, the turbine controller reduces the demanded pitch angle in order to regulate the rotor speed back to the desired value.

Choice of the Sliding Mode Gain Matrix. The gain matrix for the switching term was chosen as $\boldsymbol{\rho} = \text{diag}\begin{pmatrix}200 & 200 & 200 & 1.1 \cdot 10^9\end{pmatrix}$. With this choice, both actuator offset faults could be well reconstructed in the FAST wind turbine simulation. However, when estimating the constant $\mathcal{K}_{\text{max}}$ in the presence of faults ($\mathcal{K}_{\text{max}} \approx 10^7$), it is found that only the component $\rho_{4\times4}$ satisfies condition (20). Setting all matrix components of $\boldsymbol{\rho}$ to values greater than $\mathcal{K}_{\text{max}}$, on the other hand, leads to an inadequate observer behavior.

Apparently, (20) gives a theoretically correct, but very conservative and not necessarily practical condition for the TS SMO design.

Revisiting condition (19), which is the actually determining condition for an ideal sliding motion, one can see that (20) guarantees (19) for all possible values of the error vector $\tilde{\mathbf{e}}_y$. From a control theory standpoint, this is desirable. In practice, however, $\tilde{\mathbf{e}}_y$ assumes certain values such that condition (19) only has to be fulfilled for these certain values of $\tilde{\mathbf{e}}_y$. When estimating the values of $\tilde{\mathbf{e}}_y$ and $\mathcal{K}_{\text{max}}$ with and without the presence of the actuator faults for different operating regimes of the wind turbine, one can make out why condition (19) is probably fulfilled for the above choice of the gain matrix $\boldsymbol{\rho}$, disregarding the

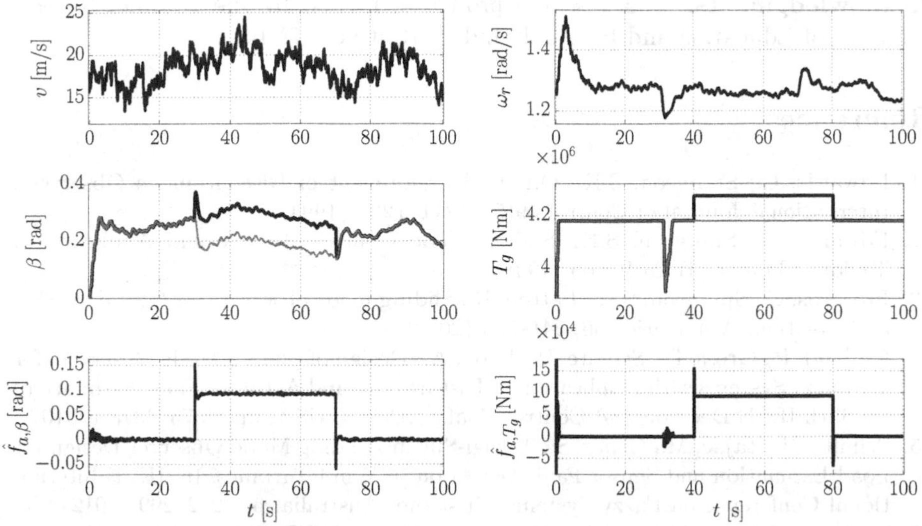

Fig. 1. FAST simulation with turbulent wind input (mean windspeed: 18 m/s) with offset actuator faults in the pitch angle and generator torque actuator. In the two middle figures, the (black) thick lines are the actual pitch angle and generator torque, whereas the (gray) thin lines are the demanded actuator values. The two lower figures show the components of the reconstructed fault vector $\hat{\mathbf{f}}_a$.

transient peaks. In a situation without faults or with only an active pitch actuator fault, $\mathcal{K}_{\max} \approx 70$, such that even condition (20) is always fulfilled. When the torque actuator fault is active, the T_g-component of $\tilde{\mathbf{e}}_y$ assumes a mean value of $\approx -1.8 \cdot 10^4$ Nm (not shown). Since all other components of $\tilde{\mathbf{e}}_y$ are close to zero, the T_g-part contributes a large positive value to the quadratic form (19), which outweighs the other small negative values, so that (19) is probably fulfilled.

5 Conclusion

In this paper, a Takagi-Sugeno sliding mode observer (TS SMO) with enhanced design possibilities was introduced. The discontinuous switching term responsible for establishing the sliding motion was modified such that the individual system outputs can be appropriately weighted according to their respective orders of magnitudes. A formal proof was given that this modified TS SMO is able to establish a stable sliding motion and that the sliding surface can be reached in finite time. As an example, the modified TS SMO was used for actuator fault diagnosis for wind turbines. While a formal sufficient, but not necessary, design condition for the individual components of the gain matrix for the switching term was obtained, this condition is very conservative and needs to be replaced in practice by a judicious choice of the gain matrix depending on estimated error bounds of the system states and outputs.

Acknowledgments. This research project is funded by the German Federal Ministry of Education and Research under grant no. 17N1411.

References

1. Edwards, C., Spurgeon, S.K.: On the Development of Discontinuous Observers. International Journal of Control 59(5), 1211–1229 (1994)
2. Edwards, C., Spurgeon, S.K.: Sliding Mode Control: Theory and Applications. Taylor & Francis, Boca Raton (1998)
3. Edwards, C., Spurgeon, S.K., Patton, R.: Sliding mode observers for fault detection and isolation. Automatica 36, 541–553 (2000)
4. Gerland, P., Gross, D., Schulte, H., Kroll, A.: Design of Sliding Mode Observers for TS Fuzzy Systems with Application to Disturbance and Actuator Fault Estimation. In: 49th IEEE Decision and Control Conf., Atlanta, USA, pp. 4373–4378 (2010)
5. Schulte, H., Zajac, M., Georg, S.: Takagi-Sugeno Sliding Mode Observer Design for Load Estimation and Sensor Fault Detection in Wind Turbines. In: IEEE International Conference on Fuzzy Systems, Brisbane, Australia, pp. 292–299 (2012)
6. Jonkman, J.M., Buhl Jr., M.L.: FAST User's Guide. NREL TP-500-38230. National Renewable Energy Laboratory, Golden, Colorado (2005)
7. Jonkman, J., Butterfield, S., Musial, W., Scott, G.: Definition of a 5-MW Reference Wind Turbine for Offshore System Development. NREL TP-500-38060. National Renewable Energy Laboratory, Golden, Colorado (2009)
8. Takagi, T., Sugeno, M.: Fuzzy Identification of Systems and Its Applications to Modeling and Control. IEEE Transactions on Systems, Man, and Cybernetics 15(1), 116–132 (1985)
9. Tanaka, K., Wang, H.O.: Fuzzy Control Systems Design and Analysis: A Linear Matrix Inequality Approach. John Wiley & Sons, Inc. (2001)
10. Bianchi, F.D., Battista, H., De Mantz, R.J.: Wind Turbine Control Systems: Principles, Modelling and Gain Scheduling Design. Springer-Verlag London Limited (2007)
11. Burton, T., Jenkins, N., Sharpe, D., Bossanyi, E.: Wind Energy Handbook, 2nd edn. John Wiley & Sons, Ltd., Chichester (2011)
12. Georg, S., Schulte, H., Aschemann, H.: Control-Oriented Modelling of Wind Turbines Using a Takagi-Sugeno Model Structure. In: IEEE International Conference on Fuzzy Systems, Brisbane, Australia, pp. 1737–1744 (2012)
13. Odgaard, P.F., Stoustrup, J., Kinnaert, M.: Fault Tolerant Control of Wind Turbines - a benchmark model. In: 7th IFAC Symposium on Fault Detection, Supervision and Safety of Technical Processes, Barcelona, Spain, pp. 155–160 (2009)
14. Georg, S., Mueller, M., Schulte, H.: Wind Turbine Model and Observer in Takagi-Sugeno Model Structure. In: Conference 'The Science of Making Torque from Wind', European Academy of Wind Energy (EAWE), Oldenburg, Germany (2012)

Control Reconfiguration for One Class of Takagi-Sugeno Fuzzy SISO Systems

Anna Filasová, Dušan Krokavec, and Vladimír Serbák

Technical University of Košice, Faculty of Electrical Engineering and Informatics,
Department of Cybernetics and Artificial Intelligence,
Letná 9, 042 00 Košice, Slovakia
{anna.filasova,dusan.krokavec,vladimir.serbak}@tuke.sk
http://web.tuke.sk/kkui

Abstract. Control reconfiguration via state feedback eigenstructure assignment, exploiting switching control principles for SISO systems, is considered in the paper. Focusing on linear SISO systems, as well as on one class of Takagi-Sugeno fuzzy systems, the design aim is to synthesize a set of state feedback control laws in such way that all closed-loop matrix constituents are sharing a common left eigenvector. The principle is based on the necessary stability condition for Hurwitz polynomials and design conditions are derived in terms of linear matrix inequalities.

Keywords: State feedback, switching systems, exponential stability, linear matrix inequality, eigenstructure assignment.

1 Introduction

Eigenstructure assignment belongs to the prominent design problems of modern control theory, concerning with the establishment of eigenvalues and corresponding eigenvectors via feedback control laws, especially in the single-input single-output (SISO) case, where a solution, when it exists, is unique. Consequently, the eigenvalues are the main factors that govern the system stability, and the eigenvectors, on the other hand, are dual factors that together determine the relative shape of the system dynamic response [1], [11], [16]. Moreover, recent advances in computational techniques, have resulted in the use of linear matrix inequality (LMI) techniques to solve the eigenvalues placement problem closest to its algebraic nature.

One of the possible ways to verify the properties of control reconfiguration is the switched systems analyze, even though that the stability of switched systems gives rise to a number of mathematical problems, since an arbitrary switching problem is concerned with obtaining verifiable conditions on the constituent set of matrices that guarantee the exponential stability of the switched system for any switching signal. Since the matrices in the constituent set have to be all Hurwitz, a number of methods relying on the construction of common quadratic functions for the constituent set, was proposed. For details on quadratic methods application in switched systems stability analysis see, e.g., [10], [12], [13].

J. Korbicz and M. Kowal (eds.), *Intelligent Systems in Technical and Medical Diagnostics*, Advances in Intelligent Systems and Computing 230,
DOI: 10.1007/978-3-642-39881-0_4,

In the paper, the problem of control reconfiguration by assigning the matrix eigenstructure through a state feedback control is reformulated to stabilization problem of switching linear SISO, as well Takagi-Sugeno fuzzy SISO systems. Utilizing algebraic methods [3], [7], the exposition of eigenstructure assignment problem is generalized to handle the common left eigenvector in the state feedback control of a constituent set of the closed-loop system matrices. Preferring methodology given in [6], a degree of freedom is used to design the closed-loop constituents of the Hurwitz type. The obtained set of LMIs, representing necessary condition for control law parameter design, provides a methodology usable in stabilization of the constituents if the state-feedback control is used.

The outline of the paper is as follows: Following Introduction, Section 2 places the preliminary results. In Section 3 and 4 the properties of linear SISO systems as well one class of Takagi-Sugeno fuzzy SISO systems are analyzed, to formulate the control law design conditions. Conforming these results, Section 5 presents the illustrative examples and Section 7 gives some concluding remarks.

Throughout the paper, the following notations are used: $\boldsymbol{x}^T$, $\boldsymbol{X}^T$ denotes the transpose of the vector $\boldsymbol{x}$ and matrix $\boldsymbol{X}$, respectively, $diag[\,\cdot\,]$ denotes a block diagonal matrix, for a square matrix $\boldsymbol{X} < 0$ means that $\boldsymbol{X}$ is a symmetric negative definite matrix, the symbol $\boldsymbol{I}_n$ indicates the n-th order unit matrix, $I\!R$ denotes the set of real numbers and $I\!R^{n \times r}$ refers to the set of $n \times r$ real matrices.

2 Basic Preliminaries

Proposition 1. *[2] (control canonical form) If a system relations take the forms*

$$\dot{\boldsymbol{q}}(t) = \boldsymbol{A}\boldsymbol{q}(t) + \boldsymbol{b}u(t), \qquad y(t) = \boldsymbol{c}^T\boldsymbol{q}(t)\ , \tag{1}$$

then there exists the coordinates change such that

$$\dot{\boldsymbol{q}}_c(t) = \boldsymbol{A}_c\boldsymbol{q}_c(t) + \boldsymbol{b}_c u(t), \qquad y(t) = \boldsymbol{c}_c^T\boldsymbol{q}_c(t)\ , \tag{2}$$

$$\boldsymbol{q}(t) = \boldsymbol{T}\boldsymbol{q}_c(t),\ \det\boldsymbol{T} \neq 0,\ \boldsymbol{A}_c = \boldsymbol{T}^{-1}\boldsymbol{A}\boldsymbol{T},\ \boldsymbol{b}_c = \boldsymbol{T}^{-1}\boldsymbol{b},\ \boldsymbol{c}_c^T = \boldsymbol{c}^T\boldsymbol{T}\ , \tag{3}$$

$$\boldsymbol{A}_c = \begin{bmatrix} 0 & 1 & 0 & \cdots & 0 \\ 0 & 0 & 1 & \cdots & 0 \\ \vdots & \vdots & \vdots & \ddots & \\ 0 & 0 & 0 & \cdots & 1 \\ -a_0 & -a_1 & -a_2 & \cdots & -a_{n-1} \end{bmatrix}, \quad \boldsymbol{b}_c = \begin{bmatrix} 0 \\ \vdots \\ 0 \\ 1 \end{bmatrix}. \tag{4}$$

Consequently,

$$P(s) = det(s\boldsymbol{I} - \boldsymbol{A}) = det(s\boldsymbol{I} - \boldsymbol{A}_c) = s^n + a_{n-1}s^{-1} + \ldots + a_1 s + a_0\ , \tag{5}$$

$$Q(s) = \begin{bmatrix} s^0\ s^1 \cdots s^{n-1} \end{bmatrix} \boldsymbol{c}_c = b_{n-1}s^{n-1} + b_{n-2}s^{-2} + \cdots + b_1 s + b_0\ , \tag{6}$$

where $P(s)$ is the characteristic polynomial and $Q(s)$ is the nominator of the system transfer function (TF). Using the state control law

$$u(t) = -\boldsymbol{k}^T\boldsymbol{q}(t)\ , \tag{7}$$

the zeros of the TFs of the closed-loop system and the plant are identical.

Lemma 1. *(left eigenvector assignment) Assigning a real closed-loop eigenvalue as $s_\circ$, $s_\circ < 0$, the corresponding left eigenvector as $\boldsymbol{m}^T \in I\!R^n$ and using the state feedback control of type (7), the closed loop system matrix $\boldsymbol{A}_{uc}$ takes the form*

$$\boldsymbol{A}_{uc} = \left(\boldsymbol{I} - \boldsymbol{b}_c\boldsymbol{m}_c^T\right)\boldsymbol{A}_c + s_\circ\boldsymbol{b}_c\boldsymbol{m}_c^T \tag{8}$$

and the closed-loop characteristic polynomial is

$$P_u(s) = det(s\boldsymbol{I} - \boldsymbol{A}_{uc}) = (s - s_\circ)(s^{n-1} + m_{c,n-1}s^{n-2} + \ldots + m_{c2}s + m_{c1}) \,, \tag{9}$$

where the normalized left eigenvector $\boldsymbol{m}_c^T$, associated with the eigenvalue $s_\circ$, is

$$\boldsymbol{m}_c^T = \left[m_{c,1}\ m_{c,2} \cdots m_{c,n-1}\ 1\right] . \tag{10}$$

Proof. If $\boldsymbol{m}_c^T$ is the left eigenvector corresponding to the eigenvalue $s_\circ$, then

$$\boldsymbol{m}_c^T\boldsymbol{A}_{uc} = \boldsymbol{m}_c^T(\boldsymbol{A}_c - \boldsymbol{b}_c\boldsymbol{k}_c^T) = s_\circ\boldsymbol{m}_c^T \,, \tag{11}$$

$$\boldsymbol{m}_c^T(\boldsymbol{A}_c - s_\circ\boldsymbol{I}) - \boldsymbol{m}_c^T\boldsymbol{b}_c\boldsymbol{k}_c^T = 0 \,, \tag{12}$$

respectively, and the feedback gain vector $\boldsymbol{k}_c^T$ can be explained as

$$\boldsymbol{k}_c^T = \frac{\boldsymbol{m}_c^T(\boldsymbol{A}_c - s_\circ\boldsymbol{I})}{\boldsymbol{m}_c^T\boldsymbol{b}_c} = \frac{\boldsymbol{m}_c^T(\boldsymbol{A}_c - s_\circ\boldsymbol{I})}{\boldsymbol{b}_c^T\boldsymbol{m}_c} . \tag{13}$$

Substituting (13) into (24) gives

$$\boldsymbol{A}_{uc} = \boldsymbol{A}_c - \frac{\boldsymbol{b}_c\boldsymbol{m}_c^T(\boldsymbol{A}_c - s_\circ\boldsymbol{I})}{\boldsymbol{b}_c^T\boldsymbol{m}_c} = \left(\boldsymbol{A}_c - \frac{\boldsymbol{b}_c\boldsymbol{m}_c^T}{\boldsymbol{b}_c^T\boldsymbol{m}_c}\boldsymbol{A}_c\right) + s_\circ\frac{\boldsymbol{b}_c\boldsymbol{m}_c^T}{\boldsymbol{b}_c^T\boldsymbol{m}_c} . \tag{14}$$

Considering $\boldsymbol{m}_c^T$ in the form (10) and $\boldsymbol{b}_c$ of the structure (4), thereafter

$$\boldsymbol{m}_c^T\boldsymbol{b}_c = 1 \tag{15}$$

and (14) implies (8). Using (4), the condition (8) gives

$$\boldsymbol{A}_{uc} = \begin{bmatrix} 1 & 0 & \cdots & 0 & 0 \\ 0 & 1 & \cdots & 0 & 0 \\ \vdots & & \ddots & & \vdots \\ 0 & 0 & \cdots & 1 & 0 \\ -m_{c1} & -m_{c2} & \cdots & -m_{c,n-1} & 0 \end{bmatrix} \boldsymbol{A}_c + \begin{bmatrix} 0 & \cdots & 0 & 0 \\ 0 & \cdots & 0 & 0 \\ \vdots & \ddots & \vdots & \vdots \\ 0 & \cdots & 0 & 0 \\ s_\circ m_{c,1} & \cdots & s_\circ m_{c,n-1} & s_\circ \end{bmatrix} , \tag{16}$$

$$\boldsymbol{A}_{uc} = \begin{bmatrix} 0 & 1 & 0 & \cdots & 0 \\ 0 & 0 & 1 & \cdots & 0 \\ \vdots & \vdots & \vdots & \ddots & \vdots \\ 0 & 0 & 0 & \cdots & 1 \\ 0 & -m_{c1} & -m_{c2} & \cdots & -m_{c,n-1} \end{bmatrix} + \begin{bmatrix} 0 & \cdots & 0 & 0 \\ 0 & \cdots & 0 & 0 \\ \vdots & \ddots & \vdots & \vdots \\ 0 & \cdots & 0 & 0 \\ s_\circ m_{c,1} & \cdots & s_\circ m_{c,n-1} & s_\circ \end{bmatrix} . \tag{17}$$

Since (17) takes the Frobenius form, it is obvious that

$$P_u(s) = s^n + (m_{c,n-1} - s_\circ)s^{n-1} + \cdots + (m_{c1} - s_\circ m_{c2})s - s_\circ m_{c1} \,, \tag{18}$$

$$P_u(s) = s(s^{n-1} + m_{c,n-1}s^{n-2} + \cdots + m_{c1}) - s_\circ(s^{n-1} + m_{c,n-1}s^{n-2} + \cdots + m_{c1}) \,, \tag{19}$$

respectively. Hence, (19) implies (9). This concludes the proof. ∎

Remark 1. Evidently, the closed-loop characteristic polynomial is stable if $s_\circ < 0$ and the polynomial

$$P_{um}(s) = s^{n-1} + m_{c,n-1}s^{n-2} + \ldots + m_{c2}s + m_{c1} = \begin{bmatrix} s^0 & s^1 & \cdots & s^{n-1} \end{bmatrix} m_c \tag{20}$$

is stable. Thus, the necessary condition to obtain a stable closed-loop characteristic polynomial is that all components of the normalized left eigenvector (10) are positive. Note, in the sense of the Routh's criterion, this condition is necessary but not sufficient.

The above generally implies that, for a prescribed $s_\circ < 0$, the stability of the closed-loop system is uniquely determined by the left eigenvector $m^T \in I\!R^n$.

3 Switched Linear SISO Systems

3.1 Problem Statement

The l-th controllable constituent of the switched linear SISO system is modeled by the state-space equation

$$\dot{q}(t) = A_l q(t) + b_l u_l(t) \tag{21}$$

with constant matrices $A_l \in I\!R^{n \times n}$ and vectors $b_l \in I\!R^n$, $l = 1, 2, \ldots, w$, where $q(t) \in I\!R^n$ is the vector of state variables, and $u_l(t) \in I\!R$ is the system input variable.

Defining the linear state feedback control law as follows

$$u_l(t) = -k_l^T q(t), \quad l = 1, 2, \ldots, w \,, \tag{22}$$

$k_l \in I\!R^n$, then applying (22) to (21) gives

$$\dot{q}(t) = (A_l - b_l k_l^T) q(t) \,, \tag{23}$$

where roots of the closed-loop characteristic polynomial are eigenvalues of the matrix $A_{ul} \in I\!R^{n \times n}$

$$A_{ul} = A_l - b_l k_l^T \,. \tag{24}$$

If a switched linear SISO system is defined as

$$\dot{q}(t) = A_{u\delta(q(t))} q(t) \,, \tag{25}$$

where $\delta(q(t))$ is a piecewise constant function (switching signal) defining the active modes $\dot{q}(t) = A_\alpha q(t)$, $A_\alpha \in \mathcal{A}$, and $\mathcal{A}$ is a collection of matrices $\mathcal{A} = \{A_{ul},\ l = 1, 2, \ldots, w\}$, the goal is to design the set of gain matrices (22) in such way that the collection of matrices $\mathcal{A}$ is Hurwitz.

3.2 Stability Analysis

Theorem 1. *The switched system (25) is stable for an arbitrary defined function $\delta(\boldsymbol{q}(t))$ if all matrices $\boldsymbol{A}_\alpha \in \mathcal{A}$ are Hurwitz and sharing the same left eigenvector $\boldsymbol{m}^T$, associated with stable real eigenvalues set $\{s_{\circ,l} < 0,\ l = 1, 2, \ldots, w\}$.*

Proof. (compare [6]) For a given switching function $\delta(\boldsymbol{q}(t))$ and a fixed initial condition $\boldsymbol{q}(t_0)$, the solution of (25) is uniquely given as

$$\boldsymbol{q}(t) = \left[e^{\boldsymbol{A}_{u\delta_k}(t-t_{k-1})} e^{\boldsymbol{A}_{u\delta_{k-1}}(t-t_{k-1})} \cdots e^{\boldsymbol{A}_{u\delta_0}(t-t_0)} \right] \boldsymbol{q}(t_0) \,, \tag{26}$$

where $t_0 < t_1 < \cdots < t_{k-1} < t_k$ refer to the switching times (switching instances), t_k is the largest switching instance smaller than t, and $\delta_j = \delta(\boldsymbol{q}(t_j))$, $j = 0, 1, 2, \ldots k$.

If all matrices in $\boldsymbol{A}_\alpha$ are sharing the same left eigenvector $\boldsymbol{m}^T$, then multiplying the left-hand side of (26) by $\boldsymbol{m}^T$ results in

$$\begin{aligned}
\boldsymbol{m}^T \boldsymbol{q}(t) &= \boldsymbol{m}^T e^{\boldsymbol{A}_{u\delta_k}(t-t_k)} e^{\boldsymbol{A}_{u\delta_{k-1}}(t-t_{k-1})} \cdots e^{\boldsymbol{A}_{u\delta_0}(t-t_0)} \boldsymbol{q}(t_0) = \\
&= e^{s_{\circ,k}(t-t_k)} \boldsymbol{m}^T e^{\boldsymbol{A}_{u\delta_{k-1}}(t-t_{k-1})} \cdots e^{\boldsymbol{A}_{u\delta_0}(t-t_0)} \boldsymbol{q}(t_0) = \cdots = \\
&= e^{s_{\circ,k}(t-t_k)} e^{s_{\circ,k-1}(t-t_{k-1})} \cdots e^{s_{\circ,0}(t-t_0)} \boldsymbol{m}^T \boldsymbol{q}(t_0) \,.
\end{aligned} \tag{27}$$

Thus, if all eigenvalues $s_{\circ,j}$, $j = 0, 1, 2, \ldots k$, in the sequence are stable, then (27) implies

$$\lim_{t\to\infty} \boldsymbol{m}^T \boldsymbol{q}(t) = \lim_{t\to\infty} \boldsymbol{m}^T \boldsymbol{q}(t_0) = 0 \tag{28}$$

and, consequently, all solutions of (26), independently of the initial value $\boldsymbol{q}(t_0)$, converge to the invariant set

$$\mathcal{Q} = \{\boldsymbol{q}(t) : \boldsymbol{m}^T \boldsymbol{q} = \boldsymbol{0}\} \,, \tag{29}$$

where $\boldsymbol{q}(t) \in I\!R^n$.

Finally, multiplying the left-hand side of (25) by $\boldsymbol{m}^T$ gives

$$\boldsymbol{m}^T \dot{\boldsymbol{q}}(t) = \boldsymbol{m}^T \boldsymbol{A}_{u\delta(\boldsymbol{q}(t))} \boldsymbol{q}(t) = s_{\circ,k} \boldsymbol{m}^T \boldsymbol{q}(t) \,, \tag{30}$$

where $s_{\circ,k}$ is the eigenvalue associated with $\boldsymbol{m}^T$ in the last active mode k, $k \in \{1, 2 \cdots, w\}$. It is obvious that (28), (30) implies

$$\lim_{t\to\infty} \boldsymbol{m}^T \dot{\boldsymbol{q}}(t) = \lim_{t\to\infty} s_{\circ,k} \boldsymbol{m}^T \boldsymbol{q}(t) = \lim_{t\to\infty} s_{\circ,k} \boldsymbol{m}^T \boldsymbol{q}(t_0) = 0 \,. \tag{31}$$

This completes the proof. ∎

Remark 2. According to (27), for all t and for all $j = 0, 1, 2, \ldots k$,

$$\left\| \boldsymbol{m}^T \boldsymbol{q}(t) \right\| \le e^{\max(s_{\circ,j})(t-t_0)} \left\| \boldsymbol{m}^T \boldsymbol{q}(t_0) \right\| \tag{32}$$

and (32) implies that the solutions (27) converge to the invariant set $\mathcal{Q}$ exponentially, i.e., the switched system (25) is exponentially stable. ∎

3.3 Control Law Parameter Design

Since there does not exist a direct way to exploit the results of Theorem 1, the design approach has to be based on a general form of the constituents state description and the common left eigenvector, exploiting results of Remark 1.

Theorem 2. *For given set of constituents (21) and for a desired set of real eigenvalues* $\{s_{\circ,l} < 0,\ l = 1,2,\ldots,w\}$, *the necessary condition for existence of a collection of Hurwitz matrices* $\mathcal{A} = \{\boldsymbol{A}_{ul}\}$ *means that there exist a positive constant* $\gamma \in I\!R$ *and a positive definite diagonal matrix* $\boldsymbol{M} \in I\!R^{n\times n}$ *such that*

$$\boldsymbol{M} > 0, \quad \gamma > 0\,, \tag{33}$$

$$\boldsymbol{b}_l^T\boldsymbol{M}\boldsymbol{v} + \boldsymbol{v}^T\boldsymbol{M}\boldsymbol{b}_l > 0, \quad l = 1,2,\ldots,w, \quad \boldsymbol{v}^T = \begin{bmatrix}1\,1\cdots 1\end{bmatrix}, \tag{34}$$

$$\begin{bmatrix} -\gamma\boldsymbol{I}_n & (\boldsymbol{A}_1 - s_{\circ,1}\boldsymbol{I})^T\boldsymbol{M}\boldsymbol{v} & \cdots & (\boldsymbol{A}_s - s_{\circ,w}\boldsymbol{I})^T\boldsymbol{M}\boldsymbol{v} \\ * & -1 & \cdots & 0 \\ \vdots & \vdots & \ddots & \vdots \\ * & * & \cdots & -1 \end{bmatrix} < 0\,. \tag{35}$$

When the above conditions hold, the design parameters are given as

$$\boldsymbol{k}_l^T = \frac{\boldsymbol{v}^T\boldsymbol{M}(\boldsymbol{A}_l - s_{\circ,l}\boldsymbol{I})}{\boldsymbol{b}_l^T\boldsymbol{M}\boldsymbol{v}}, \quad \boldsymbol{A}_{ul} = \boldsymbol{A}_l - \boldsymbol{b}_l\boldsymbol{k}_l^T, \quad \forall\, l\,. \tag{36}$$

Hereafter, $*$ denotes the symmetric item in a symmetric matrix.

Proof. Considering the performance index as follows [4]

$$\sum_{l=1}^{s} u_l(t)^T u_l(t) = \boldsymbol{q}^T(t)\sum_{l=1}^{s}\boldsymbol{k}_l\boldsymbol{k}_l^T\boldsymbol{q}(t) < \boldsymbol{q}^T(t)\sum_{l=1}^{s}\gamma\boldsymbol{I}_n\boldsymbol{q}(t)\,, \tag{37}$$

where $\gamma > 0$, $\gamma \in I\!R$ and, implying from (9), the control law gain structure as

$$\boldsymbol{k}_l^T = \boldsymbol{m}^T\boldsymbol{A}_{ul}^{\circ}, \quad \boldsymbol{A}_{ul}^{\circ} = (\boldsymbol{A}_l - s_{\circ,l}\boldsymbol{I})\,, \tag{38}$$

then, after substituting (38) into (37), it yields

$$\sum_{l=1}^{w}\boldsymbol{A}_{ul}^{\circ T}\boldsymbol{m}\boldsymbol{m}^T\boldsymbol{A}_{ul}^{\circ} - \gamma\boldsymbol{I}_n < 0\,. \tag{39}$$

Since Hurwitz necessary stability condition implies that all coefficients of a characteristic polynomial have to be positive, then defining

$$\boldsymbol{M} = \operatorname{diag}\begin{bmatrix}m_1\, m_2 \cdots m_n\end{bmatrix} > 0, \quad \boldsymbol{m}^T = \boldsymbol{v}^T\boldsymbol{M}, \quad \boldsymbol{b}_l^T\boldsymbol{M}\boldsymbol{v} > 0 \tag{40}$$

and using Schur complement property, (39) can be rewritten in the form

$$\begin{bmatrix} -\gamma\boldsymbol{I}_n & \boldsymbol{A}_{u1}^{\circ T}\boldsymbol{m} & \cdots & \boldsymbol{A}_{uw}^{\circ T}\boldsymbol{m} \\ * & -1 & \cdots & 0 \\ \vdots & \vdots & \ddots & \vdots \\ * & * & \cdots & -1 \end{bmatrix} < 0\,. \tag{41}$$

Thus, with (38) now (40), (41) implies (34), (35), respectively. ∎

Remark 3. The condition (33)–(35) is necessary but not sufficient. A stable solution can by obtained by suitable choosing the set of real eigenvalues $\{s_{\circ,l} < 0, l = 1, 2, \ldots, s\}$.

4 Switched TS Fuzzy SISO Systems Class

4.1 Problem Statement

Considering that all constituents of the switched TS fuzzy SISO system class are shackled with the same set of membership functions, i.e., the l-th constituent is

$$\dot{\boldsymbol{q}}(t) = \sum_{i=1}^{s} \mathrm{h}_i \boldsymbol{\theta}(t))\boldsymbol{A}_{li}\boldsymbol{q}(t) + \boldsymbol{b}_l u(t) \tag{42}$$

where $\boldsymbol{q}(t) \in \mathbb{R}^n$ is the vector of state variables, $u(t) \in \mathbb{R}$ is the input variable, $\boldsymbol{A}_{li} \in \mathbb{R}^{n \times n}$ is a set of constant matrices for given l and $i = 1, 2, \ldots, s$, $\boldsymbol{b}_l \in \mathbb{R}^n$ is a constant vector, $\mathrm{h}_i(\boldsymbol{\theta}(t))$ is the weight for the i-th rule, satisfying, by definition, the property

$$0 \leq \mathrm{h}_i(\boldsymbol{\theta}(t)) \leq 1, \quad \sum_{i=1}^{s} \mathrm{h}_i(\boldsymbol{\theta}(t)) = 1 \; \forall \; i \in \langle 1, \ldots, s \rangle \;, \tag{43}$$

and

$$\boldsymbol{\theta}(t) = \begin{bmatrix} \theta_1(t) \; \theta_2(t) \cdots \theta_o(t) \end{bmatrix} \tag{44}$$

is the vector of premise variables, where s, o are the numbers of fuzzy rules and premise variables, respectively. It is supposed in the next that all premise variables are measurable and independent on $u(t)$ and that all pairs $(\boldsymbol{A}_{li}, \boldsymbol{b}_l)$, $i = 1, 2, \ldots, s$, $l = 1, 2, \ldots, w$, are controllable (more details can be found, e.g., in [9], [15]).

Since $\boldsymbol{b}_l$ is constant in the l-th constituent, using the concept of parallel distributed compensation [14], the state feedback control law is defined as

$$u(t) = -\sum_{i=1}^{s} \mathrm{h}_i(\boldsymbol{\theta}(t))\boldsymbol{k}_i^T \boldsymbol{q}(t), \quad l = 1, 2, \ldots, w \;, \tag{45}$$

where $\boldsymbol{k}_{li} \in \mathbb{R}^n$ and applying (45) to (42) gives

$$\dot{\boldsymbol{q}}(t) = \sum_{i=1}^{s} \mathrm{h}_i(\boldsymbol{\theta}(t))\boldsymbol{A}_{uli}\boldsymbol{q}(t) \;, \tag{46}$$

where

$$\boldsymbol{A}_{uli} = \boldsymbol{A}_{li} - \boldsymbol{b}_l \boldsymbol{k}_{li}^T \;. \tag{47}$$

If a switched TS fuzzy SISO system is defined as

$$\dot{\boldsymbol{q}}(t) = \boldsymbol{A}_{uij\delta(\boldsymbol{q}(t))}\boldsymbol{q}(t) \;, \tag{48}$$

where, analogously, $\delta(\boldsymbol{q}(t))$ is a switching signal defining the active modes $\dot{\boldsymbol{q}}(t) = \boldsymbol{A}_\alpha \boldsymbol{q}(t)$, $\boldsymbol{A}_\alpha \in \mathcal{A}$, and $\mathcal{A}$ is a collection of matrices $\mathcal{A} = \{\boldsymbol{A}_{ulij}, \; l = 1, 2, \ldots, w, \; i = 1, 2, \ldots, s\}$, to obtain a stable switching fuzzy system, the collection of matrices $\mathcal{A}$ has to be Hurwitz.

4.2 Stability Analysis

Theorem 3. *The switched fuzzy system (42) is stable for an arbitrary defined function $\delta(\boldsymbol{q}(t))$ if all matrices $\boldsymbol{A}_\alpha \in \mathcal{A}$ are Hurwitz and sharing the same left eigenvector $\boldsymbol{m}^T$, associated with a stable real eigenvalues set $\{s_{\circ,li} < 0,\ l = 1,2,\ldots,w,\ i = 1,2,\ldots,s\}$.*

Proof. For given switching function $\delta(\boldsymbol{q}(t))$, the solution of (48) is given by

$$\boldsymbol{q}(t) = \Big[\sum_{i=1}^{s} \mathrm{h}_i(\boldsymbol{\theta}(t))e^{\boldsymbol{A}_{ui\delta_k}(t-t_{k-1})} \cdots \sum_{i=1}^{s} \mathrm{h}_i(\boldsymbol{\theta}(t))e^{\boldsymbol{A}_{ui\delta_0}(t-t_0)}\Big]\boldsymbol{q}(t_0)\,, \tag{49}$$

with the switching instances $t_0 < t_1 < \cdots < t_{k-1} < t_k$ and $\delta_j = \delta(\boldsymbol{q}(t_j))$, $j = 0,1,2,\ldots k$. If all matrices in $\boldsymbol{A}_\alpha$ are sharing the same eigenvector $\boldsymbol{m}^T$, then

$$\boldsymbol{m}^T\boldsymbol{q}(t) = \sum_{i=1}^{s} \mathrm{h}_i(\boldsymbol{\theta}(t))e^{s_{\circ,ki}(t-t_k)} \cdots \sum_{i=1}^{s} \mathrm{h}_i(\boldsymbol{\theta}(t))e^{s_{\circ,0}(t-t_0)}\boldsymbol{m}^T\boldsymbol{q}(t_0)\,. \tag{50}$$

Thus, defining a stable eigenvalues set $\{s_{\circ,ji}\}$, (50) implies

$$\lim_{t\to\infty} \boldsymbol{m}^T\boldsymbol{q}(t) = \lim_{t\to\infty} \boldsymbol{m}^T\boldsymbol{q}(t_0) = 0 \tag{51}$$

that is, all solutions of (49) converge to the invariant set (29). Alike to (28), (30),

$$\boldsymbol{m}^T\dot{\boldsymbol{q}}(t) = \boldsymbol{m}^T\sum_{i=1}^{s} \mathrm{h}_i(\boldsymbol{\theta}(t))\boldsymbol{A}_{ui\delta(\boldsymbol{q}(t))}\boldsymbol{q}(t) = \sum_{i=1}^{s} \mathrm{h}_i(\boldsymbol{\theta}(t))s_{\circ,ki}\boldsymbol{m}^T\boldsymbol{q}(t)\,, \tag{52}$$

$$\lim_{t\to\infty} \boldsymbol{m}^T\dot{\boldsymbol{q}}(t) = \lim_{t\to\infty} \sum_{i=1}^{s} \mathrm{h}_i(\boldsymbol{\theta}(t))s_{\circ,ki}\boldsymbol{m}^T\boldsymbol{q}(t_0) = 0\,. \tag{53}$$

Generalizing Remark 2, it can be proven that the switched fuzzy systems is exponentially stable. This completes the proof. ■

4.3 Parameter Design

Theorem 4. *For sets of constituents (42) and real eigenvalues $\{s_{\circ,li} < 0,\ l = 1,2,\ldots,w,\ i = 1,2,\ldots,s\}$, the necessary condition for existence of a collection of Hurwitz matrices $\mathcal{A} = \{\boldsymbol{A}_{uli}\}$ means that there exist a positive constant $\gamma \in I\!R$ and a positive definite diagonal matrix $\boldsymbol{M} \in I\!R^{n\times n}$ such that*

$$\boldsymbol{M} > 0, \quad \gamma > 0\,, \tag{54}$$

$$\boldsymbol{b}_l^T\boldsymbol{M}\boldsymbol{v} + \boldsymbol{v}^T\boldsymbol{M}\boldsymbol{b}_l, \quad l = 1,2,\ldots,w, \quad \boldsymbol{v}^T = \begin{bmatrix}1\ 1 \cdots 1\end{bmatrix}, \tag{55}$$

$$\begin{bmatrix} -\gamma\boldsymbol{I}_n & (\boldsymbol{A}_1 - s_{\circ,1i}\boldsymbol{I})^T\boldsymbol{M}\boldsymbol{v} & \cdots & (\boldsymbol{A}_s - s_{\circ,wi}\boldsymbol{I})^T\boldsymbol{M}\boldsymbol{v} \\ * & -1 & \cdots & 0 \\ \vdots & \vdots & \ddots & \vdots \\ * & * & \cdots & -1 \end{bmatrix} < 0,\ i = 1,2,\ldots,s\,, \tag{56}$$

When the above conditions hold, the sets of gain matrices and constituents are

$$\boldsymbol{k}_{li}^T = \frac{\boldsymbol{v}^T\boldsymbol{M}(\boldsymbol{A}_{li} - s_{\circ,li}\boldsymbol{I})}{\boldsymbol{b}_l^T\boldsymbol{M}\boldsymbol{v}}, \quad \boldsymbol{A}_{uli} = \boldsymbol{A}_{li} - \boldsymbol{b}_l\boldsymbol{k}_{li}^T \quad \forall\, l,i\,. \tag{57}$$

Proof. According to equations (46), (47),

$$\sum_{i=1}^{s} \mathrm{h}_i(\boldsymbol{\theta}(t))\boldsymbol{m}^T(\boldsymbol{A}_{li} - \boldsymbol{b}_l\boldsymbol{k}_{li}^T) = s_{\circ,li}\boldsymbol{m}^T \,, \tag{58}$$

where $\boldsymbol{m}^T$ is the common left eigenvector corresponding to the eigenvalue $s_{\circ,li}$. Therefore, using the property (43), then (58) implies

$$\boldsymbol{m}^T\boldsymbol{b}_l \sum_{i=1}^{s} \mathrm{h}_{li}(\boldsymbol{\theta}(t)) \left(\frac{\boldsymbol{m}^T(\boldsymbol{A}_{li} - s_{\circ,li}\boldsymbol{I}_n)}{\boldsymbol{m}^T\boldsymbol{b}_l} - \boldsymbol{k}_{li}^T \right) = \boldsymbol{0} \,, \tag{59}$$

$$\boldsymbol{k}_{li}^T = \frac{\boldsymbol{m}^T(\boldsymbol{A}_{li} - s_{\circ,li}\boldsymbol{I}_n)}{\boldsymbol{m}^T\boldsymbol{b}_l} \,. \tag{60}$$

Since (43) is satisfied, the performance index (37) can be modified as [8]

$$\begin{aligned} \sum_{l=1}^{s} u_l(t)^T u_l(t) &= \boldsymbol{q}^T(t) \sum_{l=1}^{s}\sum_{i=1}^{s}\sum_{j=1}^{s} \mathrm{h}_{li}(\boldsymbol{\theta}(t))\mathrm{h}_{lj}(\boldsymbol{\theta}(t))\boldsymbol{k}_{li}\boldsymbol{k}_{lj}^T\boldsymbol{q}(t) \leq \\ &\leq \boldsymbol{q}^T(t) \sum_{l=1}^{s}\sum_{i=1}^{s} \mathrm{h}_{li}(\boldsymbol{\theta}(t))\boldsymbol{k}_{li}\boldsymbol{k}_{li}^T\boldsymbol{q}(t) < \boldsymbol{q}^T(t) \sum_{l=1}^{s} \gamma\boldsymbol{I}_n\boldsymbol{q}(t) \,, \end{aligned} \tag{61}$$

which inquires to be satisfied the condition

$$\sum_{l=1}^{s} \boldsymbol{k}_{li}\boldsymbol{k}_{li}^T - \gamma\boldsymbol{I}_n < 0 \quad \forall\, i \,. \tag{62}$$

Accordingly to (38), the control law gain structure can be considered as

$$\boldsymbol{k}_{li}^T = \boldsymbol{m}^T\boldsymbol{A}_{uli}^{\circ}, \quad \boldsymbol{A}_{uli}^{\circ} = \boldsymbol{A}_{li} - s_{\circ,li}\boldsymbol{I}_n \tag{63}$$

and evidently, using (40), it yields

$$\begin{bmatrix} -\gamma\boldsymbol{I}_n & \boldsymbol{A}_{u1i}^{\circ T}\boldsymbol{m} & \cdots & \boldsymbol{A}_{uwi}^{\circ T}\boldsymbol{m} \\ * & -1 & \cdots & 0 \\ \vdots & \vdots & \ddots & \vdots \\ * & * & \cdots & -1 \end{bmatrix} < 0 \quad \forall\, i \,. \tag{64}$$

Thus, with (36) now (40), (64) implies (55), (56), respectively. This concludes the proof. ■

5 Illustrative Example

The linear SISO system was described by (21), where

$$\boldsymbol{A}_1 = \begin{bmatrix} 0.0 & 1.0 & 1.0 \\ -2.5 & -6.5 & -2.5 \\ 2.5 & 7.5 & 1.5 \end{bmatrix}, \; \boldsymbol{A}_2 = \begin{bmatrix} 0.0 & 0.5 & 0.5 \\ -6.0 & -8.0 & -3.0 \\ 6.0 & 9.0 & 2.0 \end{bmatrix}, \; \boldsymbol{A}_3 = \begin{bmatrix} -0.5 & -0.5 & 1.0 \\ 0.5 & 0.5 & 1.0 \\ -1.0 & -6.0 & -3.0 \end{bmatrix},$$

$$\boldsymbol{b}_1^T = \begin{bmatrix} 0.0 & 0.5 & -0.5 \end{bmatrix}, \quad \boldsymbol{b}_2^T = \begin{bmatrix} 0 & 1 & -1 \end{bmatrix}, \quad \boldsymbol{b}_3^T = \begin{bmatrix} 0 & 0 & 1 \end{bmatrix}.$$

Prescribing $s_{\circ 1} = -0.2$, $s_{\circ 2} = -0.1$, $s_{\circ 3} = -2$ and solving (33)–(35) with SeDuMi for the LMI matrix variables M, δ, the design problem was feasible and

$$\boldsymbol{M} = \operatorname{diag}\begin{bmatrix} 0.2177 & 0.2292 & 0.1731 \end{bmatrix}, \quad \gamma = 1.0559,$$

$$\boldsymbol{k}_1^T = \begin{bmatrix} -3.4472 & 2.5738 & -2.1758 \end{bmatrix}, \quad \boldsymbol{k}_2^T = \begin{bmatrix} -5.6118 & -2.5628 & -3.8377 \end{bmatrix},$$

$$\boldsymbol{k}_3^T = \begin{bmatrix} 1.5480 & -3.3189 & 1.5813 \end{bmatrix}.$$

The set of the closed-loop system eigenvalues is

$$\rho(\boldsymbol{A}_{u1}) = \{-0.2000 \; -1.3279 \; -5.8470\}, \; \rho(\boldsymbol{A}_{u2}) = \{-0.1000 \; -0.5895 \; -6.5853\},$$

$$\rho(\boldsymbol{A}_{u3}) = \{-2.0000 \; -0.0260 \; -2.5552\}$$

and the closed-loop eigenstructure satisfies the condition $\boldsymbol{m}^T \boldsymbol{A}_{iu} = s_{\circ i} \boldsymbol{m}^T$, i.e.,

$$\boldsymbol{m}^T \boldsymbol{A}_{u1} = s_{\circ 1} \boldsymbol{m}^T = \begin{bmatrix} -0.0435 & -0.0458 & -0.0346 \end{bmatrix},$$

$$\boldsymbol{m}^T \boldsymbol{A}_{u2} = s_{\circ 2} \boldsymbol{m}^T = \begin{bmatrix} -0.0218 & -0.0229 & -0.0173 \end{bmatrix},$$

$$\boldsymbol{m}^T \boldsymbol{A}_{u3} = s_{\circ 3} \boldsymbol{m}^T = \begin{bmatrix} -0.4353 & -0.4584 & -0.3462 \end{bmatrix},$$

where

$$s_{\circ 1} \boldsymbol{m}^T = \frac{s_{\circ 1}}{s_{\circ 2}} (s_{\circ 2} \boldsymbol{m}^T) = \frac{s_{\circ 1}}{s_{\circ 3}} (s_{\circ 3} \boldsymbol{m}^T).$$

Evidently, the prescribed items $s_{\circ l}$, $l = 1, 2, 3$, are placed in the set of closed-loop eigenvalues.

Using a model of TS fuzzy SISO system [5], since the variable $q_1(t) \in \langle 0.05, 0.85 \rangle$ and $\sqrt{q_2(t)} \in \langle 1.085, 4.470 \rangle$ are bounded on the prescribed sectors, the vector of premise variables was chosen as follows

$$\boldsymbol{\theta}(t) = \begin{bmatrix} \theta_1(t) & \theta_2(t) \end{bmatrix} = \begin{bmatrix} q_1(t) & \sqrt{q_2(t)} \end{bmatrix}.$$

Now, the associated set of nonlinear sector functions is

$$w_{11}(\theta_1(t)) = \frac{c_1 - \theta_1(t)}{c_1 - c_2}, \; w_{12}(\theta_1(t)) = 1 - w_{11}(\theta_1(t)), \; c_1 = 0.05, \; c_2 = 0.85,$$

$$w_{21}(\theta_2(t))) = \frac{d_1 - \theta_2(t)}{d_1 - d_2}, \; w_{22}(\theta_2(t)) = 1 - w_{21}(\theta_2(t)), \; d_1 = 1.085, \; d_2 = 4.470,$$

which implies the set of normalized membership functions

$$h_1(\boldsymbol{\theta}(t)) = w_{11}(\theta_1(t)) w_{21}(\theta_2(t)), \qquad h_2(\boldsymbol{\theta}(t)) = w_{12}(\theta_1(t)) w_{21}(\theta_2(t)),$$

$$h_3(\boldsymbol{\theta}(t)) = w_{11}(\theta_1(t)) w_{22}(\theta_2(t)), \qquad h_4(\boldsymbol{\theta}(t)) = w_{12}(\theta_1(t)) w_{22}(\theta_2(t)),$$

as well as, considering an outage of the actuator gain $\boldsymbol{b}_l$, $l = 1, 2$, the parameters

$$\boldsymbol{A}_{l1} = \begin{bmatrix} -0.1627 & 0.0000 \\ 0.1627 & -0.1446 \end{bmatrix}, \; \boldsymbol{A}_{l2} = \begin{bmatrix} -0.3873 & 0.0000 \\ 0.3873 & -0.1446 \end{bmatrix}, \; \boldsymbol{b}_1 = \begin{bmatrix} 1.1667 \\ 0.0000 \end{bmatrix},$$

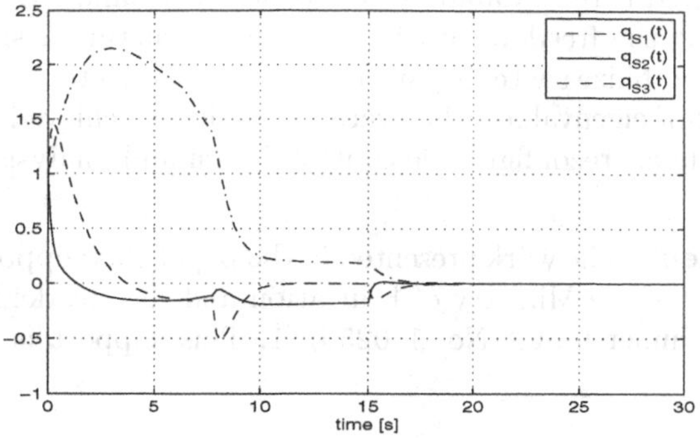

Fig. 1. A switched closed-loop state response

$$A_{l3} = \begin{bmatrix} -0.1627 & 0.0000 \\ 0.1627 & -0.3443 \end{bmatrix}, \; A_{l4} = \begin{bmatrix} -0.3873 & 0.0000 \\ 0.3873 & -0.3443 \end{bmatrix}, \; b_2 = \begin{bmatrix} 0.5834 \\ 0.0000 \end{bmatrix}.$$

Prescribing $s_{o1i} = -0.5$, $s_{o2i} = -0.3$, $i = 1, 2, 3, 4$ and solving the same problem as above using (54)–(56), the feasible solution was

$$M = \text{diag}\begin{bmatrix} 0.5672 & 0.3982 \end{bmatrix}, \quad \gamma = 1.0060\,,$$

$$k_{11}^T = \begin{bmatrix} 0.3870 & 0.2139 \end{bmatrix}, \; k_{12}^T = \begin{bmatrix} 0.3297 & 0.2139 \end{bmatrix}, \; k_{13}^T = \begin{bmatrix} 0.3870 & 0.0937 \end{bmatrix},$$

$$k_{14}^T = \begin{bmatrix} 0.3297 \, 0.0937 \end{bmatrix}, \; k_{21}^T = \begin{bmatrix} 0.4312 \, 0.1870 \end{bmatrix}, \; k_{22}^T = \begin{bmatrix} 0.3165 \, 0.1870 \end{bmatrix},$$

$$k_{23}^T = \begin{bmatrix} 0.4312 \, -0.0533 \end{bmatrix}, \; k_{24}^T = \begin{bmatrix} 0.3165 \, -0.0533 \end{bmatrix}.$$

The closed-loop eigenstructure satisfy the conditions $m^T A_{uli} = s_{oli} m^T$, i.e.,

$$m^T A_{u1i} = \begin{bmatrix} -0.2836 \, -0.1991 \end{bmatrix}, \; m^T A_{u2i} = \begin{bmatrix} -0.1702 \, -0.1195 \end{bmatrix}, \; \forall\, i = 1, 2, 3, 4$$

and all sets of the closed-loop eigenvalues are stable.

Fig. 1 shows the example of the linear unforced closed-loop switched system state response with nonzero initial state in the starting switching instant. For simplicity, it is considered that there is no time delay in detection of switching signal, and the initial state of a system in the j-th switching instant is the final state of the system in preceding interval.

6 Concluding Remarks

This paper provides a design method for memory-free controllers, where the problem of assigning the eigenstructure for state feedback switched systems is considered. The method exploits numerical optimization procedures to manipulate the common closed loop left eigenvector as a structured matrix variable. With generalization of the known algorithms, the modified exposition of the problem is presented here to handle stable SISO linear and TS fuzzy switching systems, exploiting the freedom provided by eigenstructure assignment to find controllers which stabilize switching systems. It was shown that by the appropriately assigned set of eigenvalues, the overall stability is achieved. The presented results are suitable for reconfigurable control design and analysis.

Acknowledgement. The work presented in the paper was supported by VEGA, the Grant Agency of the Ministry of Education and the Academy of Science of Slovak Republic, under Grant No. 1/0256/11. This support is very gratefully acknowledged.

References

1. Bachelier, O., Bosche, J., Mehdi, D.: On Pole Placement via Eigenstructure Assignment Approach. IEEE Trans. Automatic Control 51(9), 1554–1558 (2006)
2. Ipsen, I.C.S.: Numerical Matrix Analysis: Linear Systems and Least Squares. SIAM, Philadelphia (2009)
3. Kailath, T.: Linear Systems. Prentice-Hall, Englewood Cliffs (1980)
4. Kocsis, P., Gontkovič, D., Serbák, V.: Stabilizing Control for Switched Linear SISO Systems. In: Proc. 13th Int. Carpathian Control Conference, ICCC 2012, Podbanské, Slovakia, pp. 290–295 (2012)
5. Korba, P., Frank, P.M.: An Applied Optimization-Based Gain-Scheduled Fuzzy Control. In: Proc. American Control Conference, Chicago, USA, pp. 3383–3387 (2000)
6. Kouhi, Y., Bajcinca, N.: Nonsmooth Control Design for Stabilizing Switched Linear Systems by Left Eigenstructure Assignment. In: Prepr. 18th IFAC World Congress, Milano, Italy, pp. 380–385 (2011)
7. Krokavec, D., Filasová, A.: Dynamic Systems Diagnosis, Elfa, Košice (2006) (in Slovak)
8. Krokavec, D., Filasová, A.: Optimal Fuzzy Control for a Class of Nonlinear Systems. Mathematical Problems in Engineering, ID 481942, 29 p. (2012)
9. Krokavec, D., Filasová, A.: Stabilizing Fuzzy Output Control For a Class of Nonlinear Systems. Advances in Fuzzy Systems, ID 294971, 9p. (2013)
10. Liberzon, D.: Switching in Systems and Control. Birkhäser, Boston (2003)
11. Sobel, K.M., Lallman, F.J.: Eigenstructure Assignment for the Control of Highly Augmented Aircraft. J. Guidance, Control and Dynamics 12(3), 318–324 (1989)
12. Shorten, R., Wirth, F., Mason, O., Wulff, K., King, C.: Stability Criteria for Switched and Hybrid Systems. SIAM Review 49(4), 545–592 (2007)
13. Sun, Z., Ge, S.S.: Stability Theory of Switched Dynamical Systems. Springer, London (2011)
14. Takagi, T., Sugeno, M.: Fuzzy Identification of Systems and its Applications to Modeling and Control. IEEE Trans. Systems, Man, and Cybernetics 15(1), 116–132 (1985)
15. Tanaka, K., Wang, H.O.: Fuzzy Control Systems Design and Analysis: A Linear Matrix Inequality Approach. John Wiley & Sons, New York (2001)
16. Xu, X.H., Xie, X.K.: Eigenstructure Assignment by Output Feedback in Descriptor Systems. IMA J. Mathematical Control & Information 12, 127–132 (1995)

Efficient Predictive Fault-Tolerant Control for Non-linear Systems

Marcin Witczak and Piotr Witczak

Institute of Control and Computation Engineering, University of Zielona Góra,
ul. Pogórna 50, 65-246 Zielona Góra, Poland
M.Witczak@issi.uz.zgora.pl

Abstract. The paper deals with the problem of robust predictive fault-tolerant control for non-linear discrete-time systems. The proposed approach is based on a triple stage procedure, i.e. its starts from fault estimation, the fault is compensated with a robust controller. Finally, if the fault compensation does not provide satisfactory, which means that the current state does not belong to the robust invariant set, then a suitable predictive control actions are performed in order to enhance the invariant set. This appealing phenomenon makes it possible to enlarge the domain of attraction, which makes the proposed approach an efficient solution. The final part of the paper shows how to extend the proposed approach to the non-linear systems that can be described with the Takagi-Sugeno models.

Keywords: Fault-tolerant control, robust control, predictive control, robust invariant set.

1 Introduction

The Fault-Tolerant Control (FTC) systems are classified into two distinct classes [23]: passive and active. In the passive FTC [3], controllers are designed to be robust against a set of predefined faults, therefore there is no need for fault diagnosis, but such a design usually degrades the overall performance. In the contrast to the passive ones, active FTC schemes react to faults actively by reconfiguring control actions, and by doing so the system stability and acceptable performance is maintained. To achieve that, the control system relies on the Fault Detection and Isolation (FDI) [9,14,16,20,2,11] as well as an accommodation technique [1]. Most of the existing works treat the FDI and FTC problems separately. Unfortunately, a perfect FDI and fault identification are impossible and hence there always is an inaccuracy related to this process. Thus, there is a need for integrated FDI and FTC schemes for both linear and non-linear systems.

A number of books was published in the last decade on the emerging problem of the FTC. In particular, the book [8], which is mainly devoted to fault diagnosis and its applications provides some general rules for the hardware-redundancy-based FTC. On the contrary, the work [15] introduces the concepts of the active

DOI: 10.1007/978-3-642-39881-0_5,

and passive FTC. It also investigates the problem of performance and stability of the FTC under imperfect fault diagnosis. In particular, the authors consider (under a chain of some, not necessary easy to satisfy assumptions) the effect of a delayed fault detection and an imperfect fault identification but the fault diagnosis scheme is treated separately during the design and no real integration of the fault diagnosis and the FTC is proposed. The FTC is also treated in a very interesting work [17] where the number of practical case studies of FTC is presented, i.e., a winding machine, a three-tank system, and an active suspension system. Unfortunately, in spite of the incontestable appeal of the proposed approaches neither the FTC integrated with the fault diagnosis nor a systematic approach to non-linear systems are studied.

The proposed approach overcomes the above-mentioned difficulties and provides an elegant way of incorporating fault diagnosis (particularly the fault identification) into the fault-tolerant control framework. The proposed approach is based on a triple stage procedure, i.e. its starts from fault estimation, the fault is compensated with a robust controller. Finally, if the fault compensation does not provide satisfactory, which means that the current state does not belong to the robust invariant set, then a suitable predictive control actions are performed in order to enhance the invariant set. This appealing phenomenon makes it possible to enlarge the domain of attraction, which makes the proposed approach an efficient solution. Indeed, the presented solution can be perceived as an extension of the recent developments in this area [21] The paper also show how to extend the proposed approach to the non-linear systems that can be described with the Takagi-Sugeno [5] models.

The paper is organised as follows. Section 2 describes the robust control approach along with the fault compensation mechanism. Whilst section 3 presents the concept of the robust invariant set along with its derivation. It is shown that all the above problems can be efficiently solve with the Linear Matrix Inequalities (LMIs). The subsequent section 4 presents the proposed efficient predictive control for linear systems. Finally, the last section extends the proposed approach to be capable of handling nonlinearities that can be described in the Takagi-Sugeno modelling framework.

2 Design of a Robust Fault-Tolerant Controller

Let us consider a linear discrete-time system:

$$\boldsymbol{x}_{f,k+1} = \boldsymbol{A}\boldsymbol{x}_{f,k} + \boldsymbol{B}\boldsymbol{u}_{f,k} + \boldsymbol{B}\boldsymbol{f}_k + \boldsymbol{W}\boldsymbol{w}_k, \tag{1}$$

where $\boldsymbol{x}_k \in \mathbb{X} \subset \mathbb{R}^n$ is the state vector, $\boldsymbol{u}_k \in \mathbb{R}^r$ stands for the input, $\boldsymbol{f}_k \in \mathbb{R}^r$ is the actuator fault, and $\boldsymbol{w}_k \in \mathbb{R}^n$ is a an exogenous disturbance vector, while:

$$l_2 = \left\{ \mathbf{w} \in \mathbb{R}^n | \; \|\mathbf{w}\|_{l_2} < +\infty \right\}, \tag{2}$$

$$\|\mathbf{w}\|_{l_2} = \left(\sum_{k=0}^{\infty} \|\boldsymbol{w}_k\|^2 \right)^{\frac{1}{2}}. \tag{3}$$

Let us also assume that the system is controllable and the matrix $\boldsymbol{B}$ is a full rank one. Thus, following [6,20], it is possible to compute $\boldsymbol{H} = \boldsymbol{B}^{+}$. Subsequently, multiplying (1) by $\boldsymbol{H}$ and then extracting $\boldsymbol{f}_k$ gives:

$$\boldsymbol{f}_k = \boldsymbol{H}\boldsymbol{x}_{f,k+1} - \boldsymbol{H}\boldsymbol{A}\boldsymbol{x}_{f,k} - \boldsymbol{H}\boldsymbol{B}\boldsymbol{u}_{f,k} - \boldsymbol{H}\boldsymbol{W}\boldsymbol{w}_k, \tag{4}$$

while its estimate can be given as:

$$\hat{\boldsymbol{f}}_k = \boldsymbol{H}\boldsymbol{x}_{f,k+1} - \boldsymbol{H}\boldsymbol{A}\boldsymbol{x}_{f,k} - \boldsymbol{H}\boldsymbol{B}\boldsymbol{u}_{f,k}, \tag{5}$$

with the associated estimation error

$$\boldsymbol{\varepsilon}_{f,k} = \boldsymbol{f}_k - \hat{\boldsymbol{f}}_k = -\boldsymbol{H}\boldsymbol{W}\boldsymbol{w}_k. \tag{6}$$

Note that in order to obtain $\hat{\boldsymbol{f}}_k$ it is necessary to have $\boldsymbol{x}_{f,k+1}$. This determines the following control strategy

$$\boldsymbol{u}_{f,k} = -\hat{\boldsymbol{f}}_{k-1} - \boldsymbol{K}\boldsymbol{x}_{f,k}. \tag{7}$$

Bearing in mind that $\boldsymbol{f}_k$ is bounded, without a lost of generality, it is possible to write

$$\hat{\boldsymbol{f}}_k = \hat{\boldsymbol{f}}_{k-1} + \boldsymbol{v}_k, \quad \boldsymbol{v}_k \in l_2, \tag{8}$$

Thus, (7) can be written in an equivalent form, which will be used in further deliberations

$$\boldsymbol{u}_{f,k} = -\hat{\boldsymbol{f}}_k + \boldsymbol{v}_k - \boldsymbol{K}\boldsymbol{x}_{f,k}. \tag{9}$$

Substituting (9) into (1) gives

$$\boldsymbol{x}_{f,k+1} = \boldsymbol{A}_1\boldsymbol{x}_{f,k} + [\boldsymbol{I} - \boldsymbol{B}\boldsymbol{H}]\,\boldsymbol{W}\boldsymbol{w}_k + \boldsymbol{B}\boldsymbol{v}_k \tag{10}$$

with $\boldsymbol{A}_1 = \boldsymbol{A} - \boldsymbol{B}\boldsymbol{K}$. The equation (10) can be equivalently written as:

$$\boldsymbol{x}_{f,k+1} = \boldsymbol{A}_1\boldsymbol{x}_{f,k} + \bar{\boldsymbol{W}}\bar{\boldsymbol{w}}_k \tag{11}$$

with $\bar{\boldsymbol{W}} = [[\boldsymbol{I} - \boldsymbol{B}\boldsymbol{H}]\boldsymbol{W} \quad \boldsymbol{B}]$, $\quad \bar{\boldsymbol{w}}_k = \left[\boldsymbol{w}_k^T, \, \boldsymbol{v}_k^T\right]^T \in l_2$. Following the general $\mathcal{H}_\infty$ approach [13,22], the robust controller $\boldsymbol{K}$ is to be designed by satisfying the following inequlity

$$\Delta V + \boldsymbol{x}_{f,k}^T\boldsymbol{x}_{f,k} - \mu^2\bar{\boldsymbol{w}}_k^T\bar{\boldsymbol{w}}_k < 0, \quad \Delta V = V_{k+1} - V_k, \tag{12}$$

where the Lyapunov function has the following form

$$V_k = \boldsymbol{x}_{f,k}^T\boldsymbol{P}\boldsymbol{x}_{f,k}. \tag{13}$$

This implies that the inequality (12) becomes

$$\begin{aligned}
&\Delta V + \boldsymbol{x}_{f,k}^T\boldsymbol{x}_{f,k} - \mu^2\bar{\boldsymbol{w}}_k^T\bar{\boldsymbol{w}}_k = \\
&\boldsymbol{x}_{f,k}^T\left[\boldsymbol{A}_1^T\boldsymbol{P}\boldsymbol{A}_1 + \boldsymbol{I} - \boldsymbol{P}\right]\boldsymbol{x}_{f,k} + \boldsymbol{x}_{f,k}^T\left[\boldsymbol{A}_1^T\boldsymbol{P}\bar{\boldsymbol{W}}\right]\bar{\boldsymbol{w}}_k + \\
&+ \bar{\boldsymbol{w}}_k^T\left[\bar{\boldsymbol{W}}^T\boldsymbol{P}\boldsymbol{A}_1\right]\boldsymbol{x}_{f,k} + \bar{\boldsymbol{w}}_k^T\left[\bar{\boldsymbol{W}}^T\boldsymbol{P}\bar{\boldsymbol{W}} - \mu^2\boldsymbol{I}\right]\bar{\boldsymbol{w}}_k < 0.
\end{aligned} \tag{14}$$

It can be easily shown that (14) is equivalent to

$$\begin{bmatrix} \boldsymbol{A}_1^T \boldsymbol{P} \boldsymbol{A}_1 + \boldsymbol{I} - \boldsymbol{P} & \boldsymbol{A}_1^T \boldsymbol{P} \bar{\boldsymbol{W}} \\ \bar{\boldsymbol{W}}^T \boldsymbol{P} \boldsymbol{A}_1 & \bar{\boldsymbol{W}}^T \boldsymbol{P} \bar{\boldsymbol{W}} - \mu^2 \boldsymbol{I} \end{bmatrix} \prec 0. \tag{15}$$

The following lemma can be perceived as the generalisation of one presented in [4].

Lemma 1. *The following statements are equivalent*

1. *There exists* $\boldsymbol{X} \succ 0$ *such that*

$$\boldsymbol{V}^T \boldsymbol{X} \boldsymbol{V} - \boldsymbol{W} \prec 0 \tag{16}$$

2. *There exists* $\boldsymbol{X} \succ 0$ *such that*

$$\begin{bmatrix} -\boldsymbol{W} & \boldsymbol{V}^T \boldsymbol{U}^T \\ \boldsymbol{U} \boldsymbol{V} & \boldsymbol{X} - \boldsymbol{U} - \boldsymbol{U}^T \end{bmatrix} \prec 0. \tag{17}$$

Proof. Applying the Schur complement to (18) gives

$$\boldsymbol{V}^T \boldsymbol{U}^T (\boldsymbol{U}^T + \boldsymbol{U} - \boldsymbol{X})^{-1} \boldsymbol{U} \boldsymbol{V} - \boldsymbol{W} \prec 0. \tag{18}$$

Substituting $\boldsymbol{U} = \boldsymbol{U}^T = \boldsymbol{X}$ yield

$$\boldsymbol{V}^T \boldsymbol{X} \boldsymbol{V} - \boldsymbol{W} \prec 0. \tag{19}$$

Thus, (18) implies (19).

Multiplying (17) by $\boldsymbol{T} = \begin{bmatrix} \boldsymbol{I} & \boldsymbol{V}^T \end{bmatrix}$ on the left and by $\boldsymbol{T}^T$ on the right of (17) gives (16), which means that (19) implies (18) and hence the proof is completed.

Applying Lemma 1 (along with suitable transposition, cf. [4]) to (15) and then substituting $\boldsymbol{A}_1 \boldsymbol{U} = (\boldsymbol{A} - \boldsymbol{B}\boldsymbol{K})\boldsymbol{U} = \boldsymbol{A}\boldsymbol{U} - \boldsymbol{B}\boldsymbol{N}$, gives

$$\begin{bmatrix} \boldsymbol{I} - \boldsymbol{P} & \boldsymbol{0} & \boldsymbol{A}\boldsymbol{U} - \boldsymbol{B}\boldsymbol{N} \\ \boldsymbol{0} & -\mu^2 \boldsymbol{I} & \bar{\boldsymbol{W}} \boldsymbol{U} \\ \boldsymbol{U}^T \boldsymbol{A}^T - \boldsymbol{N}^T \boldsymbol{B}^T & \boldsymbol{U}^T \bar{\boldsymbol{W}}^T & \boldsymbol{P} - \boldsymbol{U} - \boldsymbol{U}^T \end{bmatrix} \prec \boldsymbol{0}. \tag{20}$$

Finally, the design procedure boils down to solving (20) with respect to $\boldsymbol{U}$, $\boldsymbol{N}$ and $\boldsymbol{P}$, and then calculating

$$\boldsymbol{K} = \boldsymbol{N} \boldsymbol{U}^{-1}. \tag{21}$$

Since the robust observer is given then the robust invariant set can be described in details.

3 Derivation of a Robust Invariant Set

In order to maintain desired system behaviour, the idea of invariant set of state variables can be employed [10,24]. In this section we use ellipsoidal bounding, which will describe the robust invariant set for

$$\boldsymbol{x}_{f,k+1} = \boldsymbol{A}_1\boldsymbol{x}_{f,k} + \bar{\boldsymbol{W}}\bar{\boldsymbol{w}}_k \tag{22}$$

with an additional assumption that:

$$\bar{\boldsymbol{w}}_k^T\boldsymbol{Q}\bar{\boldsymbol{w}}_k \leq 1, \quad \boldsymbol{Q} \succ \boldsymbol{0}. \tag{23}$$

In particular, $\boldsymbol{E}_{\boldsymbol{x}_f}$ is a robust invariant set for (22) if

$$\boldsymbol{x}_k \in \boldsymbol{E}_{\boldsymbol{x}_f} \Longrightarrow \boldsymbol{x}_{k+1} \in \boldsymbol{E}_{\boldsymbol{x}_f}. \tag{24}$$

Thus, the ellipsoidal robust invariant set is given by

$$\boldsymbol{E}_{\boldsymbol{x}_f} = \{\boldsymbol{x}_f | \boldsymbol{x}_f^T\boldsymbol{P}\boldsymbol{x}_f \leq 1\} \tag{25}$$

The above definition implies the following constraints:

$$\boldsymbol{x}_{f,k}^T\boldsymbol{P}\boldsymbol{x}_{f,k} \leq 1 \tag{26}$$

$$\boldsymbol{x}_{f,k+1}^T\boldsymbol{P}\boldsymbol{x}_{f,k+1} \leq 1 \tag{27}$$

which by applying the S-procedure along with (23) yield the following coupled constraint

$$\begin{aligned}
&(\boldsymbol{x}_{f,k}^T\boldsymbol{A}_1^T + \bar{\boldsymbol{w}}_k^T\bar{\boldsymbol{W}}^T)\boldsymbol{P}(\boldsymbol{A}_1\boldsymbol{x}_{f,k} + \bar{\boldsymbol{W}}\bar{\boldsymbol{w}}_k) - 1 + \gamma(1 - \boldsymbol{x}_{f,k}^T\boldsymbol{P}\boldsymbol{x}_{f,k}) + \beta(1 - \bar{\boldsymbol{w}}_k^T\boldsymbol{Q}\bar{\boldsymbol{w}}_k)\\
&= \boldsymbol{x}_{f,k}^T\boldsymbol{A}_1^T\boldsymbol{P}\boldsymbol{A}_1\boldsymbol{x}_{f,k} + \boldsymbol{x}_{f,k}^T\boldsymbol{A}_1^T\boldsymbol{P}\bar{\boldsymbol{W}}\bar{\boldsymbol{w}}_k + \bar{\boldsymbol{w}}_k^T\bar{\boldsymbol{W}}\boldsymbol{P}\boldsymbol{A}_1\boldsymbol{x}_{f,k} + \bar{\boldsymbol{w}}_k^T - 1 + \gamma - \gamma\boldsymbol{x}_{f,k}^T\boldsymbol{P}\boldsymbol{x}_{f,k}\\
&\qquad + \beta - \beta\bar{\boldsymbol{w}}_k^T\boldsymbol{Q}\bar{\boldsymbol{w}}_k \leq 0
\end{aligned}$$

with $\gamma > 0$ and $\beta > 0$. The above inequality can be written in a more compact form:

$$\begin{aligned}
&\boldsymbol{x}_{f,k}^T\left[\boldsymbol{A}_1^T\boldsymbol{P}\boldsymbol{A}_1 - \gamma\boldsymbol{P}\right]\boldsymbol{x}_{f,k} + \boldsymbol{x}_{f,k}^T\boldsymbol{A}_1^T\boldsymbol{P}\bar{\boldsymbol{W}}\bar{\boldsymbol{w}}_k + \bar{\boldsymbol{w}}_k^T\bar{\boldsymbol{W}}^T\boldsymbol{P}\boldsymbol{A}_1\boldsymbol{x}_{f,k}\\
&+ \bar{\boldsymbol{w}}_k^T\left[\bar{\boldsymbol{W}}^T\boldsymbol{P}\bar{\boldsymbol{W}} - \beta\boldsymbol{Q}\right]\bar{\boldsymbol{w}}_k - 1 + \gamma + \beta \leq 0,
\end{aligned} \tag{28}$$

or described in a matrix form:

$$\begin{bmatrix} \boldsymbol{A}_1^T\boldsymbol{P}\boldsymbol{A}_1 - \gamma\boldsymbol{P} & \boldsymbol{A}_1^T\boldsymbol{P}\bar{\boldsymbol{W}} & 0 \\ \bar{\boldsymbol{W}}^T\boldsymbol{P}\boldsymbol{A}_1 & \bar{\boldsymbol{W}}^T\boldsymbol{P}\bar{\boldsymbol{W}} - \beta\boldsymbol{Q} & 0 \\ 0 & 0 & \gamma + \beta - 1 \end{bmatrix} \preceq \boldsymbol{0} \tag{29}$$

From (29), it is obvious that

$$\gamma + \beta \leq 1 \Rightarrow \beta = 1 - \gamma \Rightarrow 0 \leq \gamma \leq 1. \tag{30}$$

This leads directly to:

$$\begin{bmatrix} \boldsymbol{A}_1^T \\ \bar{\boldsymbol{W}}^T \end{bmatrix} \boldsymbol{P} \left[\boldsymbol{A}_1 \; \bar{\boldsymbol{W}}\right] + \begin{bmatrix} -\gamma \boldsymbol{P} & 0 \\ 0 & -(1-\gamma)\boldsymbol{Q} \end{bmatrix} \preceq \boldsymbol{0}, \quad 0 \leq \gamma \leq 1, \tag{31}$$

which using Lemma 1 can be written as

$$\begin{bmatrix} -\gamma \boldsymbol{P} & \boldsymbol{0} & \boldsymbol{A}_1^T \boldsymbol{U}^T \\ \boldsymbol{0} & -(1-\gamma)\boldsymbol{Q} & \bar{\boldsymbol{W}}^T \boldsymbol{U}^T \\ \boldsymbol{U}\boldsymbol{A}_1 & \boldsymbol{U}\bar{\boldsymbol{W}} & \boldsymbol{P} - \boldsymbol{U} - \boldsymbol{U}^T \end{bmatrix} \preceq \boldsymbol{0}. \tag{32}$$

Note that for a fixed $0 \leq \gamma \leq 1$, the inequality (32) becomes the usual LMI. Another strategy is to formulate (32) as a generalised eigenvalue optimisation problem. Both of them can be efficiently solved with the numerical packages like Matlab.

4 Efficient Predictive FTC

The robust fault-tolerant control presented in Sec. 3 is based on the idea of estimating the fault, and then compensating it with a suitable increase or decrease of the control feeding the faulty actuator. In spite of the incontestable appeal of the proposed approach, its main drawback is that it does not take into account the fact that all actuators obey some saturation rules. Thus, the idea behind the approach presented in this section is as follows: when a saturation of a faulty actuator appears then perturb (or modify) the control strategy of the remaining actuators in such a way as to increase the robust invariant set and to make the overall control problem feasible. The subsequent part of this section is devoted to the implementation of such a strategy.

The system is assumed to be prestabilised (in the $\boldsymbol{H}_\infty$ sense) by a feedback controller $\boldsymbol{K}$ designed in the unconstrained case. Following [12], the degrees of freedom are expressed through $\boldsymbol{c}_k$ changing the original control strategy associated with $\boldsymbol{K}$. Thus, the modified control strategy is

$$u_j = \begin{cases} -\boldsymbol{K}\boldsymbol{x}_j - \hat{\boldsymbol{f}}_{k-1} + \boldsymbol{c}_j, \; j = k, \ldots, k + n_c - 1, \\ -\boldsymbol{K}\boldsymbol{x}_j - \hat{\boldsymbol{f}}_{k-1}, \qquad\qquad\qquad j \geq k + n_c. \end{cases} \tag{33}$$

Note that the fault estimate does not change within the prediction horizon n_c. Indeed, it is the only logical setting, since it is impossible to predict a future fault behaviour. Thus, the on-line optimisation is to be realised with the new available $\boldsymbol{c}_j$. Beyond the control horizon n_c, $\boldsymbol{c}_i$ is set to zero, which denotes the feasibility of the $\boldsymbol{H}_\infty$ control. This implies that (11) has to be suitably modified, which means that

$$\boldsymbol{x}_{f,k+1} = \boldsymbol{A}_1 \boldsymbol{x}_{f,k} + \boldsymbol{B}\boldsymbol{c}_k + \bar{\boldsymbol{W}}\bar{\boldsymbol{w}}_k \tag{34}$$

The robust controller proposed in Sec. 2 apart from its unquestionable appeal regarding the $\boldsymbol{H}_\infty$-based optimisation, does not take into account the constraints related to every real system, e.g. an actuator saturation. Such a control limits can be described as follows:

$$-\bar{\boldsymbol{u}}_i \leq \boldsymbol{u}_{i,k} \leq \bar{\boldsymbol{u}}_i, \quad i = 1, \ldots, r. \tag{35}$$

where $\bar{\boldsymbol{u}} > 0$ is a given control limit. Thus, the objective of the subsequent part of this section is to develop a suitable control strategy that takes into account the actuator saturation. For this purpose, the efficient predictive control scheme introduced by [12] is utilised. In particular, the proposed scheme is suitably extended to cope with the external disturbances, and hence, achieving robustness.

It can be easily shown that the input constraints (35) for the ith input are (cf. (33))

$$-\bar{\boldsymbol{u}}_i \leq \left[-\boldsymbol{K}_i^T \; 1\right] \begin{bmatrix} \boldsymbol{x}_{f,k} \\ \boldsymbol{c}_{i,k} - \hat{\boldsymbol{f}}_{i,k-1} \end{bmatrix} \leq \bar{\boldsymbol{u}}_i, i = 1, \ldots, r \tag{36}$$

where $\boldsymbol{K}_i$ stands for the ith row of $\boldsymbol{K}$. Thus, predictions at time k are generated as follows [12]:

$$\boldsymbol{z}_{k+1} = \boldsymbol{Z}\boldsymbol{z}_k + \tilde{\boldsymbol{W}}\bar{\boldsymbol{w}}_k. \tag{37}$$

where

$$\tilde{\boldsymbol{W}} = \begin{bmatrix} \bar{\boldsymbol{W}} \\ \boldsymbol{0} \end{bmatrix}, \; \boldsymbol{Z} = \begin{bmatrix} \boldsymbol{A} - \boldsymbol{B}\boldsymbol{K} & \boldsymbol{B}\boldsymbol{T} \\ \boldsymbol{0} & \boldsymbol{M} \end{bmatrix}, \; \boldsymbol{M} = \begin{bmatrix} \boldsymbol{0}_{(n_c-1)r\times r} & \boldsymbol{I} \\ \boldsymbol{0}_{r\times r} & \boldsymbol{0}_{r\times(n_c-1)r} \end{bmatrix},$$

$$\boldsymbol{z}_k = \begin{bmatrix} \boldsymbol{x}_{f,k} \\ \boldsymbol{\omega}_k \end{bmatrix}, \quad \boldsymbol{\omega}_k = \begin{bmatrix} \boldsymbol{c}_k \\ \boldsymbol{c}_{k+1} \\ \cdots \\ \boldsymbol{c}_{k+n_c-1} \end{bmatrix}, \qquad \boldsymbol{T} = \left[\boldsymbol{I}_{r\times r} \; \boldsymbol{0} \ldots \boldsymbol{0}\right].$$

Note that the stability of the autonomous system in the $\boldsymbol{H}_\infty$ (37) is guaranteed by the stability of $\boldsymbol{A} - \boldsymbol{B}\boldsymbol{K}$. Following [12], it can be pointed out that if there exists robust invariant set $\boldsymbol{E}_{\boldsymbol{x}_f}$ (cf. (25)) for (22), then there must exist at least one robust invariant set $\boldsymbol{E}_{\boldsymbol{z}}$ for (37). Thus, (32) can be easily adapted for (37), which gives the robust invariant set for the proposed fault-tolerant predictive scheme:

$$\begin{bmatrix} -\gamma\boldsymbol{P} & \boldsymbol{0} & \boldsymbol{Z}^T\boldsymbol{U}^T \\ \boldsymbol{0} & -(1-\gamma)\boldsymbol{Q} & \tilde{\boldsymbol{W}}^T\boldsymbol{U}^T \\ \boldsymbol{U}\boldsymbol{Z} & \boldsymbol{U}\tilde{\boldsymbol{W}} & \boldsymbol{P} - \boldsymbol{U} - \boldsymbol{U}^T \end{bmatrix} \preceq \boldsymbol{0}, \quad 0 \leq \gamma \leq 1. \tag{38}$$

Since the robust invariant set for (37) is given then it is possible to introduce the input constraints (36). The easiest way to do this is to suitably scale $\boldsymbol{\omega}_k$ in (37) as follows, i.e. $\boldsymbol{\omega}_k$ replaced by:

$$\bar{\boldsymbol{\omega}}_k = \begin{bmatrix} \boldsymbol{c}_k - \hat{\boldsymbol{f}}_{k-1} \\ \boldsymbol{c}_{k+1} \\ \cdots \\ \boldsymbol{c}_{k+n_c-1} \end{bmatrix}, \tag{39}$$

Thus, constraints (36) can be described as follows

$$-\bar{\boldsymbol{u}}_i \leq \left[-\boldsymbol{K}_i^T \ \boldsymbol{e}_i^T\right] \boldsymbol{z}_k \leq \bar{\boldsymbol{u}}_i, i = 1, \ldots, r \tag{40}$$

where $\boldsymbol{e}_i$ is the ith column of the identity matrix, or equivalently

$$\left| \left[-\boldsymbol{K}_i^T \ \boldsymbol{e}_i^T\right] \boldsymbol{z}_k \right| \leq \bar{\boldsymbol{u}}_i, i = 1, \ldots, r \tag{41}$$

For $\boldsymbol{z}_k \in \boldsymbol{E}_{\boldsymbol{z}}$, the above inequality implies

$$\begin{aligned} \left| \left[-\boldsymbol{K}_i^T \ \boldsymbol{e}_i^T\right] \boldsymbol{z}_k \right| = & \left| \left[-\boldsymbol{K}_i^T \ \boldsymbol{e}_i^T\right] \boldsymbol{P}^{-\frac{1}{2}} \boldsymbol{P}^{\frac{1}{2}} \boldsymbol{z}_k \right| \leq \\ & \left\| \left[-\boldsymbol{K}_i^T \ \boldsymbol{e}_i^T\right] \boldsymbol{P}^{-\frac{1}{2}} \right\| \left\| \boldsymbol{P}^{\frac{1}{2}} \boldsymbol{z}_k \right\| \leq \\ & \left\| \left[-\boldsymbol{K}_i^T \ \boldsymbol{e}_i^T\right] \boldsymbol{P}^{\frac{1}{2}} \right\| \end{aligned} \tag{42}$$

which is equivalent to

$$\left[-\boldsymbol{K}_i^T \ \boldsymbol{e}_i^T\right] \boldsymbol{P}^{-1} \left[-\boldsymbol{K}_i \ \boldsymbol{e}_i\right] - \bar{\boldsymbol{u}}_i^2 \leq \boldsymbol{0}, \quad i = 1, \ldots, r, \tag{43}$$

where $\boldsymbol{e}_i$ stands for the ith column of the identity matrix. Finally, using the Schur complements, inequalities (43) can be written in an LMI form

$$\begin{bmatrix} -\bar{\boldsymbol{u}}_i^2 & [-\boldsymbol{K}_i^T \ \boldsymbol{e}_i^T]\boldsymbol{P} \\ \boldsymbol{P}[-\boldsymbol{K}_i \ \boldsymbol{e}_i] & -\boldsymbol{P} \end{bmatrix} \preceq \boldsymbol{0}, \quad i = 1, \ldots, r. \tag{44}$$

Thus, to maximise the volume of the $\boldsymbol{E}_{\boldsymbol{x}_f}$ the following determinant should be maximised [12]

$$\max \det(\boldsymbol{T}_z \boldsymbol{P} \boldsymbol{T}_z^T), \tag{45}$$

with

$$\boldsymbol{x}_{f,k} = \boldsymbol{T}_z \boldsymbol{z}_k, \tag{46}$$

under the constraints (38) and (44). The algorithm for computing $\boldsymbol{c}_k$ in (34) is also inspired by [12] and boils down to perform, at each sampling time, the following minimisation

$$\boldsymbol{\omega}_k^* = \min_{\boldsymbol{\omega}_k} \boldsymbol{\omega}_k^T \boldsymbol{\omega}_k, \qquad s.t. \ \boldsymbol{z}_k^T \boldsymbol{P} \boldsymbol{z}_k \leq 1, \tag{47}$$

which can be equivalently written as:

$$\boldsymbol{\omega}_k^* = \min_{\boldsymbol{\omega}_k} \boldsymbol{\omega}_k^T \boldsymbol{\omega}_k, \tag{48}$$

$$s.t. \ \boldsymbol{x}_{f,k}^T \boldsymbol{P}_{1,1} \boldsymbol{x}_{f,k} + 2\boldsymbol{x}_{f,k}^T \boldsymbol{P}_{1,2} \bar{\boldsymbol{\omega}}_k + \bar{\boldsymbol{\omega}}_k^T \boldsymbol{P}_{2,2} \bar{\boldsymbol{\omega}}_k \leq 1, \tag{49}$$

where $\boldsymbol{P}_{1,1}$, $\boldsymbol{P}_{1,2}$ and $\boldsymbol{P}_{2,2}$ are block partitions of $\boldsymbol{P}$ conformal to the partition of $\boldsymbol{z}_k = [\boldsymbol{x}_{f,k}^T \ \bar{\boldsymbol{\omega}}_k^T]^T$. Thus, if the $\mathcal{H}_\infty$ control is feasible then $\boldsymbol{\omega} = \boldsymbol{0}$, otherwise the solution lies on the boundary of $\boldsymbol{E}_{\boldsymbol{z}}$ described by (49). This means that when $\boldsymbol{\omega} = \boldsymbol{0}$ is contained in $\boldsymbol{E}_{\boldsymbol{z}}$ described by (49), then there is no need for optimisation and the opitimal solution is $\boldsymbol{\omega} = \boldsymbol{0}$. Otherwise, as indicated in [12], the above optimisation problem has a unique solution and can be very efficiently solved with, e.g., the Newton-Raphson algorithm [7].

5 Extension to Non-linear Systems

A non-linear dynamic system can be described in a simple way by a Takagi-Sugeno fuzzy model, which uses series of locally linearised models from the nonlinear system, parameter identiffcation of an a priori given structure or transformation of a nonlinear model using the nonlinear sector approach (see, e.g. [18]). According to this model, a non-linear dynamic systems can be linearised around a number of operating points. Each of these linear models represents the local system behaviour around the operating point. Thus, a fuzzy fusion of all linear model outputs describes the global system behaviour. A T-S model is described by fuzzy IF-THEN rules which represent local linear I/O relations of the non-linear system. It has a rule base of M rules, each having p antecedents, where ith rule is expressed as

$$R^i: \text{ IF } w_k^1 \text{ is } F_1^i \text{ and } \dots \text{ and } s_k^p \text{ is } F_p^i, \text{ THEN } \left\{ \boldsymbol{x}_{k+1} = \boldsymbol{A}^i \boldsymbol{x}_k + \boldsymbol{B}^i \boldsymbol{u}_k, \right. \tag{50}$$

where $i = 1, \dots, M$, F_j^i $(j = 1, \dots, p)$ are fuzzy sets and $\boldsymbol{w}_k = [w_k^1, w_k^2, \dots, w_k^p]$ is a known vector of premise variables ([18,9]).

Given a pair of $(\boldsymbol{s}_k, \boldsymbol{u}_k)$ and a product inference engine, the final output of the normalized T-S fuzzy model can be inferred as:

$$\boldsymbol{x}_{k+1} = \sum_{i=1}^{M} h_i(\boldsymbol{s}_k)[\boldsymbol{A}^i \boldsymbol{x}_k + \boldsymbol{B}\boldsymbol{u}_k], \tag{51}$$

where $h_i(\boldsymbol{s}_k)$ are normalized rule firing strengths defined as

$$h_i(\boldsymbol{s}_k) = \frac{\mathcal{T}_{j=1}^p \mu_{F_j^i}(s_k^j)}{\sum_{i=1}^M (\mathcal{T}_{j=1}^p \mu_{F_j^i}(w_k^j))} \tag{52}$$

and $\mathcal{T}$ denotes a t-norm (e.g., a product). The term $\mu_{F_j^i}(w_k^j)$ is the grade of membership of the premise variable s_k^j. Moreover, the rule firing strengths $h_i(\boldsymbol{s}_k)$ $(i = 1, \dots, M)$ satisfy the following constraints:

$$\begin{cases} \sum_{i=1}^M h_i(\boldsymbol{s}_k) = 1, & \\ 0 \leqslant h_i(\boldsymbol{s}_k) \leqslant 1, & \forall i = 1, \dots, M. \end{cases} \tag{53}$$

The objective of the subsequent part of this section is to extend the proposed robust predictive fault-tolerant control scheme to non-linear systems described by (51). Note that (51) can be written in an equivalent form

$$\boldsymbol{x}_{k+1} = \sum_{i=1}^{M} \alpha_i [\boldsymbol{A}^i \boldsymbol{x}_k + \boldsymbol{B}\boldsymbol{u}_k] = \boldsymbol{A}(\alpha)\boldsymbol{x}_k + \boldsymbol{B}\boldsymbol{u}_k, \quad \alpha_i = h_i(\boldsymbol{s}_k). \tag{54}$$

with

$$\mathbb{A} = \left\{ \boldsymbol{A}(\alpha) : \quad \boldsymbol{A}(\alpha) = \sum_{i=1}^{M} \alpha_i \boldsymbol{A}_i, \sum_{i=1}^{M} \alpha_i = 1, \, \alpha_i \geq 0 \right\}. \tag{55}$$

Using Lemma 1 and proceeding in a similar way as the one for (20), it can be shown that the robust controller for (54) can be obtained by solving:

$$\begin{bmatrix} \boldsymbol{I} - \boldsymbol{P}_i & \boldsymbol{0} & \boldsymbol{A}_i\boldsymbol{U} - \boldsymbol{B}\boldsymbol{N} \\ \boldsymbol{0} & -\mu^2\boldsymbol{I} & \bar{\boldsymbol{W}}\boldsymbol{U} \\ \boldsymbol{U}^T\boldsymbol{A}_i^T - \boldsymbol{N}^T\boldsymbol{B}^T & \boldsymbol{U}^T\bar{\boldsymbol{W}}^T & \boldsymbol{P}_i - \boldsymbol{U} - \boldsymbol{U}^T \end{bmatrix} \prec \boldsymbol{0}, \quad i = 1, \ldots, M, \tag{56}$$

with respect to $\boldsymbol{U}$, $\boldsymbol{N}$ and $\boldsymbol{P}_i$ $(i = 1, \ldots, M)$, and then calculating

$$\boldsymbol{K} = \boldsymbol{N}\boldsymbol{U}^{-1}. \tag{57}$$

Proceeding in a similar way, the robust invariant set for (54) can be described with

$$\begin{bmatrix} -\gamma\boldsymbol{P}_i & \boldsymbol{0} & \boldsymbol{A}_{i,1}^T\boldsymbol{U}^T \\ \boldsymbol{0} & -(1-\gamma)\boldsymbol{Q} & \bar{\boldsymbol{W}}^T\boldsymbol{U}^T \\ \boldsymbol{U}\boldsymbol{A}_{i,1} & \boldsymbol{U}\bar{\boldsymbol{W}} & \boldsymbol{P}_i - \boldsymbol{U} - \boldsymbol{U}^T \end{bmatrix} \preceq \boldsymbol{0}, \ 0 \leq \gamma \leq 1 \quad i = 1, \ldots, M.$$

The resulting robust invariant set is then given by

$$\boldsymbol{E}_{\boldsymbol{x}_f} = \{\boldsymbol{x}_f | \boldsymbol{x}_f^T\boldsymbol{P}(\alpha)\boldsymbol{x}_f \ \leq 1\}, \boldsymbol{P}(\alpha) = \sum_{i=1}^{M} \alpha_i\boldsymbol{P}_i. \tag{58}$$

Since the robust invariant set is given then it is possible to design a predictive fault tolerant control. Note that the LMIs corresponding to constraints are now defined as:

$$\begin{bmatrix} -\bar{u}_j^2 & [-\boldsymbol{K}_j^T \ \boldsymbol{e}_j^T]\boldsymbol{P}_i \\ \boldsymbol{P}_i[-\boldsymbol{K}_j \ \boldsymbol{e}_j] & -\boldsymbol{P}_i \end{bmatrix} \preceq \boldsymbol{0}, \quad j = 1, \ldots, r, \quad i = 1, \ldots, M. \tag{59}$$

Note that the vector α $(\alpha_i = h_i(\boldsymbol{s}_k), i = 1, \ldots, M)$is known for the Takagi-Sugeno description. This means that at every k the matrix $\boldsymbol{P}(alpha)$ can be easily computed, which means that the predictive control obey the same rules with the final optimisation problem

$$\boldsymbol{\omega}_k^* = \min_{\boldsymbol{\omega}_k} \boldsymbol{\omega}_k^T\boldsymbol{\omega}_k, \qquad s.t. \ \boldsymbol{z}_k^T\boldsymbol{P}(\alpha)\boldsymbol{z}_k \leq 1. \tag{60}$$

6 Conclusions

The main contribution of the paper was to propose a robust predictive fault-tolerant control scheme for both linear and non-linear systems. Indeed, the contribution can be divided into a few important points: extension of the efficient predictive control to the robust case with exogenous external disturbances acting on the system, development of robust fault estimation and compensation scheme, and an integration of the developed schemes within a unified robust predictive fault-tolerant control framework. It is worth to note that the framework was also suitably extended to non-linear systems that can be described with the Takagi-Sugeno models. All the proposed approaches can be efficiently implemented, i.e. the off-line computations boils down to solving a number of linear matrix inequalities while the on-line computation reduces to the application of the Newton-Raphson method.

Acknowledgments. The work was financed as a research project with the science funds for years 2011-2014 with the kind support of the National Science Centre in Poland under the grant NN514678440 *Predictive fault-tolerant control for non-linear systems.*

References

1. Blanke, M., Kinnaert, M., Lunze, J., Staroswiecki, M.: Diagnosis and Fault-Tolerant Control. Springer, New York (2003)
2. Chen, W., Khan, A.Q., Abid, M., Ding, S.X.: Integrated design of observer-based fault detection for a class of uncertain non-linear systems. International Journal of Applied Mathematics and Computer Science 21(4), 619–636 (2011)
3. De Oca, S., Puig, V., Witczak, M., Dziekan, L.: Fault-tolerant control strategy for actuator faults using lpv techniques: application to a two degree of freedom helicopter. International Journal of Applied Mathematics and Computer Science 22(1), 161–171 (2012)
4. de Oliveira, M.C., Bernussou, J., Geromel, J.C.: A new discrete-time robust stability condition. Systems and Control Letters 37(4), 261–265 (1999)
5. Dziekan, L., Witczak, M., Korbicz, J.: Active fault-tolerant control design for takagi-sugeno fuzzy systems. Bulletin of the Polish Academy of Sciences: Technical Sciences 59(1), 93–102 (2011)
6. Gillijns, S., De Moor, B.: Unbiased minimum-variance input and state estimation for linear discrete-time systems. Automatica 43, 111–116 (2007)
7. Imsland, L., Bar, N., Foss, B.A.: More efficient predictive control. Automatica 41(8), 1395–1403 (2005)
8. Isermann, R.: Fault Diagnosis Applications: Model Based Condition Monitoring, Actuators, Drives, Machinery, Plants, Sensors, and Fault-tolerant Systems. Springer, Berlin (2011)
9. Korbicz, J., Kościelny, J., Kowalczuk, Z., Cholewa, W. (eds.): Fault diagnosis. Models, Artificial Intelligence, Applications. Springer, Berlin (2004)
10. Kouvaritakis, B., Lee, Y.I., Cannon, M.: Extended invariance and its use in model predictive control. Automatica 41 (2005)
11. Luzar, M., Czajkowski, A., Witczak, M., Mrugalski, M.: Actuators and sensors fault diagnosis with dynamic, state-space neural networks. In: MMAR 2012: Proceedings of the 17th IEEE International Conference on Methods and Models in Automation and Robotics, pp. 196–201 (2012)
12. Kouvaritakis, B., Rossiter, J.A.: Schuurmans J. Efficient robust predictive control. IEEE Transactions on Automatic Control 45(8), 1545–1549 (2000)
13. Li, H., Fu, M.: A linear matrix inequality approach to robust h_∞ filtering. IEEE Trans. Signal Processing 45(9), 2338–2350 (1997)
14. Li, H., Zhao, Q., Yang, Z.: Reliability modeling of fault tolerant control systems. International Journal of Applied Mathematics and Computer Science 17(4), 491–504 (2007)
15. Mahmoud, M., Jiang, J., Zhang, Y.: Active Fault Tolerant Control Systems: Stochastic Analysis and Synthesis. Springer, Berlin (2003)
16. Mrugalski, M.: An unscented kalman filter in designing dynamic gmdh neural networks for robust fault detection. International Journal of Applied Mathematics and Computer Science 23(1), 157–169 (2013)

17. Noura, H., Theilliol, D., Ponsart, J., Chamseddine, A.: Fault-tolerant Control Systems: Design and Practical Applications. Springer, Berlin (2003)
18. Takagi, T., Sugeno, M.: Fuzzy identification of systems and its application to modeling and control. IEEE Trans. Systems, Man and Cybernetics 15(1), 116–132 (1985)
19. Tanaka, K., Wang, H.O.: Fuzzy Control Systems Design and Analysis: A Linear Matrix Inequality Approach. Wiley Interscience, New York (2001)
20. Witczak, M.: Modelling and Estimation Strategies for Fault Diagnosis of Non-linear Systems. Springer, Berlin (2007)
21. Witczak, M., Puig, V.: Design of an extended unknown input observer with stochastic robustness techniques and evolutionary algorithms. International Journal of Control 86 (in print, 2013)
22. Zemouche, A., Boutayeb, M., Iulia Bara, G.: Observer for a class of Lipschitz systems with extension to $\mathcal{H}_\infty$ performance analysis. Systems and Control Letters 57(1), 18–27 (2008)
23. Zhang, Y., Jiang, J.: Bibliographical review on reconfigurable fault-tolerant control systems. In: IFAC Symposium Fault Detection Supervision and Safety of Technical Processes, Safeprocess, Washington, D.C., USA, pp. 265–276 (2003)
24. Zongli, L., Liang, L.: Set invariance conditions for singular linear systems subject to actuator saturation. IEEE Transactions on Automatic Control 52(12), 2351–2355 (2007)

Design of Maximum Power Fuzzy Controller for PV Systems Based on the LMI-Based Stability

Elkhatib Kamal and Abdel Aitouche

LAGIS, University Lille Nord de France
elkateb.kamal@gmail.com, abdel.aitouche@hei.fr

Abstract. This paper deals with a design methodology for stabilization of a class of nonlinear Photovoltaic (PV) systems. First, a nonlinear PV plant with a Takagi-Sugeno (TS) fuzzy model is presented. Then a model based fuzzy controller design utilizing the concept of the Parallel Distributed Compensation (PDC) is employed. The proposed algorithm depends on the fuzzy reference model. The fuzzy control system consists of observer and fuzzy control components. The state observer does not require the system states to be available for measurement. Sufficient conditions are derived for stabilization and are formulated in Linear Matrix Inequalities (LMIs). An application to PV system is given in this paper to illustrate the effectiveness of the proposed TS fuzzy controller and observer design methodology. The simulation results demonstrate that the proposed fuzzy control system can guarantee the system stability and also maintain a good tracking performance.

Keywords: Fuzzy observer, Fuzzy control, MPPT, TS fuzzy model, PV.

1 Introduction

Nonlinear control has emerged as a research area of rapidly increasing activity. Especially, the theory of explicitly linearizing the input-output response of nonlinear systems to linear systems using state feedback which has received great attention [1]. There are several approaches to control of a nonlinear system. A typical approach is the feedback stabilization of nonlinear systems where a linear feedback control is designed for the linearization of the system about a nominal operating point. This approach, however, generally only renders a local result. Other approaches [2] such as feedback linearization are rather involved and tend to result in rather complicated controllers.

In recent years, there have been significant advances in the study of the stability analysis and controller synthesis for Takagi-Sugeno (TS), which have been used to represent certain complex nonlinear systems. In the TS fuzzy model, the local dynamics in different state-space regions are represented by linear models. The overall model of the system is obtained by the fuzzy blending of these local models.

Stability and design issues of fuzzy control systems have been discussed by many researchers [3-5]. The control design is based on the fuzzy model by the

J. Korbicz and M. Kowal (eds.), *Intelligent Systems in Technical and Medical Diagnostics*, Advances in Intelligent Systems and Computing 230,
DOI: 10.1007/978-3-642-39881-0_6, © Springer-Verlag Berlin Heidelberg 2014

Parallel Distributed Compensation (PDC) [6]. A linear feedback control is designed for each local linear model. The resulting overall controller is again a fuzzy blending of the individual linear controllers. Originally, sufficient conditions for the stability of the TS fuzzy systems in the sense of Lyapunov are given in [2], [6]. The conditions for the existence of a common Lyapunov function are obtained by solving Linear Matrix Inequalities (LMIs).

In addition, Photovoltaic (PV) power generation has stimulated considerable interest over the past two decades. While solar energy is the only available energy source in space, it is an important alternative for many terrestrial applications. At a given temperature and insolation level, PV cells supply maximum power at one particular operation point called the Maximum Power Point (MPP). Unlike conventional energy sources, it is desirable to operate PV systems at its MPP. However, the MPP locus varies over a wide range, depending on PV array temperature and insolation intensity. Maximum Power Point Trackers (MPPTs) play an important role in PV power systems because they maximize the power output from a PV system for a given set of conditions, and therefore maximize the array efficiency. Thus, an MPPT can minimize the overall system cost. MPPTs find and maintain operation at the maximum power point, using an MPPT algorithm. Many MPPT techniques have been proposed, the Perturb and Observe (PO) [7], modified PO, incremental conductance methods [8], [9], Constant Voltage (CV) [10], incremental conductance and CV combined [10], Short Current Pulse [11], Open Circuit Voltage [12].

In this paper, we propose a fuzzy controller based on the reference fuzzy model and fuzzy observer. The designed fuzzy controller will drive the system states of the nonlinear system to follow those of the reference fuzzy model. Moreover, simulation of the whole system is presented focusing on DC/DC converter's control strategy so that the system operates in MPP and converter's output voltage remains constant. Incremental conductance algorithm [8], [9] is used for MPPT implementation. A MPPT is obtained by controlling the duty cycle D and photovoltaic array's voltage. Simulation results are shown and analyzed.

This paper is organized as follows. In section 2, TS fuzzy model, reference model, fuzzy observer and the proposed fuzzy controller are presented. The stability analysis condition is given in section 3. Section 4 presents the electric characteristic of PV array and modeling of general solar PV power control systems. Operation principle of the proposed MPPT control method is given in section 5. Simulations results are carried out in sections 6 to show the control performance.

2 Reference Model, TS Fuzzy Model, Fuzzy Observer and Fuzzy Controller

In this section, we will present a TS Fuzzy multivariable nonlinear control system which consists of a nonlinear plant, fuzzy observer, fuzzy reference model and a nonlinear controller connected in closed-loop.

2.1 TS Fuzzy Model

Consider a nonlinear system that can be described by the TS fuzzy model [14]. The inferred fuzzy model system is given by,

$$\dot{x}(t) = \sum_{i=1}^{q} \mu_i(Z(t)) \left[A_i x(t) + B_i u(t)\right]$$

$$y(t) = \sum_{i=1}^{q} \mu_i(Z(t)) \left[C_i x(t)\right] i = 1, \cdots q \tag{1}$$

where q is the number of rules of the TS fuzzy mode, $Z(t) = [Z_1(t), \cdots, Z_k(t)]$are measurable variables, i.e., the premise variables,$x(t) \in R^{n*1}$ is the state vector, $u(t) \in R^{m*1}$ is the control input vector, $y(t) \in R^{g*1}$ is the output vector, $A_i \in R^{n*n}$, $B_i \in R^{n*m}$ and $C_i \in R^{g*n}$ are system, input and output matrices, respectively. where

$$\mu_i(Z(t)) = \frac{h_i(Z(t))}{\sum_{i=1}^{q} h_i Z(t)} h_i(Z(t)) = \prod_{s=1}^{k} M_{si}(Z(t))$$

where M_{si} is a fuzzy set, $s = 1, \cdots, k$. k is a positive integer. It is clear that

$$0 \leq \mu_i(Z(t)) \leq 1, \sum_{i=1}^{q} \mu_i(Z(t)) = 1 \forall i = 1, 2, \cdots, q \tag{2}$$

2.2 Fuzzy Reference Model

Consider the stable reference fuzzy model with rules given by [13]. Then the dynamics of the reference model are defined as the following

$$\dot{\bar{x}}(t) = \sum_{i=1}^{q} \mu_i(Z(t)) \left[A_{ri}\bar{x}(t) + B_{ri} r_r(t)\right]$$

$$\bar{y}(t) = \sum_{i=1}^{q} \mu_i(Z(t)) \left[C_{ri}\bar{x}(t)\right] i = 1, \cdots q \tag{3}$$

where $\bar{x}(t) \in R^{n*1}$ is the state vector of the reference model, $r_r(t) \in R^{m*1}$ is the bounded reference input, $A_r \in R^{n*n}$ is the constant stable system matrix, $B_r \in R^{n*m}$ is the constant input vector, $C_r \in R^{g*n}$ is the constant output matrix and $\bar{y}(t) \in R^{g*1}$ is the reference output. The reason of using this fuzzy model instead of a linear model [14] as the reference is that it can give a better performance.

2.3 Fuzzy Observer Design

For the fuzzy observer design, it is assumed that the fuzzy system (1) is locally observable. First, due to the fact that only partial states are available in measurement, the following fuzzy observer is designed to complete the state feedback. The defuzzified output of the fuzzy observer for TS fuzzy model (1) is formulated as follows [18]:

$$\dot{\hat{x}}(t) = \sum_{i=1}^{q} \mu_i(Z(t)) \left[A_i \hat{x}(t) + B_i u(t) + K_i(y(t) - \hat{y}(t))\right]$$

$$\hat{y}(t) = \sum_{i=1}^{q} \mu_i(Z(t)) \left[C_i \hat{x}(t)\right] i = 1, \cdots q \tag{4}$$

where $K_i \in R^{n*g}$ is constant observer gain to be determined.

2.4 Proposed Fuzzy Controller

For the fuzzy model (1) we construct the following fuzzy controller via PDC [6]. It is used when the membership functions are known and the rule antecedents of the TS fuzzy plant model and the fuzzy controller are the same [6],[13]. From (4), then the output of the fuzzy controller is

$$u(t) = \sum_{j=1}^{q} \mu_j(Z(t)) \left[-G_j \hat{x}(t) + r(t)\right] \tag{5}$$

Where $G_j \in R^{m*n}$ is the feedback gains vectors and r(t) is the reference input.

2.5 Stability Analysis and Proposed Fuzzy Controller Design

In this section, we present the stability analysis and design of the fuzzy controller of (5) based on the fuzzy plant model of (1) such that the system states of the nonlinear system will follow the fuzzy reference model of (3). Writing $\mu(Z(t))$ as μ and let,

$$e_1 = x(t) - \bar{x}(t) \tag{6}$$

$$e_2 = x(t) - \hat{x}(t) \tag{7}$$

From (1),(3),(5) and (6), the dynamics of $e_1(t)$ are given by :

$$\dot{e_1}(t) = \sum_{i=1}^{q} \mu_i A_{ri} e_1(t) + \sum_{i=1}^{q} \mu_i \left[(A_i - A_{ri}\right] x(t)$$

$$-\sum_{j=1}^{q} \mu_j G_j x(t) + \sum_{j=1}^{q} \mu_j B_j G_j e_2(t) + \sum_{j=1}^{q} \mu_j B_j r(t) - B_{ri} r_r(t) \tag{8}$$

From (8), letting

$$\sum_{j=1}^{q} \mu_j B_j G_i x(t) = A_i - A_{ri}, \sum_{j=1}^{q} \mu_j B_j r(t) = B_{ri} r_{ri}(t) \tag{9}$$

We obtain,

$$\dot{e_1}(t) = \sum_{i=1}^{q} \mu_i A_{ri} e_1(t) + \sum_{j=1}^{q} \mu_j B_j G_i e_2(t) \tag{10}$$

From(1),(4),(5) and (7), the dynamics of $e_2(t)$ are also given by

$$\dot{e_2}(t) = \sum_{i=1}^{q} \mu_i \left[(A_i - K_i C_i) e_2(t)\right] \tag{11}$$

Combining (10) and (11) yields the following augmented fuzzy system error.

$$e(t) = \sum_{i=1}^{q} H_i e(t) \tag{12}$$

Where $e(t) = \begin{pmatrix} e_1(t) \\ e_2(t) \end{pmatrix}$, $H_i = \begin{pmatrix} A_{ri} & B_i G_i \\ 0 & (A_i - K_i C_i) \end{pmatrix}$

If there exists a symmetric positive definite matrix $P \in R^{n*n}$ such that the following conditions are satisfied, $H_i^T P + P H_i < 0$, the reference fuzzy model, fuzzy observer and the system of (12) are guaranteed to be stable.

Lemma: The nonlinear system of (1) is guaranteed to be globally asymptotically stable and follows those of the reference fuzzy model (3), if the gains of the nonlinear fuzzy controller of (5) are designed such that,

$$\sum_{j=1}^{q} \mu_j B_j G_i x(t) = (A_i - A_{ri}) \tag{13}$$

$$\sum_{j=1}^{q} \mu_j B_j r(t) = B_{ri} r_{ri}(t) \tag{14}$$

and there is exists a symmetric positive definite matrix $P \in R^{n*n}$ such that the following conditions are satisfied, $H_i^T P + P H_i < 0$,

Theorem. The TS fuzzy system (2) is asymptotically stabilizable if there exist symmetric and positive definite matrix P, some matrix G_j, $Ki(j, i = 1, 2, \cdots, q)$, such that the following LMIs are satisfied

$$A_{ri}^T P_1 + P_1 A_{ri} < 0 \tag{15}$$

$$A_{ri}^T P_2 + P_2 A_i - (O_i C_i)^T - (O_i C_i) < 0 \tag{16}$$

Where $P = diag(P_1, P_2), O_i = P_2 K_i$.

3 PV Model System and PV Model

To show the effectiveness of the proposed controller design techniques, PV Model System [16]-[18] are simulated.

3.1 PV Electric Characteristics

The power generation of the PV array [17]:

$$P_{pv} = i_{pv}v_{pv} = n_pI_{ph}v_{pv} - n_pI_{rs}v_{pv}(exp(qv_{pv}/n_s\varphi KT) - 1) \tag{17}$$

where n_p and n_s are the number of the parallel and series cells, respectively, K is the Boltzmanns' constant, T is the cell temperature, φ is the $p-n$ junction characteristic, q is the electron charge, i_{pv} and v_{pv} are the output current and the v_{pv} array voltage on the capacitance C_1, I_{ph} and I_{rs} are the photocurrent and reverse saturation current, respectively and are given by

$$I_{ph} = (I_{sc} + K_I(T - T_r))\lambda/100, I_{rs} = I_{or}(T/T_r)^3exp(qE_{gp}(1/T_r - 1/T)/KT) \tag{18}$$

whereI_{sc} and I_{or} is the short circuit cell current at reference temperature and insolation and the reverse saturation current at the reference temperature T_r, E_{gp} is the semiconductor bandgap energy, K_I is the short circuit current temperature coefficient; and λ is the insolation in mW/cm^2. The relationship of P_{pv} versus v_{pv}, for different values of the insolation and temperature= $25C$ is shown in Fig.4 [18].

3.2 PV Modelling

In order to adapt the array photovoltaic to a large voltage range, the PV MPPT system adopts Buck-Boost DC-DC converter topology as shown in Fig.1 [18]. When the buck-boost converter is used in PV applications the input voltage changes continuously with the atmospheric conditions. Therefore, the duty cycle should change to track the maximum power point of photovoltaic array. The sate equation and output equation can be expressed by the following [16],[18]:

$$\dot{x}(t) = A(x)x(t) + B(x)u$$

$$y(t) = Cx(t), y_m(t) = Ex(t) \tag{19}$$

Where $x(t) = \begin{pmatrix} v_{pv} \\ i_L(t) \\ V_0 \end{pmatrix}$, $A(x) = \begin{pmatrix} i_{ri}/C_1v_{pv} & 0 & 0 \\ 0 & 0 & -1/L \\ 0 & 1/C & -1/CR \end{pmatrix}$,

$$B(x) = \begin{pmatrix} 0 & i_L/C & 0 \\ v_{pv}/L & 0 & v_0/L \\ 0 & -i_L/C & 0 \end{pmatrix} C^T = \begin{pmatrix} 1 \\ 0 \\ 0 \end{pmatrix}, E^T = \begin{pmatrix} 1\,0 \\ 0\,1 \\ 0\,0 \end{pmatrix}$$

Where i_L is the current across the inductor L, V_0 is the voltage in the capacitor C, u is the duty ratio of the Pulse Width Modulated (PWM) signal to control the switching MOSFET, $u \in \{0, 1\}$defines the switch position and supposed that the pparameters R, L, C_1 and C are supposed to be known constants.

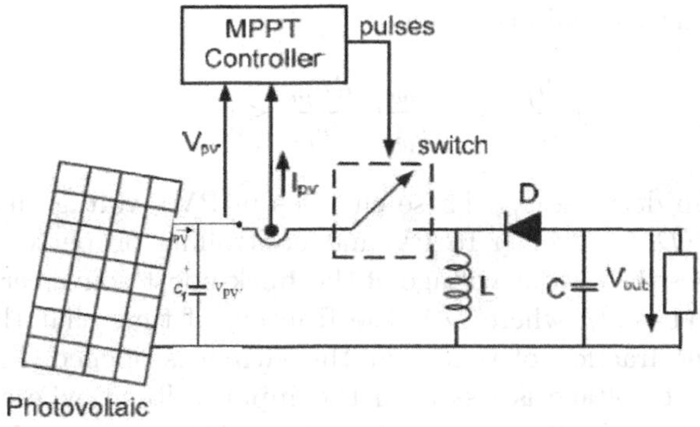

Fig. 1. PV power control system using a Buck-Boost DC-DC converter

4 Operation Principle of the Proposed MPPT Control Method

In this section we will present the MPPT control method and TS fuzzy model to design the proposed fuzzy controller.

4.1 MPPT Using Incremental Conductance Method

The maximum power point satisfies the condition $dP_{pv}/dv_{vp} = 0$. However, due to the high nonlinearity, the maximum power point is difficult to be solved from (17). This is the reason why the MPPT cannot be achieved easily [17], [18]. The incremental conductance algorithm is derived by differentiating the PV array power with respect to voltage and setting the result equal to zero. Therefore,

$$\frac{dp_{pv}}{dv_{pv}} = 0 \Rightarrow i_{pv} + v_{pv}\frac{di_{pv}}{dv_{pv}} = 0 \Rightarrow -\frac{i_{pv}}{v_{pv}} = \frac{di_{pv}}{dv_{pv}} \tag{20}$$

at the MPP (20)

Where $-\frac{i_{pv}}{v_{pv}}$ is the opposite of instant conductance of PV cells and $\frac{di_{pv}}{dv_{pv}}$ is the incremental conductance. From (20), at the MPPT, there are two quantities equal in magnitude, but opposite in sign. If the operating point is off of the MPPT, a set of inequalities can be derived from (20) that indicate whether the operating voltage is above or below the MPPT voltage. These rela-tionships are summarized as the following equations.

$$if\frac{di_{pv}}{dv_{pv}} > -\frac{i_{pv}}{v_{pv}}; (\frac{dP_{pv}}{dv_{pv}} > 0) \tag{21}$$

left of MPP, then increase v_{pv}

$$if\frac{di_{pv}}{dv_{pv}} = -\frac{i_{pv}}{v_{pv}}; (\frac{dP_{pv}}{dv_{pv}} = 0) \tag{22}$$

at MPP, then v_{pv} must remain constant

$$if\frac{di_{pv}}{dv_{pv}} < -\frac{i_{pv}}{v_{pv}};(\frac{dP_{pv}}{dv_{pv}} < 0) \tag{23}$$

Right of MPP, then decrease v_{pv} These changes in PV's voltage may be done by coupling a DC/DC converter to PV and controlling properly its duty cycle. Neglecting losses the output voltage of the buck-boost converter is given by the $V_o = v_{pv} * D/(1-D)$, where D is the fraction of time that the switch is closed, $1-D$ is the fraction of time that the switch is opened. For values of $D < 0.5$, the output voltage is less than the input voltage, whereas the output voltage is greater than the input voltage for values of $D > 0.5$. Therefore, $(V_o/v_{pv}) = v_{pv} * D/(1-D) = fd$. By using f_d, the reference MPPT is determined from the incremental conductance method [8], [9]. The fuzzy logic controller is constructed with the fuzzy rule and controller structure as in section 2. Fig.2 shows a flowchart of the proposed maximum power point tracking.

4.2 TS Fuzzy Model of PV for Design Fuzzy Controller

The TS fuzzy model of the PV is presented as the following. Frist, the system (19) is described by a TS fuzzy representation with the $Z_1(t) = i_{pv}/v_{pv}$, $Z_2(t) = i_L$, $Z_3(t) = v_{pv}$, and $Z_4(t) = V_o$, as the premise variables. Next, calculate the minimum and maximum values of $Z_j(t), Z_{jmin} \leq Zj(t) \leq Z_{jmax}(j = 1,2,3,4)$, where Z_{jmax} and Z_{jmin} are the upper and the lower bounds of the variable $Z_j(t)$. From the maximum and minimum values $Z_j(t)$, can obtain the fuzzy term of rule i as the following

$$M_{ji}(q_j(t)) = \frac{Z_{jmax} - Z_j(t)}{Z_{jmax} - Z_{jmin}}, \bar{M}_{ji}(Z_j(t)) = 1 - M_{ji}(Z_j(t)) \tag{24}$$

Where $-\frac{i_{pv}}{v_{pv}}$is the opposite of instant conductance of PV cells and $\frac{di_{pv}}{dv_{pv}}$ is the incremental conductance. From (20), at the MPPT, there are two quantities equal in magnitude, but opposite in sign. If the operating point is off of the MPPT, a set of inequalities can be derived from (20) that indicate whether the operating voltage is above or below the MPPT voltage. These rela-tionships are summarized as the following equations.

$$if\frac{di_{pv}}{dv_{pv}} > -\frac{i_{pv}}{v_{pv}};(\frac{dP_{pv}}{dv_{pv}} > 0) \tag{25}$$

left of MPP, then increase v_{pv}

$$if\frac{di_{pv}}{dv_{pv}} = -\frac{i_{pv}}{v_{pv}};(\frac{dP_{pv}}{dv_{pv}} = 0) \tag{26}$$

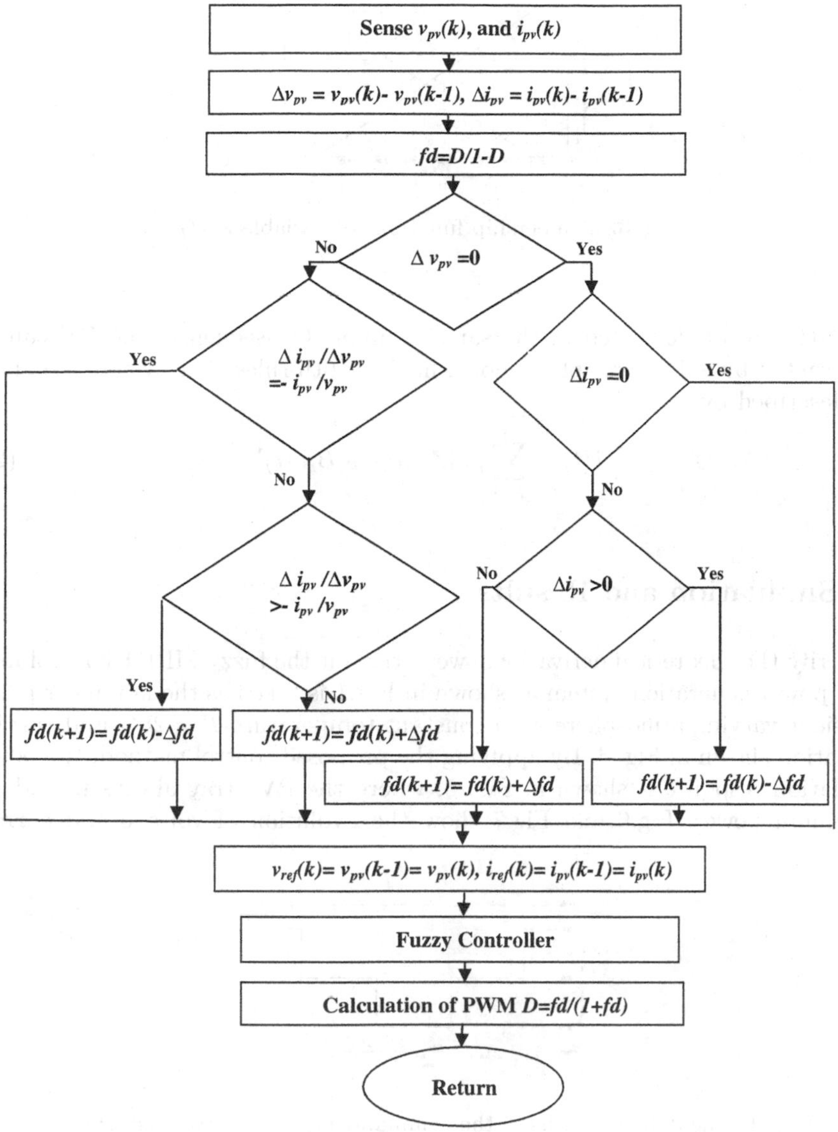

Fig. 2. Incremental conductance algorithm and fuzzy controller flowchart

at MPP, then v_{pv} must remain constant

$$if \frac{di_{pv}}{dv_{pv}} < -\frac{i_{pv}}{v_{pv}}; (\frac{dP_{pv}}{dv_{pv}} < 0) \tag{27}$$

Right of MPP, then decrease v_{pv} The degree of membership function for $Z_1(t)$ is depicted in Fig.3. The degree of membership function for state $Z_2(t), Z_3(t)$

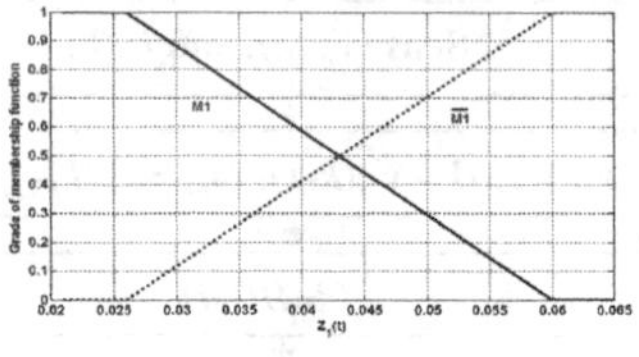

Fig. 3. Membership functions of variable $Z_1(t)$

and$Z_4(t)$ are implemented in the same manner. Consequently the PV can be represented by a TS-fuzzy plant model having 16th rules. The system dynamics are described by

$$\dot{x}(t) = \sum_{i=1}^{16} \mu_i \left[A_i x(t) + B_i u(t) \right] \tag{28}$$

5 Simulation and Results

To verify the theoretical derivations, we carry out the fuzzy MPPT control for a solar power generation system as shown in Fig.1 is taken as the nonlinear plant. Consider varying atmosphere with constant temperature $T = 25C$ and varying insolation shown in Fig. 4. By applying the proposed control method, the power regulation response is shown in Fig. 5, where the PV array always provides a maximum power. Fig.6 and Fig.7 show the evolution of voltage and current

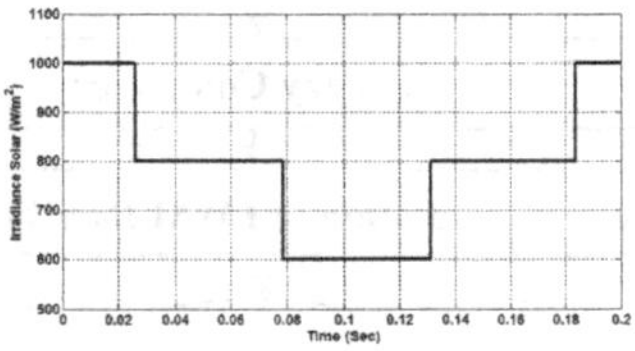

Fig. 4. The different paths of the command u_{tcom} with two disturbance

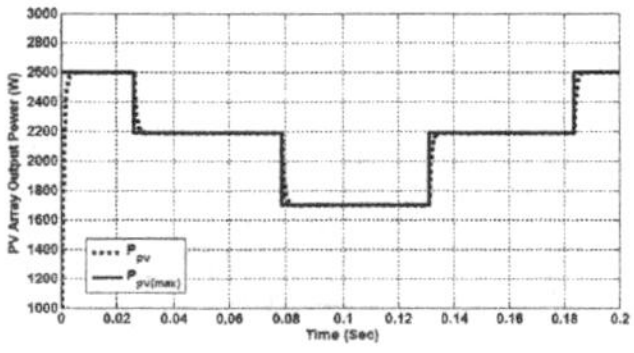

Fig. 5. The different paths of the command u_{tcom} with two disturbance

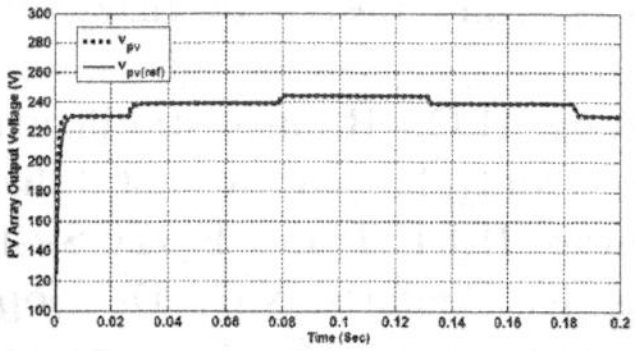

Fig. 6. PV Array Output Voltage (V)

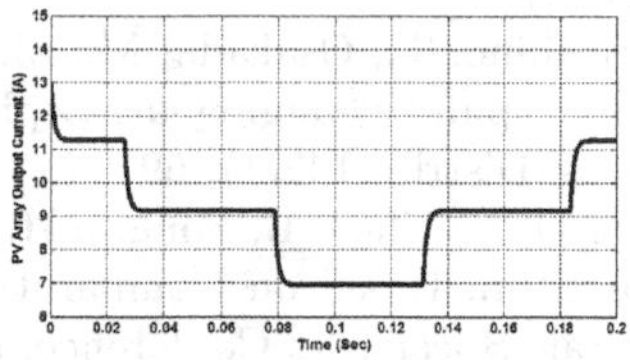

Fig. 7. PV Array Output Current (A)

PV array for different values of irradiance steps applied at different instants, respectively.

From the simulation results, it can be seen that the proposed controller method has the ability to drive the system states to follows those of the reference model. Moreover, the maximum power tracking is obtained.

6 Conclusion

A fuzzy controller was designed to drive the system states of the nonlinear system to follow those of a reference model. Stability conditions were derived to guarantee the system stability. The sufficient conditions are formulated in the format of LMIs to obtain the observer and controller gains. An application example on applying the proposed approach to PV energy was given. The fuzzy control is proposed to control the MPPT for a PV system. Simulation results illustrate the merits of the proposed fuzzy controller.

References

1. Isidori, A.: Nonlinear Control System. Springer, New York (1989)
2. Slotine, J.-J.E., Li, W.: Applied Nonlinear Control. Prentice-Hall, Englewood Cliffs, New Jersey (1991)
3. Tanaka, K., Sugeno, M.: Stability analysis and design of fuzzy control systems. Fuzzy Sets Syst. 45(2), 135–156 (1992)
4. Sugeno, M.: On stability of fuzzy systems expressed by fuzzy rules with singleton consequents. IEEE Trans. Fuzzy Syst. 7, 201–224 (1999)

5. Chen, C.-L., et al.: Analysis and design of fuzzy control systems. Fuzzy Sets Syst. 57, 125–140 (1993)
6. Tanaka, K., Ikeda, T., Wang, H.O.: Robust stabilization of a class of uncer-tain nonlinear systems via fuzzy control: Quadratic stabilizability, H_∞ control theory, and linear matrix inequalities. IEEE Trans. Fuzzy Syst. 4, 1–13 (1996)
7. Koutroulis, E., Kalaitzakis, K., Voularis, N.C.: Development of a microcon-troller-based, photovoltaic maximum power point tracking control system. IEEE Trans. Power Electron 16(1), 46–54 (1995)
8. Salas, V., Olias, E., Barrado, A., Lazaro, A.: Review of the maximum power point tracking algorithms for stand-alone photovoltaic systems. Solar Energy Mater. Solar Cells 90, 1555–1578 (2006)
9. Hussein, K.H., Muta, I., Hoshino, T., Osakada, M.: Maximum photovoltaic power tracking: An algorithm for rapidly changing atmospheric condition. Proc. Inst. Electr. Eng. Gen. Transmiss. Distrib. 142(1), 59–64 (1995)
10. Yu, G.J., Jung, Y.S., Choi, J.Y., Choy, I., Song, J.H., Kim, G.S.: A Novel Two-Mode MPPT Control Algorithm Based on Comparative Study of Existing Algorithms. In: Proc. Photovoltaic Specialists Conference, pp. 1531–1534 (2002)
11. Noguchi, T., Togashi, S., Nakamoto, R.: Short-Current Pulse-Based Maxi-mum-Power-Point Tracking Method for Multiple Photovoltaic-and-Converter Module System. IEEE Trans. Ind. Electron 49(1), 217–223 (2002)
12. Lee, D.Y., Noh, H.J., Hyun, D.S., Choy, I.: An Improved MPPT Converter Using Current Compensation Method for Small Scaled PV-Applications. Proc. APEC, 540–545 (2003)
13. Kamal, E., Koutb, M., Sobaih, A.A., Abozalam, B.: An Intelligent Maximum Power Extraction Algorithm for Hybrid Wind-Diesel-Storage System. Int. J. Electr. Power Energy Syst. 32(3), 170–177 (2010)
14. Lam, H.K., Hung, F.H.F., Tam, P.K.S.: On Design of a Switching Controller for Nonlinear Systems with Unknown Parameters based on a Model Refer-ence Approach. In: IEEE International Fuzzy Systems Conference, pp. 1263–1266 (2001)
15. Kamal, E., Aitouche, A., Ghorbani, R., Bayart, M.: Robust Fuzzy Fault Tolerant Control of Wind Energy Conversion Systems Subject to Sensor Faults. IEEE Transactions on Sustainable Energy 3(2), 231–241 (2012)
16. Kamal, E., Aitouche, A.: Robust fault tolerant control of DFIG wind energy system with unknown inputs. Renewable Energy 56, 2–15 (2013)
17. Kim, I.-S., Kim, M.-B., Youn, M.-J.: New maximum power point tracker using sliding mode observer for estimation of solar array current in the grid-connected photovoltaic system. IEEE Trans. Ind. Electron 53(4), 1027–1035 (2006)
18. Chiu, C.-S.: T-S Fuzzy Maximum Power Point Tracking Control of Solar Power Generation Systems. IEEE Trans. Energy Convers. 25(4), 1123–1132 (2010)

Design of a Fuzzy System Performing Data Analysis for Pressure Equipment Service Life-Span Estimation

Leszek Koszałka[1], Piotr Franz[2], and Karol Gęga[2]

[1] Department of Electronics, Wroclaw University of Technology, Wroclaw, Poland
leszek.koszalka@pwr.wroc.pl
[2] Metegrity, Met Inc., Edmonton, Alberta, Canada
{pfranz,kgega}@metegrity.com
http://www.metegrity.com

Abstract. The paper deals with the problem of risk based inspection of petroleum refineries various pressure equipment. Describing a real off-shelf product designed to help in maintaining equipments integrity authors present briefly their studies on a fuzzy expert system, which can deal with data analysis for risk assessment purposes.

Keywords: risk based inspeciton, fuzzy expert systems, rules extraction, data analysis.

1 Introduction

The need of equipment life-cycle management comes from the desire of profit. As the sudden equipment breakdowns generates great loses, it is often to expensive or even impossible to use robust materials in equipment's constructions. That is why the condition of equipment at each processing plant needs to be constantly monitored.

Looking at petroleum refineries - complex plants with extensive piping carrying streams liquid fluids between different chemical processing units in order to refine pure petroleum into different products, e.g., oil fuel, gasoline, asphalt base, kerosene, heating oil (and many others) - possible equipment failures, if not spotted in time, can generate great loses. One of the common factors which damages the equipment is corrosion. Additionally corrosion processes also dictates cleaning schedules, during which whole plant needs to be closed. In year 2001 the lost of petroleum industry in U.S. only due the corrosion was estimated at level of 3.7 billion US [1]. Apart from corrosion problems, there are many minor faults possibilities (e.g., material cracking) that need to be considered. This show how crucial is constant equipment monitoring.

However, to maintain the integrity of various equipments (tanks, pipes, vessels, etc.) used in the refining and petrochemical industry it is desirable to move to risk based inspection programs, not only base on known history of equipment condition [2]. In order to predict failures probability, a well understanding of process operating conditions and resulting damage mechanisms are required.

J. Korbicz and M. Kowal (eds.), *Intelligent Systems in Technical and Medical Diagnostics*, Advances in Intelligent Systems and Computing 230,
DOI: 10.1007/978-3-642-39881-0_7, © Springer-Verlag Berlin Heidelberg 2014

Those are not the only requirements for reliable risk assessment to make. Without any data analysis method or mechanism making any predictions would be difficult or even impossible.

The goal of the paper is to present authors studies on design of a fully efficient expert system, which is capable of performing precise analysis of signal gathered from equipment's sensors in order to track various process plants equipment integrity violations. The system is targeted to decrease the amount of work that the human operator needs to spend on integrity violations analysis.

2 Offshelf Risk Assesment Solutions

All further considerations will be focused around the data analysis module that is part of the *Visions* platform developed by *Metegrity*, a *Met Inc.* division [3].

2.1 *Visions* Platform

A simplified process of obtaining data needed by the Visions systems for risk assessment purposes can be easily shown on a flowchart (Fig. 1). The refinery or petrochemical institute uses its own probes system to gather the data (process parameters) from pressure equipments and stores it in so called data historian system. *Visions'* has its own scheduler, so that trough developed interface it can connect with data historian database and obtain stored data. Those data are subject to analysis (*CBIM* [4]) aimed at tracking the integrity violations. The output of *CBIM* are variables, that can be directly used for risk assessment (*RBI* module) - equipment's damage (like corrosion) prediction. To track violations (exceeding the process' variable limit for predefined time) and excursions (exceeding the process' variable limit that lasts too short to classify it as a violation) efficiently, the algorithms used in *CBIM* module are oriented on using defined Integrity Operating Windows (*IOWs* [5]). *IOW* is an established boundary for process parameters. Operating outside those boundaries for predefined time can affect the integrity of the equipment involved in process.

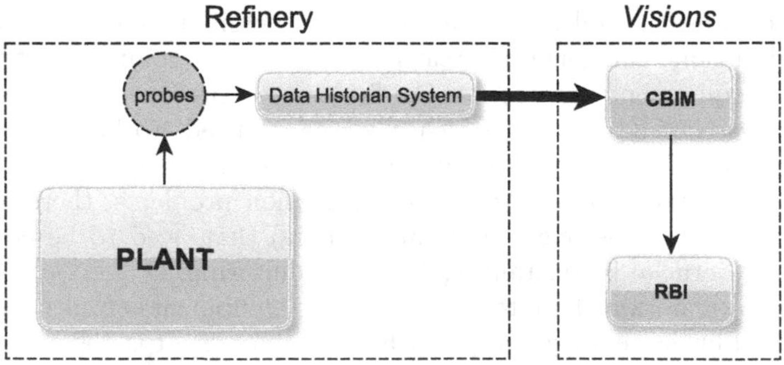

Fig. 1. Simplified data transfer schema from equipment probes to the *Visions*

2.2 Designed System and Visions Interdependence

At the moment any model or algorithm that could calculate input variables for *RBI* module does not exist, so they are calculated and entered manually by the process operator. Operator basing on historical measurements and supported by *CBIM* analysis, specifications and his knowledge decides about the parameters connected with integrity violations.

The designed system aims to extend the *CBIM* functionality by performing the analysis of the violations notified by the module, in order to at least speed up the process of estimating the integrity violations parameters. Ideally it should be able to calculate them precisely, so that the system output could be directly inputted into the RBI module.

2.3 General System Concept

The conceptual scheme of a designed system can be seen in Fig. 2. As it can be noticed, establishing a fully efficient system needs also establishing mechanisms for signal analysis and feature extraction.

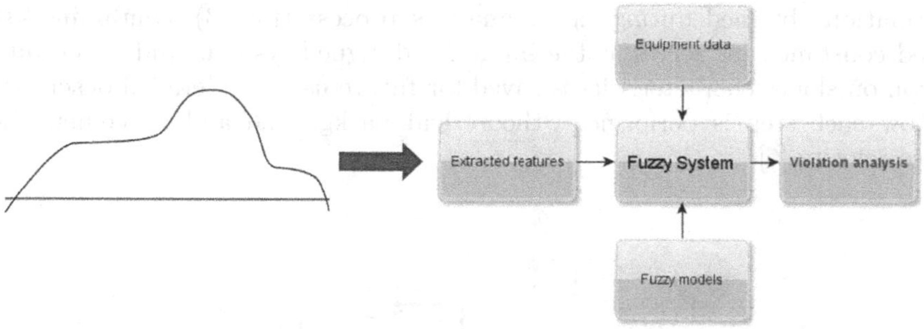

Fig. 2. Designed system conceptual scheme

The system itself is designed to be a net of specialized fuzzy inference modules, related with each other. Such structure implies significant benefits, e.g.:

- Error tracking during design and implementation process is less time consuming, thus less costly.
- Maintenance of system is simplified.
- Functionality of a system can be easily increased with new modules incorporated.

3 Signal Analysis for Feature Extraction

Several steps are needed, so that from the violation signal features can be extracted and inputted for designed system. There are:

1. Segmentation of the violation due to the signal monotonicity changes. First step is to find points in which the signal monotonicity changes. *Spearmans* rank coefficient is used to select them from signal samples.
2. Segmentation of the violation due to the signal concavity changes. At this point inflection points in the signal characteristics are found. Found points are places of further signal cuts.
3. Extraction of constant component from each of obtained violation segment. Signal part on given segment is substituted by constant run, as most of thermodynamics and kinetics e.g. corrosion rate estimation purposes are well established for constant environments. It is assumed that the constant run on a given segment is equal to the lowest real signal value on that segment if the upper IOW border is exceeded, otherwise it is equal to the highest real signal value.
4. Extraction of slope characteristics for each obtained violation segment. The information about slope characteristics on each segment is extracted for future use.
5. Adjust the constant component on each segment by the signal change on that segment. Constant run on each of the segments is adjusted so that integra of constant run and integra of real signal are equal.

Information obtained during signal analysis process (Fig. 3), containing estimated constant runs serves as the input for designed system. Additional information on slopes characteristics is saved for future use. For detailed description on how each step is performed theoretical background and implementation please refer to [6].

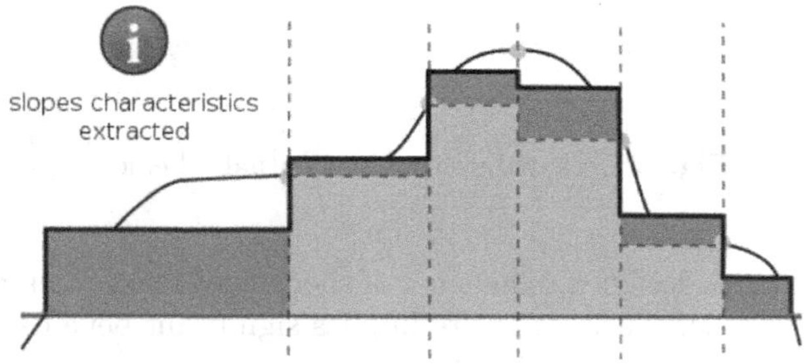

Fig. 3. Information and output signal obtained after signal analysis

4 Exemplary Module: Corrosion Risk Estimation

As a part of a bigger system for data analysis, one of important module is corrosion risk estimator. On the basis of analysis received from the module, the decision about further corrosion rate estimates will be conducted.

4.1 Module Inputs

Module is taking as an input two process parameters:

- E - electrode potential (here: equipments material potential),
- pH - measure of the activity of the (solvated) hydrogen ion.

4.2 Module Output

The module's output is a single label, identifying the condition, thus the module is dealing with the classification problem. There are three possible output classes to which the environmental conditions (due to E and pH) can be assigned:

- passive - indicating high probability of formatting protective oxide layer.
- immune - with high probability that no corrosion occurs.
- corrosive - indicating release of ions into solution - corrosion process.

4.3 Rule Base

Rule base is the most important part of the module. It is impossible to create universal rule base, which would give high performance for each equipment case. Also manual modeling of rule bases for different equipment could be annoying and often difficult process.

Taking above into consideration for the module purpose, an pattern for rule extraction from data is introduced, based on idea presented in [7]. Main steps of the rule extraction are:

1. Manual split of the $E - pH$ diagram [8] into rectangular regions according to the obtained stable material species.
2. Extraction of the region centers and influence ranges.
3. Build of rules using extracted data.
4. Rules calibration using the calibration test points set.

First two steps are well described in [7]. Third and fourth steps are more interesting.

In order to build initial module rules it is necessary to extract center of each region, denoted x^*. More precisely, as it is intended to model the input space to output space dependence, it is needed to obtain the central value of each feature for each region, denoted x^*_{ij}, where i is the region number and j is the jth feature.

What is more, also the range for each region center need to be selected denoted as r. This parameter specifies range of influence that the region center has in each of the input data dimensions, so again it is calculated for each region feature, as follows:

$$r_{ij} = \frac{1}{2} S_{ij} \qquad (1)$$

where r_{ij} is influence range of jth feature of ith region, S_{ij} is the range of jth feature in region i.

After information extraction, each region is described as an individual system rule:

$$\text{IF } X_1 = \mu_{i1} \text{ AND } X_2 = \mu_{i2} \text{ THEN } Y = Y_n \tag{2}$$

where μ_{ij} is membership function that shows the membership degree of X_j in region i and is calculated as an Gaussian function as follows:

$$\mu_{ij}(X_j) = e^{(-\frac{1}{2}(\frac{X_j - x^*_{ij}}{\sigma_{ij}})^2)} \tag{3}$$

The standard deviation σ_{ij} is calculated considering the influence range:

$$\sigma_{ij} = \sqrt{\frac{r^2_{ij}}{8}} \tag{4}$$

Such created rules are further calibrated to improve classification performance. In calibration process following error criterion is used to establish sample classification performance:

$$E = \frac{1}{2}(1 - \mu_{c,max} + \mu_{-c,max})^2 \tag{5}$$

where $\mu_{c,max}$ is the maximum degree of membership from all rules that infers class C and $\mu_{-c,max}$ is the maximum degree of membership from all rules that do not infer class C. The membership function can be adjusted be selecting new centers and influence ranges according to the gradient:

$$x^*_{ij} = x^*_{ij} \pm \lambda\mu_i \frac{(1 - \mu_{c,max} + \mu_{-c,max})(X_j - x^*_{ij})}{\sigma^2_{ij}} \tag{6}$$

$$\sigma^*_{ij} = \sigma^*_{ij} \pm \lambda\mu_i \frac{(1 - \mu_{c,max} + \mu_{-c,max})(X_j - x^*_{ij})^2}{\sigma^3_{ij}} \tag{7}$$

where μ_i is the degree of membership for the ith rule and λ is positive constant - the learning rate. Adjusted are only two rules: the one which gave $\mu_{c,max}$ and the one which gave $\mu_{-c,max}$. For rule which gave $\mu_{c,max}$ Eq. 6 and 7 are used with the $+$ sign. Rule which gave $\mu_{-c,max}$ is adjusted using equations with $-$ sign.

During the calibration process, to further improve of the performance it is allowed that the membership function can be "two-sided" Gaussian - in other words it can be a convolution of two independent Gaussian functions. During calibration only parameters of the peak closer to the X_j sample are adjusted (either left or right x^* and σ), however if the sample is located in the function plateau region the membership function is not adjusted and left as it is.

4.4 Inference Mechanism

At the moment module's decision depend only on the rule that returns the highest membership degree for given input sample. To resolve ties, as the rules are processed sequentially, the first rule that returned the highest membership degree wins.

For modeling of the AND conjunction used inside the rules the Mamdani operator is used.

4.5 Test Models

Below examples of module performance for two different environments are presented. Examples show also process of building specific base rules.

Aluminum-Water System. The $E - pH$ diagram for aluminum-water ($Al - H_2O$) system is shown in Fig. 4. The areas worth noticing that can be seen on the diagram are:

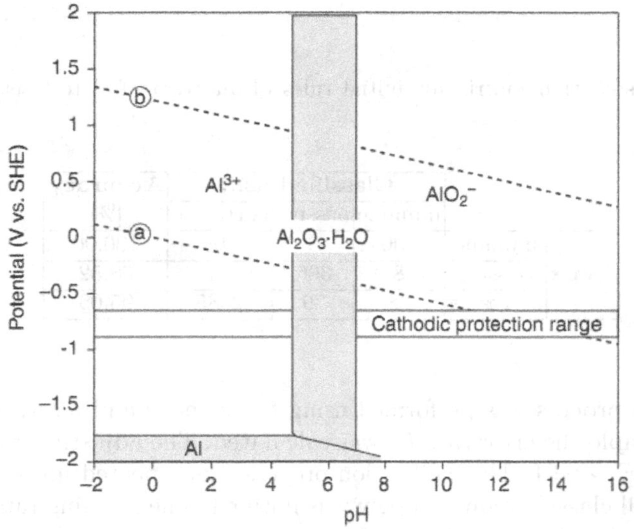

Fig. 4. $E - pH$ diagram of $Al - H_2O$ system at $25°C$

- Passivity area, where present is solid $Al_2O_3 \cdot H_2O$.
- Immunity area, where present is solid Al.
- Two corrosion areas, where present are ionic Al^{3+} and AlO_2^- aluminum species.

According to the species considered on the diagram, four initial rectangular regions were created (Tab. 1). Those four regions are the absolute minimum that needs to be considered in order to obtain fully efficient module.

Table 1. Initial region x^* (centers) and r (range) parameters with assigned classes ($Al - H_2O$ system)

No.	Stable species	pH		E		class
		x^*	r	x^*	r	
1	Al^{3+}	1.125	3.125	0.150	1.750	corrosive
2	Al_2^-	11.250	4.750	0.000	2.000	corrosive
3	Al	3.000	5.000	-1.850	0.150	immune
4	$Al_2O_3 \cdot H_2O$	5.375	1.125	0.000	2.000	passive

Initial rules obtained from regions were tested on set of 3000 samples, generated uniformly using Matlab environment, with each sample labeled with corresponding class. Results are presented in form of classification matrix in Tab. 2.

Table 2. Classification matrix for initial rules obtained on full test set ($Al - H_2O$ system)

		Classified as...			Accuracy
		immune	passive	corrosive	[%]
Class	immune	100	0	0	100.00
	passive	8	306	84	68.52
	corrosive	38	79	2385	95.09

Calibration process was performed using the same samples set. After classifying each sample the error rate E_r was calculated. The adjustment of rules was triggered for $E > 0.01$. The calibration process was repeated until no improvement in overall classification accuracy was notified. The learning rate was set at $\lambda = 0.1$. The classification matrix after the calibration process is presented in Tab. 3.

Table 3. Classification matrix for tuned rules obtained on full test set ($Al - H_2O$ system)

		Classified as...			Accuracy
		immune	passive	corrosive	[%]
Class	immune	63	16	21	63.00
	passive	1	393	4	98.74
	corrosive	14	12	2476	98.72

Iron-Water System. The $E - pH$ diagram for iron-water ($Fe - H_2O$) system is shown in Fig. 5 and, as it can be noticed its complexity is higher than for aluminum species. The areas worth noticing that can be seen on the diagram are:

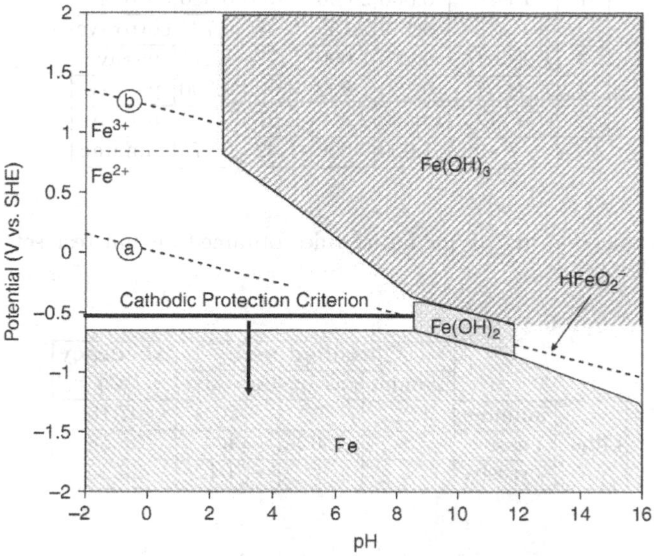

Fig. 5. $E - pH$ diagram of $Fe - H_2O$ system at 25^oC

- Two passivity areas, where as the most stable species $Fe(OH)_2$ and $Fe(OH)_3$ are present.
- Immunity area, where present is solid Fe.
- Three corrosion areas, where present are ionic Fe^{3+}, Fe^{2+} and $HFeO_2^-$ iron species.

According to the species considered on the diagram, six initial rectangular regions were created (Tab. 4). Those six regions are the absolute minimum that needs to be considered in order to obtain fully efficient module.

Initial rules were tested on set of 3005 samples, generated uniformly using Matlab environment with each sample labeled with corresponding class. The results are presented in form of classification matrix in Tab. 5.

Calibration process was performed using the same samples set, however because initial rules obtained high accuracy on test set, tuning was difficult. Finally an slight improvement was found for learning rate $\lambda = 0.01$ and error rate Er changing from 0.58 to 0.53 over each calibration iteration with step equal to 0.01. For values $E_r < 0.53$ deterioration of accuracy was noticed. Classification matrix for calibrated rules is presented in Tab. 6.

Table 4. Initial region x^* (centers) and r (range) parameters with assigned classes ($Fe - H_2O$ system)

No.	Stable species	pH		E		class
		x^*	r	x^*	r	
1	Fe^{3+}	0.000	2.000	1.400	0.600	corrosive
2	Fe^{2+}	3.150	5.150	0.075	0.725	corrosive
3	$Fe(OH)_3$	9.000	7.000	0.725	1.275	passive
4	$Fe(OH)_2$	10.100	1.800	-0.600	0.200	passive
5	$HFeO_2^-$	14.450	2.550	-0.975	0.425	corrosive
6	Fe	7.000	9.000	-1.325	0.675	immune

Table 5. Classification matrix for initial rules obtained on full test set ($Fe - H_2O$ system)

		Classified as...			Accuracy
		immune	passive	corrosive	[%]
Class	immune	922	7	20	97.15
	passive	0	1302	46	96.58
	corrosive	1	3	704	99.43

Table 6. Classification matrix for calibrated rules obtained on full test set ($Fe - H_2O$ system)

		Classified as...			Accuracy
		immune	passive	corrosive	[%]
Class	immune	934	7	8	98.41
	passive	0	1319	29	97.84
	corrosive	5	10	693	97.88

5 Summary

The paper covered mechanisms of briefly signal analysis and first of the system modules: the corrosion risk estimation module for designed data analysis system.

The main task of the analysis process is to substitute the violation signal with an constant run that impacts the equipment integrity in similar manner (however preserving information on each signal changes/slopes for future use).

Having the constant estimation of the violation signal it is possible to further process it and resolve the violation impact on considered equipment and that is done in presented corrosion risk estimation module. The module leverages fuzzy logic and uses the relationship between the electrode potential E (measured in

Volts) and solution's pH to estimate if for given material-solution environment an material corrosion may occur.

The performance of the module was presented on two test models: aluminum-water ($Al - H_2O$) and iron-water ($Fe - H_2O$) environments. As in each case the module needs unique rule base to operate, also the mechanism of extracting fuzzy rules from data was covered. Rules extraction is very simple and straightforward and, as it could be noticed, even module using minimum feasible number of rules can achieve high classification accuracy. What is more, using simples heuristic and gradient methods it is possible to further increase overall classification accuracy.

References

1. NACE International: Corrosion Costs and Preventive Strategies in the United States (2001)
2. American Petroleum Institute: API RP 581 Risk Based Inspection (unpublished)
3. Metegrity Visions, `www.visions-enterprise.com`
4. Gega, K.: Containment Boundary Integrity Monitoring (CBIM) - VisPl Server (unpublished)
5. American Petroleum Institute: API RP 584 Integrity Operating Window (unpublished)
6. Franz, P.: Design of a Fuzzy System performing data analysis for pressure equipment service life-span estimation, msc thesis, Wroclaw University of Technology, Wroclaw (2012)
7. Chiu, S.L.: Extracting Fuzzy Rules from Data for Function Approximationand Pattern Classification, Fuzzy Information Engineering: A Guided Tour of Applications, ch.9. John Wiley & Sons (1997)
8. Roberge, P.R.: Corrosion Engineering: Principles and Practice. McGraw-Hill (2008)

Robust Fault Detection Using Zonotope-Based GMDH Neural Network

Marcin Mrugalski

Institute of Control and Computation Engineering,
University of Zielona Góra,
ul. Podgórna 50, 65–246 Zielona Góra, Poland
M.Mrugalski@issi.uz.zgora.pl

Abstract. The paper deals with the problem of determination of model uncertainty during non-linear systems identification via application of the self-organising dynamic Group Method of Data Handling (GMDH) neural network. The Zonotope-Based Algorithm (ZBA) was applied to calculate the uncertainty of the parameters and the model output. The mathematical description of the model uncertainty enabled designing the robust fault detection system which effectiveness was verified by tunnel furnace.

Keywords: Robust fault detection, GMDH neural networks, zonotope-based algorithm.

1 Introduction

The application of the Artificial Neural Networks (ANNs) in the Fault Detection and Isolation (FDI) [1,2,3,4,5,6] and Fault Tolerant Control (FTC) [7,8,9,10] systems follows from their effectiveness in the modelling of the non-linear systems. Unfortunately, this effectiveness is limited by inefficient quality of the neural model following from the inappropriate selection of the network architecture and the errors following from inaccurate estimation of the neurons parameters. The high quality of the neural model is crucial because it is used in the FDI systems to generate the residuals which should be close to zero in the fault-free case, and it should be distinguishably different from zero in the faulty case. Under such an assumption, the faults are detected by the application of a fixed threshold on the residual signal. Unfortunately, appearing of the neural model uncertainty or measurements noise can lead to false alarms or undetected faults. For this reason it is necessary to develops a new robust fault detection methods. In order to achieve this goal the methodology of calculation of the neural model uncertainty have to be provided. This result can be used during obtaining of the output adaptive thresholds and their application in robust fault diagnosis.

To overcome this problem a methodology based on the GMDH neural networks [11,12,13,14,15] has been proposed. The synthesis process of this network is based on the iterative processing of a sequence of operations what results in the evolution of the resulting model structure in such a way so as to obtain

J. Korbicz and M. Kowal (eds.), *Intelligent Systems in Technical and Medical Diagnostics*, Advances in Intelligent Systems and Computing 230,
DOI: 10.1007/978-3-642-39881-0_8, © Springer-Verlag Berlin Heidelberg 2014

the best quality approximation of the identified system what results in the reduction of the neural model uncertainty. Apart from the contribution to the model uncertainty of the errors caused by the inappropriate structure selection also the parameter estimates inaccuracy influences on the model quality. This problem was widely presented in the paper [16,17], where the potential sources of the model uncertainty following from the application of improper parameters estimation algorithm were described. In particular, it was shown that the usual parameters estimation algorithms work based on the incorrect assumptions concerning the properties of the noises and disturbances which affect on the data used during system identification.

In order to overcome this problem the Bounded Error Approach (BEA) [18,19,20] can be applied. This approach is based on more realistic assumption that the noises and disturbances lie between given prior bounds. Unfortunately, in spite of many advantages this algorithm has also significant memory and time expenditure. In the case of neural model, which consists of a great number of neurons, determination of parameters estimations will not be possible considering significant memory and time expenditure. In order to overcome this disadvantage the ZBA [21,22] can be used. The application of this algorithm allows to estimate the parameters of the GMDH neural network and the model uncertainty in a form of the zonotope which is an approximation of the exact admissible parameter set. Based on this result it is possible to define the model uncertainty in the form of the output adaptive thresholds, which can be applied in the robust fault detection tasks.

2 ZBA in GMDH Network Synthesis for Robust Fault Detection

The GMDH neural model is gradually increasing (cf. Fig. 1) by adding new layers of neurons according following repeated steps [5,11,23,24,25]:

- Creation of a neuron layer on the basis of system inputs combinations,
- Parameters estimation of each neuron,
- Calculation of neurons responses and its uncertainty,
- Quality evaluation of neurons,
- Selection of neurons,
- Termination condition testing.

In the GMDH network it is assumed that each neuron should reflect the behaviour of the dynamic system being identified. Moreover, the neurons parameters are estimated seriately for each neuron in such a way that its outputs are the best approximation of the real system outputs. In the case of non-linear dynamic system identification the dynamic neuron with an Infinite Impulse Response (IIR) filter can be used [4,13]. Such a neuron model consists of the filter module and the activation module. The behaviour of the filter module is described by the following equation:

$$y_{n,k}^{(l)} = \left(\boldsymbol{r}_{n,k}^{(l)}\right)^T \boldsymbol{p}_n^{(l)}. \tag{1}$$

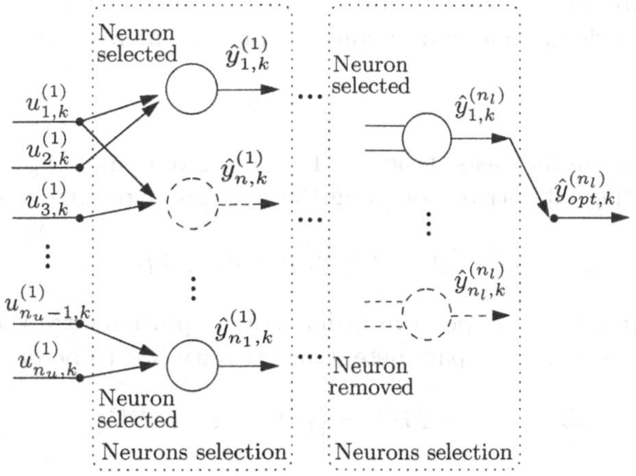

Fig. 1. Synthesis of the GMDH model

where $\boldsymbol{r}_{n,k}^{(l)} = [-y_{n,k-1}^{(l)}, \ldots, -y_{n,k-n_a}^{(l)}, \boldsymbol{u}_{n,k}^{(l)}, \boldsymbol{u}_{n,k-1}^{(l)}, \ldots, \boldsymbol{u}_{n,k-n_b}^{(l)}]$ and $\hat{\boldsymbol{p}}_n^{(l)} = [a_1, \ldots, a_{n_a}, v_0, v_1, \ldots, v_{n_b}]$ are the regressor and the filter parameters respectively and l is the layer number of the GMDH network and n is the neuron number in the l-th layer. The filter output is used as the input for the activation module:

$$\hat{\boldsymbol{y}}_{n,k}^{(l)} = f(y_{n,k}^{(l)}). \tag{2}$$

In order to estimate the unknown parameters $\boldsymbol{p}$ the techniques for the parameter estimation of linear-in-parameter models can be used, e.g. Least Mean Square method (LMS) [11]. It follows from the facts that the parameters of the each neuron in the GMDH network are estimated separately and it is assumed that the neuron's activation function $f(\cdot)$ is an invertible, i.e. exists there $f^{-1}(\cdot)$. Unfortunately, the usual statistical parameter estimation framework such as LMS which can be applied to the parameters of GMDH network estimation, assumes that the data are corrupted by errors which can be modelled as the realisations of independent random variables, with a known or parameterised distribution:

$$\mathcal{E}\left[\varepsilon_n^{(l)}\right] = 0, \quad \mathrm{cov}\left[\boldsymbol{\varepsilon}_n^{(l)}\right] = \left(\sigma_n^{(l)}\right)^2 \boldsymbol{I}. \tag{3}$$

A more realistic approach is to assume that the errors lie between given prior bounds. Such reasoning leads directly to the BEA [18,19]. The application of this algorithm can lead to obtain unbiased parameters estimates. Unfortunately, the main drawback of the BEA is associated high computational cost and complexity of the representation of the exact feasible parameters set. In order to overcome these limitations the approaches based on the approximation of the exact feasible parameters set by the zonotopes can be applied [21,22]. In this algorithm the following parameterized form of the identified system is assumed:

$$y_k = \boldsymbol{r}_k^T \boldsymbol{p} + \varepsilon_k. \tag{4}$$

In this algorithm the following form of the bounding of the errors ε_k, which represents the modeling uncertainty and the measurements noise, is assumed:

$$\varepsilon_k \in \mathbb{R} : |\varepsilon| \leq \sigma. \tag{5}$$

On the beginning, on the basis of the $k = 1, ..., n_D$ given inputs $\boldsymbol{r}_k$ and outputs y_k measurements, the data strips bounding the consistent parameters are defined:

$$\mathbb{S}_k = \{\boldsymbol{p} : -\sigma \leq \boldsymbol{r}_k^T \boldsymbol{p} - y_k \leq \sigma\}. \tag{6}$$

In the next step, the zonotope of m order which approximates the feasible parameter set representing the parameters uncertainty has to be defined as:

$$\mathbb{Z}_{m,k} = \boldsymbol{p} \oplus \mathbb{P}\boldsymbol{B}^m = \{\boldsymbol{p} + \mathbb{P}z : z \in \boldsymbol{B}^m\}, \tag{7}$$

where:

- $\boldsymbol{p} \in \mathbb{R}^{n_p}$ – given vector of parameters,
- $\mathbb{P} \in \mathbb{R}^{n_p \times m}$ – matrix of parameters uncertainty,
- $\boldsymbol{B}^m \in \mathbb{R}^{m \times 1}$ – the unitary box composed of m unitary intervals $\boldsymbol{B} = [-1, 1]$,
- $\oplus$ – the Minkowski sum of two sets (e.g. $\mathcal{X} \oplus \mathcal{Y} = \{x + y : x \in \mathcal{X}, y \in \mathcal{Y}\}$).

In the fact the zonotope of m order is defined as the Minkowski sum of segments defined by columns of matrix $\mathbb{P} : \boldsymbol{p} \oplus \mathbb{P}\boldsymbol{B}^m = \boldsymbol{p} \oplus \mathbb{P}_1\boldsymbol{B} \oplus \mathbb{P}_2\boldsymbol{B} \oplus, ..., \oplus \mathbb{P}_m\boldsymbol{B}$ where $\mathbb{P}_i$ denotes the i-th column of matrix $\mathbb{P}$ and $\mathbb{P}_i\boldsymbol{B} = \{\mathbb{P}_i b : b \in [-1, 1]\}$. It should be underlined that if $m < n_p$ or $\mathbb{P}$ is not a full-rank matrix then the zonotope $\mathbb{Z}_m$ has zero volume. Moreover, the zonotope is always a bounded set provided that the entries of matrix $\mathbb{P}$ are bounded. Furthermore, the order of zonotope m represents its geometrical complexity.

The intersection of the defined zonotope $\mathbb{Z}_{m,k}$ with the subsequent data strips $\mathbb{S}_k$ allows to obtain the approximation feasible solution set containing the parameters estimates:

$$AFSS_{k+1} = \mathbb{S}_k \cap \mathbb{Z}_k, \quad \text{for} \quad k = 1, ..., n_d. \tag{8}$$

Unfortunately, the application of the above identification algorithm can lead to increase of the zonotope complexity for each subsequent sample time k. For this reason, it is necessary to use an algorithm of overbounding of the higher-order zonotope by a lower-order zonotope. In order to achieve this goal, it is advisable to define so-called zonotope support strip $\mathbb{S}_{s,k}$ for a given zonotope $\mathbb{Z}_{m,k}$ and a vector $\boldsymbol{r}_k \in \mathbb{R}^{n_r}$:

$$\mathbb{S}_{s,k} = \{\boldsymbol{p} : q_b \leq \boldsymbol{r}_k^T \boldsymbol{p} \leq q_a\}, \tag{9}$$

where q_a and q_b satisfy:

$$q_a = \max_{\boldsymbol{p} \in \mathbb{Z}} \boldsymbol{r}_k^T \boldsymbol{p}, \tag{10}$$

and

$$q_b = \min_{\boldsymbol{p} \in \mathbb{Z}} \boldsymbol{r}_k^T \boldsymbol{p}. \tag{11}$$

Knowing that $\mathbb{Z}_{m,k} = \boldsymbol{p} \oplus \mathbb{P}\boldsymbol{B}^m$ and vector $\boldsymbol{r}_k \in \mathbb{R}^{n_r}$, the values of q_a and q_b can be calculated according to the following equations:

$$q_a = \boldsymbol{r}_k^T \boldsymbol{p} + \|\mathbb{P}^T \boldsymbol{r}_k\|_1, \tag{12}$$

and:

$$q_b = \boldsymbol{r}_k^T \boldsymbol{p} - \|\mathbb{P}^T \boldsymbol{r}_k\|_1, \tag{13}$$

where $\|\mathbb{P}^T \boldsymbol{r}_k\|_1$ represents the sum of the absolute values of the components of $\mathbb{P}^T \boldsymbol{r}_k$. Moreover, for a given zonotope $\mathbb{Z}_{m,k} = \boldsymbol{p} \oplus \mathbb{P}\boldsymbol{B}^m$ and strip $\mathbb{S}_k$, the zonotope tight strip $\mathbb{S}_{t,k}$ is obtained as:

$$\mathbb{S}_{t,k} = \mathbb{S}_k \cap \mathbb{S}_{s,k}. \tag{14}$$

The intersection of the tight strips $\mathbb{S}_{t,k} = \{\boldsymbol{p} : |\boldsymbol{r}_k^T \boldsymbol{p} - y_k| \leq \sigma\}$ and the zonotope $\mathbb{Z}_{m,k}$ leads to obtaining a new reduced order zonotope $\mathbb{Z}_{r,k} = \boldsymbol{p} \oplus \mathbb{P}\boldsymbol{B}^r = \boldsymbol{p} \oplus [\mathbb{P}_1, \mathbb{P}_2, ..., \mathbb{P}_r]\boldsymbol{B}^r$:

$$\mathbb{Z}_{r,k} = \mathbb{Z}_{m,k} \cap \mathbb{S}_{t,k} \subseteq \boldsymbol{p}_j \oplus T_j \boldsymbol{B}^r, \tag{15}$$

where $\boldsymbol{p}_j$ and T_j for every integer j where $0 \leq j \leq r$ can be calculated according the following equations:

$$\boldsymbol{p}_j = \begin{cases} \boldsymbol{p} + (\frac{y_k - \boldsymbol{r}_k^T \boldsymbol{p}}{\boldsymbol{r}_k^T \mathbb{P}_j})\mathbb{P}_j, & \text{if} \quad 1 \leq j \leq r \quad \text{and} \quad \boldsymbol{r}_k^T \mathbb{P}_j \neq 0 \\ \boldsymbol{p}, & \text{otherwise,} \end{cases} \tag{16}$$

$$T_j = \begin{cases} [T_1^j, T_2^j, ..., T_r^j], & \text{if} \quad 1 \leq j \leq r \quad \text{and} \quad \boldsymbol{r}_k^T \mathbb{P}_j \neq 0 \\ \mathbb{P}, & \text{otherwise,} \end{cases} \tag{17}$$

$$T_i^j = \begin{cases} \mathbb{P}_i - (\frac{\boldsymbol{r}_k^T \mathbb{P}_i}{\boldsymbol{r}_k^T \mathbb{P}_j})\mathbb{P}_j, & \text{if} \quad i \neq j \\ (\frac{\sigma}{\boldsymbol{r}_k^T \mathbb{P}_j})\mathbb{P}_j, & \text{if} \quad i = j. \end{cases} \tag{18}$$

In order to minimize the volume of the new zonotope $\mathbb{Z}_{r,k}$ bounding the intersection, it is necessary to chose the optimal value of j_{opt} from $r+1$ possible choices of j:

$$j_{opt} = \arg \min_{0 \leq j \leq r} vol(\boldsymbol{p}_j \oplus T_j \boldsymbol{B}^r), \tag{19}$$

and in order to make this task less computationally demanding, the choice of j_{opt} can be done by:

$$j_{opt} = \arg \min_{0 \leq j \leq r} \det(T_j T_j^T). \tag{20}$$

The description of neurons uncertainty in the form of zonotope $\mathbb{Z}$ (15) allows to obtain the GMDH neural model-based robust fault detection scheme. The proposed approach enable perform robust fault detection on the basis of the output adaptive thresholds according to the scheme presented in Fig. 2.

The proposed technique relies on the calculation of the model output uncertainty interval based on the estimated parameters which values are known at some confidence level.

$$\hat{\boldsymbol{y}}_k^m \leq \hat{\boldsymbol{y}}_k \leq \hat{\boldsymbol{y}}_k^M, \tag{21}$$

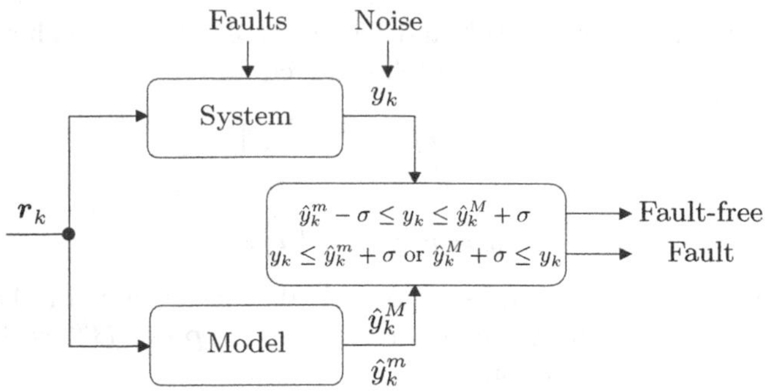

Fig. 2. Robust fault detection with the adaptive time-variant threshold

On the basis of (12-13) the neuron output uncertainty interval can be defined as follows:

$$\boldsymbol{r}_k^T \boldsymbol{p} - \|\mathbb{P}^T \boldsymbol{r}_k\|_1 \le \hat{\boldsymbol{y}}_k \le \boldsymbol{r}_k^T \boldsymbol{p} + \|\mathbb{P}^T \boldsymbol{r}_k\|_1, \tag{22}$$

Additionally, as the measurement of the diagnosed system responses $\boldsymbol{y}_k$ are corrupted by the noise, it is necessary to add the boundary values of the output error ε_k^m and ε_k^M to the model output uncertainty interval. In proposed approach these values are overabounded be σ according to relation (5). Since (22) describes a neuron output uncertainty interval, the system output adaptive threshold will satisfy:

$$\boldsymbol{r}_k^T \boldsymbol{p} - \|\mathbb{P}^T \boldsymbol{r}_k\|_1 - \sigma \le \boldsymbol{y}_k \le \boldsymbol{r}_k^T \boldsymbol{p} + \|\mathbb{P}^T \boldsymbol{r}_k\|_1 + \sigma, \tag{23}$$

and it is presented in Fig. 3. Defined by (23) output adaptive threshold should contain the real system response in the fault free mode.

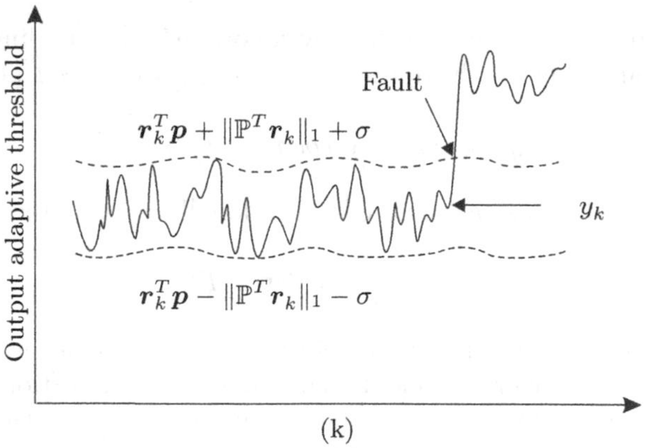

Fig. 3. Robust fault detection via output adaptive threshold

3 Robust Fault Detection of Tunnel Furnace

The objective of this section is to design a GMDH model according to the approach described in the previous section and its application to the robust fault detection of the tunnel furnace. On the beginning the effectiveness of the LMS, BEA, OBE and ZBA in the task of parameters estimation of the neurons in the GMDH network is presented. Let us consider the following static system:

$$y_k = p_1 \sin(u_k^2) + p_2 u_k + \varepsilon_k,$$

where the nominal values of the parameters are $\boldsymbol{p} = [1.5, 0.5]^T$, the input data u_k and the noise ε_k, $k = 1, \ldots, n_{\mathcal{T}}$, are generated according to the uniform distribution, i.e. $u_k \in \mathcal{U}(0, 2)$ and $\varepsilon_k \in \mathcal{U}(-0.05, 0.1)$, where $\mathcal{U}(\cdot)$ stands for the uniform distribution. It should be noticed that the noise ε_k does not satisfy (3). The problem is to obtain the parameter estimate $\hat{\boldsymbol{p}}$ and the corresponding feasible parameter set using the set of input-output measurements $\{u_k, y_k)\}_{k=1}^{n_{\mathcal{T}}=200}$. In order to obtain the feasible parameter set for the LMS, the F-test [26] was employed and a 95% confidence region was obtained, whilst in the case of the BEA that initial boundary values of the disturbances $\varepsilon_k^m = -0.1$, $\varepsilon_k^M = 0.2$ were assumed. In the case of the LMS the parameters uncertainty was obtained with the confidence level of 95%, whereas the application of the BEA, OBE and ZBA allows to calculate the parameters uncertainty with the confidence level of 100%. Parameters estimates and their minimal and maximal values of the parameter estimates for all methods are presented in Tab. 1. In Fig. 4 the parameters estimates and the feasible parameters sets obtained with the ZBA is presented. The results show that the parameters estimates obtained with the application of the BEA, OBE and ZBA are similar to the nominal parameters $\boldsymbol{p} = [1.5, 0.5]^T$, opposite to parameters estimates calculated with the LMS which is the worst method. Indeed, as it can be seen in Tab. 1, the feasible parameter set for the LMS does not even contain the nominal parameter values. It follows from the fact that the condition (3) concerning noise is not fulfilled. Moreover, from the results presented in Tab. 1 it is clear that the BEA is superior to the other methods. This method provides the parameters estimate which is the closest to the nominal parameters values. Furthermore, the obtained feasible parameter set has the smallest size. Unfortunately, this accuracy is occupied by a large computational cost. The parameters estimates obtained with the application of

Table 1. Parameters estimates obtained with the LMS, BEA, OBE and ZBA

$\hat{\boldsymbol{p}}$	$\hat{p}_1$	$[\hat{p}_1^{\min}, \hat{p}_1^{\max}]$	$\hat{p}_2$	$[\hat{p}_2^{\min}, \hat{p}_2^{\max}]$
LMS	1.5170	[1.4803, 1.5537]	0.5265	[0.5068, 0.5461]
BEA	1.5004	[1.4992, 1.5016]	0.5001	[0.4995, 0.5008]
OBE	1.5007	[1.4473, 1.5541]	0.4925	[0.4649, 0.5201]
ZBA	1.5111	[1.4751, 1.5472]	0.5153	[0.4980, 0.5326]

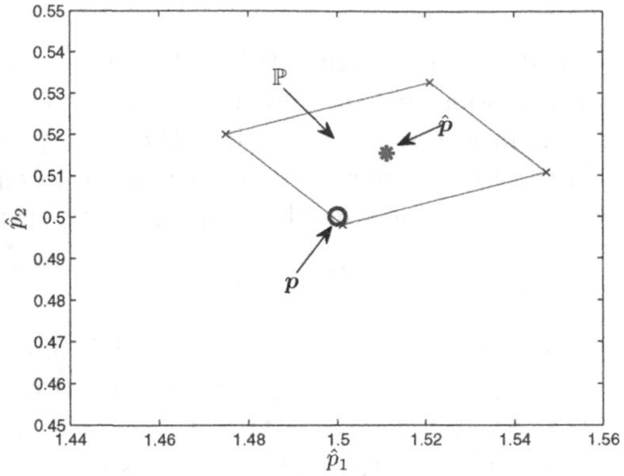

Fig. 4. Parameters estimate and feasible parameter set obtained via ZBA

the OBE and ZBA are quite accurate, however, they are not so good as those obtained with the application of the BEA. It should be emphasized that feasible parameter set obtained with the OBE algorithm is quite large. In the case of the GMDH neural network that is composed of many neurons, such uncertainty can be accumulated and hence, the performance of the network can be significantly degraded. From this reason, the ZBA seem to be attractive trade-off between the accuracy offered by the BEA and performance given by the OBE algorithm.

The obtained feasible parameter sets describes the parametrical uncertainty of the each neuron in the GMDH network. On the basis of the obtained matrixes $\mathbb{P}$ it is possible to calculate the neuron uncertainty in the form of the confidence interval of the model output. Taking into account these results and the disturbances enable calculation of the output adaptive thresholds which can be applied to robust fault diagnosis. The output adaptive thresholds were calculated for the following validation signal:

$$u_{1,k} = \sin(2\pi(k)/250) \quad \text{for} \quad k = 1, \ldots, 100,$$

$$u_{2,k} = 0.8\sin(2\pi(k/250)) + 0.2\sin(2\pi(k/25)) \quad \text{for} \quad k = 101, \ldots, 200.$$

For the LMS and ZBA the output adaptive thresholds together with system output are depicted in Fig. 5. The results obtained with the LMS indicate that the system output adaptive threshold does not contain the system output calculated based on the nominal parameters $\boldsymbol{p}$. This result is consistent with the expectations because the parameters estimate obtained with the application of the LMS was biased and the feasible parameter set does not contain the nominal parameters. These results obtained with the application of the ZBA shows that this approach provides better results than OBE algorithm.

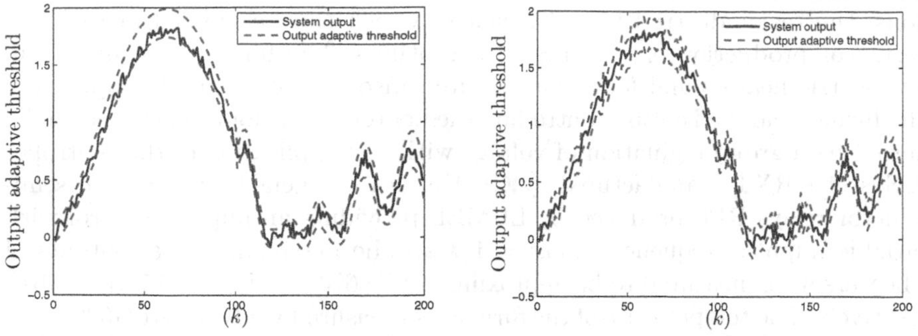

Fig. 5. System output and output adaptive threshold for LMS (left) and ZBA (right)

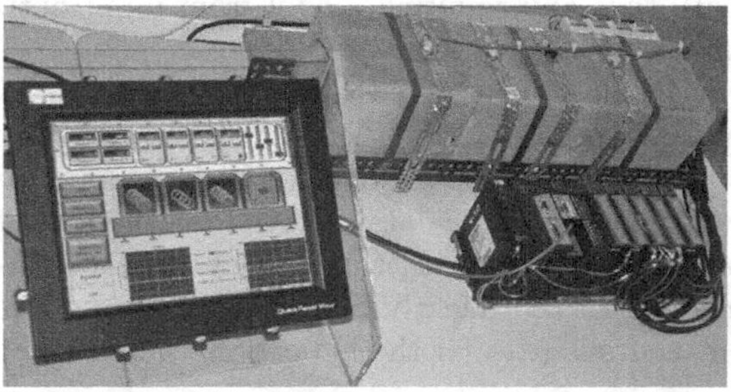

Fig. 6. Laboratory model of a tunnel furnace

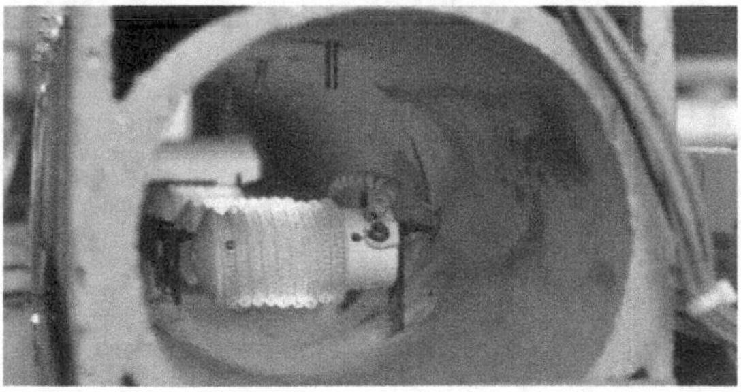

Fig. 7. Interior of a tunnel furnace

The above presented illustrative example shows that the ZBA can be successively applied to estimate parameters and uncertainty of the GMDH neural model. In the subsequent part of this section the developed approach is applied to the robust fault detection of the tunnel furnace (cf. Figs. 6-7).

The considered tunnel furnace is designed to mimic, in the laboratory conditions, the real industrial tunnel furnaces, which can be applied in the food industry or production of ceramics among others. The furnace is equipped in three electric heaters and four temperature sensors. The required temperature of the furnace can be kept by controlling the heaters behaviour. This task can be achieved by a group regulation of voltage with the application of the controller PACSystems RX3i manufactured by GE Fanuc Intelligent Platforms and semiconductor relays RP6 produced by LUMEL providing an impulse control with a variable impulse frequency $fmax = 1\ Hz$. The maximum power outputs of the heaters were measured to be approximately $686W$, $693W$ and $756W \pm 20W$, respectively. The temperature of the furnace is measured via IC695ALG600 module with Pt100 Resistive Thermal Devices (RTDs) with an accuracy of $\pm 0.7°C$. The visualisation of the behaviour of the tunnel furnace is made by Quickpanel CE device from GE Fanuc Intelligent Platforms. It is worth to note that the considered system is a distributedparameter one (i.e., a system whose state space is infinite dimensional), thus any resulting model from input-output data will be at best an approximation.

The tunnel furnace can be considered as a three-input and four-output system $(t_1, t_2, t_3, t_4) = f(u_1, u_2, u_3)$, where the $t_1, \ldots, t_4$ represent measurements of the temperatures from sensors and values $u_1, \ldots, u_3$ denote the input voltages allowing to control the heaters. The data set used for the identification consists of 2600 samples and was filtered with the Matlab Signal Processing Toolbox. The output signals were scaled linearly taking into consideration the response range of the output neurons (e.g. for the hyperbolic tangent neurons this range is $[-1, 1]$.

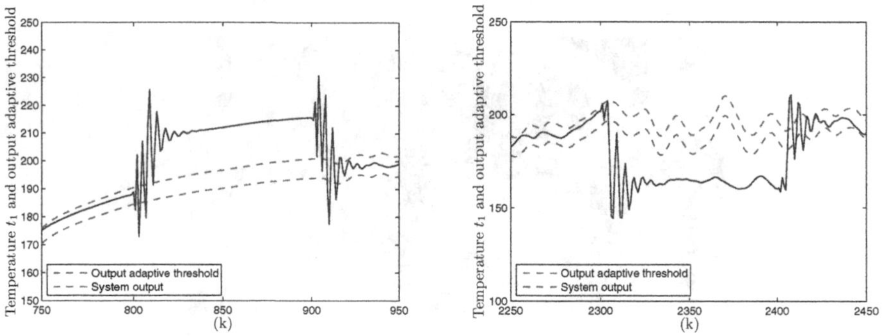

Fig. 8. Detection of faulty temperature sensor (left) and faulty electric heater (right) via output adaptive thresholds

The parameters of the dynamic neurons are estimated with the application of the ZBA presented in Sect. 2. The selection of the best performing neurons in the terms of their processing accuracy is realized with the application of the soft selection method [13] based on the sum squared error evaluation criterion. After the synthesis of the GMDH model, it is possible to employ it for robust fault detection. The detection of the faulty sensor for the temperature t_1 (simulated during 100sec.) and the faulty first electric heater of the tunnel furnace via output adaptive threshold are presented in Fig. 8. As it can be seen the faults are detected when the measurements of temperature t_1 cross the output adaptive thresholds.

4 Conclusions

One of the crucial problems occurring during system identification with the application of the ANNs is the choice of appropriate neural model architecture. The GMDH approach solves this problem and allows to chose such an architecture directly only on the basis of measurements data. Unfortunately, irrespective of the identification method used, there is always the problem of model uncertainty, i.e. the model-reality mismatch. The application of the ZBA to parameters estimation of the GMDH model also allows obtaining a neural model uncertainty. The calculation of the GMDH model uncertainty in the form of the output adaptive threshold allows perform robust fault detection of industrial systems.

Acknowledgments. The work was supported by the National Science Centre of Poland under grant: 2011-2014

References

1. Ding, S.: Model-based Fault Diagnosis Techniques: Design Schemes, Algorithms, and Tools. Springer, Heidelberg (2008)
2. Korbicz, J., Kościelny, J., Kowalczuk, Z., Cholewa, W. (eds.): Fault diagnosis. Models, Artificial Intelligence, Applications. Springer, Berlin (2004)
3. Korbicz, J., Kościelny, J. (eds.): Modeling, Diagnostics and Process Control: Implementation in the DiaSter System. Springer, Berlin (2011)
4. Mrugalski, M., Witczak, M., Korbicz, J.: Confidence estimation of the multi-layer perceptron and its application in fault detection systems. Engineering Applications of Artificial Intelligence 21(6), 895–906 (2008)
5. Mrugalski, M., Witczak, M.: State-space GMDH neural networks for actuator robust fault diagnosis. Advances in Electrical and Computer Engineering 12(3), 65–72 (2012)
6. Kościelny, J., Bartyś, M., Syfert, M.: Method of multiple fault isolation in large scale systems. IEEE Transactions on Control Systems Technology 20(5), 1302–1310 (2012)
7. Noura, H., Theilliol, D., Ponsart, J., Chamseddine, A.: Fault-tolerant Control Systems: Design and Practical Applications. Springer, London (2009)
8. Tong, S., Yang, G., Zhang, W.: Observer-based fault-tolerant control againts sensor failures for fuzzy systems with time delays. International Journal of Applied Mathematics and Computer Science 21(4), 617–628 (2011)

9. Pedro, J., Dahunsi, O.: Neural network based feedback linearization control of a servo-hydraulic vehicle suspension system. International Journal of Applied Mathematics and Computer Science 21, 137–147 (2011)
10. Krzysztof, P., Korbicz, J.: Nonlinear model predictive control of a boiler unit: A fault tolerant control study. International Journal of Applied Mathematics and Computer Science 22(1), 225–237 (2012)
11. Ivakhnenko, A., Mueller, J.: Self-organizing of nets of active neurons. System Analysis Modelling Simulation 20, 93–106 (1995)
12. Mrugalski, M., Korbicz, J., Patton, R.J.: Robust fault detection via GMDH neural networks. In: Proceedings of 16th IFAC World Congress (2005)
13. Korbicz, J., Mrugalski, M.: Confidence estimation of GMDH neural networks and its application in fault detection system. International Journal of System Science 39(8), 783–800 (2008)
14. Mrugalski, M., Korbicz, J.: GMDH neural networks. In: Wilamowski, B., Irwin, J. (eds.) Intelligent systems: The Industrial Electronics Handbook, 2nd edn., vol. 5, pp. 8-1–8-21. CRC Press, Taylor Francis Group, Boca Raton (2011)
15. Mrugalski, M.: An unscented kalman filter in designing dynamic GMDH neural networks for robust fault detection. International Journal of Applied Mathematics and Computer Science 23(1), 157–169 (2013)
16. Witczak, M., Korbicz, J., Mrugalski, M., Patton, R.: A GMDH neural network based approach to robust fault detection and its application to solve the damadics benchmark problem. Control Engineering Practice 14(6), 671–683 (2006)
17. Patan, K., Witczak, M., Korbicz, J.: Towards robustness in neural network based fault diagnosis. International Journal of Applied Mathematics and Computer Science 18(4), 443–454 (2008)
18. Walter, E., Pronzato, L.: Identification of Parametric Models from Experimental Data. Springer, Berlin (1997)
19. Cerone, V., Piga, D., Regruto, D.: Bounded error identification of hammerstein systems through sparse polynomial optimization. Automatica (2012)
20. Mrugalska, B., Kawecka-Endler, A.: Practical application of product design method robust to disturbances. Human Factors and Ergonomics in Manufacturing & Service Industries (2012)
21. Alamo, T., Bravo, J., Camacho, E.: Guaranteed state estimation by zonotopes. Automatica 41(6), 1035–1043 (2005)
22. Bravo, J., Alamo, T., Camacho, E.: Bounded error identification of systems with time-varying parameters. IEEE Transactions on Automatic Control 51(7), 1144–1150 (2006)
23. Mrugalski, M., Witczak, M.: Parameter estimation of dynamic GMDH neural networks with the bounded-error technique. Journal Applied Computer Science 10(1), 77–90 (2002)
24. Mrugalski, M., Arinton, E., Korbicz, J.: Dynamic GMDH type neural networks. In: Rutkowski, L., Kacprzyk, J. (eds.) Neural Networks and Soft Computing. Advances in Soft Computing, pp. 698–703 (2003); 6th International Conference on Neural Networks and Soft Computing, Zakopane, Poland, June 11-15 (2002)
25. Mrugalski, M., Korbicz, J.: Least mean square vs. Outer bounding ellipsoid algorithm in confidence estimation of the GMDH neural networks. In: Beliczynski, B., Dzielinski, A., Iwanowski, M., Ribeiro, B. (eds.) ICANNGA 2007. LNCS, vol. 4432, pp. 19–26. Springer, Heidelberg (2007)
26. Atkinson, A.C., Donev, A.N.: Optimum Experimental Designs. Oxford University Press, New York (1992)

Model Predictive Control of the Two Rotor Aero-Dynamical System Using State Space Neural Networks with Delays

Andrzej Czajkowski and Krzysztof Patan

Institute of Control and Computation Engineering,
University of Zielona Góra, ul. Podgórna 50, 65-246 Zielona Góra
{a.czajkowski,k.patan}@issi.uz.zgora.pl
http://www.issi.uz.zgora.pl

Abstract. This paper deals with the application of state space neural network model with delays to design a model predictive control for a laboratory stand of the Two Rotor Aero-dynamical system. The work describes approach based on the so-called instantaneous linearisation of the already trained nonlinear state space model of the system. With obtained linear model it is possible to derive a vector of future controls based on the minimisation of the cost function within one optimisation window. Repeating procedure in each step of simulation and applying the obtained control signal allows for efficiently control of the nonlinear systems. All data used in experiments is obtain from the real-time laboratory stand which is working in Matlab/Simulink RTW environment.

Keywords: State Space Neural Networks, Model Predictive Control, Two Rotor Aerodynamical System, Nonlinear System Modeling, MIMO system, Linearisation.

1 Introduction

Currently, the rapid industrial development has caused the expansion of automatic control techniques. The very well known technique of Model Predictive Control (MPC) is being increasingly used in industrial solutions. Also over the last years, many papers and applications described in the open literature are focused on MPC ([8], [15], [13], [9]). This technique is widely used in industrial processes due to its ability to handle constraints explicitly. Furthermore it can be practically implemented and can be applied to multi-variable processes naturally. Those advantages makes this method the most widely used advanced control technology in industry. The main drawback of MPC is that the computational time necessary to calculate a control signal at each sampling time can be much greater than the amount of time that is available in a real time setting. The consequence is that MPC is typically applied only to slow processes, such as those which occur in process control of refineries and petrochemical industries. Although the major customers of MPC remain such systems, the rapid growth of automation industry (especially faster CPU's and bigger memory resources)

J. Korbicz and M. Kowal (eds.), *Intelligent Systems in Technical and Medical Diagnostics*, Advances in Intelligent Systems and Computing 230,
DOI: 10.1007/978-3-642-39881-0_9, © Springer-Verlag Berlin Heidelberg 2014

allow much more often to use such techniques in very dynamic real-time objects. So the MPC becomes more attractive to much wider range of industry systems and require much more research in that area. In this paper such very fast approach based on linearised neural state space model is proposed.

The paper is organized as follows. The general description of the state space neural networks with delays, its linearisation and preparation to use in the MPC is described in Section 2. Section 3 presents a Model Predictive Control strategy and the proposed algorithm of calculating control signal. Section 4 describes laboratory installation used in experiments, while experimental results are included in Section 5.

2 State Space Neural Network with Delays

The crucial part in designing the MPC is the model of the plant. In this work the State Space Neural Network (SSNN) which are a very important class of dynamic neural networks are used. The typical structure of such network in this paper is modified by adding both delayed inputs and outputs to the overall neural model. The structure of such neural network is depicted in Fig. 1. The output of the hidden layer is fed back to the input layer through a bank of unit delays. The number of state unit delays determines the order of the model. In general, the user decides how many neurons are used to produce feedback. Let $\boldsymbol{u}(k) \in \mathbb{R}^n$ be the input vector which is equal to $\boldsymbol{u}(k) = [\boldsymbol{u}_s(k), \boldsymbol{u}_s(k-1), \ldots, \boldsymbol{u}_s(k-m), \bar{\boldsymbol{y}}(k-1), \bar{\boldsymbol{y}}(k-2), \ldots, \bar{\boldsymbol{y}}(k-n))]^T$ where $\boldsymbol{u}_s(k)$ is the standard input vector at time k, $\bar{\boldsymbol{x}}(k) \in \mathbb{R}^q$ - the output of the hidden layer at time k, and $\bar{\boldsymbol{y}}(k) \in \mathbb{R}^m$ - the output vector. The state space representation of the neural model considered is described by the equations

$$\begin{aligned} \bar{\boldsymbol{x}}(k+1) &= \bar{g}(\bar{\boldsymbol{x}}(k), \boldsymbol{u}(k)) \\ \bar{\boldsymbol{y}}(k) &= \boldsymbol{C}\bar{\boldsymbol{x}}(k) \end{aligned}, \tag{1}$$

where $\bar{g}(\cdot)$ is a nonlinear function characterizing the hidden layer, and $\boldsymbol{C}$ represents synaptic weights between hidden and output neurons. Introducing the

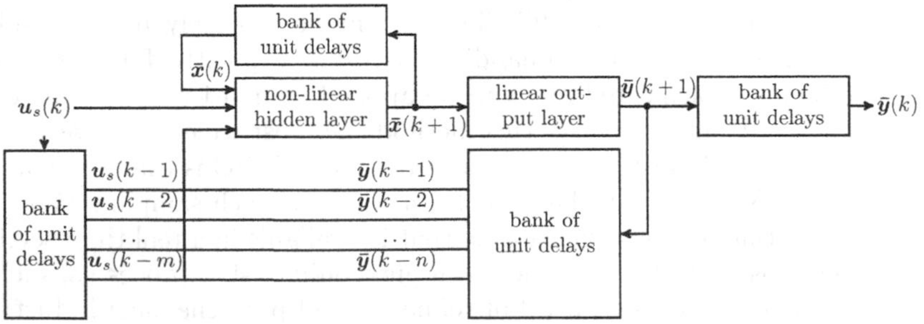

Fig. 1. Block scheme of the state space neural network

weight matrix between input and hidden layers $\boldsymbol{W}^u$ and the matrix of recurrent links $\boldsymbol{W}^x$ the representation (1) can be rewritten in the following form:

$$\begin{aligned} \bar{\boldsymbol{x}}(k+1) &= h(\boldsymbol{W}^x\bar{\boldsymbol{x}}(k) + \boldsymbol{W}^u\boldsymbol{u}(k)) \\ \bar{\boldsymbol{y}}(k) &= \boldsymbol{C}\bar{\boldsymbol{x}}(k) \end{aligned}, \tag{2}$$

where h stands for the activation function of the hidden neurons. In most cases the hyperbolic tangent activation function is selected giving pretty well modelling result. For the state space model the outputs of hidden neurons which are fed back are, in general, unknown during training. Therefore, state space neural models can be trained only by minimizing the simulation error. If state measurements are available, the training can be carried out much easier, using the series-parallel identification scheme similarly as for the external dynamic approach (the feedforward network with tapped delay lines). In spite of this inconvenience, state space models have a number of advantages, contrary to other recurrent neural networks:

- The number of states (model order) can be selected independently from the number of hidden neurons. In the recurrent networks, e.g. Williams-Zipser, Elman, locally recurrent networks, the number of neurons directly influences the model order, which significantly impedes the modelling phase.
- Model states are easily accessible from the outside (they feed the network input). This property can be useful when state measurements are available at some time instants.
- State space neural models are useful in the fault tolerant control framework as they can be used to determine the approximation of a fault effect. As the fault effect can be represented in the state space one can handle different kind of faults including multiplicative and additive ones.

2.1 Model Linearisation

In this paper, to obtain the control law in the model predictive control manner the instantaneous linearisation is applied which means that nonlinear state space model is linearised at each time instant and then obtained linear model is used for control law calculation. Linearisation can be carried out by expanding the model into Taylor series around the current operating point $(\boldsymbol{x}, \boldsymbol{u}) = (\boldsymbol{x}(\tau), \boldsymbol{u}(\tau))$, rejecting the nonlinear components. The state space model expanded into the Taylor series of first order have the form

$$\begin{aligned} h(\boldsymbol{x}(k), \boldsymbol{u}(k)) &= h(\boldsymbol{x}(\tau), \boldsymbol{u}(\tau)) + \frac{\partial h}{\partial \boldsymbol{x}}\Big|_{(x,u)} \Delta x + \frac{\partial h}{\partial \boldsymbol{u}}\Big|_{(x,u)} \Delta u \\ &= h(\boldsymbol{x}(\tau), \boldsymbol{u}(\tau)) + h' \boldsymbol{W}^x(\boldsymbol{x}(k) - \boldsymbol{x}(\tau)) \\ &\quad + h' \boldsymbol{W}^u(\boldsymbol{u}(k) - \boldsymbol{u}(\tau)) \end{aligned} \tag{3}$$

As a result, the linear state space model in the form

$$\begin{aligned} \bar{\boldsymbol{x}}(k+1) &= \boldsymbol{A}\bar{\boldsymbol{x}}(k) + \boldsymbol{B}\boldsymbol{u}(k) + \boldsymbol{E} \\ \bar{\boldsymbol{y}}(k) &= \boldsymbol{C}\bar{\boldsymbol{x}}(k) \end{aligned}, \tag{4}$$

is obtained, where $\boldsymbol{A} = h'\boldsymbol{W}^x$, $\boldsymbol{B} = h'\boldsymbol{W}^u$, $\boldsymbol{E} = \boldsymbol{x}(\tau) - \boldsymbol{A}\bar{\boldsymbol{x}}(\tau-1) - \boldsymbol{B}\boldsymbol{u}(\tau-1)$. Symbol h' represents the first derivative of the activation function. If one use hyperbolic tangent as the activation function this derivative can be simply calculated as

$$h' = 1 - \tanh^2 . \tag{5}$$

This is very useful property as the linearisation can be performed very quickly not influencing significantly on the computation burden of simulations carried out in real-time.

2.2 Augmented Linear Model

The model predictive control presented in this paper in the form presented in Section 3, cannot be used with the linearised state space model with delays because it does not take into consideration delays of inputs and outputs. Using this method for not modified model would lead to calculation of control corrections for past controls which obviously cannot be achieved in real time simulation. In this paper it was decided to modify the matrices describing linear model in such manner that delayed inputs and outputs will be part of the state vector. The original state vector $\bar{\boldsymbol{x}}(k)$ and control vector which can be presented as:

$$\boldsymbol{u}(k) = [\boldsymbol{u}_s(k)\ \boldsymbol{u}_s(k-1)\ \ldots\ \boldsymbol{u}_s(k-m)\ \bar{\boldsymbol{y}}(k-1)\ \bar{\boldsymbol{y}}(k-2)\ \ldots\ \bar{\boldsymbol{y}}(k-n)\ ]^T , \tag{6}$$

need to be modified in such way that the model input vector is $\boldsymbol{u}_s(k)$ and the new state vector take the forms:

$$\bar{\boldsymbol{x}}_m(k) = \begin{bmatrix} \bar{\boldsymbol{x}}(k) \\ \boldsymbol{u}_s(k) \\ \boldsymbol{u}_s(k-1) \\ \vdots \\ \boldsymbol{u}_s(k-m) \\ \bar{\boldsymbol{y}}(k-1) \\ \bar{\boldsymbol{y}}(k-2) \\ \vdots \\ \bar{\boldsymbol{y}}(k-n) \end{bmatrix} . \tag{7}$$

With such modification of these vectors the model matrices need to be joint as follows:

$$
A_m = \begin{bmatrix} \multicolumn{3}{c}{A \qquad B_1} \\ 0\,0\,0 \ldots 0\,0\,0 & \ldots & 0 \\ 0\,I\,0 \ldots 0\,0\,0 & \ldots & 0 \\ 0\,0\,I \ldots 0\,0\,0 & \ldots & 0 \\ \vdots & & \vdots \\ 0\,0\,0 \ldots I\,0\,0 & \ldots & 0 \\ \multicolumn{3}{c}{CA \qquad CB_1} \\ 0\,0\,0 \ldots 0\,0\,I & \ldots & 0 \\ \vdots & & \vdots \\ 0\,0\,0 \ldots 0\,0\,0 & \ldots & I \end{bmatrix}, B_m = \begin{bmatrix} B_2 \\ I \\ 0 \\ 0 \\ \vdots \\ 0 \\ CB_2 \\ 0 \\ \vdots \\ 0 \end{bmatrix}, C_m = \begin{bmatrix} I \\ 0 \\ 0 \\ 0 \\ \vdots \\ 0 \\ 0 \\ 0 \\ \vdots \\ 0 \end{bmatrix}^T, \tag{8}
$$

where $B_1 = B[0\ I \ldots I]^T$, $B_2 = B[I\ 0 \ldots 0]^T$ and $B = [B_2\ B_1]$, the length of that vectors is equal to the original input control which depends on number of the delayed inputs and outputs and number of the system inputs. With such transformed system matrices it is possible to apply the method presented in Section 3.

3 Model Predictive Control within One Optimisation Window Based on Receding Horizon Control Principle

MPC is a specific control strategy which uses a model of the process to derive the control signal by minimizing some objective function over a finite receding horizon. The process model is used to predict future plant outputs based on past and current outputs as well as future control signals. These are calculated through the minimization of a cost function. Predictive control algorithms are very interesting because they are able to consider constraints imposed on both controls and process outputs. In order to design MPC for a given problem, two elements are of crucial importance: the model of a process working in normal operating conditions and the robust optimization procedure.

In this paper the predictive control is carried out within one optimisation window based on the method descried in [16]. Described method is enhanced with the technique of instantaneous linearisation which should lead to better results for nonlinear systems.

To carry out the optimisation procedure let consider the following state-space linear model which in this paper is obtained through instantaneous linearisation of the nonlinear neural network state space model:

$$
\begin{aligned} \bar{x}_m(k+1) &= A_m \bar{x}_m(k) + B_m u_s(k) \\ \bar{y}(k) &= C_m \bar{x}_m(k) \end{aligned}, \tag{9}
$$

Taking a difference operation on both sides of (9), we obtain that

$$
\bar{x}_m(k+1) - \bar{x}_m(k) = A_m(\bar{x}_m(k) - \bar{x}_m(k-1)) + B_m(u_s(k) - u_s(k-1)) \tag{10}
$$

Now, the difference of the state can be denoted as:

$$\Delta \bar{\boldsymbol{x}}_m(k+1) = \bar{\boldsymbol{x}}_m(k+1) - \bar{\boldsymbol{x}}_m(k), \tag{11}$$

and the difference of the control variable is equal:

$$\Delta \boldsymbol{u}_s(k+1) = \boldsymbol{u}_s(k+1) - \boldsymbol{u}_s(k) \tag{12}$$

With (11) and (12) the (10) can be rewritten as:

$$\Delta \bar{\boldsymbol{x}}_m(k+1) = \boldsymbol{A}_m \Delta \bar{\boldsymbol{x}}_m(k) + \boldsymbol{B}_m \Delta \boldsymbol{u}_s(k). \tag{13}$$

Now, it is necessary to chose a new state variable vector:

$$\bar{\boldsymbol{x}}_a(k) = [\Delta \bar{\boldsymbol{x}}_m(k)^T \; \bar{\boldsymbol{y}}(k)^T]^T. \tag{14}$$

Knowing, that the following is true:

$$\begin{aligned} \bar{\boldsymbol{y}}(k+1) - \bar{\boldsymbol{y}}(k) &= \boldsymbol{C}_m(\bar{\boldsymbol{x}}(k+1) - \bar{\boldsymbol{x}}(k)) \\ &= \boldsymbol{C}_m \Delta \bar{\boldsymbol{x}}(k+1) = \boldsymbol{C}_m \boldsymbol{A}_m \Delta \bar{\boldsymbol{x}}(k) + \boldsymbol{C}_m \boldsymbol{B}_m \Delta \boldsymbol{u}_s(k), \end{aligned} \tag{15}$$

we can use it with the (13) to obtain the following state-space model:

$$\begin{aligned} \begin{bmatrix} \Delta \bar{\boldsymbol{x}}_m(k+1) \\ \bar{\boldsymbol{y}}(k+1) \end{bmatrix} &= \begin{bmatrix} \boldsymbol{A}_m & \boldsymbol{0} \\ \boldsymbol{C}_m \boldsymbol{A}_m & \boldsymbol{1} \end{bmatrix} \bar{\boldsymbol{x}}_m(k) + \begin{bmatrix} \boldsymbol{B}_m \\ \boldsymbol{C}_m \boldsymbol{B}_m \end{bmatrix} \Delta \boldsymbol{u}_s(k) \\ \bar{\boldsymbol{y}}(k) &= [\boldsymbol{0} \; \boldsymbol{1}] \begin{bmatrix} \Delta \bar{\boldsymbol{x}}_m(k) \\ \bar{\boldsymbol{y}}(k) \end{bmatrix}. \end{aligned} \tag{16}$$

From that we can form the following matrices:

$$\boldsymbol{A_a} = \begin{bmatrix} \boldsymbol{A}_m & \boldsymbol{0} \\ \boldsymbol{C}_m \boldsymbol{A}_m & \boldsymbol{1} \end{bmatrix}, \boldsymbol{B_a} = \begin{bmatrix} \boldsymbol{B}_m \\ \boldsymbol{C}_m \boldsymbol{B}_m \end{bmatrix}, \boldsymbol{C_a} = [\boldsymbol{0} \; \boldsymbol{1}] \tag{17}$$

which are called the augmented model. Assuming that N_p is prediction horizon and N_c is control horizon then using the augmented state-space model, the future state variables are calculated sequentially using the set of future control parameters:

$$\begin{aligned} \bar{\boldsymbol{x}}_a(k+1) &= \boldsymbol{A_a} \bar{\boldsymbol{x}}_a(k) + \boldsymbol{B_a} \Delta \boldsymbol{u}_s(k) \\ \bar{\boldsymbol{x}}_a(k+2) &= \boldsymbol{A_a}^2 \bar{\boldsymbol{x}}_a(k) + \boldsymbol{A_a} \boldsymbol{B_a} \Delta \boldsymbol{u}_s(k) + \boldsymbol{B_a} \Delta \boldsymbol{u}_s(k+1) \\ &\vdots \\ \bar{\boldsymbol{x}}_a(k+N_p) &= \boldsymbol{A_a}^{N_p} \bar{\boldsymbol{x}}_a(k) + \boldsymbol{A_a}^{N_p-1} \boldsymbol{B_a} \Delta \boldsymbol{u}_s(k) + \boldsymbol{A_a}^{N_p-2} \boldsymbol{B_a} \Delta \boldsymbol{u}_s(k+1) \\ &\quad + \ldots + \boldsymbol{A_a}^{N_p-N_c} \boldsymbol{B_a} \Delta \boldsymbol{u}_s(k+N_c-1) \end{aligned} \tag{18}$$

From the predicted state variables, the predicted output variables are, by substitution

$$\begin{aligned} \bar{\boldsymbol{y}}(k+1) &= \boldsymbol{C_a} \boldsymbol{A_a} \bar{\boldsymbol{x}}_a(k) + \boldsymbol{C_a} \boldsymbol{B_a} \Delta \boldsymbol{u}_s(k) \\ \bar{\boldsymbol{y}}(k+2) &= \boldsymbol{C_a} \boldsymbol{A_a}^2 \bar{\boldsymbol{x}}_a(k) + \boldsymbol{C_a} \boldsymbol{A_a} \boldsymbol{B_a} \Delta \boldsymbol{u}_s(k) + \boldsymbol{C_a} \boldsymbol{B_a} \Delta \boldsymbol{u}_s(k+1) \\ &\vdots \\ \bar{\boldsymbol{y}}(k+N_p) &= \boldsymbol{C_a} \boldsymbol{A_a}^{N_p} \bar{\boldsymbol{x}}_a(k) + \boldsymbol{C_a} \boldsymbol{A_a}^{N_p-1} \boldsymbol{B_a} \Delta \boldsymbol{u}_s(k) \\ &\quad + \boldsymbol{C_a} \boldsymbol{A_a}^{N_p-2} \boldsymbol{B_a} \Delta \boldsymbol{u}_s(k+1) + \ldots \\ &\quad + \boldsymbol{C_a} \boldsymbol{A_a}^{N_p-N_c} \boldsymbol{B_a} \Delta \boldsymbol{u}_s(k+N_c-1) \end{aligned} \quad . \tag{19}$$

Now, to complete the optimisation procedure define a vectors:

$$\bar{\boldsymbol{Y}} = [\bar{\boldsymbol{y}}(k+1)\ \bar{\boldsymbol{y}}(k+2)\ \dots\ \bar{\boldsymbol{y}}(k+N_p)]^T, \tag{20}$$

$$\Delta \boldsymbol{U} = [\Delta \boldsymbol{u}_s(k)\ \Delta \boldsymbol{u}_s(k+1)\ \dots \Delta \boldsymbol{u}_s(k+N_c-1)]^T, \tag{21}$$

The (18) and (19) with defined vectors can be presented in compact matrix form:

$$\bar{\boldsymbol{Y}} = \boldsymbol{F}\bar{\boldsymbol{x}}_a(k) + \boldsymbol{\Phi}\Delta \boldsymbol{U} \tag{22}$$

where

$$\boldsymbol{F} = \begin{bmatrix} \boldsymbol{C_a A_a} \\ \boldsymbol{C_a A_a}^2 \\ \vdots \\ \boldsymbol{C_a A_a}^{N_p} \end{bmatrix}, \boldsymbol{\Phi} = \begin{bmatrix} \boldsymbol{C_a B_a} & \boldsymbol{0} & \boldsymbol{0} \\ \boldsymbol{C_a A_a B_a} & \boldsymbol{C_a B_a} & \boldsymbol{0} \\ \vdots & & \\ \boldsymbol{C_a A_a}^{N_p-1}\boldsymbol{B_a} & \boldsymbol{C_a A_a}^{N_p-2}\boldsymbol{B_a} & \boldsymbol{C_a A_a}^{N_p-N_c}\boldsymbol{B_a} \end{bmatrix}.$$

3.1 Optimisation

The final step to obtain a control signal is to define the cost function J that reflects the control objective as:

$$J = (\boldsymbol{R}_s - \bar{\boldsymbol{Y}})^T(\boldsymbol{R}_s - \bar{\boldsymbol{Y}}) + \Delta \boldsymbol{U}^T \bar{\boldsymbol{R}} \Delta \boldsymbol{U}, \tag{23}$$

where $\bar{\boldsymbol{R}}$ is a diagonal matrix in the form that $\bar{\boldsymbol{R}} = r_w \boldsymbol{I}_{N_c \times N_c}$ where r_w is used as a tuning parameter for the desired closed-loop performance. The $\boldsymbol{R}_s$ is the reference signal vector and is equal to $\boldsymbol{R}_s = [\boldsymbol{I}_{m\times m}\ \cdots\ \boldsymbol{I}_{m\times m}]^T_{m\times N_c m}\boldsymbol{r}(k)$ where m is number of outputs and $\boldsymbol{r}(k)$ is the reference signal at the time k. By using (22), J is expressed as:

$$\begin{aligned} J = (\boldsymbol{R}_s - \boldsymbol{F}\bar{\boldsymbol{x}}_a(k))^T(\boldsymbol{R}_s - \boldsymbol{F}\bar{\boldsymbol{x}}_a(k)) - 2\Delta \boldsymbol{U}^T \boldsymbol{\Phi}^T(\boldsymbol{R}_s - \boldsymbol{F}\bar{\boldsymbol{x}}_a(k)) \\ + \Delta \boldsymbol{U}^T(\boldsymbol{\Phi}^T\boldsymbol{\Phi} + \bar{\boldsymbol{R}})\Delta \boldsymbol{U}. \end{aligned} \tag{24}$$

The necessary condition to find such $\Delta \boldsymbol{U}$ that will minimize J is:

$$\frac{\partial J}{\partial \Delta \boldsymbol{U}} = 0, \tag{25}$$

and the first derivative of the cost function J is equal:

$$\frac{\partial J}{\partial \Delta \boldsymbol{U}} = -2\boldsymbol{\Phi}^T(\boldsymbol{R}_s - \boldsymbol{F}\bar{\boldsymbol{x}}_a(k)) + 2(\boldsymbol{\Phi}^T\boldsymbol{\Phi} + \bar{\boldsymbol{R}})\Delta \boldsymbol{U}. \tag{26}$$

from which we find the optimal solution for the control signal as

$$\Delta \boldsymbol{U} = (\boldsymbol{\Phi}^T\boldsymbol{\Phi} + \bar{\boldsymbol{R}})^{-1}\boldsymbol{\Phi}^T(\boldsymbol{R}_s - \boldsymbol{F}\bar{\boldsymbol{x}}_a(k)) \tag{27}$$

with the assumption that $(\boldsymbol{\Phi}^T\boldsymbol{\Phi} + \bar{\boldsymbol{R}})^{-1}$ exists. The matrix $(\boldsymbol{\Phi}^T\boldsymbol{\Phi} + \bar{\boldsymbol{R}})^{-1}$ is called the Hessian matrix in the optimization literature.

At each step the first N-elements of vector $\Delta \boldsymbol{U}$ are used as the inputs to the system where N is the number of system inputs.

4 Two Rotor Aero-dynamical System

The Two Rotor Aero-dynamical System (TRAS) is a laboratory set-up designed for control experiments. In certain aspects its behaviour resembles that of a helicopter. From the control point of view it exemplifies a high order non-linear system with significant cross-couplings. The system is controlled from a PC. Therefore it is delivered with hardware and software which can be easily mounted and installed in a laboratory. The laboratory setup consists of the mechanical unit with power supply and interface to a PC and the dedicated RTDAC/USB2 I/O board configured in the Xilinx technology. The software operates in real time under MS Windows XP/7 32-bit using MATLAB R2009/10,11, Simulink and RTW toolboxes. Real-time is supported by the RT-CON toolbox from INTECO. Control experiments are programmed and executed in real-time in the MATLAB/Simulink environment. The real-life installation is presented in Fig.2, and the scheme of the system is presented on Fig.3.

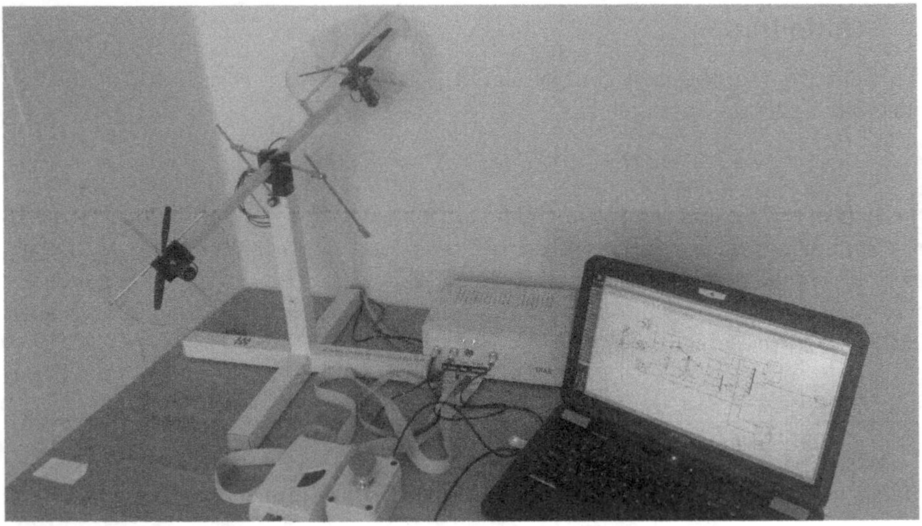

Fig. 2. Two Rotor Aero-dynamical System - laboratory setup

According to the TRAS instruction manual the equations describing the motion of the system can be written as follows:

$$\begin{aligned}\frac{d\Omega_v}{dt} =& \frac{l_m F_v(\omega_m) + g((A-B)cos\alpha_v - Csin\alpha_v) - a_1\Omega_v abs(\omega_v)}{J_v} \dots \\ &+ \frac{-\frac{1}{2}\Omega_h^2(A+B+C)sin2\alpha_v U_h k_{hv} + U_h k_{hv} - \Omega_v k_v}{J_v}\end{aligned} \tag{28}$$

$$\frac{d\alpha_v}{dt} = \Omega_v \tag{29}$$

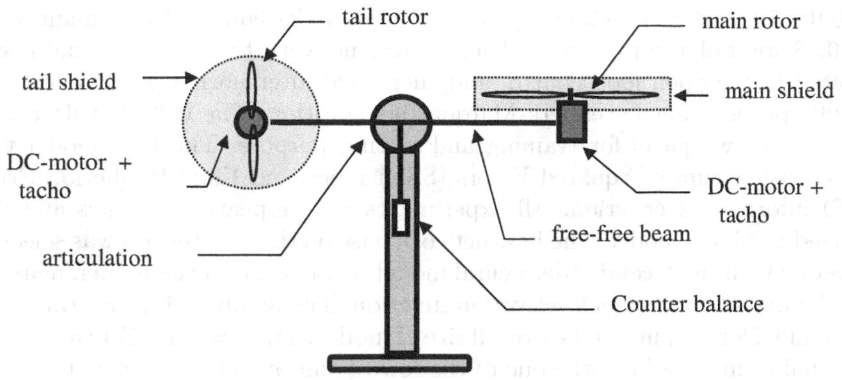

Fig. 3. Two Rotor Aero-dynamical System - parts scheme. Source: [5].

$$\frac{dK_h}{dt} = \frac{M_h}{J_h} = \frac{l_t F_h(\omega_t) cos\alpha_v - \Omega_h k_h + U_v k_{vh} - a_2 \Omega_h abs(\omega_h)}{D sin^2 \alpha_v + E cos^2 \alpha_v + F} \tag{30}$$

$$\frac{d\alpha_h}{dt} = \Omega_h, \; \Omega_h = \frac{K_h}{J_h(\alpha_v)}, \tag{31}$$

and two equations describing the motion of rotors:

$$I_h \frac{d\omega_h}{dt} = U_h - H_h^{-1}(\omega_h) \tag{32}$$

and

$$I_v \frac{d\omega_v}{dt} = U_v - H_v^{-1}(\omega_v) \tag{33}$$

where:
Ω_v - angular velocity (pitch velocity) of TRAS beam [rad/s];
Ω_h - angular velocity (azimuth velocity) of TRAS beam [rad/s];
ω_v - rotational speed of main rotor [rad/s];
ω_h - rotational speed of tail rotor [rad/s]
K_h - horizontal angular momentum [N m s];
M_h - horizontal turning torque [Nm];
I_h - moment of inertia of the main rotor.
I_v - moment of inertia of the tail rotor
The descriptions of the other symbols can be found in [5].

5 Experiments

5.1 Neural Modelling

The first step in the model predictive control design is the process modelling. To build a proper model, the training data describing the process under normal operating conditions is required. Based on spectral analysis of chirp signal from range 0 – 10 hz the sinusoidal signals with frequnecies equal 0.16 for main rotor and 0.48 for tail rotor were used for neural network training. The data was collected in the open loop control and then used after normalizing in training for 100 epochs using Levenberg-Marquardt algorithm. The collected data was divided into two parts for training and testing purpose. The best model was selected using Sum of Squared Errors (SSE) index and Final Prediction Error (FPE) information criterion. All experiments were repeated 10 times and the averaged results of model. The best network base on those criterions was selected as the one with the second order neural model consisting of seven hidden neurons with the hyperbolic tangent activation function. The number of delays was 4 for inputs and 2 for outputs so the overall size of model's input was 14. For this model both quality indexes have the one of the lowest values and also the structure is also not oversized which could burden computational cost of using such one network. The details of modelling and different neural structure comparisons are presented in other work of the authors (currently in review).

5.2 Model Predictive Control

The designed model of the plant allows to carry out experiments in which system is controlled by MPC controller. Initial experiments were carried out to choose best parameters of the MPC controller. Different values of N_p, N_c and tuning parameter r_w were tried and the best one were chosen based on average value of the cost function in that initial simulations. The parameters were chosen as follows: $N_p = 20$, $N_c = 5$ and $r_w = 0.7$. Lower value of r_w lead to unstable behaviour and higher to large control error.

Two control scenarios were considered. The results were compared with cross-coupled PID controller tuned by the manufacturer of the plant. All experiments were carried out with real-life object in real time simulation. In first scenario the sinusoid reference signal was applied and in second one the random step control signal was used. In first scenario after start-up of the simulation it was expected that the system was controlled in such way that it should quickly start to follow required sinusoidal reference trajectory. The results are presented in Fig. 5. Fig. 4 presents the calculated inputs which were later introduced to the system.

In the second scenario the random-step reference signal was used. The task of the control system was to achieve desired level and stabilise on that level minimising any oscillation. The results are presented in Fig. 7. Fig. 6 presents the calculated inputs which were later introduced to the system.

In both scenarios as one can see the control system worked quite efficiently. There was not much of oscillations which could lead to instability. During experiments two major problems were noticed. First one was not the perfect accuracy of the control system. The second problem appeared when the reference signal was close to boundaries values of the output signal. In such situations the control system could be unstable. Those two aspects will be the main tasks to solve in future work.

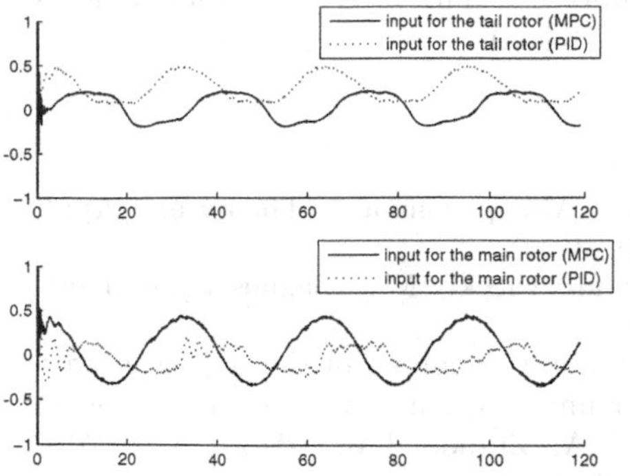

Fig. 4. Calculated control signals with the sinusoidal reference signal

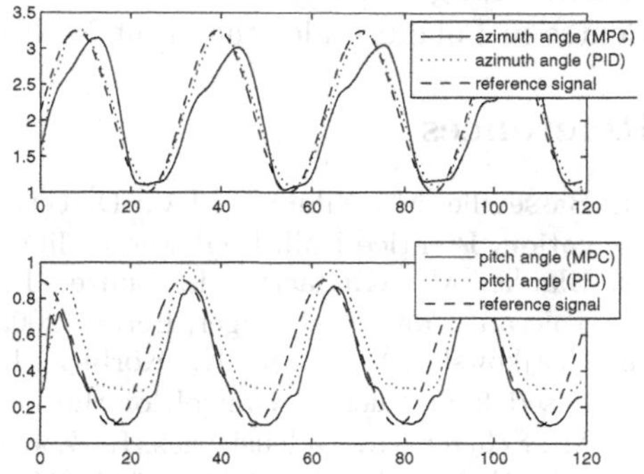

Fig. 5. Output signals with the sinusoidal reference signal

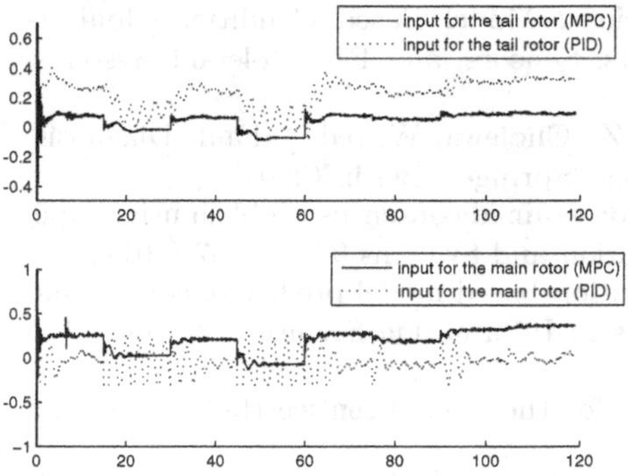

Fig. 6. Calculated control signals with the random-step reference signal

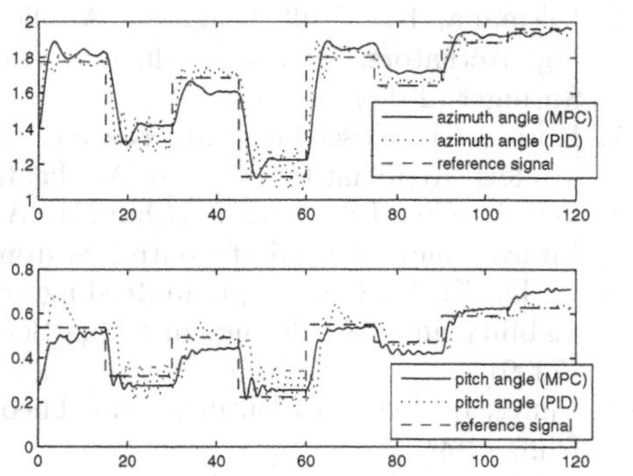

Fig. 7. Output signals with the random-step reference signal

6 Conclusion

The purpose of this work was to design MPC controller based on neural network state space model with delays for the purpose of controlling a fast dynamic system as the Two Rotor Aero-Dynamical system, where computational requirements cannot be very high. As was shown in experiments such controller based on linearisation of the model can be very easily and what is more important efficiently used in the control of the real time setting of nonlinear system. Also worth mentioning is the speed of such method which allows to achieve sampling time equal to $0.1s$. Such approach seems to be very promising and quite better when compared to classic PID controller especially when considering random-steps experiments in which the proposed control scheme characterized with much better stability of control.

Our future work will be focused on using such approach in case of fault to achieve fault tolerance.

Acknowledgements. This work was supported in part by the National Science Centre in Poland under the grant N N514 6784 40.

References

1. Basseville, M., Nikiforov, I.V.: Detection of Abrupt Changes: Theory and Application. Prentice Hall, Englewood Cliffs (1993)
2. Blanke, M., Kinnaert, M., Lunze, J., Staroswiecki, M.: Diagnosis and Fault-Tolerant Control. Springer, Berlin (2006)
3. Czajkowski, A., Patan, K., Korbicz, J.: Stability analysis of the neural network based fault tolerant control for the boiler unit. In: Rutkowski, L., Korytkowski, M., Scherer, R., Tadeusiewicz, R., Zadeh, L.A., Zurada, J.M. (eds.) ICAISC 2012, Part II. LNCS, vol. 7268, pp. 548–556. Springer, Heidelberg (2012)
4. Fletcher, R.: Practical methods of optimization, 2nd edn. Wiley-Interscience, New York (1987)
5. INTECO: Two Rotor Aero-dynamical System - Users Manual (2012), `www.inteco.com.pl`
6. Isermann, R.: Fault Diagnosis Applications: Model Based Condition Monitoring, Actuators, Drives, Machinery, Plants, Sensors, and Fault-tolerant Systems. Springer (2011)
7. Korbicz, J., Koscielny, J.M., Kowalczuk, Z., Cholewa, W. (eds.): Fault Diagnosis. Models, Artificial Intelligence, Applications. Springer, Berlin (2004)
8. Lee, J., Kim, J.S., Song, H., Shim, H.: A constrained consensus problem using mpc. International Journal of Control, Automation and Systems 9, 952–957 (2011)
9. Li, D., Xi, Y.: Quality guaranteed aggregation based model predictive control and stability analysis. Science in China Series F: Information Sciences 52, 1145–1156 (2009)
10. Ljung, L.: System Identification - Theory for the User. Prentice Hall, Englewood Cliffs (1999)
11. Luzar, M., Czajkowski, A., Witczak, M., Korbicz, J.: Actuators and sensors fault diagnosis with dynamic, state-space neural networks. In: MMAR 2012: 17th International Conference on Methods and Models in Automation and Robotics, pp. 196–201 [CD–ROM] (2012) ISBN: 978-83-7518-453-2
12. Maciejowski, J.: The implicit daisy-chaining property of constrained predictive control. Applied Mathematics and Computer Science 8(4), 101–117 (1998)
13. Milman, R., Davison, E.: A fast mpc algorithm using nonfeasible active set methods. Journal of Optimization Theory and Applications 139, 591–616 (2008)
14. Patan, K.: Artificial Neural Networks for the Modelling and Fault Diagnosis of Technical Processes. Springer, Berlin (2008)
15. Patan, K., Korbicz, J.: Nonlinear model predictive control of a boiler unit: a faul tolerant control study. International Journal of Applied Mathematics and Computer Science 22(1), 225–237 (2012)
16. Wang, L.: Model Predictive Control System Design and Implementation Using MATLAB®. Advances in Industrial Control. Springer (2010)
17. Won, W., Lee, K., Kim, I., Lee, B., Lee, S., Lee, S.: Model predictive control of condensate recycle process in a cogeneration power station: Controller design and numerical application. Korean Journal of Chemical Engineering 25, 972–979 (2008)
18. Yetendje, A., Seron, M.M., Don, J.A.D.: Robust multisensor fault tolerant model-following mpc design for constrained systems. Internationl Journal of Applied Mathematics and Computer Science 22(1), 211–223 (2012)

Robust $\mathcal{H}_\infty$ Sensor Fault Diagnosis with Neural Network

Marcel Luzar[1], Marcin Witczak[1], and Christophe Aubrun[2]

[1] Institute of Control and Computation Engineering, University of Zielona Góra, ul. Pogórna 50, 65-246 Zielona Góra, Poland
{m.luzar,m.witczak}@issi.uz.zgora.pl

[2] Centre de Recherche en Automatique de Nancy, CRAN-UMR 7039, Nancy-Université, CNRS, F-54506 Vandoeuvre-les-Nancy Cedex, France
christophe.aubrun@cran.uhp-nancy.fr

Abstract. The paper deals with the problem of a robust fault diagnosis for Linear Parameter-Varying (LPV) systems with Recurrent Neural-Network (RNN). The preliminary part of the paper describes the derivation of a discrete-time polytopic LPV model with RNN. Subsequently, a robust fault detection, isolation and identification scheme is developed, which is based on the observer and $\mathcal{H}_\infty$ framework for a class of non-linear systems. The proposed approach is designed in such a way that a prescribed disturbance attenuation level is achieved with respect to the sensor fault estimation error while guaranteeing the convergence of the observer.

Keywords: Fault diagnosis, fault identification, robust estimation, non-linear systems, observers, neural-networks.

1 Introduction

The problem of fault diagnosis (FD) of non-linear industrial systems [4,10,8] has received considerable attention during the last three decades. Indeed, it developed from the art of designing a satisfactory performing systems into the modern theory and practice that it is today. Within the usual framework, the system being diagnosed is divided into three main components, i.e. plant (or system dynamics [14]), actuators and sensors. The paper deals with the problem of full fault diagnosis of sensors, i.e. apart from the usual two steps consisting of fault detection and isolation (FDI), the fault identification is also performed. This last step is especially important from the viewpoint of Fault-Tolerant Control (FTC) [9,12], which is possible if and only if there is an information about the size of the fault being a result of fault identification (or fault estimation).

In this paper a robust fault estimation approach is proposed, which can be efficiently applied to realise the above-mentioned procedure. The proposed approach can be perceived as a combination of the linear-system strategies [2] and [11] for a class of non-linear systems [13]. The proposed approach is designed

J. Korbicz and M. Kowal (eds.), *Intelligent Systems in Technical and Medical Diagnostics*, Advances in Intelligent Systems and Computing 230,
DOI: 10.1007/978-3-642-39881-0_10,

in such a way that a prescribed disturbance attenuation level is achieved with respect to the fault estimation error while guaranteeing the convergence of the observer. The paper is organised as follows. On the beginning the derivation of a discrete-time polytopic LPV model with neural network is described. In section 3 the robust observer design procedure for sensor fault diagnosis is presented. Section 4 shows experimental results obtained with a laboratory multitank model. Finally, Section 5 conclude the paper.

2 Derivation of a Discrete-Time Polytopic LPV Model

The goal of this section is to present a neural state-space model that can be represent a general class of state-space models and can be easily transformed into a LPV one. The transformation method is derived by Lachhab et. all and presented in [6].

2.1 Recurrent Neural-Network Topology

Let us consider the following non-linear discrete-time LPV model:

$$\boldsymbol{x}_{k+1} = \boldsymbol{A}(\theta_k)\boldsymbol{x}_k + \boldsymbol{B}(\theta_k)\boldsymbol{u}_k, \tag{1}$$

$$\boldsymbol{y}_{k+1} = \boldsymbol{C}(\theta_k)\boldsymbol{x}_{k+1}, \tag{2}$$

where $\boldsymbol{A}(\theta_k)$, $\boldsymbol{B}(\theta_k)$, $\boldsymbol{C}(\theta_k)$ are continuous mappings and θ_k is a time-varying parameter. The dependence of the $\boldsymbol{A}$,$\boldsymbol{B}$,$\boldsymbol{C}$ on θ_k represents a general LPV model; imposing that any of these matrices is parameter independent, i.e. fixed, will restrict the generality of the LPV model. The latter is indeed the case when RNN is transformed into the LPV model: $\boldsymbol{A}$ is parameter dependent but $\boldsymbol{B}$ and $\boldsymbol{C}$ are fixed. A general form of state-space neural network model is

$$\boldsymbol{x}_{k+1} = \boldsymbol{A}\boldsymbol{x}_k + \boldsymbol{B}\boldsymbol{u}_k + \boldsymbol{A}_1\sigma(\boldsymbol{E}_1\boldsymbol{x}_k) + \boldsymbol{B}_1\sigma(\boldsymbol{E}_2\boldsymbol{u}_k), \tag{3}$$

$$\boldsymbol{y}_{k+1} = \boldsymbol{C}\boldsymbol{x}_{k+1} + \boldsymbol{C}_1\sigma(\boldsymbol{E}_3\boldsymbol{x}_{k+1}), \tag{4}$$

where $\boldsymbol{x} \in \mathbb{R}^n$ denotes the state vector, $\boldsymbol{y} \in \mathbb{R}^p$ the output and $\boldsymbol{u} \in \mathbb{R}^m$ the input vector. $\boldsymbol{A}$, $\boldsymbol{A}_1$, $\boldsymbol{B}$, $\boldsymbol{B}_1$, $\boldsymbol{C}$, $\boldsymbol{C}_1$, $\boldsymbol{E}_1$, $\boldsymbol{E}_2$ and $\boldsymbol{E}_3$ are real valued matrices of appropriate dimensions and represent the weights which will be adjusted during the training stage of the RNN. The non-linear activation function $\sigma(\cdot)$, which is applied elementwise in (3)–(4) is taken as a continuous, differentiable and bounded function. This RNN leads to a general form of the neural state-space model in the sense that if it is transformed into an LPV model in the form (1)–(2), the matrices $\boldsymbol{A}$,$\boldsymbol{B}$ and $\boldsymbol{C}$ will be parameter dependent.

For stability and identifiability proofs of the proposed RNN reader is refereed to [6].

2.2 Simplification and Assumptions

The Recurrent Neural-Network (RNN) model described by (3)–(4) can be simplified by removing any of the sigmoidal layers. This can be done according to *a priori* information about the identified system. The LPV model (1)–(2) of multitank system, which was derived from a physical laboratory model and used in experiments presented in Section 4, has only matrix $\boldsymbol{A}$ parameter dependent. Thus, a specific transformation proposed by [5] is used to simplify such a model. Removing the sigmoidal layers from the input and the output paths means that the resulting LPV model has parameter dependance in matrix $\boldsymbol{A}$ only. Moreover, for a practical implementation this simplified RNN is modified as shown in Fig. 1: the outputs instead of the states are taken as input to sigmoidal layer. This modification facilities the implementation of LPV controllers designed based on this model. The modified RNN is represented as

$$\boldsymbol{x}_{k+1} = \boldsymbol{A}\boldsymbol{x}_k + \boldsymbol{B}\boldsymbol{u}_k + \boldsymbol{A}_1\sigma(\boldsymbol{E}_1\boldsymbol{C}\boldsymbol{x}_k), \tag{5}$$

$$\boldsymbol{y}_{k+1} = \boldsymbol{C}\boldsymbol{x}_{k+1}. \tag{6}$$

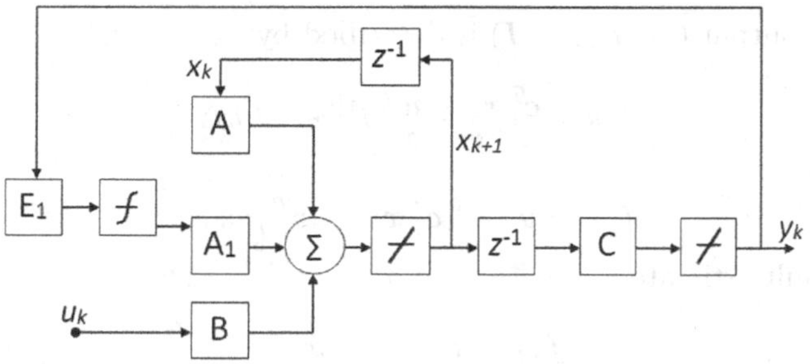

Fig. 1. Simplified state-space recurrent neural network

3 Sensor Fault Diagnosis

Neural model described by (5)–(6) can be modify to following state-space form:

$$\boldsymbol{x}_{k+1} = \boldsymbol{A}\boldsymbol{x}_k + \boldsymbol{B}\boldsymbol{u}_k + \boldsymbol{g}(\boldsymbol{x}_k) + \boldsymbol{W}_1\boldsymbol{w}_k, \tag{7}$$

$$\boldsymbol{y}_{k+1} = \boldsymbol{C}\boldsymbol{x}_{k+1} + \boldsymbol{L}_s\boldsymbol{f}_{s,k} + \boldsymbol{W}_2\boldsymbol{w}_{k+1}, \tag{8}$$

where $\boldsymbol{x}_k \in \mathbb{X} \subset \mathbb{R}^n$ is the state vector, $\boldsymbol{u}_k \in \mathbb{R}^r$ stands for the input, $\boldsymbol{y}_k \in \mathbb{R}^m$ denotes the output, $\boldsymbol{f}_{s,k} \in \mathbb{R}^m$ stands for the sensor fault. While , $\boldsymbol{w}_k \in l_2$ is a an exogenous disturbance vector with $\boldsymbol{W}_1 \in \mathbb{R}^{n\times n}$, $\boldsymbol{W}_2 \in \mathbb{R}^{m\times n}$ being its distribution matrices while

$$l_2 = \left\{\mathbf{w} \in \mathbb{R}^n \middle| \, \|\mathbf{w}\|_{l_2} < +\infty\right\}, \|\mathbf{w}\|_{l_2} = \left(\sum_{k=0}^{\infty} \|\boldsymbol{w}_k\|^2\right)^{\frac{1}{2}}. \tag{9}$$

The main objective of this section is to provide a detailed design procedure of the robust observer, which can be used for sensor fault diagnosis. In other words, the main role of this observer is to provide the information about the sensor fault. Indeed, apart from serving as a usual residual generator (see, e.g.,[14]), the observer should be designed in such a way that a prescribed disturbance attenuation level is achieved with respect to the sensor fault estimation error while guaranteeing the convergence of the observer.

Let us define the matrix $\boldsymbol{X}$ be partitioned in such a way that

$$\boldsymbol{X} = \begin{bmatrix} \boldsymbol{x}_1^T \\ \vdots \\ \boldsymbol{x}_{n_x}^T \end{bmatrix} \tag{10}$$

where $\boldsymbol{x}_j$ stands for the jth row of $\boldsymbol{X}$. Let us also denote $\boldsymbol{X}^j$ as the matrix $\boldsymbol{X}$ without the jth row and $\boldsymbol{y}^j$ as a vector $\boldsymbol{y}$ without the jth element.

The sensor fault diagnosis will be realised by a set of m observers of the form:

$$\hat{\boldsymbol{x}}_{k+1} = \boldsymbol{A}\hat{\boldsymbol{x}}_k + \boldsymbol{g}(\hat{\boldsymbol{x}}_k) + \boldsymbol{K}_j\left(\boldsymbol{y}_k^j - \boldsymbol{C}^j\hat{\boldsymbol{x}}_k\right), \quad j = 1, \ldots, m, \tag{11}$$

while the jth output (for $\boldsymbol{L}_{s,k} = \boldsymbol{I}$) is described by

$$\boldsymbol{y}_{j,k} = \boldsymbol{c}_j^T\boldsymbol{x}_k + \boldsymbol{w}_{2,j}^T\boldsymbol{w}_k + \boldsymbol{f}_{j,k}. \tag{12}$$

Thus:

$$\boldsymbol{f}_{j,k} = \boldsymbol{y}_{j,k} - \boldsymbol{c}_j^T\boldsymbol{x}_k - \boldsymbol{w}_{2,j}^T\boldsymbol{w}_k, \tag{13}$$

and an jth fault estimate is

$$\hat{\boldsymbol{f}}_{j,k} = \boldsymbol{y}_{j,k} - \boldsymbol{c}_j^T\hat{\boldsymbol{x}}_k. \tag{14}$$

The fault estimation error $\boldsymbol{\varepsilon}_{f_{j,k}}$ of the jth sensor is

$$\boldsymbol{\varepsilon}_{f_{j,k}} = \boldsymbol{f}_{j,k} - \hat{\boldsymbol{f}}_{j,k} = -\boldsymbol{c}_j^T\boldsymbol{x}_k + \boldsymbol{c}_j^T\hat{\boldsymbol{x}}_k - \boldsymbol{w}_{2,j}^T\boldsymbol{w}_k = -\boldsymbol{c}_j^T\boldsymbol{e}_k - \boldsymbol{w}_{2,j}^T\boldsymbol{w}_k, \tag{15}$$

while the state estimation error is:

$$\boldsymbol{e}_{k+1} = \boldsymbol{A}\boldsymbol{e}_k + \boldsymbol{s}_k - \boldsymbol{K}_j\boldsymbol{C}^j\boldsymbol{e}_k - \boldsymbol{K}_j\boldsymbol{W}_2\boldsymbol{w}_k + \boldsymbol{W}_1\boldsymbol{w}_k, \tag{16}$$

$$\boldsymbol{e}_{k+1} = \left(\boldsymbol{A} - \boldsymbol{K}^j\boldsymbol{C}^j\right)\boldsymbol{e}_k + \boldsymbol{s}_k - \bar{\boldsymbol{W}}\boldsymbol{w}_k, \tag{17}$$

$$\boldsymbol{e}_{k+1} = \boldsymbol{A}_1\boldsymbol{e}_k + \boldsymbol{s}_k - \bar{\boldsymbol{W}}\boldsymbol{w}_k. \tag{18}$$

where

$$\boldsymbol{s}_k = \boldsymbol{g}\left(\boldsymbol{x}_k\right) - \boldsymbol{g}\left(\hat{\boldsymbol{x}}_k\right). \tag{19}$$

Note that both $\boldsymbol{e}_k$ and $\boldsymbol{\varepsilon}_{f_s,k}$ are non-linear with respect to $\boldsymbol{e}_k$. To settle this problem within the framework of this paper, the following solution is proposed. Using the Differential Mean Value Theorem (DMVT) [15], it can be shown that

$$\boldsymbol{g}\left(\boldsymbol{a}\right) - \boldsymbol{g}\left(\boldsymbol{b}\right) = \boldsymbol{M}_x(\boldsymbol{a} - \boldsymbol{b}), \tag{20}$$

with

$$\boldsymbol{M}_x = \begin{bmatrix} \frac{\partial g_1}{\partial x}(\boldsymbol{c}_1) \\ \vdots \\ \frac{\partial g_n}{\partial x}(\boldsymbol{c}_n) \end{bmatrix}, \tag{21}$$

where $\boldsymbol{c}_1, \ldots, \boldsymbol{c}_n \in \mathrm{Co}(\boldsymbol{a}, \boldsymbol{b})$, $\boldsymbol{c}_i \neq \boldsymbol{a}$, $\boldsymbol{c}_i \neq \boldsymbol{b}$, $i = 1, \ldots, n$. Assuming that

$$\bar{a}_{i,j} \geq \frac{\partial g_i}{\partial x_j} \geq \underline{a}_{i,j}, \quad i = 1, \ldots, n, \quad j = 1, \ldots, n, \tag{22}$$

it is clear that:

$$\mathbb{M}_x = \left\{ \boldsymbol{M} \in \mathbb{R}^{n \times n} | \bar{a}_{i,j} \geq m_{x,i,j} \geq \underline{a}_{i,j}, \, i, j = 1, \ldots, n, \right\} \tag{23}$$

Thus, using (20), the term $\boldsymbol{A}_1 \boldsymbol{e}_k + \boldsymbol{s}_k$ in (18) can be written as

$$\boldsymbol{A}_1 \boldsymbol{e}_k + \boldsymbol{s}_k = (\boldsymbol{A} + \boldsymbol{M}_{x,k} - \boldsymbol{K}^j \boldsymbol{C}^j) \boldsymbol{e}_k \tag{24}$$

where $\boldsymbol{M}_{x,k} \in \mathbb{M}_x$.

From (24), it can be deduced that the state estimation error can be converted into an equivalent form

$$\begin{aligned} \boldsymbol{e}_{k+1} &= \boldsymbol{A}_2(\alpha) \boldsymbol{e}_k - \bar{\boldsymbol{W}} \boldsymbol{w}_k, \\ \boldsymbol{A}_2(\alpha) &= \tilde{\boldsymbol{A}}(\alpha) - \boldsymbol{K}^j \boldsymbol{C}^j, \end{aligned} \tag{25}$$

which defines an LPV polytopic system [1] with

$$\tilde{\mathbb{A}} = \left\{ \tilde{\boldsymbol{A}}(\alpha) : \quad \tilde{\boldsymbol{A}}(\alpha) = \sum_{i=1}^{N} \alpha_i \tilde{\boldsymbol{A}}_i, \, \sum_{i=1}^{N} \alpha_i = 1, \, \alpha_i \geq 0 \right\}, \tag{26}$$

where $N = 2^{n^2}$. Note that this is a general description, which does not take into account that some elements of $\boldsymbol{M}_{x,k}$ maybe constant. In such cases, N is given by $N = 2^{(n-c)^2}$ where c stands for the number of constant elements of $\boldsymbol{M}_{x,k}$.

The problem of $\mathcal{H}_\infty$ observer design [7] is to determine the gain matrix $\boldsymbol{K}$ such that

$$\lim_{k \to \infty} \boldsymbol{e}_k = \boldsymbol{0} \quad \text{for } \boldsymbol{w}_k = \boldsymbol{0}, \tag{27}$$

$$\|\boldsymbol{\varepsilon}_f\|_{l_2} \leq \omega \|\mathbf{w}\|_{l_2} \quad \text{for } \boldsymbol{w}_k \neq \boldsymbol{0}, \, \boldsymbol{e}_0 = \boldsymbol{0}. \tag{28}$$

The general framework for designing robust observer is:

$$\Delta V_k + \boldsymbol{\varepsilon}_{f_{j,k}}^T \boldsymbol{\varepsilon}_{f_{j,k}} - \mu^2 \boldsymbol{w}_k^T \boldsymbol{w}_k < 0, \, k = 0, \ldots \infty, \tag{29}$$

with

$$V_k = \boldsymbol{e}_k^T \boldsymbol{P}(\alpha) \boldsymbol{e}_k \tag{30}$$

Consequently, it can be shown that:

$$\begin{aligned}&\Delta V_k + \varepsilon_{f_{j,k}}^T \varepsilon_{f_{j,k}} - \mu^2 \boldsymbol{w}_k^T \boldsymbol{w}_k = \\ &\boldsymbol{e}_k^T \left(\boldsymbol{A}_2(\alpha)^T \boldsymbol{P}(\alpha)\boldsymbol{A}_2(\alpha)\right)\boldsymbol{e}_k + \\ &\boldsymbol{e}_k^T \left(\boldsymbol{A}_2(\alpha)^T \boldsymbol{P}(\alpha)\bar{\boldsymbol{W}}\right)\boldsymbol{w}_k + \\ &\boldsymbol{w}_k^T \left(\bar{\boldsymbol{W}}^T \boldsymbol{P}(\alpha)\boldsymbol{A}_2(\alpha)\right)\boldsymbol{e}_k + \\ &\boldsymbol{w}_k^T \left(\bar{\boldsymbol{W}}^T \boldsymbol{P}(\alpha)\bar{\boldsymbol{W}}\right)\boldsymbol{w}_k < 0\end{aligned} \tag{31}$$

By defining

$$\boldsymbol{v}_k = \left[\boldsymbol{e}_k^T, \boldsymbol{w}_k^T\right]^T, \tag{32}$$

the inequality (31) becomes

$$\Delta V_k + \varepsilon_{f_j,k}^T \varepsilon_{f_j,k} - \mu^2 \boldsymbol{w}_k^T \boldsymbol{w}_k = \boldsymbol{v}_k^T \boldsymbol{M}_V \boldsymbol{v}_k < 0, \tag{33}$$

where

$$\boldsymbol{M}_V = \begin{bmatrix} \boldsymbol{A}_2(\alpha)^T \boldsymbol{P}(\alpha)\boldsymbol{A}_2(\alpha) - \boldsymbol{P}(\alpha) + \boldsymbol{c}_j \boldsymbol{c}_j^T & \boldsymbol{A}_2(\alpha)^T \boldsymbol{P}(\alpha)\bar{\boldsymbol{W}} + \boldsymbol{c}_j \boldsymbol{w}_{2,j}^T \\ \bar{\boldsymbol{W}}^T \boldsymbol{P}(\alpha)\boldsymbol{A}_2(\alpha) + \boldsymbol{w}_{2,j}\boldsymbol{c}_j^T & \bar{\boldsymbol{W}}^T \boldsymbol{P}(\alpha)\bar{\boldsymbol{W}} + \boldsymbol{w}_{2,j}\boldsymbol{w}_{2,j}^T - \mu^2 \boldsymbol{I} \end{bmatrix}. \tag{34}$$

The following two lemmas can be perceived as the generalisation of those presented in [1].

Lemma 1. *The following statements are equivalent*

1. *There exists* $\boldsymbol{X} \succ 0$ *such that*

$$\boldsymbol{V}^T \boldsymbol{X} \boldsymbol{V} - \boldsymbol{W} \prec 0 \tag{35}$$

2. *There exists* $\boldsymbol{X} \succ 0$ *such that*

$$\begin{bmatrix} -\boldsymbol{W} & \boldsymbol{V}^T \boldsymbol{U}^T \\ \boldsymbol{U}\boldsymbol{V} & \boldsymbol{X} - \boldsymbol{U} - \boldsymbol{U}^T \end{bmatrix} \prec 0. \tag{36}$$

Proof. Applying the Schur complement to (*2*) gives

$$\boldsymbol{V}^T \boldsymbol{U}^T (\boldsymbol{U}^T + \boldsymbol{U} - \boldsymbol{X})^{-1} \boldsymbol{U}\boldsymbol{V} - \boldsymbol{W} \prec 0. \tag{37}$$

Substituting $\boldsymbol{U} = \boldsymbol{U}^T = \boldsymbol{X}$ yield

$$\boldsymbol{V}^T \boldsymbol{X} \boldsymbol{V} - \boldsymbol{W} \prec 0. \tag{38}$$

Thus, (*1*) implies (*2*).

Multiplying (36) by $\boldsymbol{T} = \begin{bmatrix} \boldsymbol{I} & \boldsymbol{V}^T \end{bmatrix}$ on the left and by $\boldsymbol{T}^T$ on the left of (36) gives (35), which means that (*2*) implies (*1*) and hence the proof is completed.

Lemma 2. *The following statements are equivalent*

1. *There exists* $\boldsymbol{X}(\alpha) \succ \mathbf{0}$ *such that*

$$\boldsymbol{V}(\alpha)^T \boldsymbol{X}(\alpha) \boldsymbol{V}(\alpha) - \boldsymbol{W}(\alpha) \prec \mathbf{0}, \tag{39}$$

2. *There exists* $\boldsymbol{X}(\alpha) \succ \mathbf{0}$ *such that*

$$\begin{bmatrix} -\boldsymbol{W}(\alpha) & \boldsymbol{V}(\alpha)^T \boldsymbol{U}^T \\ \boldsymbol{U}\boldsymbol{V}(\alpha) & \boldsymbol{X}(\alpha) - \boldsymbol{U} - \boldsymbol{U}^T \end{bmatrix} \prec \mathbf{0}. \tag{40}$$

Proof. The proof can be realised by following the same line of reasoning as the one of Lemma 1.

It is easy to show that that (40) is satisfied if there exist matrices $\boldsymbol{X}_i \succ \mathbf{0}$ such that

$$\begin{bmatrix} -\boldsymbol{W}_i & \boldsymbol{V}_i^T \boldsymbol{U}^T \\ \boldsymbol{U}\boldsymbol{V}_i & \boldsymbol{X}_i - \boldsymbol{U} - \boldsymbol{U}^T \end{bmatrix} \prec \mathbf{0}, \quad i = 1, \ldots, N. \tag{41}$$

Theorem 1. *For a prescribed disturbance attenuation level* $\mu > 0$ *for the fault estimation error* (15)*, the* $\mathcal{H}_\infty$ *observer design problem for the system* (7)–(8) *and the observer* (11) *is solvable if there exists matrices* $\boldsymbol{P}_i \succ \mathbf{0}$ $(i = 1, \ldots, N)$*,* $\boldsymbol{U}$ *and* $\boldsymbol{N}$ *such that the following LMIs are satisfied:*

$$\begin{bmatrix} -\boldsymbol{P}_i + \boldsymbol{c}_j \boldsymbol{c}_j^T & \boldsymbol{c}_j \boldsymbol{w}_{2,j}^T & \boldsymbol{A}_{2,i}^T \boldsymbol{U}^T \\ \boldsymbol{w}_{2,j} \boldsymbol{c}_j^T & \boldsymbol{w}_{2,j} \boldsymbol{w}_{2,j}^T - \mu^2 \boldsymbol{I} & \bar{\boldsymbol{W}}^T \boldsymbol{U}^T \\ \boldsymbol{U}\boldsymbol{A}_{2,i} & \boldsymbol{U}\bar{\boldsymbol{W}} & \boldsymbol{P}_i - \boldsymbol{U} - \boldsymbol{U}^T \end{bmatrix} \prec 0. \tag{42}$$

where

$$\boldsymbol{U}\boldsymbol{A}_{2,i} = \boldsymbol{U}(\hat{\boldsymbol{A}}_i - \boldsymbol{K}\boldsymbol{C}) = \boldsymbol{U}\hat{\boldsymbol{A}}_i - \boldsymbol{N}\boldsymbol{C}. \tag{43}$$

Proof. Observing that the matrix (34) must be negative definite and writing it as

$$\begin{bmatrix} \boldsymbol{A}_2(\alpha)^T \\ \bar{\boldsymbol{W}}^T \end{bmatrix} \boldsymbol{P}(\alpha) \begin{bmatrix} \boldsymbol{A}_2(\alpha)^T & \bar{\boldsymbol{W}}^T \end{bmatrix} + \begin{bmatrix} -\boldsymbol{P}(\alpha) + \boldsymbol{c}_j \boldsymbol{c}_j^T & \boldsymbol{c}_j \boldsymbol{w}_{2,j}^T \\ \boldsymbol{w}_{2,j} \boldsymbol{c}_j^T & \boldsymbol{w}_{2,j} \boldsymbol{w}_{2,j}^T - \mu^2 \boldsymbol{I} \end{bmatrix} \prec 0. \tag{44}$$

and then applying Lemma 2 and (41) leads to (42), which completes the proof.

Finally, the design procedure boils down to solving LMI (42) and then $\boldsymbol{K} = \boldsymbol{U}^{-1}\boldsymbol{N}$.

4 Experimental Results

To verify proposed approach, a sensor fault detection was implemented for multitank system. The considered multi-tank system (Fig. 2) is designed for simulating the real industrial multi-tank system in the laboratory conditions [3]. The multi-tank system can be efficiently used to practically verify both linear and non-linear control, identification and diagnostics methods. The multi-tank system consists of three separate tanks placed each above other and equipped with drain valves and level sensors based on a hydraulic pressure measurement. Each of them has a different cross-section in order to reflect system nonlinearities. The lower bottom tank is a water reservoir for the system. A variable speed water pump is used to fill the upper tank. The water outflows the tanks due to gravity. The considered multi-tank system has been designed to operate with an external, PC-based digital controller. The control computer communicates with the level sensors, valves and a pump by a dedicated I/O board and the power interface. The I/O board is controlled by the real-time software, which operates in a Matlab/Simulink environment.

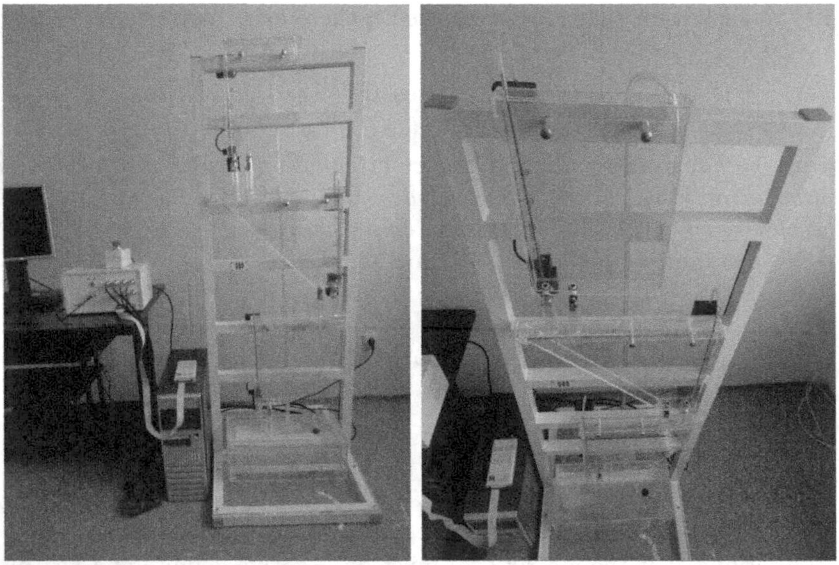

Fig. 2. Multi-tank system

The simplified RNN presented in Section 2 is implemented in Neural Network Time Series Tool (*ntstool*) provided by MATLAB and trained with classical Levenberg-Marquardt algorithm.

Let the initial condition for the system and the observer be:

$$\boldsymbol{x}_0 = [0.1, 0.2, 0.3]^T, \quad \hat{\boldsymbol{x}}_0 = \mathbf{0.00001}, \tag{45}$$

while the input and the exogenous disturbance are:

$$\boldsymbol{u}_k = 1, \quad \boldsymbol{w}_k \sim \mathcal{N}(\mathbf{0}, 0.01\boldsymbol{I}). \tag{46}$$

Although measurements from all water level sensors are available, in experiments measurements from two of three sensor were taken into account in order to simulate insensitivity of the system for third faulty sensor. To present the results of sensor fault diagnosis, the following set of faults scenarios for i^{th} sensor were introduced:

1. Incipient fault

$$\boldsymbol{f}_{s_1,k} = \begin{cases} -0.005(k-300), \text{ for } 500 \geq k \geq 300, \\ 0, \quad \text{otherwise}, \end{cases}$$

$$\boldsymbol{f}_{s_2,k} = \begin{cases} -0.005(k-1000), \text{ for } 1500 \geq k \geq 1000, \\ 0, \quad \text{otherwise}, \end{cases}$$

$$\boldsymbol{f}_{s_3,k} = \begin{cases} -0.005(k-1800), \text{ for } 2100 \geq k \geq 1800, \\ 0, \quad \text{otherwise}, \end{cases}$$

2. 20% decrease in the accuracy

$$\boldsymbol{f}_{s_1,k} = \begin{cases} -0.2\boldsymbol{y}_k, \text{ for } 500 \geq k \geq 300, \\ 0, \quad \text{otherwise}. \end{cases}$$

$$\boldsymbol{f}_{s_2,k} = \begin{cases} -0.2\boldsymbol{y}_k, \text{ for } 1500 \geq k \geq 1000, \\ 0, \quad \text{otherwise}. \end{cases}$$

$$\boldsymbol{f}_{s_3,k} = \begin{cases} -0.2\boldsymbol{y}_k, \text{ for } 2100 \geq k \geq 1800, \\ 0, \quad \text{otherwise}. \end{cases}$$

Figure 3 presents results of fault identification for each type of fault for the first sensor for the nominal case ($\hat{\boldsymbol{x}}_0 \neq \boldsymbol{x}_0$ and $\boldsymbol{w}_k \neq \boldsymbol{0}$). Moreover, figure 6 shows the evolution of $\|\boldsymbol{e}_k\|$ for first and second sensor which confirms that condition (28) is satisfied.

Similar results of fault identification were obtained for second and third sensor and shown in figures (4)-(5).

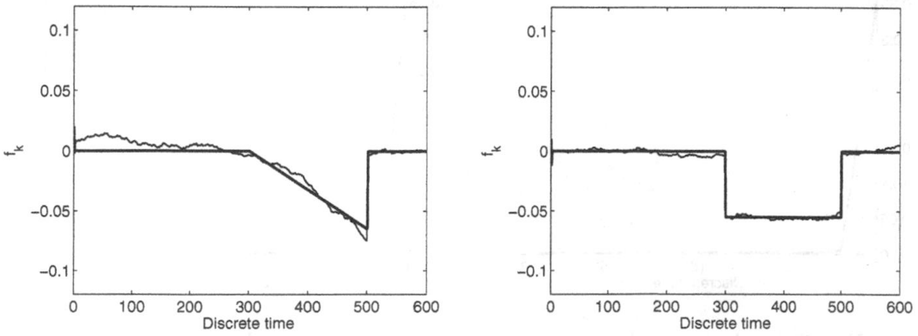

Fig. 3. Fault identification for incipient and decrease fault in first sensor

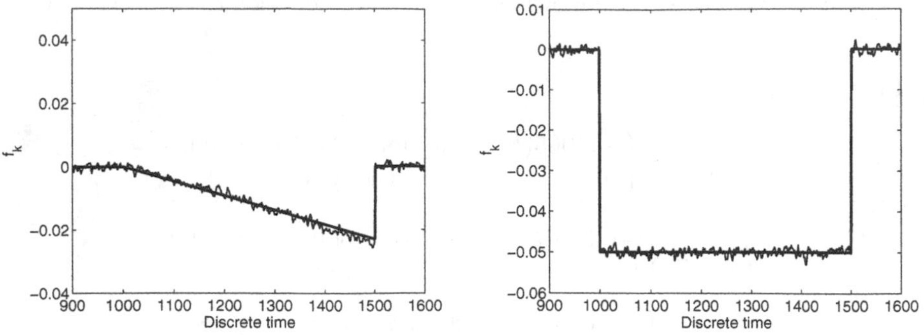

Fig. 4. Fault identification for incipient and decrease fault in second sensor

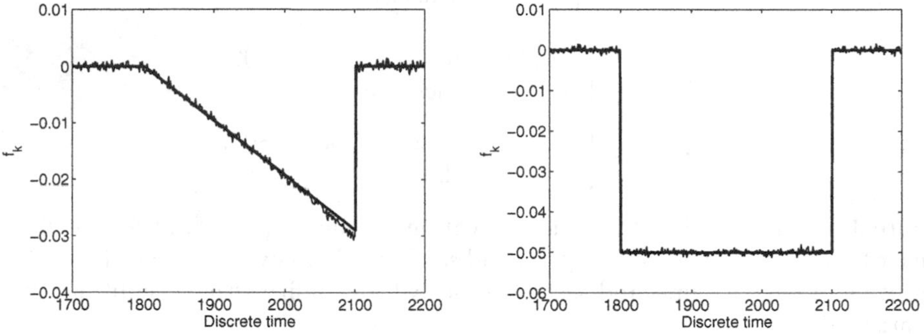

Fig. 5. Fault identification for incipient and decrease fault in third sensor

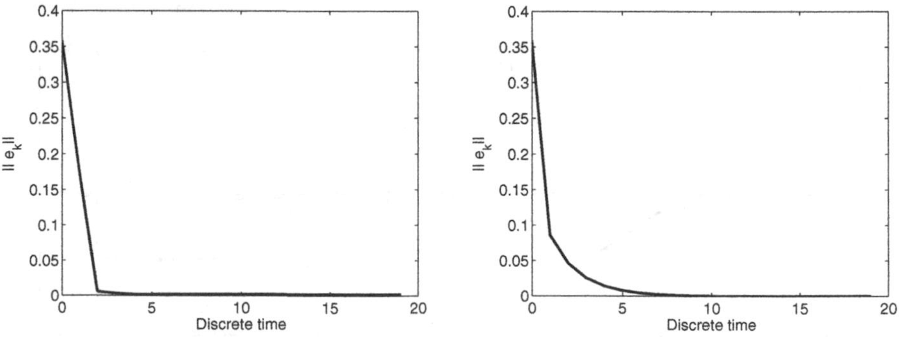

Fig. 6. Evolution of $\|\boldsymbol{e}_k\|$ (for $k = 0, \ldots, 20$) for first and second sensor

5 Conclusions

The paper deals with the problem of robust sensor fault estimation with neural networks. In particular, a combination of the celebrated generalised observer scheme with the robust $\mathcal{H}_\infty$ approach is proposed to settle the problem of robust fault diagnosis. The proposed approach is designed in such a way that a prescribed disturbance attenuation level is achieved with respect to the sensor fault estimation error while guaranteeing the convergence of the observer. The final part of the paper is concerned with a comprehensive case study regarding the multi-tank system. The achieved results show the performance of the proposed approach, which confirm its practical usefulness.

References

1. de Oliveira, M.C., Bernussou, J., Geromel, J.C.: A new discrete-time robust stability condition. Systems and Control Letters 37(4), 261–265 (1999)
2. Gillijns, S., De Moor, B.: Unbiased minimum-variance input and state estimation for linear discrete-time systems. Automatica 43, 111–116 (2007)
3. INTECO. Multitank System - User's manual (2013), `www.inteco.com.pl`
4. Iserman, R.: Fault Diagnosis Applications: Model Based Condition Monitoring, Actuators, Drives, Machinery, Plants, Sensors, and Fault-tolerant Systems. Springer, Berlin (2011)
5. Kajiwara, H., Apkarian, P., Gahinet, P.: LPV techniques for control on an inverted pendulum. IEEE Control Systems Magazine 19(1), 44–54 (1999)
6. Lachhab, N., Abbas, H., Werner, H.: A neural-network based technique for modelling and LPV control of an arm-driven inverted pendulum. In: Proceedings of the 47th IEEE Conference on Decision and Control, Cancun, Mexico, pp. 3860–3865 (2008)
7. Li, H., Fu, M.: A linear matrix inequality approach to robust $\mathcal{H}_\infty$ filtering. IEEE Trans. Signal Processing 45(9), 2338–2350 (1997)
8. Luzar, M., Czajkowski, A., Witczak, M., Mrugalski, M.: Actuators and sensors fault diagnosis with dynamic, state-space neural networks. In: MMAR 2012: Proceedings of the 17th IEEE International Conference on Methods and Models in Automation and Robotics, pp. 196–201 (2012)
9. Montes de Oca, S., Puig, V., Witczak, M., Dziekan, Ł.: Fault-tolerant control strategy for actuator faults using LPV techniques: Application to a two degree of freedom helicopter. International Journal of Applied Mathematics and Computer Science 22(1), 161–171 (2012)
10. Mrugalski, M.: An unscented kalman filter in designing dynamic gmdh neural networks for robust fault detection. International Journal of Applied Mathematics and Computer Science 23(1), 157–169 (2013)
11. Nobrega, E.G., Abdalla, M.O., Grigoriadis, K.M.: Robust fault estimation of unceratain systems using an LMI-based approach. International Journal of Robust and Nonlinear Control 18(7), 1657–1680 (2008)
12. Puig, V.: Fault diagnosis and fault tolerant control using set-membership approaches: Application to real case studies. International Journal of Applied Mathematics and Computer Science 20(4), 619–635 (2010)

13. Stipanovic, D.M., Siljak, D.D.: Robust stability and stabilization of discrete-time non-linear: the LMI approach. International Journal of Control 74(5), 873–879 (2001)
14. Witczak, M.: Modelling and Estimation Strategies for Fault Diagnosis of Non-linear Systems. Springer, Berlin (2007)
15. Zemouche, A., Boutayeb, M.: Observer design for Lipschitz non-linear systems: the discrete time case. IEEE Trans. Circuits and Systems - II:Express Briefs 53(8), 777–781 (2006)

Neural Network Simulator's Application to Reference Performance Determination of Turbine Blading in the Heat-Flow Diagnostics

Anna Butterweck and Jerzy Głuch

Gdańsk University of Technology, Faculty of Ocean Engineering and Ship Technology, Narutowicza 11/12, 80-233 Gdańsk Poland
anna.butterweck@gmail.com, jgluch@pg.gda.pl

Abstract. In the paper, the possibility of application of artificial neural networks to perform the fluid flow calculations through both damaged and undamaged turbine blading was investigated. Preliminary results are presented and show the potentiality of further development of the method for the purpose of heat-flow diagnostics.

Keywords: Artificial neural networks, heat-flow diagnosis, turbine stage.

1 Introduction

Modern operating systems of technical objects increase usage of diagnostic methods and diagnostic models [2,4,6,7,9,10,12]. That leads to safer and more efficient operation of technical objects. Diagnostic methods are mainly based on individual safety (e.g. vibration) [3] or on efficiency [6,10,12]. Improving indicators of efficiency in operating the technical objects lead to the better protection of the environment by reducing noise and the emission of harmful substances into the environment.

Diagnostic models can also be applied to energy facilities, including power plants, marine power turbines and aircraft turbine engines. The efficiency, of mentioned objects, is determined mainly by efficient work of each part of thermal cycle [10]. In particular it applies to a turbine, which is the key element of thermal circuit.

Turbine itself is a complex thermo-mechanical device. Its effectiveness, in turn, is determined by the quality of the flow of the working medium through turbine blades. Contemporary research methods have led to the high efficiency of turbine blades [11]. Further improvement, with today's technology development, can be ascribed only to maintain these values throughout the whole time of operation [10]. This is one of the most important tasks of thermal and flow diagnostics, which focuses primarily on operational aspects of efficiency [6,10,12].

Thermal and flow diagnostics can apply various methods. Determination of symptoms of inefficient work is its task. Symptoms, often gathered in signatures,

J. Korbicz and M. Kowal (eds.), *Intelligent Systems in Technical and Medical Diagnostics*, Advances in Intelligent Systems and Computing 230,
DOI: 10.1007/978-3-642-39881-0_11, © Springer-Verlag Berlin Heidelberg 2014

differ from parameters and efficiency indicators for the current state of the machine and the corresponding reference parameter. Parameters (pressure, temperature, mass flows, etc.) of the object with undamaged, geometry are taken as a reference state.

The reference state parameters create a model of object's efficient operation. Symptom-based diagnostics is an example of diagnostic method based on the application of the reference state. Symptom value greater than measurement uncertainty indicates degradation in the object [6,10].

In diagnostics there are few critical steps. The first of them is the task of fast determining of the reference. Therefore, the reference state heat-flow diagnostics yet uses simplified computational models to determine the reference state [6,10] - mainly due to the diagnostic system requirement of a short time for determining the parameters of the reference.

Shortening the calculation time is possible when the numerical procedures are not too complicated. However, their disadvantage is the inability to determine and take into account all the details of the characteristics that while using of simpler models (e.g., 1 dimension, 1D) are averaged and do not show all the needed details [4,10]. Mentioned disadvantages do not occur while the usage of the diagnostics of three-dimensional models to calculate spatial distribution the thermal-flow parameters. Furthermore, one could consider additional phenomena already used in other types of diagnostics such as acoustics. This would vastly extend time of calculations.

The perspective of application detailed multivariate methods in the diagnostics must, therefore, lead to their modification. In the modification the listed advantages of the method should be kept while the computation time should be significantly reduced. It can be done based on the method of artificial neural networks (ANN). If practical solutions turn out to be working fast enough, the proposed models could be used even in the on-line diagnostics.

Replacing detailed 3D methods with neural network has already been reported in the literature. However, it is applied only to assist in the design of thermal turbines [5,8,11]. Due to lack of the geometrical data, in the design of turbine systems dimensionless geometrical characteristics were used. Similar modelling for diagnostic purposes must be done differently. The reference geometry is known. During operating of the turbine, it degrades. Therefore, the input data for neural model must consist both of reference undamaged characteristics of geometric of turbine system, and the model and the degraded one.

In the paper the preliminary results of the neural modelling of the steam turbine blade row of known geometry are presented. Investigation of possibility of replacement of CFD model by ANN model is the main aim of the paper. The purpose of neural model application is the close fit reconstruction of distribution of velocity, pressure, temperature and flow losses in a channel of complex geometry. The presented results are encouraging for further work and apply to 2-dimensional model at the desired diameter of the blade.

2 Physics and Numerical Model of the Turbine Flow

The task of turbine blade rows is to transform a stream of heat into stream of kinetic energy. In order to complete the energy transformation the speed up and working channels are used. These channels are formed by the turbine blades, in the channels fluid flows. Usually the working fluid is steam (steam turbines) or exhaust gas (gas turbines). Efficient operation of blade channels requires inter alia the blades smooth surfaces, good condition of trailing edge and the absence of sediment or erosion cavities. The example of undamaged and damaged blade is presented in Fig. 1. Degradation results in energy dissipation, and hence losses.

Actual systems can be modelled mathematically with respect to all geometric complexity. Such advanced flow simulations are commonly used and allow determining three-dimensional distributions of: velocity, pressure temperature and energy losses. Calculations also allow to take into account at least some of operational degradation of geometry. Integration of flow fields gives the general averaged characteristics of turbine stage. The results of the calculations can be applied both to the design and diagnostics.

Fig. 1. Example of a new (undamaged) turbine rotor blade and the possible degradation in service

The accuracy of the simulation depends on both modeling and computational methods. The methods are based on nonlinear balance equations of mass, energy and momentum (and closing equations taken into account) in small finite volumes elements. The space of fluid flow channel is divided into vast (for a single turbine channel in 2D about 16000 elements and about $2 \cdot 10^6$for 3D) amount of elements (creating a mesh). Mentioned equations are iterated in every node of the mesh. The CFD requires a significant amount of time to perform the calculations. In the case of diagnostic applications computational time should be as short as possible. Therefore, attention of authors was brought to the method of artificial neural networks (ANN) that provide an opportunity to meet this requirement. Simultaneously the ANN method is expected to provide satisfying level of accuracy and its ability to reconstruct energy and flow phenomena. This requires a study of the conditions for obtaining good quality. Considerations of the mentioned problems of ANN model are the goal of the paper.

Fluid flow of compressible media is described by [13]:

– the continuity equation

$$\frac{\partial \rho}{\partial t} + div\,(\rho \mathbf{u}) = 0, \tag{1}$$

where: ρ - density, $\mathbf{u}$- vector of velocity.

– conservation equation of momentum:

$$\rho \frac{d\mathbf{u}}{dt} = \rho \mathbf{f} - grad\,p - grad\left(\frac{2}{3}\mu div\,\mathbf{u}\right) + div\,(2\mu \mathbf{D}), \tag{2}$$

where: $\mathbf{f}$ - density of distribution of mass forces, p - pressure, μ- dynamic viscosity, $\mathbf{D}$ - tensor of velocity of deformation

– conservation of energy:

$$\rho \frac{de}{dt} = T\dot{s}_m + \frac{p}{\rho}\frac{d\rho}{dt} + \lambda \Delta T. \tag{3}$$

where: e - internal energy, T - temperature, $\dot{s}_m$ - intensity of mechanical entrophy sources, λ - thermal conductivity

In order to close system of equations (1-3) it is necessary to involve equation of fluid (4)

$$e = \int_{T_0}^{T} c_v(T)dT, \tag{4}$$

and state equation (5)

$$\frac{p}{\rho} = Z(p,T)RT, \tag{5}$$

where: c_v - heat at constant volume, Z - compressibility function, R - gas constant.

As a reference state, to train neural network, computational fluid dynamics (CFD) calculations were made. To apply equations (1)-(5) the discretisation with following schemes were proceeded:

- pressure-velocity coupling - SIMPLE
- density - second order Upwind
- momentum - second order Upwind
- turbulent kinetic energy - first order upwind
- turbulet dissipation rate - first order upwind
- energy - second order upwind

Fig. 2. Non-structural mesh of flow channel

In order to proceed the calculations the geometry of flow channel was built and meshed. In this case flow channel was created by two blades of steam turbine stator. In Fig. 2 the geometry of the flow channel and the mesh is shown. Boundary layer was meshed structurally the rest of the channel was meshed non-structurally.

In Table 1 inlet and outlet values of pressure and temperature for both desig and off-design fluid flow channel were shown. As a first step towards complex application of neural network modeling in diagnostics calculations at range values of inlet and outlet parameters were made. Inlet and outlet pressure varied from 93% to 100% of design values.

Table 1. Design and off-design boundary conditions

	Design	Off-design
Inlet pressure [MPa]	7,93	7,37
Inlet temperature [K]	746	746
Outlet pressure [MPa]	7,22	6,71
Outlet temperature [K]	732	732

In Fig. 3 and Fig. 4 results of two simulations (with lowest and highest values of inlet and outlet pressure) were presented. In Fig. 3 the pressure distributions were shown, for better presentation of difference between cases the comparing figure was made. In Fig. 4 distribution of velocity magnitudes were shown.

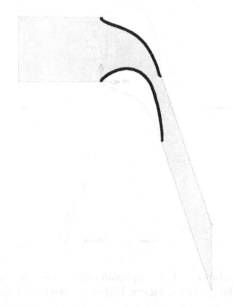

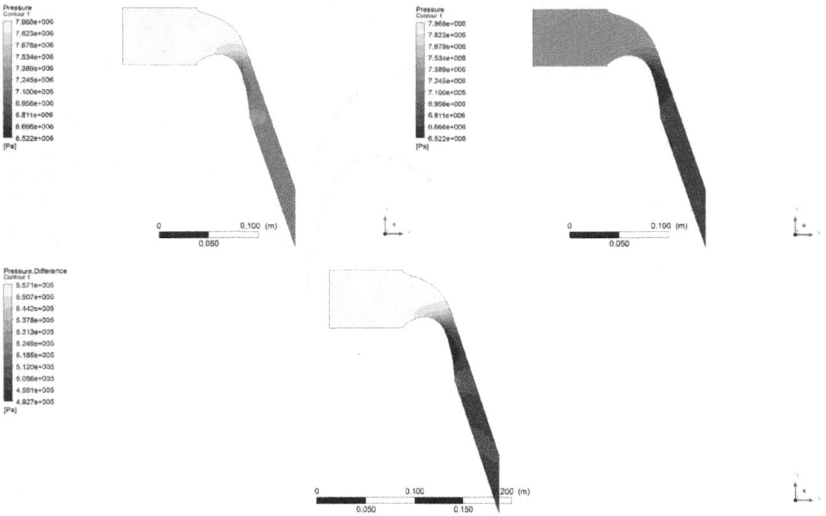

Fig. 3. Pressure distributions. Left – up: Nominal values of thermodynamic parameters, Right up: 93% of inlet pressure; Down: Difference between both cases

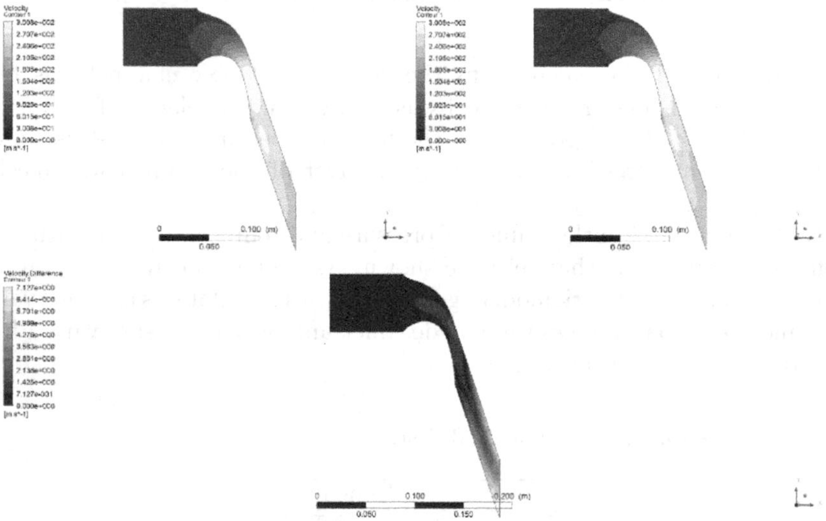

Fig. 4. Velocity magnitude distributions. Left – up: Nominal values of thermodynamic parameters, Right up: 93% of inlet pressure; Down: Difference between both cases

The advantages of proceeding CFD calculations with commercial solvers are: solvers were validated and these are well known and well described in literature. On the other hand using CFD commercial solver is highly time consuming and cost of licensing is high. Due to the disadvantages of CFD solvers it is impossible to apply such a calculations in diagnostics. The authors foresee the chance of shortening time of calculation in applying artificial neural networks.

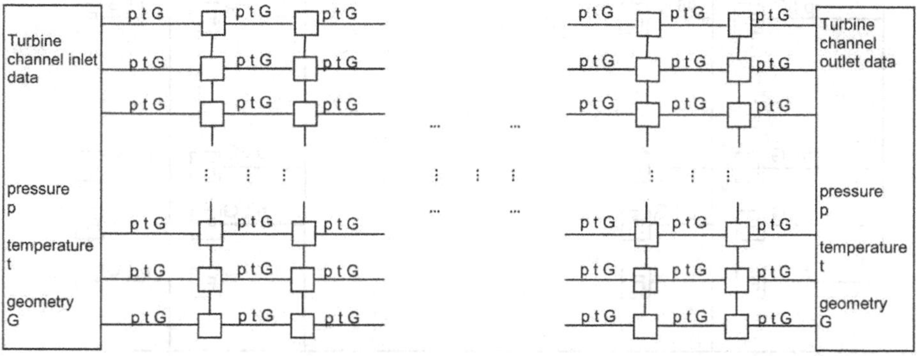

Fig. 5. Set of single finite volume cells modelling flow channel

3 Description of Neural Model

The proposed neural model is designed to replace the analytical calculation model. As is well known, neural models are not physical models. It is author's responsibility to set such an input values for the neural model to reproduce the results of the overlap physical phenomena. It is one of the first objectives of research presented in the paper.

To restore the flow phenomena and processes of energy conversion it was assumed that at the input data used in the analytical and numerical models should be given. The working fluid is a gas with known physical and chemical properties; constructed neural model applies only to that gas. Based on authors experience with other research on the application of artificial neural networks methods to diagnostic purposes, individual neural models were built for each of the distributions describing the flow field, such as pressure, velocity, energy loss, etc. This provides a better quality model [6,12].

Finally the following assumptions for each of the individual neural models were given:

- the simple feedfoward neural network (widley described in the literature) applied for modeling each individual mesh element,
- thermodynamic and flow parameters: temperature and pressure were treated as each individual ANN input data,
- inlet and outlet pressure on the blade modeled for the variable load conditions,
- physical boundaries defined by the geometrical parameters of turbine blades.

Simplified diagram illustrating the individual graphical computational model of the flow channel is presented in Fig. 6.

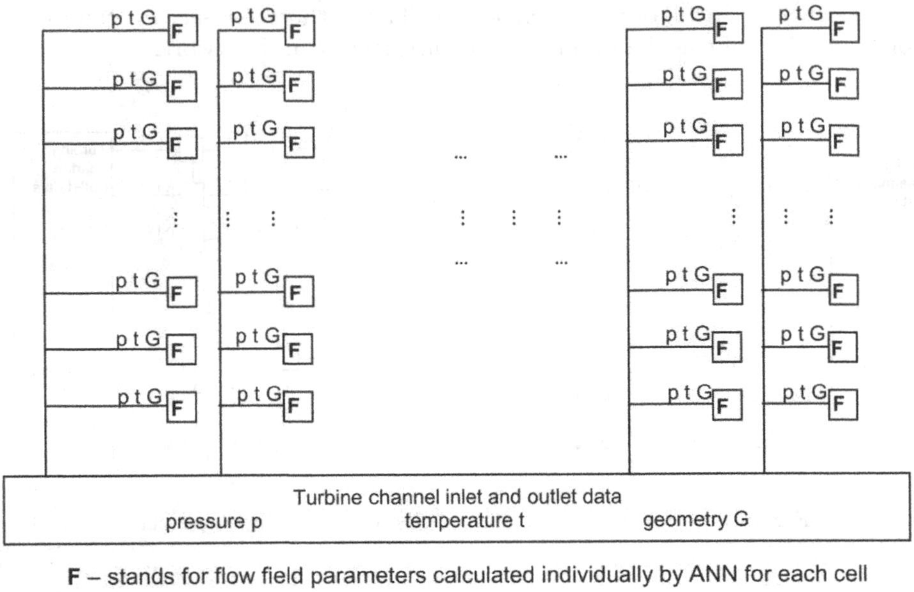

Fig. 6. Set of single ANN modeling individual finite volume cells

The proposed neural model differs from the models of neural networks presented in the literature [8,11]. In [8,11] authors used, instead of applying geometry data, applied generalized, dimensionless design characteristics for turbine blade rows. This fact determined that neural models were prepared only for nominal flow and thermodynamic parameters describing load conditions.

However, diagnostic applications require the fullest possible description of geometric turbine blades. Possitive effect of presented calculations conditions further steps in development of the method. In a further research the geometry of blades will be changed to simulate selected geometric degradation. The next aim is to prepare a diagnostic neural model applied to changing load conditions by variation of thermal-hydraulic parameters.

4 Application of Neural Network Model to Determine the Reference Flow

The described neural model was trained under supervision by using the results of numerical calculations based on analytical models of the flow channels. In earlier part of the article graphic presentation of numerically calculated distributions of chosen variables was made.

In the first stage of the research backpropagation algorithm was used as a training method. Backpropagation algorithm is one of the older methods of training and in future work will be replaced with more developed alghoritm.

However even the application of backpropagation algorithm leads to satisfactory results.

In the following figures (Fig. 7), the results of the neural model training are satisfactory. Relative percentage error was calculated as:

$$err = \frac{p_{CFD} - p_{ANN}}{p_{CFD}} 100\%. \quad (6)$$

Values of the error varies from 0 to 0.25 % (shown in Fig. 7). In Fig. 7 the distribution of relative percentage error near the trailing edge was show. Area of the trailing edge is area of the strongest variation of the pressure and persumably is the place of the highest error value. In further course of the research the authors plan to apply more sophisticated methods of training to minimise the time and effectivnes of training. Distribution of pressure calculated both with ANN and ANSYS Fluent (Fig. 7) indicate the high accuracy of the training of ANN. The time of the calculation, with CFD solver, of 2-dimensional (2D) model composed of 16200 elements is 90 seconds. While time of computation of the same case with ANN is significantly shorter - about 1 second. That is a premise of the development of the ANN method for future application in heat-flow diagnostics. For example full 3-dimensional (3D) model of the investigated

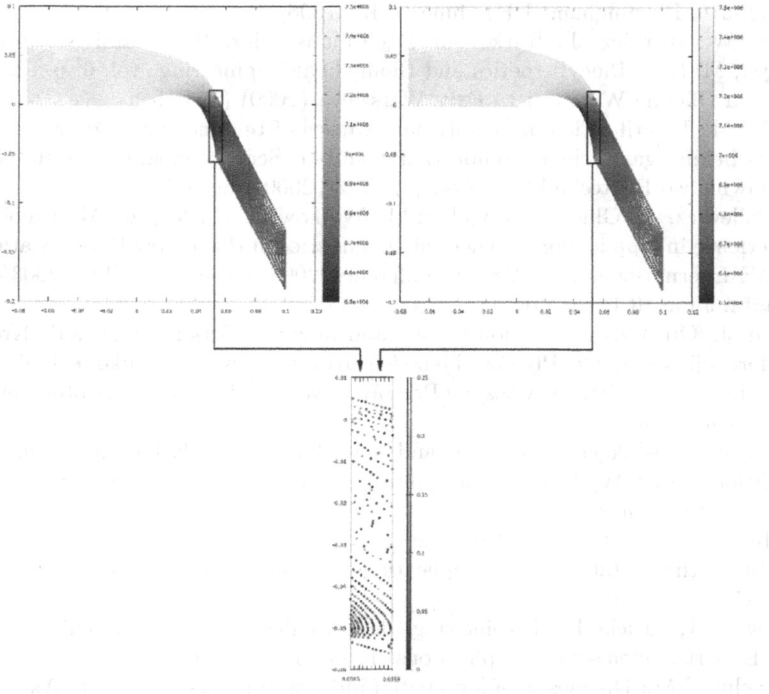

Fig. 7. Distribution of pressure (case: 93% of inlet pressure): left: calculated with neural network; right: calculated with ANSYS Fluent

turbine channel consists of about $2 \cdot 10^6$ elements and time of CFD calculations is considered in hours. On the other hand the time of calculating the same 3-D case with ANN is still expected to be in seconds.

5 Final Conclusions

Application of artificial neural networks (ANN) to perform the fluid flow calculations through both damaged and undamaged turbine blading was investigated. Presented results confirm high accuracy of the method. Percentage relative error values, related to value of thermodynamical variables resulted from CFD, were found in range from 0 to 0,25%. Due to shortening time of computation (in investigated case - 90 times) and high accuracy the ANN based method is applicable to detailed heat-flow turbine diagnostics.

Further work on the method assume application of more developed training methods of ANN and expanding the method to 3-dimentional space.

References

1. Chbat, N.W., Rajamani, R., Ashley, T.A.: Estimating Gas Turbine Internal Cycle Parameters Using a Neural Network, ASME-TURBO-EXPO, ASME Paper No 96-GT-316, Birmingham (UK), June 10-13 (1996)
2. Duch, W., Korbicz, J., Rutkowski, L., Tadeusiewicz, R.: Neural Networks. In: Nałęcz, M. (ed.) Biocybernetics and Biomedical Engineering, vol. 6, p. 833. Akademicka Oficyna Wydawnicza Exit, Warszawa (2000) (in Polish)
3. Dzida, M.: Identification of unstationary causes of temperature and pressure variations behind gas turbines combustion chamber. Seria Monografie Nr 16, Gdańsk Wydawnictwo Politechniki Gdańskiej, p. 123 (2000) (in Polish)
4. Gardzilewicz, A., Gluch, J., Bogulicz, M., Walkowiak, R., Najwer, M., Kiebdój, J.: Experience in Application of Thermal Diagnostics in the Turów Power Station. In: ASME International Joint Power Conference 2003, Paper No. IJPGC2003-40017, Atlanta, June 16-19 (2003)
5. Głuch, J.: On ANN application to heat and flow calculations. Mat. XIV Krajowej Konferencji Mechaniki Płynów, Łódź-Arturówek , Zeszyty Naukowe Politechniki Łódzkiej Nr 851, Cieplne Maszyny Przepływowe 117/2000, Łódź, September 18-22 (2000) (in Polish)
6. Głuch, J.: – redakcja i inni, Heat and flow diagnostic relations under industrial conditions. Wyd. Wydział Oceanotechniki i Okrętownictwa Politechniki Gdańskiej, Gdańsk (2007) (in Polish)
7. Korbicz, J., Kościelny, J.M., Kowalczuk, Z., Cholewa, W.: Fault diagnostics. Models. In: Artificial Intelligence. Applications, p. 865. Springer (2004) ISBN 3-540-40767-7
8. Kosowski, K., Tucki, K.: Turbine stage design aided by artificial intelligence methods. Expert Systems with Applications, 11536–11542 (2009)
9. Kościelny, J.M.: Diagnostic of automated industrial processes, p. 418. Akademicka Oficyna Wydawnicza EXIT, Warszawa (2001) (in Polish)
10. Krzyżanowski, J., Głuch, J.: Heat and flow diagnostics of energetics objects, p. 302. Wydawnictwo Inst. Masz. Przepł, Gdańsk (2004) (in Polish)

11. Pierret, S.: Designing Turbomachinery Blades by Means of the Function Approximation Concept Based on Artificial Neural Network, Genetic Algorithm, and the Navier-Stokes Equation, PhD Thesis, Faculte Polytechnique de Mons, Von Karman Institute for Fluid Dynamics, Turbomachinery Department, p. 176 (1999)
12. Ślęzak-Żołna, J., Głuch, J.: Towards Symptoms of Degradation In On-Line Thermal and Flow Diagnostics of Power Objects, W: CD, Mat. Konferencji SAFEPROCESS, Pekin, (2006)
13. Puzyrewski, R., Sawicki, J.: Fundamentals of fluid mechanics and hydraulics, Wydawnictwo Naukowe PWN, Warszawa (2000)

Clustering of Delaminations in Composite Rotors Using Self-Organizing Maps

Marcin Amarowicz and Andrzej Katunin

Silesian University of Technology, Institute of Fundamentals of Machinery Design, Konarskiego 18A, 44-100 Gliwice, Poland
{marcin.amarowicz,andrzej.katunin}@polsl.pl

Abstract. The study presented in this paper concerned with the evaluation of delamination presence and its type basing on the analysis of vibration data of composite rotors using Kohonen self-organizing maps. The approach used for the evaluation considered axial displacements of modal shapes of rotated composite rotors with simulated delaminations and measurement points on the surface of a rotor in order to detect and classify the delaminations with respect to their geometric properties and location. A large number of numerical models with simulated delaminations were considered in order to create representative training sets. Obtained results show that the proposed approach allows for detection of delaminations with unknown shapes and locations. The approach could be further applied in practical solutions using advanced non-contact or embedded measurement systems.

Keywords: self-organizing maps, artificial neural networks, damage clustering, composite rotors, delamination.

1 Introduction

An increased development of novel composite materials and technologies cause that the composite structures become more and more common in engineering applications. Application of lightweight composites allows for significant reduction of mass-to-strength ratio, which stated a useful property in machine's design applications, primarily in the aircraft engineering. However, in comparison with traditional metal alloys, the degradation processes occurred in the structures made of polymer composites are different, i.e. some new types of damages may occur in this structures, e.g. delamination, decohesion, debonding, etc. The latter damage types cause insignificant changes in the structure, which cause that many of testing methods remain unsensitive to them, especially in the early phases of their propagation. Consequently, it is essential to develop appropriate structural health monitoring (SHM) systems, which allow for detection and characterization of the occurred damages in possible early stage of its propagation.

Colaterally SHM technologies, which provide reliable and economically efficient analysis of such structures, were developed. The greatest increase of interest in SHM technologies is observable in the past decade. Due to the tendency of

J. Korbicz and M. Kowal (eds.), *Intelligent Systems in Technical and Medical Diagnostics*, Advances in Intelligent Systems and Computing 230,
DOI: 10.1007/978-3-642-39881-0_12,

miniaturization of electronic devices used in SHM systems and actual needs and requirements of industrial applications many different techniques and methods were proposed. The numerous SHM methods are based on strain measurements and acoustic emission (AE) [15]. Advanced measurement devices in SHM systems are usually non-contact or embedded in the investigated structure. The measurement systems are often used advanced signal processing techniques, which should provide non-destructive damage evaluation and continuous monitoring of its propagation.

In order to classify and group the types of damages occurred in the structures several methods are commonly used. Due to the ability of processing of the great quantities of information in comparatively short time and great performance and accuracy in classification problems the techniques based on artificial neural networks (ANN) seems to be the leading tools applied for such a class of problems. Three types of ANN architectures are applied for the clustering problems most often [4]: the multilayer perceptron (MLP), radial basis function network (RBF) and Kohonen self-organizing maps (SOM). The first two architectures provide supervised training, which implies that the classes are a priori known. SOM uses unsupervised training approach, which gives a possibility to identify the specific classes in the clustering task. This advantage is conducted by numerous studies, including SHM problems. The authors of [5] used SOM in the on-line damage detection in smart composite structures based on strain measurements by fibre-optic sensors, whereas the authors of [15] presented a study of damage detection based on strain measurements by piezoelectric transducers. The machine condition monitoring problem using SOM was presented in [13]. The applications of SOM in AE-based SHM systems for composite structures have been reported in [2,6]. Several applications of SOM-based techniques could be found in rotating machinery diagnostics: the authors of [7,10] investigated the problem of damage detection in roller bearings.

The main objective of the presented study is to develop a SOM-based method, which allows for detection and clustering of internal damages in rotated composite rotors with respect to their shape and location. The developed method is based on the evaluation of strain magnitudes of modal shapes of rotors with delaminations. Results of several tests with different organization of training and validation data were presented and discussed. Furthermore, the ability of application of the proposed method in industrial conditions was pointed out.

2 Description of the Clustering Method

In the following section the idea of SOM and its application in the delamination clustering problem was discussed. A brief description of numerical models used as input in the analyses was presented.

2.1 Self-Organizing Maps

As it was mentioned before, the Kohonen SOM are the most suitable form of unsupervised learning of ANN. The architecture of the Kohonen SOM consists of two layers: the input layer, which is a vector of input neurons, and the output layer, which is commonly a 2D map of neurons (Fig. 1). In some applications the form of output layer could be a vector of neurons or a set of neurons in 3D space [1]. The neurons on a 2D map could constitute the hexagonal or rectangular net. The number of neurons is usually defined as $5\sqrt{N}$, where N is the number of input samples [11]. Each neuron in the input layer is connected with all of the neurons in the output layer. The eventual similarities between the subsequent vectors of input data are visualized on the map.

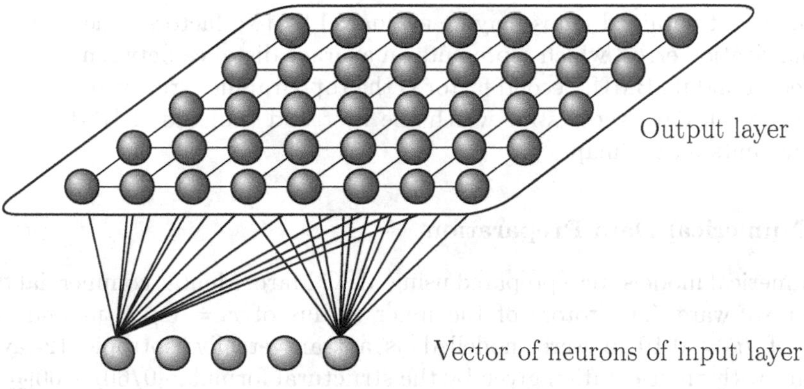

Fig. 1. The structure of the self-organizing map

SOM are commonly used in the problems of analysis of large data structures, primarily in the problems of classification and clustering. The objective of these networks is to reduce the multi-dimensional data to 2D (or sometimes 3D) data structures.

The process of learning of SOM is based on updating of weights of particular neurons of a map (so-called codebook) [8]. In this process the key role has the neighborhood of neurons. For the arbitrary neuron n_i the map has a set of neighbor neurons located in the area defined by the radius r with the origin in the considered neuron. For the map of hexagonal network considered neuron has up to 6 neighbor neurons, while in the case of rectangular map each neuron has 8 neighbor neurons (except the neurons located on the boundaries of the networks). The neighborhood of the neuron could be described by the neighborhood function $h_{ci}[\bullet] \in [0,1]_{\mathrm{R}}$ distributed along the radius r from the neuron n_i. In the next iterations the sequence of input data vectors were considered. For each of them the best neuron, so-called winner or best matching unit (BMU), is determined from the set of all neurons in the map. For this purpose some of the distance metrics e.g. Euclidean, Manhattan, Chebyshev, etc., could be used. The learning consists of two steps. Firstly the Winner Takes Most (WTM) algorithm is executed. During this step the weights of winner neuron and its neighbors are updated following the formula:

$$w_i(k+1) = w_i(k) + \eta(k)h_{ci}[x(k) - w_i(k)], \tag{1}$$

where k is the number of iterations of the learning process, $\eta(k)$ is the learning coefficient, h_{ci} is the value of the neighborhood function of the neuron n_i with respect to the neighborhood of the winner neuron n_c and $x(k)$ is the input vector inputted to the network during k-th iteration.

In the second step the Winner Takes All (WTA) algorithm is executed. During this step only the weights of the winner neuron are updated (i.e. $r = 0$) following the next formula:

$$w_c(k+1) = w_c(k) + \eta(k)[x(k) - w_c(k)]. \tag{2}$$

The quality of obtained maps may be evaluated by two factors. The first one is the quantization error, which represents an average distance between each input data vector and its BMU. Second factor is the topographic error, which describes the ratio of all data vectors for which the first and the second BMUs are not adjacent units on the map.

2.2 Numerical Data Preparation

The numerical models were prepared using MSC Marc/Mentat commercial finite element software. The rotors of the inner radius of $r = 0.038$ m and outer radius of $R = 0.19$ m were modelled as a transversally isotropic 12-layered laminate with an orientation given by the structural formula: $[0/60/-60]_{2S}$ and material properties of the layer in the origin: Young's moduli of $E_1 = 38.283$ GPa, $E_2 = 10.141$ GPa, shear modulus of $G_{12} = 3.533$ GPa and Poisson's ratio of $\nu_{12} = 0.366$ [14]. The thickness of each layer was constant and equals $h_i = 0.0002$ m. Each layer was modelled as a single deformable solid body with perfect contact in between. The model was meshed by 6000 3D hexagonal eight-node elements (500 elements per layer).

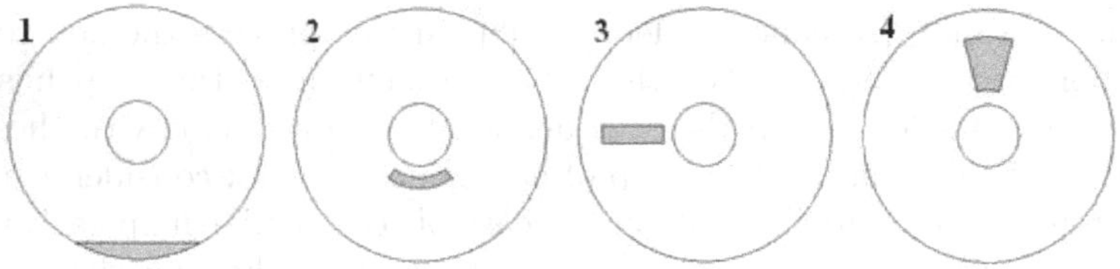

Fig. 2. Groups of the delaminations shapes

Two types of boundary conditions have been applied to the model: the fixing in the inner radius (with free rotation degree of freedom in axial direction) and a centrifugal loading with the frequency of 50 Hz applied to the whole model. The structural analysis was performed in order to determine first ten modal shapes of the rotated rotor. The undamaged case was used further as a reference.

The delaminations were modelled by deactivation of contact constraints in the selected areas. The damaged models were initially divided into four groups according to the shape of delamination (see Fig. 2). The delamination positions were shifted in angular and axial directions, which results in 924 damaged cases: 143 cases in the first group, 440 cases in the second group, 275 cases in the third group and 66 cases in the last one.

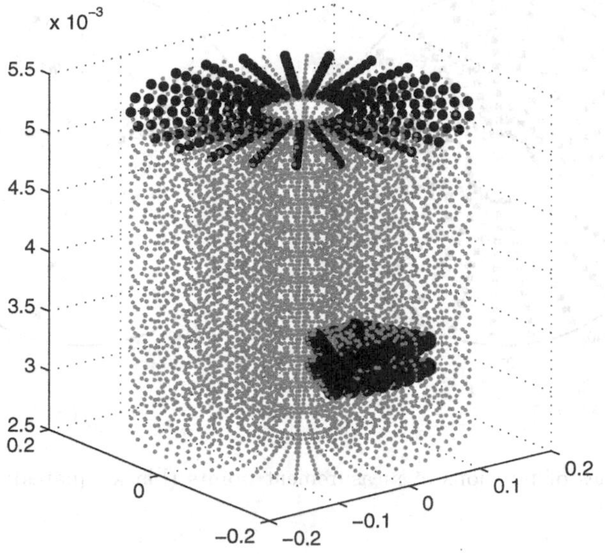

Fig. 3. Example of locations of measurement points and delamination

After performing the calculations, the displacements in radial, angular and axial directions were considered in further analyzes. An exemplary location of measurement points and the delamination was presented in Fig. 3. Obtained natural frequencies do not show any regularity of changes (as it was noticed also in [14]).

2.3 The Clustering Problem

The clustering of the considered damage types consisted of the following steps. The numerical data were prepared as an input data for SOM as follows: the displacements in all directions of measurement points on the top surface of the rotor were considered, thus the dimensions of the input matrix are $924 \times 3k$ for a single modal shape and $924 \times 3km$ for more than one modal shapes taken into consideration, where k is a number of measurement points and m is the number of considered modal shapes. The following cases of the number of points located on the top of a plate (275, 125, 50, 30 and 15 points) considered in the analyzes. An exemplary locations were shown in Fig. 4. The generation of SOM was performed using standard settings implemented in the SOM Toolbox [12], which was used in the analysis. These settings were as follows: hexagonal structure of maps, batch training algorithm, Gaussian function as the neighborhood function, Euclidean metric as the distance measure.

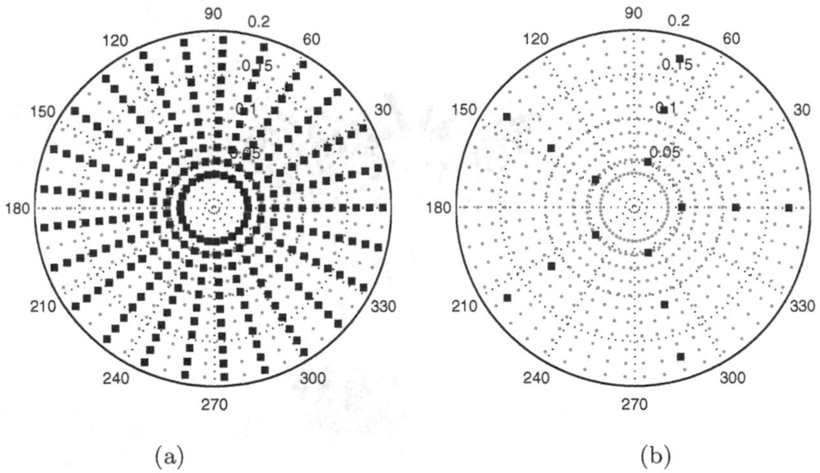

Fig. 4. Examples of positions of measurement points (black squares): (a) 275 points and (b) 15 points

During the analysis several sizes of maps were applied, from 10×10 for a single mode of vibration till 120×120 for all considered 10 modes of vibration. The number of steps in the training process (WTM, WTA) was selected depending on the size of a map and the number of considered modes and changes from (5, 10) for a single mode till (650, 550) for the largest maps and for all of the considered modes.

For each investigated case the maps were generated in the form of U-matrices. For each of them the numbers of adequate input vectors and the information about the number of cases assigned to the given BMU were defined in the nodes, which represented these BMUs. The quality of obtained results was verified basing on the inspection of the maps division and basing on the values of quantization error determined for the successive steps of a training process. In order to carry out the calculations the MATLAB environment with the SOM Toolbox was selected.

3 Results and Discussion

The simulations were carried out for numerous combinations of number of considered modes, maps sizes and the number of training cycles.

3.1 Clustering Results

Basing on the obtained results for single modes (see examples of results in Tab. 1) it is possible to conclude that:

- each of the damaged case (of 924 considered) is assigned to the one of four groups obtained from the clustering,
- each class consisted of one type of four occurred modes with changes in phase of vibration. An example of changes of phase of vibration for 9-th mode was presented in Fig. 5,
- there is no strict correlation between the types of delaminations (defined previously – see Fig. 2) and the obtained classes from the clustering,
- it could be observed that each of the class from clustering consisted 3 of 4 types of delaminations. Moreover, the modes changed by the delaminations of group 1 and 3 occurred only in two classes. Other types of delaminations (2 and 4) were assigned to all of the obtained classes, which influences much on efficiency of the clustering (Fig. 6),
- the number of considered measurement points does not influences much on the clustering results. This fact could be used in the further studies of the sensors placement problem,
- the best clustering to four classes was obtained for 8 and higher modes (see Fig. 7a-b), which could be explained by the higher number of extrema of the displacements for these modes,
- the similarity between obtained classes for various modes was about 30%.

Table 1. Obtained results for selected single modal shapes

Modal shape	Groups of delaminations	Number of cases	Classes 1	2	3	4
2	1	143	0	49	0	94
2	2	440	127	98	56	159
2	3	276	198	0	77	0
2	4	66	24	13	19	10
6	1	143	83	60	0	0
6	2	440	143	114	122	61
6	3	276	0	0	174	101
6	4	66	12	11	25	18
9	1	143	0	58	0	85
9	2	440	89	150	94	107
9	3	276	135	0	140	0
9	4	66	23	14	20	9

The examples of maps achieved for input data in the form of displacements for few modes were shown on Fig. 7c-f. Analysing obtained results it could be concluded that each additional mode considered in the analysis multiplies the number of classes twice. Considering a fact that for a single mode considered the number of classes was four, then for k considered modes 2^{k+1} classes could be obtained. The classes for k modes are created by division from the classes obtained for $k-1$ considered modes.

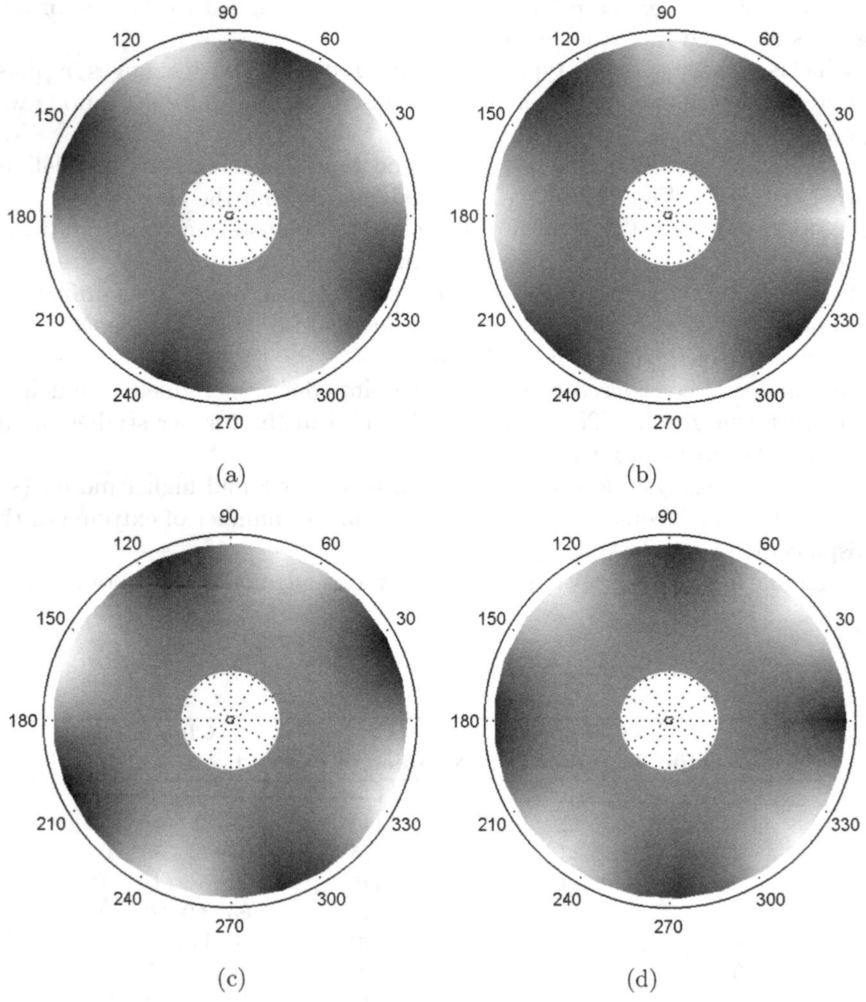

Fig. 5. Phase shifts of ninth modal shape changed by the delaminations

3.2 Ability of Hardware Implementation

Novel measurement systems allow for hardware implementation of the proposed method in order to monitor the rotors condition continuously. Due to the large number of considered measurement points on the surface of the rotor the traditional piezoelectric sensors could not be used, because of the great mass influence on the structure. The promising non-contact technique for measuring

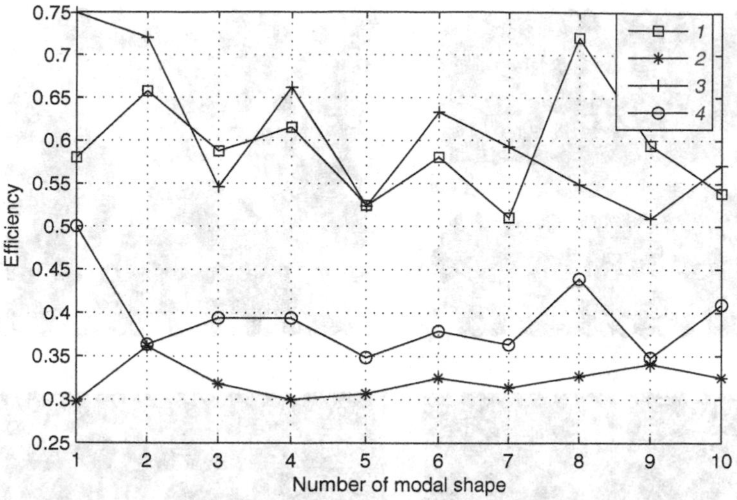

Fig. 6. The dependence of the clustering efficiency on the number of considered modes, $1 \div 4$ – number of groups of delaminations

displacements is based on application of laser Doppler vibrometer together with derotator, which allow for optical freezing of rotated rotor [3]. Another applicable approach could be based on the fiber Bragg gratings optical sensors embedded in the structure [9], however such a solution will be much more expensive.

The measured strain magnitudes in the selected points could be processed by the processor with implemented clustering procedure, which allows for obtaining dynamic Kohonen maps.

4 Conclusions

An approach based on SOM used in this study allows for precise clustering of multidimensional data. The results of considered problem of the delamination clustering with respect to its type and position show that the classes of damaged regions are well recognized, even for such tight damages as delaminations. The tendency of subdividing of classes during the increase of number of considered modes proves that the presence and position of a delamination could be determined with relatively high precision. Moreover, the analyzes show that the decreasing number of the measurement points has insignificant influence on the clustering efficiency, which allows for reducing the number of sensors used for monitoring of rotated rotors in practical applications. The possibilies of practical realizations of measurement systems for rotated rotors were presented and discussed. Proposed approaches could be used for continuous monitoring of composite rotated rotors without the necessity of shutdown the machines. In further

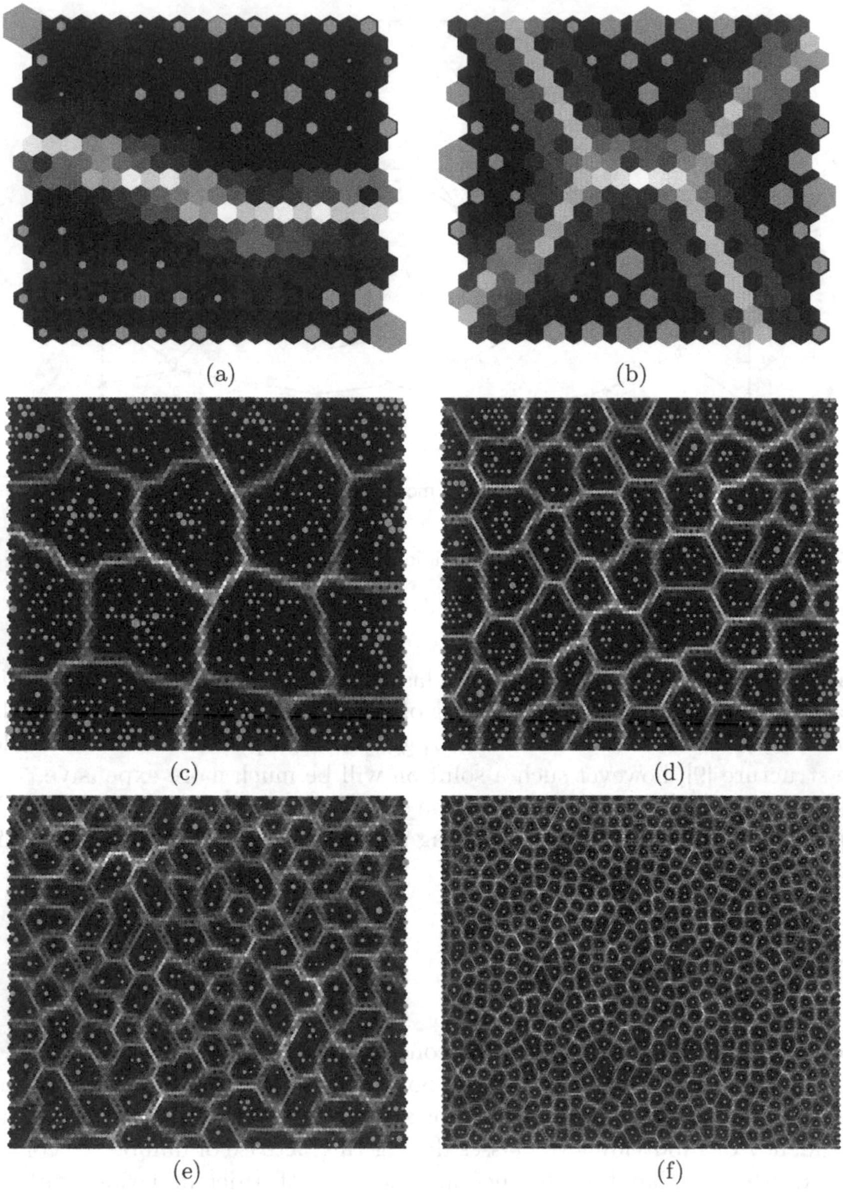

Fig. 7. Examples of obtained maps (a gray hexagons in particular classes represents a BMUs, size of this hexagons is related to the number of input data for which this hexagons is BMU): (a) second modal shape, number of measurement points – 275, size of map – 10×10, training steps – (5, 10); (b) ninth modal shape, 275, 10×10, (5, 10); (c) 5, 7 and 9 modal shapes, 275, 50×50, (40, 15); (d) 6, 7, 8, 9 and 10 modal shapes, 15, 60×60, (350, 250); (e) 1, 2, 3, 4, 5, 6 and 7 modal shapes, 50, 60×60, (150, 45); (f) all modal shapes, 15, 120×120, (650, 550)

studies the development of isolation algorithms for various types of damages in fibrous composites is planned. Moreover, the effectiveness of clustering will be examined for data with a presence of measurement noise.

References

1. Andreu, G., Crespo, A., Valiente, J.M.: Selecting the toroidal self-organizing feature maps (TSOFM) best organized to object recognition. In: International Conference on Neural Networks, vol. 2, pp. 1341–1346 (1997)
2. Bar, H.N., Bhat, M.R., Murthy, C.R.L.: Identification of failure modes in GFRP using PVDF sensors: ANN approach. Compos. Struct. 65, 231–237 (2004)
3. Beeck, M.-A., Hentschel, W.: Laser methodology – a diagnostic tool for automotive development process. Opt. Laser. Eng. 34, 101–120 (2000)
4. Bishop, C.M.: Neural networks for pattern recognition. University Press, Oxford (2000)
5. de Oliveira, R., Franão, O., Santos, J.L., Marques, A.T.: Optic fibre sensor for real-time damage detection in smart composite. Comput. Struct. 82, 1315–1321 (2004)
6. de Oliveira, R., Marques, A.T.: Health monitoring of FRP using acoustic emission and artificial neural networks. Comput. Struct. 86, 367–373 (2008)
7. Hoffman, A.J., van der Merwe, N.T.: The application of neural networks to vibrational diagnostics for multiple fault conditions. Comput. Stand. Inter. 24, 139–149 (2002)
8. Kohonen, T.: Self-Organizing Maps, 3rd edn. Information Sciences, vol. 30. Springer, New York (2001)
9. Loutas, T.H., Panopoulou, A., Roulias, D., Kostopoulos, V.: Intelligent health monitoring of aerospace composite structures based on dynamic strain measurements. Expert Syst. Appl. 39, 8412–8422 (2012)
10. Saxena, A., Saad, A.: Evonving an artificial neural network classifier for condition monitoring of rotating mechanical systems. Appl. Soft Comput. 7, 441–454 (2007)
11. Vesanto, J.: SOM implementation in SOM Toolbox. SOM Toolbox Online Help (2005), `http://www.cis.hut.fi/projects/somtoolbox/documentation/somalg.shtml`
12. Vesanto, J., Himberg, J., Alhoniemi, E., Parhankangas, J.: SOM Toolbox for Matlab 5. SOM Toolbox Team, Helsinki University of Technology, Espoo (2000)
13. Wong, M.L.D., Jack, L.B., Nandi, A.K.: Modified self-organising map for automated novelty detection applied to vibration signal monitoring. Mech. Syst. Signal Pr. 20, 593–610 (2006)
14. Wysogląd, B., Katunin, A.: Optimization of sensors location in rotating disc made of polymer-based laminate for diagnostic needs. In: Burczyński, T., Cholewa, W., Moczulski, W. (eds.). AI-METH Series, pp. 317–326 (2009)
15. Yuan, S., Wang, L., Peng, G.: Neural network method based on a new damage signature for structural health monitoring. Thin Wall. Struct. 43, 553–563 (2005)

Gender Approach to Multi-Objective Optimization of Detection Systems with Pre-selection of Criteria

Zdzisław Kowalczuk and Tomasz Białaszewski

Gdańsk University of Technology, Faculty of Electronics,
Telecommunications and Informatics, Department of Decision Systems, Gdańsk, Poland
kova@pg.gda.pl, bialas@eti.pg.gda.pl

Abstract. Novel idea of performing evolutionary computations for solving highly-dimensional multi-objective optimization (MOO) problems is proposed. The information about individual genders is applied. This information is drawn out of the fitness of individuals and applied during the parental crossover in the evolutionary multi-objective optimization (EMO) processes. The paper introduces the principles of the genetic-gender approach (GGA) and illustrates its performance by means of examples of multi-objective optimization tasks.

Keywords: GA, multi-objectives, Pareto-optimality, applications.

1 Introduction

Designing optimal systems can be performed by following evolutionary mechanisms existing in nature, which allow eliminating detrimental features and inheriting or developing desirable features in the course of genetic computations. Elaborated evolutionary algorithms simulate the natural laws connected with inheritance, crossover and mutation. They make highly efficient tools for gaining optimal solutions of practical meaning. An important determinant in the use of evolutionary algorithms EA in optimization can be seen in intelligent universal realization of iterative procedures of stochastic exploration [11, 23].

The evolutionary algorithms [11, 23, 22] have found a great number of applications [16, 2, 22, 14, 4, 9]. The significance of optimization methods emulating the evolution of biological systems is approved by their usefulness and effectiveness. The features of biological systems are their ability to re-generate, perform self-control, re-product and adapt to the variable conditions of existence. On a similar basis, we also require that analogous features characterize technical systems designed in terms of adaptation, optimality, immunity etc.

It appears clear that in such decision-making and design processes it is essential to globally optimize several objectives at the same time [11, 23, 2, 22]. Such MOO tasks are difficult to be performed, as the notion of optimality is not obvious.

In order to join a number of objectives together, it is necessary to define relations between partial objectives being considered. For the purpose of solving such problems of optimality, various methods are proposed, including: weighted profits [23],

J. Korbicz and M. Kowal (eds.), *Intelligent Systems in Technical and Medical Diagnostics*, Advances in Intelligent Systems and Computing 230,
DOI: 10.1007/978-3-642-39881-0_13, © Springer-Verlag Berlin Heidelberg 2014

distance functions [23], sequential inequalities [17], lexicographic ordering [3], or Pareto-optimality ranking [11, 22, 17]. The substance of the first three methods lies in direct integration of many objectives into one criterion submitted to optimization by using an arbitrary choice of the weighting vector, demand vector or limit values for partial objective functions. Such choices are not straightforward, and they restrict the MOO problem. The method of lexicographic ordering relies on optimization with respect to each objective function in a sequence, starting with the most important one, and proceeding according to the assigned order of importance of partial objectives.

In contrast to the above, the Pareto-optimality ranking avoids the arbitrary weighting of objectives. Instead, a constructive classification of solutions is applied that takes into account particular objectives more objectively. Though this idea of optimality does not give any hints as to the choice of a single solution from a generated set of Pareto-optimal solutions, the designer has always a chance to make an independent judgment of all the 'best' offers. The above-mentioned methods of qualifying MOO solutions can be utilized in the genetic algorithms (GAs) [11, 23].

There are two basic reasons and consequences of the evolution of gender in nature:

- long-term – in search of new mutations, beneficial improvements, and adaptation,
- short-term – for the genetic variation, significant in terms of resistance to parasites.

In the above context, this paper presents a new method, referred to as the genetic-gender approach (GGA) initiated in [17], of solving MOO problems by evolutionary search, with the use of Pareto-optimal ranking, where the information about a degree of membership to a given gender [18] is attributed to each newly generated solution under examination. This information is utilized in the process of parental crossover, in which only individuals of different genders are allowed to create their offspring.

A practical control-design example of the application of the proposed approach to MOO problems, complex FDI design issue [24, 16, 17] of a detection observer [2, 19, 27], which serve as a principal element in the procedures of detecting and isolating faults is considered. This design technique is illustrated with the use of a benchmark problem based on a ship propulsion system [15]. The complex engineering design effect given in the form of a robust optimal detection observer [17, 19] demonstrates both usefulness and effectiveness of the proposed GGA optimization method. Such an optimal system design tool – in accordance with suitable project prerequisites – allows designing systems, which perform their basic task, while having sufficient sensitivity (to errors in sensors and actuators, for instance) and simultaneously showing robustness (to certain modeling uncertainties).

2 GA Multi-Objective Optimization

There are many forms of life that have been created as a result of natural evolution. On the basis of the existing variety of life, we infer that each of the species is optimal with respect to a certain subset of 'survival' criteria.

An analogy can be found within different products of human activity and productivity. There are goods and their variants, many kinds of constructions, bridges, buildings, automotive vehicles, airplanes, home appliances and diverse equipment. We usually make out a specified set of technical criteria with respect to which a given product should be optimal, and this simplifies our choice from amongst of all 'equivalent' solutions. Frequently, we trade off just between price, reliability and safety. Thus considering equally-optimal solutions and making final decisions are parts of human nature. Quite often, based on a given (non-weighted) set of criteria, we are stuck with a selection of solutions which are merely 'mutually non-inferior' [17].

From a formal viewpoint, a MOO task [11, 17] can be defined by means of the following m-dimensional vector of objective functions

$$\boldsymbol{f}(\boldsymbol{x}) = \left[f_1(\boldsymbol{x}) \quad f_2(\boldsymbol{x}) \quad \dots \quad f_m(\boldsymbol{x})\right]^{\mathrm{T}} \in \mathbf{R}^m \tag{1}$$

where $\boldsymbol{x} = \left[x_1 \quad x_2 \quad \dots \quad x_n\right]^{\mathrm{T}} \in \mathbf{R}^n$ means a vector of the parameters searched for, and $f_j(\boldsymbol{x})$, $j = 1,2,\dots,m$, denotes a given partial-objective function. Assuming that all co-ordinates of the criterion vector (1) are profit functions, the MOO problem can be formulated as a multi-profit maximization task without constrains:

$$\max_{\boldsymbol{x}} \boldsymbol{f}(\boldsymbol{x}) \tag{2}$$

A great number of various evolutionary MOO methods [11, 12, 13, 3, 4, 16, 32, 5, 6, 7, 1] have been proposed for solving MOO problems in multi-dimensional spaces, including: VEGA, LOGA, VOES, WBGA, MOGA, NPGA/NPGA2, NSGA/NSGA2, DPGA, TDGA, MOMGA, PAES, PESA(2), SPEA(2), μGA(^{2}), MOBOA, and GGA.

3 The Genetic-Gender Approach

In spite of the range of mechanisms proposed for evolutionary algorithms, there are few isolated attempts of applying sexual categories in genetic reproduction: MSGA [21], G-GA [25], GAGS [26], HRAGA [31], GASS [29], GSGA [30], MOGASS [28]. In contrast to the other approaches [21, 25, 26, 28, 29, 30, 31], in this section we present a novel method for solving MOO tasks based on the EMO approach with the use of a genetic-gender (GG) idea, which consists in assigning each individual a specific gender relating to a degree of its membership to certain sub-populations associated with respective sub-criteria. The gendered individuals are submitted to a 'natural' crossover process. To deliver a proper view on our GGA concept and the resulting GGA algorithm, basic issues and mechanisms of the MOO machinery applied (including rank estimation and global optimality concepts) are explained, before other details of the GGA approach are given.

3.1 High Dimensionality Problem and Study Motivations

While considering MOO problems one always have to be aware of the issue of dimensionality. It is a well-known fact that when the space of the optimized objectives has a higher dimension many individuals fall within the category of being P-optimal, i.e. mutually equivalent in the Pareto sense (forming a P-front). As such, they are assigned the same rank, which, in turn, implies that they are indistinguishable from the P-optimality viewpoint. An outer effect of such a state of the Pareto assessment is a low number of Pareto fronts that obstructs distinction, estimation and ordering of solutions in evolutionary cycles. This also means that during the process of selecting individuals to new generations the Pareto-based ranking is ineffective leading to a strongly stochastic behavior of the genetic algorithm with no 'conscious' movements.

On the other hand, when one confines the scope of optimality by reducing (in some way) the dimension of the analyzed objectives space, the ability of the Pareto optimality method to differentiate between different individuals is also facilitated.

Another motivation for the development proposed for EC, is the fact that in most cases (but for some sporadic attempts mentioned in the introduction to this section) there have been only integrate estimates of the fitness functions applied in the reproduction process of the GA algorithms. This also means using the concept of one 'unisex' parental pool, unlike the pattern followed amongst the species in the nature.

By estimating a global optimality level of solutions we would easily find a difference, if we were looking only for isolation of the solutions of the same P-front. Yet, this need not lead to whatever qualitative change in the EC considered.

3.2 Concept of the Genetic Gender

In nature the gender division of a species appears to differentiate individuals with reference to reproductive functions. According to this division, our concept of an artificial genetic gender (GG) consists in dividing the set of objective functions into several subsets, each of which has a genetic gender Xj, $j = 1,2,...,s$, (Fig. 1) and portrays an assumed partial-scope sub-optimality value for design purposes [18]: GENDERS={X1, X2, ... Xj, ..., Xs}. For example, XX and XY symbolize the native two-element set of the chromosomes which are associated with some distinguishable characteristics of the individuals estimated in terms of the objectives considered.

A gender (set) and a variant (label) are associated with both a subset of criteria and a subset of individuals (best in this context). Basically, the applied division of the set of attributes should be rational and distinctive to achieve the realization of the global optimization goal in a synergy of different genders.

In this way, one gender set (Xj) can embrace objectives of a 'similar' character that are only in a kind of internal (secondary) rivalry in terms of an approximately equal meaning to the user from some pertinent point of view. Such an assortment can thus effectively discharge the designer from the task of isolating a single solution from all the Pareto-sub-optimal ones, obtained in the course of MOO.

On the other hand, different gender sets (Xj) can express various groups of 'interests' that are difficult to be judged by the user in advance. In general, this division can be employed to represent an external (primary) rivalry. On a common basis, in such cases the best method is to use the notion of Pareto-optimality.

Thus we propose a mechanism of the gender allotment during the whole evolution for the purpose of creating parental pools of different genders and generating new offspring by mating only dissimilar individuals (and a set-fitting P-sub-optimality).

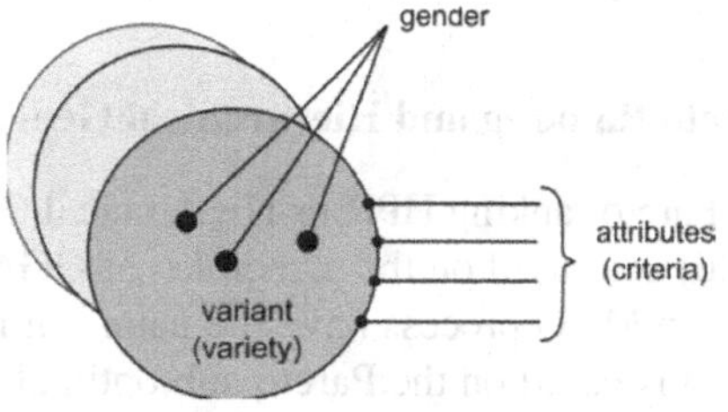

Fig. 1. Concept of the GGA method: a gender (set) with a variant (name) and its attributes

The vector of profits (1) can therefore be divided into s subvectors [18]

$$\boldsymbol{f}(\boldsymbol{x}) = \left[\boldsymbol{f}_1(\boldsymbol{x}) \quad \boldsymbol{f}_2(\boldsymbol{x}) \quad \dots \quad \boldsymbol{f}_s(\boldsymbol{x})\right]^{\mathrm{T}} \in \mathbf{R}^m \tag{3}$$

$\boldsymbol{f}_j(\boldsymbol{x})^{\mathrm{T}} \in \mathbf{R}^{m_i}$, describes the *j*-th subvector ($j = 1,2,\dots,s$) defining the genetic-gender perspective, which will be used to specify the gender set of individuals labeled by X*j*. P-sub-optimality ranking is applied which leads to a vector of ranks:

$$\boldsymbol{r}(\boldsymbol{x}_i) = \left[r_1(\boldsymbol{x}_i) \quad r_2(\boldsymbol{x}_i) \quad \dots \quad r_s(\boldsymbol{x}_i)\right]^{\mathrm{T}} \in \mathbf{R}^s \tag{4}$$

where $r_j(\boldsymbol{x}_i),\ j = 1,2,\dots, s,$ represents the rank of an *i*-th individual within X*j*.

According to GGA, the assignment of the genetic gender l_i to each individual $\boldsymbol{x}_i$ in the population is performed by computing the following procedural quantities:

$$\varphi_i = \max_{j=1,2,\dots,s} \varphi_i^j, \quad l_i = \arg \max_{j=1,2,\dots,s} \varphi_i^j \tag{5}$$

$$\varphi_i^j = r_j(\boldsymbol{x}_i) / r_{j\max}, \qquad r_{j\max} = \max_{i=1,2,\dots,N} \left\{ r_j(\boldsymbol{x}_i) \right\} \tag{6}$$

where φ_i is the highest degree of sub-optimality, a fuzzy measure of the memberships of the *i*-th individual to the l_i -th variant of the genetic gender, while the symbol $r_{j\max}$ denotes the maximum rank with respect to the *j*-th sub-criterion X*j*.

The index of the maximal degree of sub-optimality φ_i determines the gender.

Only individuals of different genders create their offspring. Selecting the parental pool is carried out according to the stochastic-remainder method [11, 17] based on the highest degree of membership to a considered gender set (X*j*). Assuming that the number of individuals in each gender can change in the process of evolution, we monitor a minimum power of the gender sets (e.g. $N/(3s)$). The lacking positions can be augmented by individuals from the lowest Pareto front of another gender set.

We presume that the sizes of the gendered parental pools are similar. This means that in the case of a small ($N/3s$) gender the rivalry between these individuals is weak. Weaker individuals will then have a greater chance of survival. In larger subpopulations the rivalry is much stronger (similarly to the mechanism of niching [17]).

3.3 Hierarchical Pareto Ranking and Hierarchical Genetic Gender

The idea of Hierarchical Pareto ranking HPR or Hierarchical Genetic Gender HGG of the analyzed solutions [20] are based on the principles of GGA and observations that human decisions made in an MOO process have the nature of hierarchical evaluation. Such an approach to MOO is based on the Pareto sub-optimal ranking. During evolutionary cycles the dynamic assignment of genders is not considered, as well as no gender restrictions are imposed on the crossover process.

In the process of creating hierarchy, a primary partition of the complete set of criteria into disjoint subsets, referred to as primary variants, is first performed. As before, the variants are marked with labels, called primary genders. Next, a similar treatment is carried out with respect to the obtained set of primary genders. As a result, we obtain a master collection of secondary genders, each representing a group of primary genders. A resulting structure can be suitably described by a tree structure.

Fig. 2(a) shows an example of division of 12 primary objective functions into four primary variants. In the second step, a complete division of the primary genders (low level) into 2 disjoint subsets of secondary 'genders' (high level) is performed. Suitable labels can be applied here to name the secondary genders. Another simple case is shown in Fig. 2(b), illustrating a 'gender distribution' of criteria using solely three genders (corresponding to the one-level gender resolution applied in GGA).

The vector (4) is applied as the representation of primary genders, which are next subject to distribution into a set of secondary variants. At both levels, the assessment of individuals is completed in the Pareto sense. It is performed at level I with respect to the degree φ_i^j of sub-optimality (5) within each original variant and get ranks. At level II, the ranks of genders (computed at the first level and next treated as fitness) selected to a secondary (aggregate) variant, are normalized and Pareto-assessed.

This process of derivative P-assessments can be generalized to higher levels, if such a hierarchical Pareto ranking (HPR) with multi-levels will better match the nature of the analyzed set of criterion functions. Exemplary hierarchies matching the way of distribution of the criterion functions of Fig. 2 are shown in Fig. 3.

Fig. 3(a) describes the optimization task consisting of 12 criteria, which are divided into four groups of Fig. 2(a). Inside each of them, the Pareto ranking (I-level) is carried out for all individuals in the population. As a result, a corresponding assessment vector of four normalized ranks is obtained. These ranks, based on P-assessment of two twofold ranks, are next used in the construction of level II of variants/genders.

In the course of the second-level HPR evaluation performed on secondary variants (II-level sub-criteria), each individual achieves a scalar rank (needing normalization).

Fig. 3(b) presents an exemplary case of a simple bi-level analysis, where nine criterion functions are divided into three groups associated with three variants. With the adopted hierarchy of multi-criteria, the II-level Pareto re-ranking assessment of all the solutions concerns a three-dimensional criterion, and leads to scalar evaluation.

Using of HPR can be extended to the highest level of aggregation (III and II for Fig. 2 and Fig. 3) meaning a scalar global estimation (a challenger for GOL).

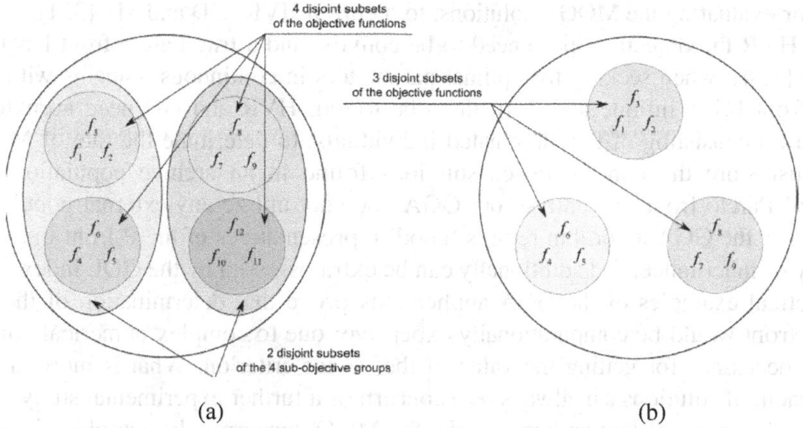

(a) (b)

Fig. 2. HPR: Exemplary distribution of objectives into 3-level (a) and 2-level (b) hierarchies

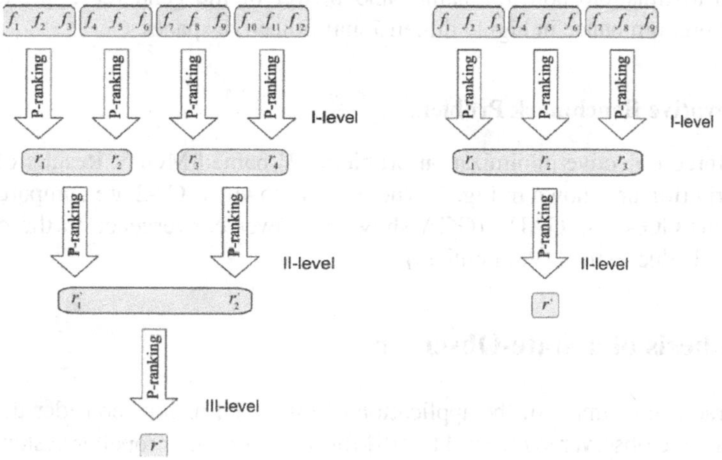

(a) (b)

Fig. 3. Exemplary ranking according to: (a) 3-level HPR, (b) 2-level HPR

The scalar ranks of individuals achieved in HPR, can be used in a number of ways for selecting individuals into the parental pool. This process can be supported, for example, by means of the method of stochastic remainder choice [17].

A global optimality level (GOL) based on the max-min principle, prefers solutions that are located in the middle of the Pareto front, and turns down the boundary solutions typical for high dimensional objective spaces being searched. It is an important feature, as from a practical viewpoint, solutions which are "excellent" with respect to a single criterion are unacceptable. The main point of using GOL during the evolutionary computations lies in measuring the progress of the optimization process of finding the Pareto-optimal solutions. Additionally, we apply GOL as a supporting measure for selecting the ultimate solution (as a final optimality criterion).

Certainly, there are a number of standard performance indices, which can be applied for evaluating the MOGA solutions, to mention HVR, GD and MS [3, 1].

For HVR the objective space need to be convex, and a true Pareto front has to be known [1, 4]; when seeking for optimal parameters in continuous domains with GD, the P-front [4] is infinite and difficult to be found; HVR and GD need an external population consisting of non-dominated individuals; to determine the rate of MS the GA must store the non-dominated solutions found in an archive population as a 'known' Pareto front. In contrast, our GGA does not utilize any external population. Moreover, the GGA algorithm returns 'good' representatives of the P-front in a natural way of inheritance, and additionally can be extra-assessed by the GOL index.

Practical examples of the GGA applications prove that determination of the true Pareto front would be computationally expensive: due to complex numerical computations necessary for getting the value of the vector criterion. What is more, a final assessment of solutions can always be a concern of a further experimental study.

Note that most of known benchmarks for MOO concern only two objectives and they are not interesting patterns in the context of GGA. On the contrary, our examples are intended to illustrate both the nature and power of the gendered approach that reveals their predominance in highly dimensional objective spaces.

3.4 Illustrative Benchmark Problem

Consider a three-objective minimization problem [3] named MOP5. Results of minimizing its criterion are shown in Fig. 4, where the renown NSGA2 are compared with our algorithms GGA and GAHP (GGA shows a slower convergence to the optimal solutions that is due to the lack of elitism).

4 Synthesis of a State-Observer

To give a practical example of the application of GGA, let us now consider the problem of linear state observer synthesis [17, 19] for a ship engine-propeller system [15].

FDI systems are founded on two principal operations: (a) detecting the occurrence of faults, and (b) isolating them. Such systems ensure a reliable operation of systems [16], where the presence of errors in the system components is unacceptable.

FDI systems compare the measurements of the plant with their predictions based on mathematical models of the monitored object. The differences between the corresponding signals are called residues. They are influenced by disturbances, noise and modeling errors. Fault detection can be achieved by appropriate filtration of the residues, and diagnostic decisions can be made on the basis of their appropriate evaluation. A practical diagnosis system is thus founded on a residue generator [24, 16, 17, 27], composed of a state observer and an additional filter.

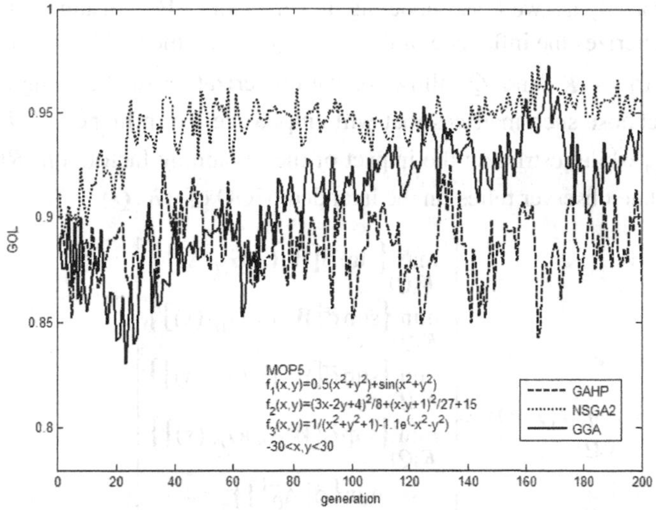

Fig. 4. Global optimality level GOL in a simulation run for a benchmark problem MOP5

4.1 Design of Residue Generators

The state observer can be expressed in the following form [24, 16, 17, 27]

$$\dot{\hat{\boldsymbol{x}}}(t) = (\boldsymbol{A} - \boldsymbol{KC})\hat{\boldsymbol{x}}(t) + (\boldsymbol{B} - \boldsymbol{KD})\boldsymbol{u}(t) + \boldsymbol{Ky}(t) \tag{7}$$

$$\hat{\boldsymbol{y}}(t) = \boldsymbol{C}\hat{\boldsymbol{x}}(t) + \boldsymbol{Du}(t) \tag{8}$$

where $\hat{\boldsymbol{x}}(t) \in \mathbf{R}^n$ is a state-vector estimate, $\hat{\boldsymbol{y}}(t) \in \mathbf{R}^m$ constitutes an estimated system output, while $\boldsymbol{K} \in \mathbf{R}^{n \times m}$ stands for a matrix observer gain, $\boldsymbol{u}(t) \in \mathbf{R}^p$ is a control vector, $\boldsymbol{y}(t) \in \mathbf{R}^m$ stands for a measurement vector. The matrices $\boldsymbol{A} \in \mathbf{R}^{n \times n}$, $\boldsymbol{B} \in \mathbf{R}^{n \times p}$, $\boldsymbol{C} \in \mathbf{R}^{m \times n}$, $\boldsymbol{D} \in \mathbf{R}^{m \times p}$, describe an underlying state-space model of the plant.

The vector fault $\boldsymbol{f}(t)$ is represented by an unknown time function and the influence of this fault on the state evolution of the system and on the measurements.

The weighted residual signal is generated as follows:

$$\boldsymbol{r}(t) = \boldsymbol{Q}(\boldsymbol{y}(t) - \hat{\boldsymbol{y}}(t)) \tag{9}$$

where the matrix $\boldsymbol{Q} \in \mathbf{R}^{r \times m}$ of weights makes an additional design parameter. The Laplace transformation of the residual signal results in an operator function:

$$\begin{aligned} \boldsymbol{R}(s) = {} & \boldsymbol{G}_{rf}(s) \cdot \boldsymbol{F}(s) + \boldsymbol{G}_{rd}(s) \cdot \boldsymbol{D}(s) + \\ & + \boldsymbol{G}_{rw}(s) \cdot \boldsymbol{W}(s) + \boldsymbol{G}_{rv}(s) \cdot \boldsymbol{V}(s) + \boldsymbol{G}_{re}(s) \cdot \boldsymbol{e}(0) \end{aligned} \tag{10}$$

Particular matrix transfer functions describe the influence of all critical factors: faults, external disturbances, state and measurement noise, initial conditions of the state estimation process, as well as modeling uncertainties. For instance, the function $\boldsymbol{G}_{rf}(s)$ characterizes the influence of the faults $\boldsymbol{f}(t)$ on the residue $\boldsymbol{r}(t)$.

Certain matrices $\boldsymbol{K}$ and $\boldsymbol{Q}$ allow for parameterization of the designed detector. We can thus choose such matrices that will emphasize the influence of $\boldsymbol{F}(s)$ on the residue $\boldsymbol{R}(s)$, while restricting the impact of the remaining factors on $\boldsymbol{R}(s)$. A synthesis of the state observer relies on the intended MOO of $(\boldsymbol{K},\boldsymbol{Q})$:

$$\underset{(\boldsymbol{K},\boldsymbol{Q})}{opt}\ \boldsymbol{J}(\boldsymbol{K},\boldsymbol{Q}) \doteq \begin{bmatrix} \max\limits_{(\boldsymbol{K},\boldsymbol{Q})} \left\{\sup\limits_{s} \overline{\sigma}[\boldsymbol{W}_1(s)\boldsymbol{G}_{rf}(s)]\right\} \\ \min\limits_{(\boldsymbol{K},\boldsymbol{Q})} \left\{\sup\limits_{s} \overline{\sigma}[\boldsymbol{W}_2(s)\boldsymbol{G}_{rd}(s)]\right\} \\ \min\limits_{(\boldsymbol{K},\boldsymbol{Q})} \left\{\sup\limits_{s} \overline{\sigma}[\boldsymbol{W}_3(s)\boldsymbol{G}_{rw}(s)]\right\} \\ \min\limits_{(\boldsymbol{K},\boldsymbol{Q})} \left\{\sup\limits_{s} \overline{\sigma}[\boldsymbol{W}_4(s)\boldsymbol{G}_{rv}(s)]\right\} \\ \min\limits_{(\boldsymbol{K})} \left\{\overline{\sigma}[\boldsymbol{A}_0^{-1}]\right\} \\ \min\limits_{(\boldsymbol{K})} \left\{\overline{\sigma}[\boldsymbol{A}_0^{-1}\boldsymbol{K}]\right\} \end{bmatrix} \tag{11}$$

where $\overline{\sigma}[\cdot]$ denotes a maximal singular value of matrix, while $\boldsymbol{W}_i(s)$ are weighting functions, allowing for separating the effects of faults from disturbances and noises.

The first co-ordinate of (11), measuring the influence of faults $\boldsymbol{f}(t)$ on $\boldsymbol{r}(t)$, is maximized. The next three co-ordinates describe of the impact of disturbances, input noises and measurement noises on state evolution. The last two co-ordinates, describing the effect of deviations from the nominal plant model, give a robustness measure.

The method of sequential inequalities [3] was utilized by [2] in an EMO optimization procedure. In this approach, the cost indices are expressed in the frequency domain and transformed into a set of inequality constraints, which are tested for a finite set of frequencies. A genetic algorithm is then applied to seek for optimal solutions satisfying all inequality constraints, and the eigenstructure assignment approach is employed to get an appropriate parameterization of the matrix $\boldsymbol{K}$.

The analyzed MOO problem is solved by a method that incorporates both Pareto optimality and genetic search in the frequency domain. The design of residue generators is based on optimization of $\boldsymbol{J}(\boldsymbol{K},\boldsymbol{Q})$. The ranking derived from P-optimality is employed to assess the solutions generated by GA operating on multi-alleles [17].

To guarantee that genetic optimization yields only permissible solutions, we search for eigenvalues formed in an n-vector $\lambda \subset \mathbf{C}^n$ ($\mathrm{spectr}[\boldsymbol{A}_0] \subset \mathbf{C}_-$), on the basis of which the matrix $\boldsymbol{K}$ is calculated by means of the pole placement method [17].

By setting the matrix $\boldsymbol{Q}$ to identity and accordingly fixing the frequency weighting matrices $\boldsymbol{W}_i(s)$, the problem of optimization can be reduced to the following task:

$$\underset{(\boldsymbol{K},\boldsymbol{Q})}{opt}\ \boldsymbol{J}(\boldsymbol{K},\boldsymbol{Q}) = \underset{\boldsymbol{K}}{opt}\, \boldsymbol{J}(\boldsymbol{K}) = \underset{\lambda}{opt}\, \boldsymbol{J}(\boldsymbol{K}(\lambda)) = \underset{\lambda}{opt}\, \boldsymbol{J}(\lambda) \tag{12}$$

4.2 Ship Propulsion System

To detail our discussion let us consider the ship propulsion system of a low-speed marine vehicle [15, 17] that consists of one engine and one propeller. Such a system is fundamental for a maneuvering ship. A failure of the propulsion unit may easily cause various dangerous events (collisions, drifting to shallows, financial and ecological losses). Such circumstances imply the necessity of monitoring the propulsion system.

A linearized model of the ship propulsion is described in the continuous-time domain by the state-space model with the above-mentioned matrices $\boldsymbol{A}$, $\boldsymbol{B}$, $\boldsymbol{C}$, and $\boldsymbol{D}$.

The faults in the observed object are associated with the potentiometer sensor of the pitch angle of the propeller, the tachometer sensor of the angular velocity of the shaft, and the diesel engine itself. Another fault, indirectly associated with this sensor, is hydraulic leakage, which brings about a slow change of a propeller pitch angle [15].

4.3 Results of Evolutionary Optimization

The criteria applied in the optimization task (11) were divided into three gender sets. The first gender is composed of the performance criterion (the impact of faults). The second gender embraces the three insensitivity criteria (the influence of disturbances and noise). The third gender consists of the two robustness measures.

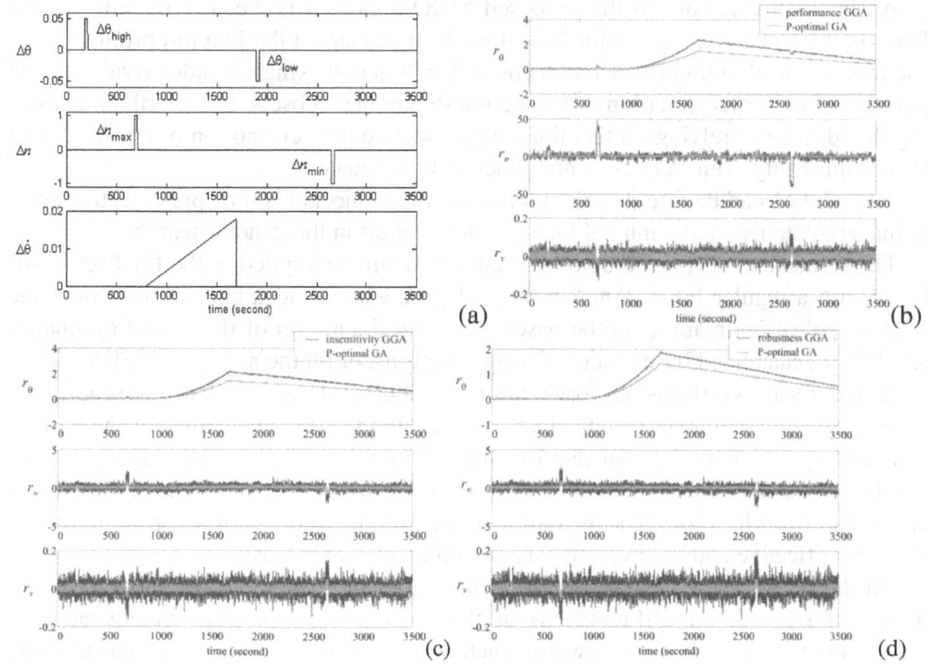

Fig. 5. Additive faults (a) along with residuals for the performance (b), insensitivity (c), and robustness (d) GG-observers against a classical GA solution

Verifying simulations of the performance of the state observers designed with the use of GGA were performed in the presence of faults, noise and other external disturbances [15, 17]. The additive faults considered are shown in Fig. 5a.

As can be easily seen, practically all the faults of Fig. 5a give distinctive symptoms in at least one of the residues. What is more, the residuals demonstrate changes analogous to the generic fault signal applied. It is obvious that with the use of appropriate filtration the symptom information included in the residues makes it possible to detect and isolate the faults (though, the temporary pitch-sensor fault effects are less clear).

The three different gender solutions and a total P-optimal solution are compared in 5b-d. All the three gender solutions are more efficient than the classical GA ones. The performance-gender state-observer reveals a greater ability to detect faults as compared to the other state observers (refer, for instance, the scale of r_n in Fig. 5b).

5 Conclusions

The proposed method of solving MOO problems is based on the evolutionary search with the recognition of genetic genders. Information about the degree of membership to a given gender set is extracted in the Pareto-sub-optimal process of ranking the fitness functions of solutions. This information is exploited in the crossover process of mating, in which only individuals of different genders are allowed to create offspring.

An instructive feature of the proposed MOO approach is the way of utilizing the Pareto-optimization results. Namely, within each gender set the Pareto-optimization is used as a tool of sub-optimal judgment of the 'internal' single-gender rivals for the purpose of uniform estimation and selection to parental subsets. It is worth emphasizing that despite of relying on this limited perspective the very notion of the set-fitting P-sub-optimality is entirely clear and practically adequate.

The method can be interpreted in terms of a new mechanism of pre-selecting individuals (solutions), and a mutual inter-gender support in the genetic search.

The standard concept of Pareto-optimality can still be applied to the final set of solutions on a regular basis. Another way of processing, and taking into account the 'full scope' of optimality, can be based either on the notion of the global optimality level GOL, calculated based on the fitness functions, or on the notion of HPR/HGG.

A major success of this approach can be attributed to the fact that it appropriately deals with a great number of objectives by reducing the dimensionality of the Pareto-analyzed spaces. Observe that due to a high dimension of the objectives space, the number of Pareto fronts is limited. This means that many individuals are estimated as equivalent (they have of the same rank). As a result, the process of selecting individuals is not effective and the evolutionary search is overly stochastic.

On the contrary, by introducing the gender approach we solve the above issue by means of restricting the dimensions of the objective spaces and bringing about a greater number of Pareto fronts within each gender population analyzed in a subspace of a restricted dimension. This, in turn, brings about diversity among the individuals that can be easily estimated and used in effectively pushing the evolutionary exploration into the desired directions on the basis of the achievable distinctive ordering.

As compared to other methods our approach is fairly simple, sticks to the very basics of the GA methodology, more profoundly utilizes hints from the nature, fulfills the requirements of technical design and ultimate decision making, and, clearly, is completely open to other amendments and developments proposed in the literature (see [11, 12, 11, 13, 32, 63, 3, 17], for instance).

Our approach is also entirely different from other propositions. Although showing several instrumental consequences, the GGA method has a conceptual nature consisting in the objective space decomposition of the initial problem. It is interesting to see that the VEGA algorithm has few limited similarities. Namely, the VEGA partial parental pools can be assigned genders from a maximum m-element set, though it does not make use of the Pareto optimality conception, its genders are not exclusive (one individual can have a number of genders), and crossover has no restrictions.

References

1. Bader, J., Zitzler, E.: A hypervolume-based optimizer for high-dimensional objective spaces. In: Jones, D., Tamiz, M., Ries, J. (eds.) Conference on Multiple Objective and Goal Programming. LNEMS, vol. 638, pp. 35–54. Springer, Heidelberg (2009)
2. Chen, J., Patton, R.J., Liu, G.: Optimal residual design for fault diagnosis using multi-objective optimization and genetic algorithms. International Journal of Systems Science 27(6), 567–576 (1996)
3. Coello, C., Lamont, G., Van Veldhuizen, D.: Evolutionary algorithms for solving multi-objective problems. In: Genetic and Evolutionary Comutation. Springer (2007)
4. Cotta, C., Schaefer, R.: Special Issue on Evolutionary Computation. International Journal of Applied Mathematics and Computer Science 14(3), 279–440 (2004)
5. Deb, K.: Current trends in evolutionary multi-objective optimization. Intern. Journal for Simulation and Multidisciplinary Optimisation 1(1), 1–8 (2007)
6. Deb, K., Gupta, H.: Introducing robustness in multi-objective optimization. Evolutionary Computation Journal 14(4), 463–494 (2006)
7. Deb, K., Mohan, M., Mishra, S.: Evaluating the domination based multiobjective evolutionary algorithm for a quick computation of Pareto-optimal solutions. Evolutionary Computation Journal 13(4), 501–525 (2005)
8. Deb, K., Pratap, A., Argarwal, S., Meyarivan, T.: A fast and elitist multi-objective genetic algorithm: NSGA- II. Tech. Report (200001) GA Lab., Kanpur, India (2000)
9. Dridi, M., Kacem, I.: A hybrid approach for scheduling transportation networks, Intern. Journal of Applied Mathematics and Computer Science 14(3), 397–409 (2004)
10. Fonseca, C.M., Fleming, P.J.: Genetic algorithms for multi-objective optimization: Formulation, discussion and modification. In: Forrest, S. (ed.) Proc. 5th Int. Conf. on Genetic Algorithms, pp. 416–423. Morgan Kaufmann, San Mateo (1993)
11. Goldberg, D.E.: Genetic Algorithms in Search, Optimization and Machine Learning. Addison-Wesley, Reading (1989)
12. Hajela, P., Lin, C.Y.: Genetic search strategies in multicriterion optimal design. Structural Optimization 4, 99–107 (1992)
13. Horn, J., Nafpliotis, N., Goldberg, D.E.: A niched Pareto genetic algorithm for multi-objective optimization. In: IEEE World Congr. on Comput. Computation, vol. 1, pp. 82–87 (1994)

14. Huang, Y., Wang, S.: The identification of fuzzy grey prediction systems by genetic algorithms. International Journal of Systems Science 28(1), 15–24 (1997)
15. Izadi-Zamanabadi, R., Blanke, M.: A ship propulsion system model for fault-tolerant control. Technical Report, Aalborg University, Denmark (4262) (1998)
16. Korbicz, J., Kościelny, J.M., Kowalczuk, Z., Cholewa, W.: Fault Diagnosis. Models, Artificial Intelligence, Applications. Springer, Berlin (2004)
17. Kowalczuk, Z., Białaszewski, T.: Genetic algorithms in multi-objective optimization of detection observers. In: [16], pp. 511–556 (2004)
18. Kowalczuk, Z., Bialaszewski, T.: Improving evolutionary multi-objective optimization using genders. In: Rutkowski, L., Tadeusiewicz, R., Zadeh, L.A., Żurada, J.M. (eds.) ICAISC 2006. LNCS (LNAI), vol. 4029, pp. 390–399. Springer, Heidelberg (2006)
19. Kowalczuk, Z., Białaszewski, T.: Designing FDI observers by improved evolutionary multi-objective optimization. In: Proc. 6th IFAC Symposium on Fault Detection, Supervision and Safety for Techn. Processes, Beijing, China, pp. 601–606 (2006)
20. Kowalczuk, Z., Białaszewski, T.: Gender selection of a criteria structure in multi-objective optimization of decision systems (in Polish). PAK 57(7), 810–814 (2011)
21. Lis, J., Eiben, A.: A multi-sexual genetic algorithm for multiobjective optimization. In: Proc. of IEEE International Conference on Evolutionary Computation, pp. 59–64 (1997)
22. Man, K.S., Tang, K.S., Kwong, S., Lang, W.A.H.: Genetic Algorithms for Control and Signal Processing. Springer, London (1997)
23. Michalewicz, Z.: Genetic Algorithms + Data Structures = Evolution Programs. Springer, Berlin (1996)
24. Patton, R.J., Frank, P.M., Clark, R.N. (eds.): Fault Diagnosis in Dynamic Systems. Theory and Application. Prentice Hall, New York (1989)
25. Rejeb, J., AbuElhaija, M.: New gender genetic algorithm for solving graph partitioning problems. In: 43rd IEEE Midwest Symposium on Circuits and Systems, vol. 1, pp. 444–446 (2000)
26. Sanchez-Velazco, J., Bullinaria, J.A.: Sexual selection with competitive/cooperative operators for genetic algorithms. In: Proc. the IASTED Intern. Conf. on Neural Networks and Computational Intelligence, pp. 191–196. ACTA Press (2003)
27. Suchomski, P., Kowalczuk, Z.: Robust H¥-optimal synthesis of FDI systems. In: [16], pp. 261–298 (2004)
28. Sodsee, S., Meesad, P., Li, Z., Halang, W.: A networking requirement application by multi-objective genetic algorithms with sexual selection. In: 3rd International Conference Intelligent System and Knowledge Engineering, vol. 1, pp. 513–518 (2008)
29. Song, G.K., Lim, A., Rodrigues, B.: Sexual selection for genetic algorithms. Artificial Intelligence Review, 123–152 (2003)
30. Vrajitoru, D.: Simulating gender separation with genetic algorithms. In: Proceedings of the Genetic and Evolutionary Computation Conference, pp. 634–641 (2002)
31. Yan, T.: An improved genetic algorithm and its blending application with neural network. In: 2nd International Workshop Intelligent Systems and Applications, pp. 1–4 (2010)
32. Zitzler, E., Thiele, L., Bader, J.: On set-based multi-objective optimization. IEEE Transactions on Evolutionary Computation 14(1), 58–79 (2010)

Importance Discounting as a Technique of Expert Knowledge Incorporation into Diagnostic Decision-Making Process

Wojciech Jamrozik

Institute of Fundamentals of Machinery Design, Silesian University of Technology
Konarskiego 18a 44-100 Gliwice, Poland
wojciech.jamrozik@polsl.pl

Abstract. In the paper the possibility of the use of importance discounting as a method of taking into account preferences of experts in a diagnostic system based on Desert-Smarandache theory is discussed. Reliability discounting is a well-known technique for belief assignments conditioning, having regard to objective properties of source of evidence. Indeed, in real applications, it is needed to differentiate between objective properties and subjective preferences of experts having proper knowledge, experience or intuition connected with the machine or the process being diagnosed. Way of importance discounting incorporation into a diagnostic system, considering gathering of expert's opinions and the conversion from opinions to values of discounting factors is described. Theoretical reflections are supported by a practical example.

Keywords: Technical diagnostics, Classifier fusion, DSmT, Importance discounting.

1 Introduction

Solution of problems in the field of technical diagnostic is often found on the way of classification. In the last years many classification algorithms as well as classification schemes were elaborated. A more and more popular concept called classifier fusion is based on the assumption that joining of several classifier outputs can lead to better results as in the case of a single classifier. There are numerous methods, providing different attempts to the fusion process. All known methods can be divided into three basic groups: fusion on the abstract level (label outputs), fusion on the rank level and fusion on the measurement level (continuous outputs). Most sophisticated methods are working with continuous classifier outputs.

In real-life applications, classification is mostly made on basis of imprecise, uncertain and incomplete data. According to that classifier answers are also affected by low quality of input data. To overcome such drawbacks theories of belief functions, originally known as Dempster-Shafer Theory (DST) [1] and developed in the framework of Desert-Smarandache Theory (DSmT) [2] are applied

J. Korbicz and M. Kowal (eds.), *Intelligent Systems in Technical and Medical Diagnostics*, Advances in Intelligent Systems and Computing 230,
DOI: 10.1007/978-3-642-39881-0_14, © Springer-Verlag Berlin Heidelberg 2014

in classifier fusion. Those methods are well situated and applied in many areas, just to mention diagnosis of induction engines [3], machine condition monitoring [4], railway track circuit diagnosis [5],[6], target recognition [7] and in many others decision support applications.

In diagnostic systems data comes from different sources, mostly from various physical sensors or human expert report. Data and/or information gathered for classification purposes in diagnostic systems describes some phenomena being observed in order to recognize technical object or process state. Properly processed data constitutes a basis (training and testing feature set) on which further classification is performed. When applying DST or DSmT for fusion of classifier outputs there are two main problems, with which a designer of fusion systems has to face. First is the choice of best fusion rule, that allows aggregation of classifier outputs expressed in form of basic belief assignments (BBAs) or belief masses. Second issue is connected to the fact that various sensors involved in the data acquisition have different reliability. This property describes objectively the ability of sensor (physical or other) and in wider context ability of source of evidence to generate the correct assessment for certain considered application. It is often necessary to incorporate some knowledge about the whole path of diagnosis making process that begins in the measurement point (sensor) through signal processing stage to classifier responsible for providing diagnostic decisions. Human experts, operators familiar with an object being the subject of a diagnostic task, could have experience or only preferences according to importance of sensors, signal processing methods, etc. Thus in contrary to the reliability, the importance of source reflects subjective opinions expressed by fusion system designer (taking into consideration opinions of other staff involved in the elaboration of the diagnostic system). Counter-intuitive decision results could be obtained if unequal evidence sources are fused and both these factors are not taken into consideration.

Differentiation between reliability and importance is a relatively new concept [8], thus number of practical application is limited. Importance discounting was applied in multi-criteria decision system for prediction of natural hazards in mountains [9] or to aid designer in aerospace industry [10]. No applications in the field of technical diagnostics are known to the author.

As reliability and importance are interpreted in different ways, so the fusion systems should allow to differentiate between them. Efficient application method of both concepts through discounting operation into DSmT based fusion system using Proportional Conflict Redistribution rules no 5 or its extension denoted as no 6 (PCR5 or PCR6) was presented in [8].

There are several methods dedicated to the reliability discounting factors assessment, mainly based on analysis of conflict between classifier outputs or performance of each classifier. In the case of importance, values of discounting factors are determined directly by human experts [10], resulting in the lack of possibility to recognize expert motivations. Additionally values assigned directly can be arbitrary, without reflecting experts intention. Thus more formal technique of opinion gathering is desirable. In the paper application of multi-criteria decision making method called Analytic Hierarchy Process (AHP) [11] for importance discounting factor is presented. In this approach sources of evidence are compared pairwise according to several named and prioritized criteria. The result of AHP are values of priorities computed for all alternatives (sources).

The paper is organized as follows. In the section 2 a brief reminder of basics of DSmT for information fusion and PCR6 fusion rule is presented, followed by the description of reliability and importance discounting techniques. Next the application of AHP method for assessing importance discounting factors is presented. Finally in section 3 example of presented method usage is shown. Conducted research is concluded in section 4.

2 Belief Functions Theory

Dempster-Shafer Theory [1] is a generalization of fundamental probability theory. In the framework of DS mathematical apparatus for aggregating beliefs and making decision basing on them is available. The frame of discernment $\Theta = \{\theta_1, \ldots, \theta_n\}$ contains a finite set of n exclusive and exhaustive hypotheses representing object or process technical states (or faults). The set of subsets of Θ is denoted by the power set 2^Θ, where $\emptyset, \theta_1, \ldots, \theta_n \in 2^\Theta$ and if subsets $F_1, F_2 \in 2^\Theta$, than $F_1 \cup F_2 \in 2^\Theta$.

Dezert-Smarandache Therory [2] is an extension of classic DS theory, which can deal with both paradoxical and uncertain information. It is based on generalized frame of discernment $\Theta = \{\theta_1, \ldots, \theta_n\}$, where elements θ_i form an exhaustive set of prepositions. In this case exclusivity of elements is not required. Hyper power-set D^Θ being the free Dedekind's lattice is defined as the set of subsets of Θ, where $\emptyset, \theta_1, \ldots, \theta_n \in D^\Theta$ and if $F_1, F_2 \in D^\Theta$, than $F_1 \cup F_2 \in D^\Theta$ and $F_1 \cap F_2 \in D^\Theta$. In real-world problems some hypothesis (states) can not occur simultaneously, e.g. normal operation and state representing any fault. Thus integrity constraints have to be put on the D^Θ, generating reduced set, D_r^Θ with lower cardinality as the original D^Θ.

A mass function or basic belief assignment, m is defined by the mapping of the hyper power set D^Θ onto $[0, 1]$ with: $m(\emptyset) = 0$, $\sum_{X \in D^\Theta} m(X) = 1$, where X is a subset of Θ defined in D^Θ. Element $X \in D^\Theta$ having $m(X) > 0$ is called focal element of $m(\cdot)$. The credibility Bel and plausibility Pl functions are defined as follows:

$$Bel\,(F) = \sum_{X \in D^\Theta, X \subseteq F, X \not\equiv \emptyset} m\,(X)\,, \qquad Pl\,(F) = \sum_{X \in D^\Theta, F \cap X \not\equiv \emptyset} m\,(X)\,. \quad (1)$$

Credibility shows total degree of trust in all subsets of F. Plausibility represents in contrast total degree of trust for not denying the subset F.

Proportional conflict redistribution rules To make the decision only on the elements in Θ, several rules propose to redistribute the conflict on these elements. The most accomplished is the PCR5 given in [2] for two sources of evidence and its generalization for M sources of evidence PCR6 presented by Martin and Osswald in [12]. PCR6 rule for M sources of evidence, when $F \in D^{\Theta}$, $F \neq \emptyset$ is defined by following expression:

$$m_{PCR6}(F) = m_c(F) + \sum_{i=1}^{M} m_i(F)^2 \sum_{\substack{\bigcap_{k=1}^{M-1} B_{\sigma_i(k)} \cap F \equiv \emptyset \\ \left(B_{\sigma_i(1)},\ldots,B_{\sigma_i(M-1)}\right) \in \left(D^{\Theta}\right)^{M-1}}} \left(\frac{\prod_{j=1}^{M-1} m_{\sigma_i(j)}\left(B_{\sigma_i(j)}\right)}{m_i(F) + \sum_{j=1}^{M-1} m_{\sigma_i(j)}\left(B_{\sigma_i(j)}\right)} \right), \tag{2}$$

where m_c is the conjunctive rule, and $\sigma_i = 1, \ldots, M$, is a factor applied to omit special case where $j = i$ and B_i is a subset of focal elements, to which belief mass was assigned by the classifier i.

Decision making In the most of applications, the decision is taken on Θ and not 2^{Θ} or D^{Θ}, thus the decided element is given by:

$$A = \underset{X \in \Theta}{\operatorname{argmax}} f_d(X), \tag{3}$$

where f_d is the decision fusion. In practice it can be credibility, plausibility, pignistic probability, etc.

Reliability discounting The reliability is directly connected with the quality of information in form of the evidence obtained from a certain source. Reliability is characterized by a discounting factor, usually denoted $\alpha \in [0, 1]$. It is usually assumed that $\alpha = 1$ when the source is fully reliable and $\alpha = 0$ if the source is totally unreliable. Discounting factor are introduced in the decision making process is defined as follows:

$$\begin{cases} m^{\alpha}(A) = (1-\alpha)\, m(A), & \text{for } A \neq \Theta \\ m^{\alpha}(\Theta) = (1-\alpha)\, m(\Theta) + \alpha \end{cases} \tag{4}$$

where $m^{\alpha}(\cdot)$ is the BBA after discounting and A is a subset of Θ.

Concept of classical Shafer's discounting was extended to a contextual one [13]. In that case for each source a vector of discounting factors $\boldsymbol{\alpha}$ is applied. Elements of $\boldsymbol{\alpha}$ are discounting factors representing reliability of source for each element of Θ in other words for each identified technical state (or fault). Contextual discounting factors are introduced as follows:

$$m^{\alpha}(A) = \prod_{\theta_l \in \overline{A}} 1 - \alpha_l \sum_{B \subseteq A} \left(\prod_{\theta_k \in (A \setminus B)} \alpha_k \right) m(B). \tag{5}$$

Contextual discounting factors are assessed using optimization a fitness function described in [13].

Importance discounting. Importance discounting is intended to bring out individual, sometimes subjective experts' preferences, supported by experience and knowledge in particular area. In contrary to the reliability discounting, part of the BBA $m(\cdot)$ is transferred to the element $\emptyset$, according to the following formula [2]:

$$\begin{cases} m^{\beta}(A) = (1-\beta)\, m(A), & \text{for } A \neq \emptyset \\ m^{\beta}(\emptyset) = (1-\beta)\, m(\emptyset) + \beta \end{cases} \tag{6}$$

It can be noticed that in the case of importance discounting, one must deal with non-normal BBA. According to the importance factor β ($\beta = 1$ means that a source has full importance), a case can occur when $m_{\beta}(\emptyset) > 0$. Classical interpretation of non-normal BBA (conflicting information from the source) is not valid when importance discounting was conducted. It is also not connected to the presence of mass assigned to unknown elements. In this case it is only a context dependant discounted mass of a source.

Importance discounting can not be applied, when aggregation of BBAs is made with Dempster's rule of combination. It is the result of how mass assigned to the empty set is normalized and redistributed with that rule. Application of one of Proportional Conflict Redistribution rule provides solution for fusion of sources with different reliabilities.

To use PCR6 within described discounting technique it is only needed to do some simple adoption. New rule PCR6$_{\emptyset}$ was defined for M sources, where $X \in D^{\Theta}$ oraz $\exists X = \emptyset$. This assumption allows to deal with $m_{PCR6_{\emptyset}}(\emptyset) \geq 0$, when $m_{PCR6}(\emptyset) = 0$. BBA $m_{PCR6_{\emptyset}}(\cdot)$ needs to be normalized, because most of decision rules uses normal BBA. Normalization was made though redistribution of non zero mass assigned to the empty set, to all other non zero focal elements, proportionally to their masses [2].

Discounting of reliability and importance. In [2] possibility of applying both, reliability discounting with values α_i (or $\boldsymbol{\alpha}_i$) and importance discounting with factors β_i, for M sources of evidence $m_i(\cdot)$, $i = 1, 2, \ldots, M$ was explained. Reliability and importance discounting are alternate operations, thus $m^{\alpha_i\beta_i}(\cdot) \neq m^{\beta_i\alpha_i}(\cdot)$, when $\alpha_i \neq 1$ i $\beta_i \neq 1$. According to that order of indications is important and $m^{\alpha_i\beta_i}(\cdot)$ means that first reliability discounting of $m_i(\cdot)$ with factor α_i, followed by importance discounting of $m^{\alpha_i}(\cdot)$ with factor β_i. Two schemes of joint use of both methods were elaborated [2]:

Scheme 1:
apply reliability discounting followed by reliability discounting, to obtain $m^{\alpha_i\beta_i}(\cdot)$, $i = 1, \ldots, M$; after that aggregation is made with PCR6$_{\emptyset}$ rule and normalization of aggregated BBA; to obtain $m^{\beta_i\alpha_i}(\cdot)$, $i = 1, \ldots, M$, first importance discounting and next reliability discounting is made and than combine BBAs with rule PCR6$_{\emptyset}$ and normalize the result; apply any combination rule (eg. averaging or PCR6) to join BBAs from two previous steps.

Scheme 2:
apply importance discounting to obtain $m^{\alpha_i}(\cdot)$, $i = 1, \ldots, M$; combine discounted BBAs with PCR6; apply reliability discounting to obtain $m^{\beta_i}(\cdot)$,

$i = 1, \ldots, M$; combine discounted BBAs with $PCR6_{\emptyset}$ and normalize result; apply any combination rule (eg. averaging or PCR6) to join BBAs from two previous steps.

In [2] only the first scheme was considered, so there was no comparative no comparative analysis made. Therefore it was decided to test both schemes during the research.

3 Assessment of the Importance Discounting Factors

Importance discounting factors are a representation of certain preferences of experts for selecting sources of signals, how signals are acquired and what is the suitability of signals for objects' state recognition. Expert opinions are often fuzzy and a direct conversion from preferences to quantitative factor values is impossible. To overcome such inconvenience, a method for systematic and unambiguous collecting of expert opinions and preferences. Analytic Hierarchy Process (AHP) seems to be an appropriate tool for repeatable interviewing of experts. It is an expert method, the result of which is a model that reflects the understanding of the problem by an expert, and not always the true problem nature [11]. The main task which had to be resolved that the AHP could be used to determine the discounting factors is the elaboration of criteria. It was assumed, that to ease the task of subject matter experts, each source of evidence will be expressed only in terms of sensors (physical or virtual). Therefore, it is required that criteria concern the sensor's properties.

For example, diagnostic system, it was assumed, that three criteria at one level of hierarchy are sufficient for method presentation purposes. However, there are no contraindications for an increase of criteria number. On the other hand, excessive increase of criteria number, can lead to problems in feedback from experts.

Identifying a set of criteria, various factors associated with the measuring system, having an impact on diagnostic decisions were analysed. These criteria should express the effect of certain properties of the measuring system, especially properties of the measurement point. Following criteria were identified:

– sensor type criterion, related mainly to a situation where the same measurand is intended to be measured with several sensors of different type, sensors using different measurement principle for the same physical quantity. Comparison of alternatives with respect to this criterion provides subjective information about the best sensor im the group of sensors performing redundant measurements, regarding immunity to interference, repeatability and accuracy;
– measured quantity criterion related to the physical quantity, measured by the individual sensors. Comparison of alternatives with respect to this criterion provides the information about the usefulness of measurements from certain sensor considering their sensitivity to a change of object/process state;

– measurement point criteria bounded with the location of the sensor. Different types of sensors can be compared, as this criterion describes the vulnerability of the sensor failure or the occurrence interference only due to the location of measurement point.

All criteria are of the same priority, because the influence of criteria on the value of importance discounting factors has to be equal.

In the AHP method, experts make pairwise comparisons of individual sensors using standard rating scale from 1 to 9.

After collecting opinions from experts and assessing weights, priorities for each decision variant is calculated according to the weight arithmetic. To change calculated priorities to discounting factors they need to be normalized. After normalization vector of factors $\boldsymbol{\beta}$ consisted in β_i values, that are importance discounting factors for source of evidence representing by sensor i, $i = 1, \ldots, N_S$, where N_S is the number of sensors in the diagnostic system.

4 Application Example

A practical example of importance discounting for incorporating experts opinions into a task of welding process state recognition is presented.

Experiment. Plates made of steel S235JR (EN 10025-2) with dimensions of 300 × 150 × 5 mm were welded on the welding stand for rectilinear MAG welding equipped in a microprocessor controlled welding machine Castolin TotalArc 5000 (Fig. 1). The edges of the joined plates were bevelled on V at an angle of 60° and the offset between them was 1 mm. A solid electrode wire with a diameter of 0,2 mm (Castolin CastoMag 45255) and a shield gas M21 (82%Ar+18%CO_2) were used. Parameters of welding process, used for joining plates were gathered in tab. 1.

Table 1. Optimal MAG welding parameters

Welding current [A]	Arc voltage [V]	Welding speed [cm/min]	Wire feeding rate [m/min]	Shield gas flow [l/min]	Electrode outlet [mm]
240	25	32	7,4	15	15

Laboratory experiment consisted in making several welded joints under conditions simulating several various states of the welding process ($Z_0 - Z_7$):

Z_0 - Correct welding process.
Z_1 - Welding with decay of the shielding gas flow.
Z_2 - Welding of the plates with distinct outbreaks of atmospheric corrosion on the welded surfaces.
Z_3 - Welding of plates with irregularities of the plate edges from side of the weld root.
Z_4 - Welding with deviation of current.

Fig. 1. View of laboratory stand

Z_5 - Welding of plates with different offset intervals.
Z_6 - Welding with deviation of voltage.
Z_7 - Welding of the plates with improper welding groove geometry.

For each state several (8-12) samples were joined. Several process output parameters were acquired: welding current, arc voltage, shield gas flow rate, wire feed rate, welding speed as well as acoustic emission with frequency of 100 kHz. From acquired signals additionally two signals were calculated: welding linear energy E and welding resistance R. All signals were low pass filtered and down sampled to frequency 10 kHz. After this simple processing procedure several signal sets were created. For further studies following sets were chosen: $\boldsymbol{I}$ - welding current, $\boldsymbol{U}$ - arc voltage $\boldsymbol{P}$ - acoustic pressure, $\boldsymbol{R}$ - resistance and $\boldsymbol{E}$ - welding linear energy. Acquired signals were assessed with use of several point estimators, widely used in the technical diagnostic: pf_1 - mean, pf_2 - absolute mean, pf_3 - squared mean, pf_4 - RMS, pf_5 - variance, pf_6 - standard deviation, pf_7 - absolute peak, pf_8 - maximum, pf_9 - minimum, pf_{10} - peak to peak, pf_{11} - form factor, pf_{12} - crest factor, pf_{13} - peak-to-average ratio (PTA), pf_{14} - peak-to-sqrt ratio (PTS), pf_{15} - asymmetry and pf_{16} - kurtosis.

After estimation stage a multi dimensional feature space was obtained. To reduce the space dimensionality, a feature selection using Fisher criterion was made. After that operation a 12 dimensional reduced feature space was created. The reduced space was consisted in following features: $pf_5\boldsymbol{R}$, $pf_1\boldsymbol{E}$, $pf_{11}\boldsymbol{E}$, $pf_{14}\boldsymbol{E}$, $pf_{15}\boldsymbol{E}$, $pf_{16}\boldsymbol{E}$, $pf_1\boldsymbol{U}$, $pf_3\boldsymbol{U}$, $pf_{16}\boldsymbol{U}$, $pf_1\boldsymbol{P}$, $pf_{11}\boldsymbol{P}$, $pf_1\boldsymbol{R}$.

Importance discounting. Importance discounting was preceded by taking experts' preferences. On this basis numerical values of discounting factor were determined.

Expert, with wide knowledge and many years of experience, was asked to do a pairwise comparison of three sensors: $\boldsymbol{E}$, $\boldsymbol{P}$, $\boldsymbol{U}$. It can be seen, that physical ($\boldsymbol{P}$, $\boldsymbol{U}$), as well as virtual sensor ($\boldsymbol{E}$) were taken into account. Expert opinions are gathered in tab. 2. It can be noticed, that because of the differences in the type measured quantities expert is not honoured any sensor due to the sensor type criterion. In the context of the measuring point and the type of measured quantity the domination of virtual sensor allowing the acquisition of welding linear energy is clearly observable. The voltage sensor is also more preferred than a microphone recording the sound pressure emitted during welding process.

Table 2. Alternatives compared with respect to all criteria

	Sensor type			Measurement point			Measured quantity		
	E	***U***	***P***	***E***	***U***	***P***	***E***	***U***	***P***
E	1	1	1	1	1	3	1	3	5
U	1	1	1	1	1	3	1/3	1	3
P	1	1	1	1/3	1/3	1	1/5	1/3	1

As the result of applying expert weights was the AHP hierarchy (tree). For each node values of priorities were calculated (Fig. 2). For each decision alternative (sensor) in the bottom of the hierarchy, final, but not normalized, priorities were obtained: $\beta_{\boldsymbol{E}_{nz}} = 0,47$, $\beta_{\boldsymbol{U}_{nz}} = 0,34$, $\beta_{\boldsymbol{P}_{nz}} = 0,19$. After normalization, priorities can be treated as importance discounting factors, having following values: $\beta_{\boldsymbol{E}} = 0,77$, $\beta_{\boldsymbol{U}} = 0,56$, $\beta_{\boldsymbol{P}} = 0,32$.

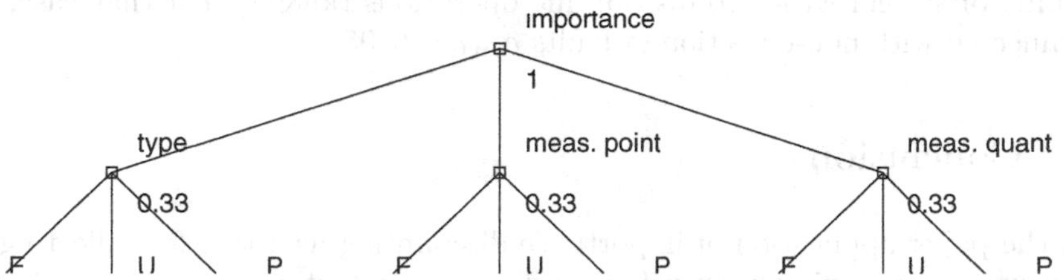

Fig. 2. AHP tree with priorities

Factors characterizing the importance of each sensor (one of main elements describing source of evidence), were applied to discount BBAs before the aggregation step. Single classifiers were trained and tested over one dimensional patterns, build from individual elements of feature space consisted of following elements: $pf_{15}\boldsymbol{E}$, $pf_1\boldsymbol{U}$, $pf_1\boldsymbol{P}$. In tab. 3 accuracies of recognition reached for each state. The overall classification accuracy (e_{acc}), and the error describing the rate decisions suggesting state with no fault, when actually the state reflecting abnormal process took place (e_{fnf}). Four scenarios, differing in the type of applied discounting method, were taken into consideration:

- aggregation without discounting (none),
- contextual reliability discounting (cnt),
- contextual reliability discounting and importance discounting (Scheme 1), (cnt&i1),
- contextual reliability discounting and importance discounting (Scheme 2), (cnt&i2).

Table 3. Accuracy of state recognition with three classifier, trained over features $cp_{15}\boldsymbol{E}$, $cp_1\boldsymbol{U}$, $cp_1\boldsymbol{P}$

Discounting type	State Z_0	Z_1	Z_2	Z_3	Z_4	Z_5	Z_6	Z_7	e_{acc}	e_{fnf}
none	0,91	0,67	0,86	0,86	1,00	0,83	0,71	0,85	0,84	0,12
cnt	0,82	1,00	0,67	1,00	1,00	1,00	1,00	0,67	0,90	0,18
cnt&i1	0,91	0,67	0,71	0,86	1,00	0,83	0,71	0,86	0,84	0,15
cnt&i2	0,73	0,67	1,00	0,86	1,00	0,83	0,71	1,00	0,85	0,05

Analysis of obtained results allows to state, that introduction of importance discounting procedure into the fusion systems can lead to increase of recognition accuracy for certain states of the welding process, decreasing the accuracy for other at the same time. Subjectivity of experts' opinions can be pointed as the main reason of variations of the classification accuracy. Experience, but also some kind of intuition, can led to depreciation of sensors, that in general are responsible for most informative and sensitive diagnostic signals. In considered case, joint use of reliability and importance discounting, gave better results, when the scheme 2 was used. It is worth to pay attention to the ability to detect faults, of system with two discounting operations (kn&i2). For that case, error connected with non-detection of faults $e_{fnf} = 0,05$.

5 Conclusions

In the paper application of importance discounting as a tool for reflecting subjective experts' opinions in values of discounting factors is presented. Analytic Hierarchy Process (AHP) was used to make the transformation from qualitative preferences to quantitative importance discounting factors. In the example use joint of reliability discounting and importance discounting, according to two described schemes.

Clear distinction between reliability and importance being two different concepts is assured, thus application of both in a diagnostic system is possible. There is a great potential for incorporating valuable, but sometimes somehow fuzzy knowledge. Designing simple, but exhaustive system of criteria with properly given priorities can lead to increase of overall accuracy of state recognition made by solving a classification task. Based on the research it is not possible to draw more general conclusions about the influence of importance discounting factors on the results of the fusion and next on the quality of the state recognition process. In presented example importance differentiation was impeded, due to lack of process data taken for example with corrupted sensor, in non-standard environmental conditions, with redundant sensors, etc. Extending the simulation range, should lead to extension of possible scenarios, on which the human expert can put his attention.

Acknowledgments. Scientific work partially financed from resources assigned to science in the years 2009-2012 as a research project N504281937 and from resources assigned to statutory activity of Institute of Fundamentals of Machinery Design, Silesian University of Technology at Gliwice.

References

1. Shafer, G.: A mathematical theory of evidence. Princeton University Press (1976)
2. Smarandache, F., Dezert, J.: Advances and applications od DSmT for information fusion (Collected works), vol. 1-3. American Research Press (2004-2009)
3. Yang, B.S., Kim, K.J.: Application of DempsterShafer theory in fault diagnosis of induction motors using vibration and current signals. Mechanical Systems and Signal Processing 20(2), 403–420 (2006)
4. Parikh, C.R., Pont, M.J., Jones, N.B.: Application of DempsterShafer theory in condition monitoring applications: a case study. Pattern Recognition Letters 22(6-6), 777–785 (2001)
5. Oukhellou, L., Debiolles, A., Denoeux, T., Aknin, P.: Fault diagnosis in railway track circuits using DempsterShafer classifier fusion. Engineering Applications of Artificial Intelligence 23(1), 117–128 (2010)
6. Aihua, L.: A method for condition evaluation based on DSmT. In: 2nd IEEE International Conference on Information Management and Engineering (ICIME), pp. 263–266 (2010)
7. Zhuang, M., Yongmei, C., Quan, P., Jun, H., Zhunga, L.: An intelligent fusion method of sequential images based on improved DSmT for target recognition. In: International Conference on Computational Aspects of Social Networks (CASoN), pp. 369–373 (2010)
8. Smarandache, F., Dezert, J., Tacnet, J.: Fusion of sources of evidence with different importances and reliabilities. In: 13th Conference on Information Fusion (FUSION), pp. 1–8 (2010)
9. Dezert, J., Tacnet, J., Batton-Hubert, M., Smarandache, F.: Multi-criteria decision making based on DSmT-AHP. In: 1st Workshop on the Theory of Belief Functions (2010)
10. Browne, F., Bell, D., Liu, W., Jin, Y., Higgins, C., Rooney, N., Wang, H., Müller, J.: Application of evidence theory and discounting techniques to aerospace design. In: Greco, S., Bouchon-Meunier, B., Coletti, G., Fedrizzi, M., Matarazzo, B., Yager, R.R. (eds.) IPMU 2012, Part III. CCIS, vol. 299, pp. 543–553. Springer, Heidelberg (2012)
11. Saaty, T.L.: How to make a decision: the analytic hierarchy process. Interfaces 24(6), 19–43 (1994)
12. Martin, A., Osswald, C.: A new generalization of the proportional conflict redistribution rule stable in terms of decision. In: Smarandache, F., Dezert, J. (eds.) Applications and Advances of DSmT for Information Fusion, Book 2, pp. 69–88. American Research Press, Rehoboth (2006)
13. Mercier, D., Denoeux, T.: Masson. M.-H.: Refined sensor tuning in the belief function framework using contextual discounting. In: Proceedings of IPMU 2006, vol. 2, pp. 1443–1450 (2006)

Simulators for Defining of Requirements for Diagnostic Systems

Marcin Amarowicz

Silesian University of Technology, Institute of Fundamentals of Machinery Design
Konarskiego 18a, 44-100 Gliwice, Poland
marcin.amarowicz@polsl.pl http://www.ipkm.polsl.pl

Abstract. The paper deals with the problem of designing diagnostic systems. It was noticed that the expected functionalities of such systems can be described by sets of requirements. In many cases, different solutions of diagnostic systems will be described by these requirements. One important task is the problem of selecting optimal solution from these sets of likely solutions. This task could be difficult and time-consuming in case of large set of likely solutions of the diagnostic systems. Therefore, it is necessary to limit this set to some subset of solutions, which contains useful examples from the diagnostic point of view. The relations *technical state* → *symptom* can be used for this purpose. For simple technical objects these relations can be elaborated based on the available literature data, which describes the process of their utilization. This approach will be difficult for more complex objects which contain e.g. innovative solutions. As the answer to this problem the author proposes diagnostic simulators.

Keywords: diagnostic systems, requirement engineering, expert system, statement network, diagnostic simulator.

1 Introduction

Permanent advances in technology result in the fact that new, more and more complex systems and technical objects emerge. Their complexity and innovative solutions which are used within these objects create a necessity to develop appropriate diagnostic systems. In particular, it is very important for critical machines, whose potential failure may result in the decrease of efficiency or total break in industrial process. Unfortunately, these situations result in some financial losses.

The process of designing a diagnostic system, like each design process, is multistage. It begins from recognition of needs and description of functionalities which the expected product should possess. Next, it is necessary to determine existing or likely limitations and develop possible solutions of a diagnostic system being developed. The last stage of this process includes setting selection criteria and identifying optimal solution which would meet all the assumptions and elaborated or forced limitations.

J. Korbicz and M. Kowal (eds.), *Intelligent Systems in Technical and Medical Diagnostics*, Advances in Intelligent Systems and Computing 230,
DOI: 10.1007/978-3-642-39881-0_15,

The designing proces can be aided by various methods and techniques, which allow to accelerate its implementation and reduce possible bugs or inconveniences.

2 Methods of Diagnostic Systems Representation

The diagnostic system may be interpreted as a set of single subsystems. These subsystems are used for the purpose of monitoring selected part of technical object e.g. subsystem for engine state or bearings monitoring etc. Each subsystem is a functionality that characterizes an overall, global diagnostic system. Consequently, the diagnostic system can be interpreted as a set of functionalities. Additionally, this set can be interpreted as a hierarchical set of functionalities because single functionality may force a necessity to use successive functionalities. Without these additional functionalities a comprehensive subsystem will not operate correctly.

Each functionality may be realized by a technical solution, which includes a measurement technique and appropriate methods of recorded data analysis. It should be noted that in many cases few solutions can be used for fulfilling single functionality. Taking into account this fact it is possible to assume that a design of diagnostic systems is a combination of functionalities and simultaneously a combination of solutions for each functionality. A very useful way that can be used to present this concept is a morphological table [3].

Particular elements of diagnostic system representation i.e. functionalities as well as diagnostic methods and techniques may be described by sets of requirements, which are widely used in software engineering.

3 Requirement Engineering

Various definitions of the term *requirement* are available in literature [12, 14]. In order to generalize, it is possible to assume that this term is a *formalized description of function which the solution should meet or realize.* One example of a requirement for the design of a diagnostic system can be the following statement: *Temperature field acquisition in the vicinity of element X of this object is necessary to positively determine technical state of the object.* The branch of science called requirement engineering deals with the overall issues of requirements i.e. acquisition, management etc.

For the purpose of carrying out of various projects the requirements may be acquired from numerous sources, e.g. principal customer, end-user, domain expert. Additionally, the requirements may be defined based on the analyzes of the available solutions or information from literature, norms, standards and also recommendations. For the design of a diagnostic system, it is possible to define requirements based on the knowledge from object operators and supervisors. In most cases the requirements definition process is based on negotiations between client and future project developer. In this process, the client states the requirements which describe the expected functionality of developed solution. At the same time, requirements are accepted, modified or deleted by a developer. During these tasks it is necessary to take into account the existing project limitations which sometimes have a huge influence on the final solution [8]. All the accepted requirements are recorded into a document called requirement specification. This document describes the final solution of the developed object.

4 Methods of Requirements and Diagnostic Knowledge Representation

The set of requirements and set of diagnostic knowledge are gathered in the knowledge acquisition process. These sets should be saved accordingly in order to use them for the purpose of developing design of diagnostic system. The method of data recording must be very universal, on the one hand it must enable to save knowledge which is uncertain or inaccurate, on the other hand it must provide easy access to this knowledge.

One of the graphical models called statement network may be used for this purpose. Statement network is a directed graph in which a finite and not empty set of vertices (nodes) N as well as a finite and not empty set of directed edges E connecting selected vertices can be distinguished. Particular nodes can be interpreted as statements. Single statement is a sentence that describes observed facts or expresses some opinions. Each of these sentences is called statement contents c and it is possible to assign some values $v \in < 0, 1 >$ for particular contents. This value gives us information on belief of truthfulness of this statement. Statements whose values are not equal 0 (false statement) or 1 (true statement) are called approximate statements. It is possible to assign as a value of statement some interval of values $< b_{min}, b_{max} >$. This interval can be interpreted as information about imprecise point value of this statement but limited to this interval. It should be noted, that it is possible to distinguish two extreme examples. The first one is given as $b_{min} = 0$ and $b_{max} = 1$. In this case it does not represent any information about truthfulness of considered statement. The second example is given as $b_{min} = b_{max}$, in this case it represents precise information about the value of considered statement.

One important advantage of using statement network is the possibility of carrying out a reasoning process. Unknown values of certain nodes are determined based on known values of others nodes. Each of the defined requirements can be interpreted as a statement and also like a node in the statement network. During the reasoning process unknown values of nodes which represent particular requirements are determined based on known values of nodes which represent diagnostic knowledge and limiting criteria. The value which is assigned to each requirements may be interpreted as a preference factor. The necessity to fulfil a functionality described by particular requirements is evaluated by this factor.

Relationships between particular nodes of statement network can belong to different classes. Bayesian networks (belief networks) are one of the widely used statement network, in which the relationships between nodes are expressed in the form of conditional probability tables (CPT) [9,11]. The necessary and sufficient conditions can be also used to describe these relationships [1].

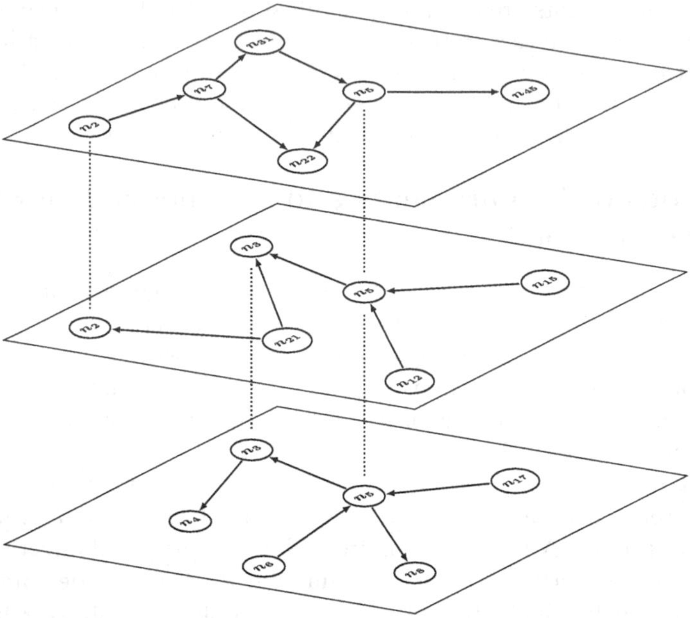

Fig. 1. Example of multimodal statement network

Single statement networks can be span over common nodes (see Fig. 1). This structure is called multimodal statement network (MSN) [2]. Particular networks of MSN can be build by independent experts. Different aspects e.g. type of control, reasoning process, etc. can be also used [13]. An important issue is a choice of appropriate method for determination of value of common nodes. Aggregation (particular networks are solved independently, final value is determined as one of statistical metrics e.g. average, max, etc.) and reconciliation (MSN is flattened, all networks are solved simultaneously) are two widely used methods for this task.

The specialized environment can be used to design and carry out a reasoning process with multimodal statement network [5, 7].

5 Designing of Diagnostic System

The realization of design of diagnostic system is connected with the necessity to perform many complex tasks. Particular tasks can be assigned to the following stages of the design process:

- identification of the project needs,
- identification of the technical object,
- identification of constrains and selection criteria,
- acquisition of knowledge which describes a technical object,
- gathering diagnostic system requirements,

- design of appropriate graphical models,
- evaluation of particular requirements based on diagnostic knowledge, constrains and selection criteria,
- analyzing and optimization of obtained results,
- elaborating requirement specification for a diagnostic system.

The main target of the identification of project needs stage is recognition and description of needs that characterize main targets of realized project. The obtained needs cannot impose the solutions of expected final product. The stage of identification of technical object consists of identification of elementary information about considered object, including a set of possible relations *diagnostic signal* $\rightarrow$ *symptom* ($Z \rightarrow S$). The correct identification of technical object (i.e. acquisition of full set of relations $Z \rightarrow S$) is necessary to carry out the remaining stages of design process. The method of this stage realization depends on the characteristics of object for which the diagnostic system will be developed. Generally, it is possible to distinguish two different cases that define the initial conditions for the needs of this process realization.

First of them describes the situations when the technical object occurs in the "familiar form". Based on the available literature, design documentation, experts knowledge, etc. it is technically possibile to define a set of appropriate relations $Z \rightarrow S$. These situations refer to the objects for which an appropriate operating data was obtained in a greater period of time. Based on these data it is possible to define suitable remarks and conclusions, which would be next verified based on new operating data.

For the purpose of elaborating a full set of relations $Z \rightarrow S$, it is necessary to have a set of data that includes cases for different technical states and faults. In many circumstances this task would not be possible to realize due to the consequences of some faults occurring during the operation of the technical object. Simultaneously, new technical objects are often characterized by innovative solutions. For this group of objects, usually there is no data available that describes precisely process of their operation and maintenance. Therefore, there is no technical premise to create an appropriate relations $Z \rightarrow S$. This situation forms the second possible case of initial conditions for the needs of realization of technical object identification process.

In practise technical objects have attributes of these two above mentioned cases. However, in many situations the mathematical model of the object is known. This model describes an activity of considered object. It's possible to developed an appropriate simulator based on this mathematical model. Based on the data generated by a simulator it is possible to elaborate a suitable relations $Z \rightarrow S$. A detailed way of using simulators for diagnostic systems implementation is shown in Section 7.

When a technical object is identified it is possible to carry out the remaining stages of design process. These stages include the acquisition of sets of limiting and selecting criteria, knowledge that describes a technical object as well as a set of requirements that describes the functionalities of the diagnostic system. In the last stage of the design process, the evaluation of the obtained set of requirements is performed based on selecting and limiting criteria as well as available diagnostic knowledge. In this manner, the subset of requirements that describe an optimal solution of diagnostic system is isolated from the whole set of requirements.

The above mentioned stages of design process can be carried out according to two scenarios which are shown in Fig. 2. First of them includes the sequential execution of the subsequent tasks which allow to obtain the necessary set of data, whereas the second scenario takes into account a possibility of existing feed-backs between particular stages. An important advantage of this approach is the possibility of beginning the whole process based on a given initial set of requirements. The elements of this set may be gathered from general requirement specification document for diagnostic systems. The content of this set can be updated based on the obtained set of limitations, knowledge and requirements. Additionally in this scenario it is possible to use diagnostic simulators to reproduce events and technical states of the object.

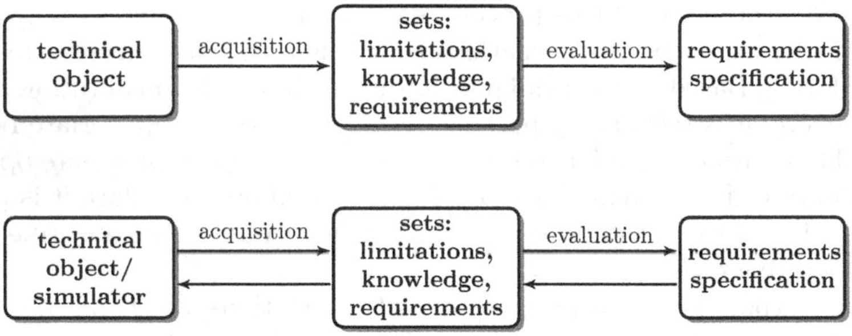

Fig. 2. Possible scenarios for design process

During the process of requirements acquisition, it is necessary to take into account various information, including the specialized diagnostic knowledge. The execution of this stage in a form of negotiation (mentioned in the Section 3) will be significantly difficult. The possible solution is a contractual assumption that the technical object will be considered as a virtual client during this process. The virtual client will be capable of establishing requirements to be met by a diagnostic system. At the same time the virtual client can be represented by an expert system with a knowledge base as a multimodal statement network [3,4].

6 Modeling the Set of Requirements for the Diagnostic Systems

Each N_i need which was identified during the initial stages of design process can be transformed into some set of requirements. This set can be expressed as a formula which consists of simple requirements. The following expression is an example

$$N_4 = (req_{11} \cup req_{26}) \cap req_{145}. \tag{1}$$

from which it results that the requirements req_{11} or req_{26} and requirement w_{145} are sufficient to fulfill this need. Each requirement req_i can be recorded as an expression built with expected functionality fu_i as well as the set of criteria

$$req_i = [fu, (k_w, k_o)]. \tag{2}$$

In the whole set of criteria, it is possible to distinguish the selection criteria k_w which are used for the optimization process as well as the limiting criteria k_o, which describe the conditions which the solutions of functionality should meet. Therefore, the expected functionalities are described by requirements and existing criteria. Each functionality fu_i can be obtained by using a subsystem $subs(fu_i)$, which has a structure s_{subs} and performs specified actions act

$$subs(fu_i) = (s_{subs}, act) = [(elem, con), act]. \tag{3}$$

The structure of this subsystem can be interpreted as a set of elements and connections between them. Simultaneously, actions can be considered as a set of activities which should be done on the structure of the subsystem which fulfill this functionality. The expression given as 2 can be expanded also to the form

$$req_i = [\{subs_1(fu_i) \cdots subs_n(fu_i)\}, (k_w, k_o)]. \tag{4}$$

Each functionality can be considered as a set of elements $subs_i(fu_i)$ in the three dimensional (elements, connections, actions) space of possible solutions. A functionality which is described by only one combination of elements, connections and actions is a trivial example. In most cases the sets of possible solutions in earlier mentioned three-dimensional space are available. Among these solutions, the optimal solution (elements, connections, actions) is searched. It should fulfill a considered functionality as well as meet the established criteria (k_w, k_o).

7 Simulators for Requirements Modeling

In many cases, including the problem of designing a diagnostic system for the complex object, the set of possible solutions can be large. Additionally, the selected solutions of diagnostic systems may contain contradictory partial solutions of particular functionalities. The selection of optimal solution may be carried out by various criteria (selection and limiting ones). Due to this problem it is necessary to limit the set of possible solutions of a diagnostic system to a subset of solutions. This subset should contain the solutions which are useful from the diagnostic point of view.

The process of reducing the set of possible solutions of particular functionalities should be performed based on acquired diagnostic knowledge about technical object. This useful knowledge is usually gathered in the form of relations *technical state* → *symptom*. For a typical technical object these relations can be elaborated based on the available literature data (see also Section 5).

If much more complicated object is considered for which this type of data is not available, it is possible to use diagnostic simulators. They allow to reproduce the activity of object in the function of possible technical states. The output of this simulator gives information that allows recognition of the technical state [6]. The unquestionable advantage of using simulators is a possibility to reproduce the situations which are impossible to perform in real technical objects. As examples of these situations we can mention the bug in the control system, faults occur during the utilization of the object, etc. In general case it is possible to develop a diagnostic simulator that belongs to one of three classes which differ as far as form of output data is concerned. The output data describes the possible symptoms of particular diagnostic states. For the subsequent classes of simulators they can be determined as elements of the following sequence of processing: *diagnostic signal* $\rightarrow$ *feature of signal* $\rightarrow$ *symptom*. In the case of first class simulators, the waveform of signals which are related to the considered technical state, are received on the outputs. The relationships between features of particular signal and adequate technical states are determined by using simulators of second class. And finally the simulators of third class allow to reproduce the full dependencies between technical state and symptoms. Exemplary output values from simulators of particular classes can be given as follows: waveform of vibration signal, average value of vibration and the information provides data about the exceeding the allowable vibration levels.

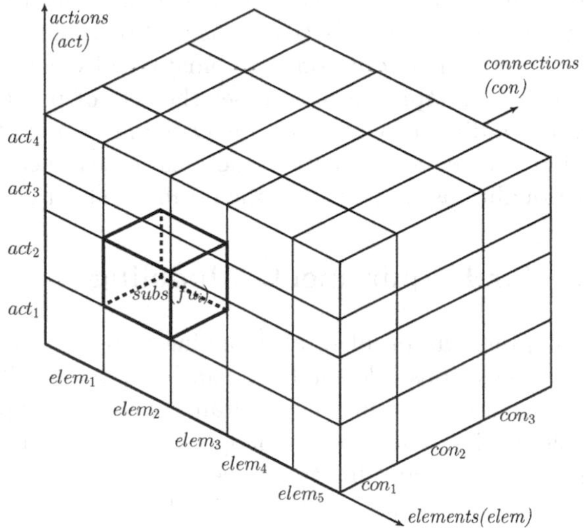

Fig. 3. Three-dimensional space of solutions and the best solution (subsystem) which fulfill the functionality fu_i

The sets of relations $Z \rightarrow S$ are determined based on the inputs (technical states) and outputs (symptoms) of the simulator. Particular relations can be used for the purpose of limiting the set of possible solutions of selected functionalities. The appropriate measurement systems and operations on the signals can be used in order to record some diagnostic signals or determine features or signals respectively. The measurement systems and operations on the signals can be interpreted as elements, connections and actions $[(elem, con), act]$ of subsystem that fulfill selected functionality. Considering the set of possible solutions that allow to fulfill functionality fu_i (the elements in the three-dimensional space, see Fig. 3) and the information obtained from diagnostic simulator it is possible to highlight the subsystem $subs(fu_i)$ (marked object on the Fig. 3). This subsystem will be useful for the purpose of identification of a considered technical state as well as for the determination of the functionality fu_i. Optimal solutions of particular functionalities can be determined by simulations of subsequent technical states with the use of a diagnostic simulator. It should be noted, that the obtained results will depend on the considered technical state of technical object. Therefore, it is possible that for two different technical states the same functionality will be achieved by two different subsystems described by $[(elem, con), act]$. At the same time a situation can occur where a few functionalities will be fulfilled by one solution (or its elements) given as $[(elem, con), act]$. In these cases it is necessary to carry out an additional analysis (e.g. based on suitable graphical models) for the proposes of eliminating coverings between particle solutions of the whole diagnostics system.

8 Example of Modeling of Requirements

Let us consider an example of simple rotating machine. For this machine it is possible to define the following need N_1 – *Diagnose the current state of the rotating machine.* Based on general knowledge about design of machines as well as specialized diagnostic knowledge, it is possible to create a database, that will be contain an information about possible technical states, faults and diagnostic methods and techniques that are related to typical elements of machines. Based on information from this database it is possible to conclude that for fulfill need N_1 it is necessary to diagnose such elements of considered machine as rotor, journal bearings, etc.

Therefore the need N_1 can be recorded as an expression built of requirements in the following form:

$$N_1 = req_1 \cap req_2 \cap req_3 \cap req_4 \cap ... \tag{5}$$

The selected requirements which are included in this expression can be considered as follows: req_1 – necessity to monitor vibration of the machine body, req_2 – necessity to monitor a vibration of the shaft, req_3 – necessity to monitor a temperature of oil in journal bearings, req_4 – necessity to store the results of the technical state monitoring of the machine, etc.

The requirement req_1 can be characterized as follows:

$$req_1 = [fu_1, (k_{w1}, k_{o1})], \tag{6}$$

where fu_1 – measurement of vibration is carry out by the vibration sensor which is mounted on the machine body, k_{w1} – minimal cost of solution, k_{o1} – measurement with an accuracy of 0.1% of the nominal value. The functionality of fu_1 can be fulfilled by a subsystem given as:

$$subs(fu_1) = [(elem_1, con_1), act_1], \tag{7}$$

where: $elem_1, con_1, act_1$ – are respectively possible elements, connections and actions describing a subsystem which fulfills the considered functionality fu_1. They can be considered as follows:
$elem_1$ – (absolute vibration sensor, DAQ card, specialized software),
con_1 – ({mounting of sensor: screw, connecting cables}, {mounting of sensor: glue, connecting cables}, {mounting of sensor: magnet, connecting cables}, {mounting of sensor: wax, connecting cables},
act_1 – analyze of recorded signals.

The next requirement, req_2 can be characterized as follows:

$$req_2 = [fu_2, k_{w2}], \tag{8}$$

where fu_2 – it is possible to measure the vibration of the shaft, k_{w2} – easiness of the measurement. The functionality of fu_2 can be fulfilled by a subsystem, which is described by $elem_2, con_2, act_2$. They can be considered as follows:
$elem_2$ – (relative vibration sensor, DAQ card, specialized software),
con_2 – connecting cables,
act_2 – analyze of recorded signals.

In the same way, it is possible to define the remaining requirements req_i. For each of them there are some sets of subsystems $subs_i$ which allow to fulfill particular functionality. The size of particular sets depends on the technical possibilities of performing considered functionality. In the standard case, it is possible to develop the set of

$$\prod_i card[subs_i(fu_i)] \tag{9}$$

solutions of diagnostic system that is able to fulfill the need N_1. This set of possible solutions may include solutions that would not be technically feasible, incomplete or simply irrational.

The optimal solution (the all criteria are met by this one) is determined from this set of solutions. For this purpose it is necessary to establish a set of criteria. For considered example, the selected criterion k_w – *minimal cost of overall system* and limiting criterion k_o – *simplicity of diagnostic system construction* can be defined. The process of selecting the optimal solution can be carried out in two stages. At the beginning the best solutions for particular functionalities are determined. In the next stage the overall solution, which fulfills the need N_1, is evaluated. Taking into account these assumptions it is possible to select an optimal solution for particular functionalities. In the considered example, for functionality fu_1 it will be $subs(fu_1)$ = [({absolute vibration sensor, DAQ card, specialized software}, {mounting of sensor: glue, connecting cables}), analyze of recorded signals], and for functionality fu_2 it will be $subs(fu_2)$ = [({relative vibration sensor, DAQ card, specialized software}, connecting cables), analyze of recorded signals]. Next, based on these results an optimal solution which fulfills the need N_1 is evaluated. It can be seen, that some of the element (DAQ card, specialized software) of subsystems $subs_i$ are repeated. So, in the final solution of diagnostic system it will be necessary to use only one specimen of DAQ card and specialized software.

The example which was presented above is relatively easy, but it presents a general assumptions that are connected with the modelling of requirements for diagnostic systems. For more complicated technical object, it will be necessary to use simulators (according to the assumptions described in Section 7) for limiting the set of possible solutions of selected functionalities. This conception are currently being developed.

9 Summary

The process of designing a diagnostic system is a very complex task. In many cases these systems are prepared for the critical machines e.g. turbines of power plant, etc. For this reason, these systems must be characterized by great reliability it would allow them to monitor technical object with high accuracy. During the development process it is necessary to take into account variety of factors that have significant influence on the final solution of the diagnostic system. One of the options is to use requirements as the description of the expected functionalities of designed diagnostic system.

The obtained set of requirements can be characterized by different solutions of designed system. This set can contain solutions that are not technically feasible, incomplete or simply irrational. So it is necessary to limit the number of possible solutions (particular elements in the three-dimensional space) to a subset of solutions. This subset would describe examples of solutions that are useful from the diagnostic point of view. Diagnostic simulators can be used for this purpose. The obtained subset of useful solutions can be optimized in the next stage of the designing process. Some additional information e.g. diagnostic knowledge or limiting criteria can be used for this task.

Acknowledgement. Described herein are selected results of study, supported partly from the budget of Research Task No. 4 implemented under the strategic program of The National Centre for Research and Development in Poland entitled *Advanced technologies of generating energy.*

References

1. Cholewa, W.: Mechanical analogy of statement networks. International Journal of Applied Mathematics and Computer Science 18(4), 477–486 (2008)
2. Cholewa, W.: Multimodal statement networks for diagnostic applications. In: S.P., Bergen, B. (eds.) Proceedings of the International Conference on Noise and Vibration Engineering, ISMA 201, pp. 817–830. Katholieke Universiteit Leuven, Leuven (2010)

3. Cholewa, W., Amarowicz, M.: Acquisition of requirements for diagnostic systems. Diagnostyka - Applied Structural Health, Usage and Condition Monitoring 62(2), 23–30 (2012)
4. Cholewa, W., Amarowicz, M.: Management of requirements for developed diagnostic systems. In: Mączak, J. (ed.) Proceedings of XI International Technical Systems Degradation Conference, XI ITSDC 2012, pp. 135–137. The Polish Maintenance Society, Warsaw (2012)
5. Cholewa, W., Chrzanowski, P., Rogala, T., Amarowicz, M.: Environment for statement network design (in Polish). Measurement Automation and Monitoring 57(9), 1079–1082 (2011)
6. Cholewa, W., Rogala, T.: Inverse models and diagnostic modeling. *Zeszyty* series, vol. 136. Silesian University of Technology, Department of Fundamentals of Machinery Design, Gliwice (2008) (in Polish)
7. Cholewa, W., Rogala, T., Chrzanowski, P., Amarowicz, M.: Statement Networks Development Environment *REx*. In: Jędrzejowicz, P., Nguyen, N.T., Hoang, K. (eds.) ICCCI 2011, Part II. LNCS, vol. 6923, pp. 30–39. Springer, Heidelberg (2011)
8. Grünbacher, P., Seyff, N.: Requirements Negotiation. In: Aurum, A., Wohlin, C. (eds.) Engineering and Managing Software Requirements, pp. 143–162. Springer, Heidelberg (2005)
9. Jensen, F.V.: Introduction to Bayesian Networks. Springer, Berlin (1997)
10. Korbicz, J., Kościelny, J.M., Kowalczuk, Z., Cholewa, W. (eds.): Fault Diagnosis. Models, Artificial Intelligence, Applications. Springer, Berlin (2004)
11. Koski, T., Noble, J.: Bayesian Networks. An Introduction. John Wiley & Sons, New York (2011)
12. Leffingwell, D., Widrig, D.: Managing Software Requirements: A Use Case Approach, 2nd edn. Pearson Education, Boston (2003)
13. Skupnik, D.: Multiacpect diagnostic models. *Zeszyty* series, vol. 140. Silesian University of Technology, Department of Fundamentals of Machinery Design, Gliwice (2009) (in Polish)
14. Sommerville, I., Sawyer, P.: Requirement Engineering: A Good Practice Guide. John Wiley & Sons, New York (1997)

Discovering Rules for Production Speed Prediction in Fasteners Manufacturing

Tomasz Mączka, Tomasz Żabiński, and Jacek Kluska

Faculty of Electrical and Computer Engineering,
Rzeszow University of Technology,
35-959 Rzeszow, Al. Powstancow Warszawy 12, Poland
{tmaczka,tomz}@kia.prz.edu.pl, jacklu@prz.edu.pl

Abstract. In this paper, an approach of evaluating efficiency for cold forging process in fasteners manufacturing is described. The purpose of the research is to find factors causing decrease in the efficiency of the cold forging production process. The proposed method allows us to discover a knowledge from the data collected in the Manufacturing Execution System (MES) operating in a fasteners manufacturing company. The structure and preparation phase of analyzed data, concerning daily work data of pushing machines for cold forging are presented. Conducted experiments detect relationships between pushing machines production speed, material and product type, in the form of transparent "If-then" rules. The results received a positive opinion of the company management board and give promising prospects for the future works.

Keywords: Intelligent Manufacturing System, C5 algorithm, decision support system.

1 Introduction

Production processes should be constantly investigated and optimized by enterprises in order to cope with many significant challenges that are currently being faced by manufacturing companies. The challenges result from the market expectations and economic trends, e.g. higher quality requirements and smaller production batches. Customers' expectations for short delivery times and low prices are growing and simultaneously more products have to be customized. Therefore, the complexity of production processes is significantly increasing, which results in the necessity for continuous optimization. Taking into account these determinants, production management requires an effective support from computer systems and continuous monitoring of manufacturing resources, e.g. processes, machines and employees.

In order to provide reliable and accurate data for a factory management staff, the computer system should be integrated with production resources located on a factory floor level. The system should enable on-line monitoring, control and maximization of the total use of manufacturing resources and should support human interactions with the system, especially on a factory floor level.

J. Korbicz and M. Kowal (eds.), *Intelligent Systems in Technical and Medical Diagnostics*, Advances in Intelligent Systems and Computing 230,
DOI: 10.1007/978-3-642-39881-0_16,

Computer systems for enterprises collect enormous amount of data related to production processes and resources, e.g. machines and operators, production orders realization, quality control, production planning, scheduling and control. Typical information in such systems is easily accessible for enterprise through various kinds of statistics or reports. However, there is a necessity to continuously discover complex relationships and knowledge hidden in the data, e.g. rules to foresee and prevent failures or efficiency decrease of the manufacturing process. Computational Intelligence (CI) methods clearly fit in this area for automated extraction of useful information and knowledge. These techniques will have a profound impact on current practices in manufacturing [1], [2]. They can be applied in manufacturing domains for association and patterns finding, classification, clustering, prediction, etc. Interesting relationships or sequences in manufacturing data can be discovered using association rules mining algorithms, e.g. Apriori, Apriori Predictive or Generalized Sequential Patterns. Classification methods can predict a class to which a new data record belongs. They require a set of training records, each associated with one of predefined classes, in order to build a classification model. A group of classification algorithms is based on decision trees, e.g. C5.0 (ID3), CART or CHAID. Support Vector Machine (SVM), Bayes trees, Gene Expression Programming (GEP) or artificial neural networks can also be applied for classification. Clustering algorithms, e.g. k-Means, enable us to partition a set of data records into clusters. Clusters are determined on the basis of similarity measures or probability density models [1], [2].

A promising perspective, especially in distributed manufacturing environments, is the use of collective intelligence techniques, like multi-agent systems or swarm intelligence [3]. Collective intelligence concepts clearly fit into new organizational structures for manufacturing industries, such as Holonic Manufacturing Systems (HMS), where an organizational structure can be treated as a conglomerate of distributed and autonomous units which operate as a set of cooperating entities [4].

The new generation of manufacturing systems, which utilizes artificial and computational intelligence techniques, is referred to as Intelligent Manufacturing System (IMS). IMS industrial application requires computer and factory automation systems characterized by a distributed structure, direct communication with manufacturing resources and the application of sophisticated embedded devices on a factory floor level [5], [6], [7], [8], [13].

2 Hardware and Software Structure of the IMS Platform

The platform for IMS has been created under the auspices of Green Forge Innovation Cluster. The platform has already been successfully implemented in four small and medium Polish metal component production companies. The most important testbed is located in the fasteners production company located in southern Poland [9]. The testbed has been working in the factory since 2009. Further experiments and platform descriptions mentioned in the paper will be referenced to this testbed.

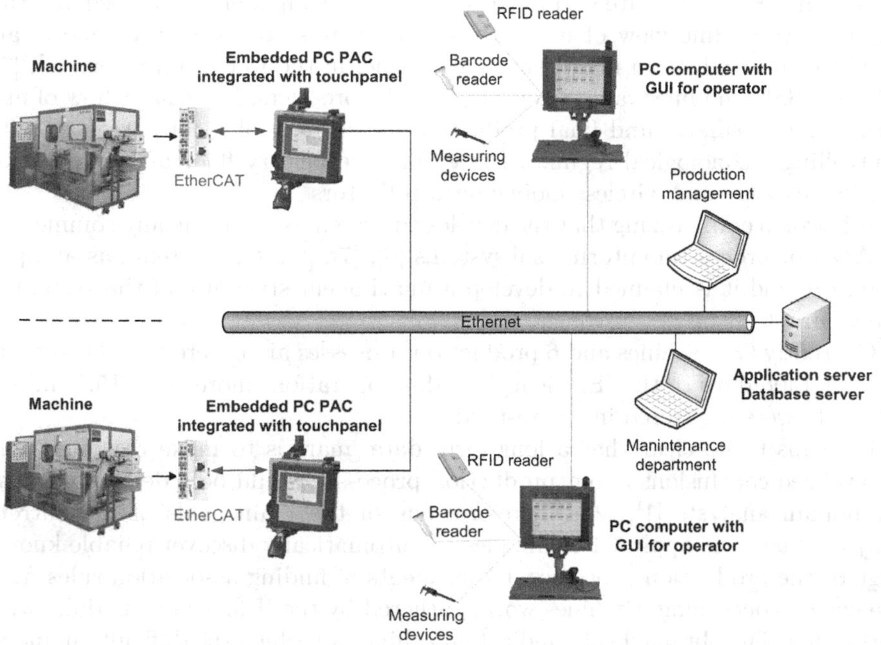

Fig. 1. IMS platform hardware structure

The system consists of embedded devices installed on the factory floor and equipped with custom made software as well as business logic components and Graphical User Interfaces (GUIs) for system end users, i.e. machine operators and production management personnel.

Programmable Automation Controllers (PACs) with touchable panels were selected for the hardware platform on the factory floor in the fasteners production company. PACs are equipped with an operating system and meet the demands of modern manufacturing systems [13]. Such devices are installed for each machine, and perform diverse tasks simultaneously, i.e. acquiring, processing and collecting data concerning machine work; communication with information systems; providing GUI and communication with peripheral devices, e.g. barcode and RFID readers or electronic calipers. An Ethernet network is used for communication between PACs and a data server layer. Besides PACs, mobile data collectors with the wireless communication are applied for tracing product genealogy on the manual work places or work places with simple machines, as well as for transportation services. The IMS platform hardware structure is presented in Fig. 1. Two modules important for our considerations are distinguished in system's software layer used on the factory floor, i.e. 'Efficiency' and 'SSK'.

The 'Efficiency' module is responsible for monitoring machines' work, i.e. for acquiring signals from machine control systems, transferring them into production events and registering in a database. It also provides GUI for machine operators which allows inputting stoppages reasons and registering operators' work time. PAC dedicated software and server components are used for this module. An on-line view of machine working states and historical reports are provided for production management personnel in the web applications [6], [7], [8]. The 'SSK' module traces a genealogy of the products. It registers flow of materials, intermediates and final products between technological processes, with controlling technological regimes. Data from the factory floor are provided by PAC controllers and wireless mobile data collectors.

It is worth emphasizing that the developed system is not using any commercial SCADA or process monitoring subsystems [6], [7], [8]. The system has an open structure and it is planned to develop a multi-agent structure of the system in the future [6].

Currently 68 machines and 6 production processes are covered by the system. Within 1408 days of the 'Efficiency' module operation, more than 15.5 million of events were registered in the testbed.

It seems to be clear that a long-term data analysis to make reasoning and generalized conclusions about production processes would be a demanding task for human analysts [1], [2]. Therefore, one of the main goals of the current project stage is to apply CI techniques to automatically discover reliable knowledge of the production processes. Experiments of finding association rules from the events concerning machines work, gathered by the 'Efficiency' module, were performed. The obtained rules indicate machines for which the difficulty in maintaining stable production occurs. Further details can be found in [10].

The experiments described in this paper are a continuation of applying CI methods for discovering knowledge from the production events. The main purpose is to detect particular factors, which have an influence on machines efficiency and unstoppable work, i.e. production speed.

3 Discovering Rules for Production Speed Prediction

Pushing machines for cold forging used in the fasteners production company have adjustable production speed. Technically, it is achieved electrically by inverters or mechanically by different gears. Therefore, machine operators have to set up a proper speed when starting production of a particular type of products. In case of machines with inverters, speed can be changed during products realization in order to obtain expected performance.

According to the factory production management board, appropriate production speed plays an important role in machine and production tools durability. In consequence, it determines the whole production process, as cold forging is the first process in the fasteners production technology used in the factory. Too high production speed can lead to more frequent machine stoppages because of improper quality of intermediate products detected by process monitoring devices. On the other hand, too low speed leads to decrease of utilization level of the factory productive capacities.

In a technological processes definition for a particular product type, there is defined a nominal production speed for all technologically adequate machines.

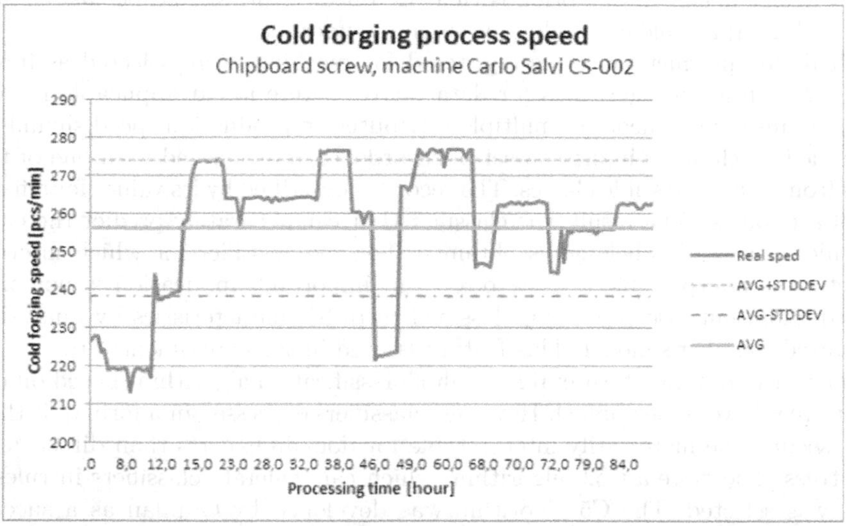

Fig. 2. Fluctuation of machine production speed during single order realization

The value of the nominal production speed in the fasteners production company is defined in the Enterprise Resource Planning (ERP) system, typically in pieces per minute for pushing machines. It is also used to estimate processing time of production orders and in consequence to predict a delivery date for customers.

However, such definition does not include factors like the type of material or material supplier used for production or difficulties that occurred during the historical realization of the assortment. Historical and current data concerning realization of production orders is stored in a factory's production monitoring system database. To detect factors that influence on the production speed, the usage of CI techniques is needed. In the next project phase, an on-line updated decision support system is going to be developed in order to suggest appropriate machines for particular order realization as well as the production speed for machines.

The example of the production speed fluctuations for a selected pushing machine during chipboard screw production is shown in Fig. 2.

The nominal speed for the product and given machine is 280 pieces per minute, whereas the real speed varied in the range 213 – 278 pieces per minute with average value 258 pieces per minute during 87 hours of the order realization.

3.1 Methodology

In order to find factors that influence the production speed, it is necessary to analyze data registered in the production monitoring system concerning machine production speed associated with selected factors. The analysis should automatically generate relationships between machine, product, material, production speed, etc. The main assumption is that obtained results should be easily readable and interpretable by production domain experts [1].

Therefore, production events registered for machines were selected as input data. Classification algorithms for data analysis were found applicable in this case, because the influence of multiple attributes on production speed should be examined. A classifier input is a set of records, each associated with one of the class from a set of possible classes. The record is described by its values for a fixed set of attributes. As a result, the classifier that can accurately predict the class to which a new case belongs is obtained. For the classifier for which internal structure is interpretable ("white-box"), an important application is not only a prediction, but also retrieving classified records' characteristics by analyzing generated classifier's model. This feature is used in the experiment [11].

In order to obtain interpretable results, classification algorithms based on decision trees have to be applied. However, classifiers expressed in a form of if-then rules seem to be more easily interpretable for domain experts than direct decision trees. Therefore a C5.0 algorithm, which can generate classifiers in ruleset form was selected. The C5 algorithm was developed by Quinlan as a successor of C4.5 a descendant of ID3. The algorithm grows an initial tree using the divide-and-conquer algorithm [12].

3.2 Data Preparation

Following consultation with the factory management board, events from one modern pushing machine Carlo Salvi CS 0002 was chosen for further analysis. The data for experiments were prepared in the form of intervals built from production events, like in experiment with association rules described in [10]. A Python script for data preparation was developed in order to transfer events into production intervals and to attach additional attributes from the company ERP system. Preprocessed data are returned in CSV format. The list of attributes of production intervals for production speed classification is as follows:

1. *length* – length of continuous production [min],
2. *stoppage_time_before* – time of stoppage period before starting continuous production [min],
3. *stoppage_time_after* – time of stoppage period after stopping continuous production [min],
4. *engine_runs_before_prod* – quantity of machine's main engine runs tries before starting continuous production,
5. *prod_speed* - value of production speed (in pieces per minute) during production interval,
6. *product_id* – type of produced assortment (from factory ERP system),
7. *material_id* – identifier of material used for production (from factory ERP system),
8. *material_supply_id* – identifier of material supplier (from factory ERP system),

where the variables 1 – 5 are continuous, and 6 – 8 are categorical. The production speed attribute was calculated on the basis of production start and production stop events, and production interval length:

$$prod\_speed = \frac{pieces_{prod\_stop} - pieces_{prod\_start}}{length\,(interval)}.$$

For experiments, events collected during 2 years of the selected machine operation time were used. Events were automatically transferred into 5266 production intervals with attributes 1 – 8 listed above. In the analyzed period for the selected machine, 205 production orders concerning 71 different assortment types were performed. For the classification purposes, three classes were proposed on the basis of a speed average ($\overline{speed}$) and a standard deviation (σ_{speed}):

$$Class = \begin{cases} -1 & \text{for} \quad prod_{speed} < \overline{speed} - \sigma_{speed} \\ 0 & \text{for} \quad \overline{speed} - \sigma_{speed} < prod_{speed} < \overline{speed} + \sigma_{speed} \\ 1 & \text{for} \quad prod_{speed} > \overline{speed} + \sigma_{speed} \end{cases}.$$

The script developed for data preparation assigns automatically classes for every input data record. The speed average and its standard deviation were computed differently for experiments described in the next Subsections: 3.3 and 3.4.

3.3 Speed Classes for Machines

In the first experiment, the production speed classes for input records were designated using the speed average value and the standard deviation value computed globally for the analyzed machine, regardless of product types or other attributes of intervals. For all analyzed production intervals, statistical values for production speed are (in pieces per minute): $\overline{speed} = 189.85$, $\sigma_{speed} = 35.8$, $speed_{\min} = 101.57$, $speed_{\max} = 260.35$. The experiment was performed using IBM SPSS Modeler software [14]. After loading the input data, the records are randomly divided into training and testing data subsets, in proportion of 80% and 20%, respectively. Then, the production speed classifier is built on the basis of data from the training subset, using the C5 algorithm. The classifier's accuracy is measured using data from the testing subset. There were 77 classifier rules generated for the production speed classes during the experiment. Rules with confidence greater than 70% are as follows:

1. (50; 0.981): if product_id = BL09-S55-0550 and material_supplier_id = 947, then -1, [normative class: 0]
2. (47; 0.98): if product_id = 1148-M050-0100, then -1, [normative class: 1]
3. (12; 0.857): if product_id = 1207-S48-0500 and material_supplier_id = 7064, then -1, [normative class: 0]
4. (142; 0.993): if product_id = 0194-M060-0160, then 0, [normative class: 0],
5. (44; 0.978): if material_supplier_id = 5895, then 0
6. (309; 0.987): if product_id = 0947-F045-0300, then 1, [normative class: 1].

The first number in brackets before each rule indicates the number of records from the testing subset, which match both the antecedent and the consequent of the rule. The second number is the rule confidence, i.e. proportion of number of records that match both the antecedent and the consequent to number of records that match only the antecedent. The rule antecedent contains concrete values of data attributes. The rule consequent contains the predicted production class for attributes from the antecedent. For the rules containing a product identifier in the antecedent, a normative production speed class was designated on the basis of a normative speed retrieved from ERP subsystem. This value is included in the rules list, right after the rule's consequent (in square brackets). Rules with different nominal and real speed classes were highlighted as especially interesting for further analysis.

The evaluation of the generated classifier was performed in order to estimate its error rate. The percentage value and the number of misclassified records is presented in Table 1 (a). The testing error was equal to 21.25%. The confusion matrix in Table 1 (b) shows the number of records for each class classified correctly and misclassified. Columns of the matrix represent classes assigned by classifier, and rows represent real classes known from the learning data set. For this experiment, most misclassified cases exist for records with real class 1, classified as 0.

Table 1. (a) - Classification accuracy for the first experiment, (b) - the confusion matrix, (RCC - records classified correctly, RM - records misclassified, Total - total number of records, RC - real class, PC - predicted class).

(a)

	Training set	Testing set
RCC	3433 (81.51%)	830 (78.75%)
RM	779 (18.49%)	224 (21.25%)
Total	4212	1054

(b)

RC\PC	-1	0	1
-1	136	78	1
0	37	590	18
1	1	89	104

3.4 Speed Classes for Products

In the second experiment, the production speed classes for input records were designated separately for each type of the assortment (product_id), on the basis of the speed average and the standard deviation, computed for the particular product locally. In consequence, obtained rules could be helpful when finding regularities or characteristics for product types. There were 57 classifier rules generated for production speed classes during this experiment. Rules with confidence greater than 70% are as follows:

1. (12; 0.929): if product_id = 1186-F060-0350 and material_id = 2077, then -1, [normative class: 1],
2. (12; 0.929): if product_id = 1186-F060-0350 and material_id = 214, then -1, [normative class: 1],
3. (98; 0.72): if product_id = 0711-M050-0110 and material_id = 1328, then -1, [normative class: 1],
4. (39; 0.976): if material_id = 494, then 0,

5. (15; 0.941) if product_id = 0711-M050-0110 and material_id = 1812, then 1, [normative class: 1],
6. (8; 0.9) if product_id = 0060-F060-0550 and material_id = 694, then 1, [normative class: 0].

The number of misclassification records presented in Table 2 (a) is higher than in the previous experiment. However, it should be noted that the constructed classifier is more specific than in the first experiment, because of locally designated speed classes. Like in the previous experiment, misclassification mostly occurred for records with the real class 1 classified as 0 (see Table 2 (b)).

Table 2. (a) - Classification accuracy for the second experiment, (b) - the confusion matrix, (RCC - records classified correctly, RM - records misclassified, Total - total number of records, RC - real class, PC - predicted class)

(a)

	Training set	Testing set
RCC	3026 (71.84%)	705 (66.89)
RM	1186 (28.16)%)	349 (33.11)
Total	4212	1054

(b)

RC\PC	-1	0	1
-1	28	140	4
0	23	655	12
1	15	155	22

3.5 Evaluation of the Rules by the Production Domain Experts

The obtained results were consulted with the factory production management board in order to evaluate usefulness of the rules. Full products descriptions, material types and material suppliers' names were substituted instead of identifiers in order to make rules more readable. It turned out that useful knowledge was discovered. Production domain experts confirmed that the usage of three types of materials for two types of products indicated in the experiments had been causing intensive tools wear and in consequence more frequent machine stoppages, which led to lower real production speed value. In consequence, material types were changed in the technology definition for the products. Moreover, the overestimated normative production speed value for one product was correctly detected. The one rule from the first experiment and three rules from the second experiment were highlighted, as those which provided new information. Therefore, an assumption could be made that rules with the production speed class computed for particular product types separately could be more useful in practice.

4 Conclusions

In this paper, the methodology for knowledge discovery on efficiency of the cold forging process using CI methods was investigated. The data preparation phase, and the experimental results, and their evaluation by production domain experts were described. The C5 classifier algorithm was used in order to generate interpretable rules for the factory management board.

Two experiments of classifier construction were performed for real data regarding two years' work of one machine. Raw data from the production monitoring system were automatically transferred to production intervals and additional attributes concerning: production speed, type of assortment and material were added. Production speed classes were defined on the basis of production speed average value and its standard deviation. In the first experiment, obtained rules reflect globally to machine, as the speed average value and standard deviation were designated regardless of the product type. In the second experiment, speed class was designated for the speed average value and standard deviation computed for particular assortments, so the classifier provides rules that identify relationships for particular product types.

The production management board evaluated the obtained rules and confirmed that real and valuable knowledge was discovered using the proposed methodology.

It is planned to implement the evaluated methodology as an element of IMS platform in the fasteners production company. This will require continuous updating of the classifier model on the basis of data acquired from the factory floor. The rules database will be used not only by the factory management personnel, but also by other IMS platform modules. For example, production planning module could automatically retrieve suggestion of the best machine selection for performing particular order in terms of production speed, production continuity, etc. Furthermore, data from other production monitoring system modules will be included, i.e. concerning production tools or quality control. Those aspects will extend classification mechanism target not only to machines' production speed, but also to quality control measurements results, different influences for tools wear etc. However, it is planned to incrementally deploy and examine developed solutions in the real production environment, starting with production speed goal. An open-source data mining solutions like Waikato Environment for Knowledge Analysis (WEKA) will be considered for implementation. In the future, it is planned to develop a data analysis module as continuously working part of a platform for Intelligent Manufacturing System.

The long-term goal of the project includes testing and applying different solutions in the field of CI and collective intelligence methods. It is planned to develop a multi-agent software platform for a holonic based IMS. The introductory research in the field will be devoted to the analysis of the system requirements and to the design of an agent system topology. It seems to be clear that in the system two different approaches for an agent encapsulation will be combined, i.e. the functional and the physical decomposition. Due to the functional decomposition, agents responsible for system support, process modeling and task scheduling (contact agent, order agent, supply agent) will be defined. Due to the physical decomposition, agents related to factory floor equipment, e.g. machine agent, machine section agent, etc. will be also defined. Due to open and extensible system structure, collective intelligence solutions can be placed in particular system modules, which supports the production path beginning from customer order, to its planning and tracing the realization. Hardware installed on factory floor gives possibility to run software agents directly on the controller in parallel with the low-level PLC or CNC programs [13].

Acknowledgments. This work was supported in part by the Rzeszow University of Technology, Poland, funds for young researchers; No U-175/DS/M. Personal computers used for performing computations were purchased in the projekt No POPW.01.03.00-18-012/09 from the Structural Funds, The Development of Eastern Poland Operational Programme co-financed by the European Union.

References

1. Wang, K.: Applying data mining to manufacturing: the nature and implications. Journal of Intelligent Manufacturing 18, 487–495 (2007)
2. Choudhary, A.K., Harding, J.A., Tiwari, M.K.: Data mining in manufacturing: a review based on the kind of knowledge. Journal of Intelligent Manufacturing 20(5), 501–521 (2009)
3. Ayhan, M.B., Aydin, M.E., Oztemel, E.: Collective Intelligence for Monitoring Innovation and Change in Manufacturing Industry. In: Proceedings of the World Congress on Engineering, London, U.K, July 4-6, vol. III, pp. 1395–1400 (2012)
4. Colombo, A., Schoop, R., Neubert, R.: An Agent-Based Intelligent Control Platform for Industrial Holonic Manufacturing Systems. IEEE Trans. Ind. Elect. 53(1), 322–337 (2006)
5. Oztemel, E.: Intelligent manufacturing systems. In: Benyoucef, L., Grabot, B. (eds.) Artificial Intelligence Techniques for Networked Manufacturing Enterprises Management, pp. 1–41. Springer, London (2010)
6. Żabiński, T., Mączka, T.: Implementation of human-system interface for manufacturing organizations. In: Hippe, Z.S., Kulikowski, J.L., Mroczek, T. (eds.) Human - Computer Systems Interaction: Backgrounds and Applications 2. AISC, vol. 98, pp. 13–31. Springer, Heidelberg (2012)
7. Żabiński, T., Mączka, T.: Human System Interface for Manufacturing Control - Industrial Implementation. In: Proceedings of 3rd IEEE International Conference on Human System Interaction, Rzeszow, Poland, May 13-15, pp. 350–355 (2010)
8. Maczka, T., Czech, T.: Manufacturing Control and Monitoring System – Concept and Implementation. In: Proceedings of IEEE International Symposium on Industrial Electronics, Bari, Italy, July 4-7, pp. 3900–3905 (2010)
9. Gawel Zaklad Produkcji Srub S.A., `http://www.gzps.pl/` .
10. Maczka, T., Zabinski, T.: Platform for Intelligent Manufacturing Systems with elements of knowledge discovery. In: Aziz, F.A. (ed.) Manufacturing System, pp. 183–203. InTech Open Access Publisher (2012)
11. Wu, X., Kumar, V., Quinlan, J.R., et al.: Top 10 algorithms in data mining. Knowledge Information Systems 14, 1–37 (2008)
12. Quinlan, J.R.: C4.5: Programs for machine learning. Morgan Kaufmann Publishers, San Mateo (1993)
13. Zabinski, T.: Implementation of Programmable Automation Controllers – Promising Perspective for Intelligent Manufacturing Systems. Management and Production Engineering Review 1(2), 56–63 (2010)
14. IBM SPSS Modeler 14.1 Algorithms Guide. IBM Corporation (2011)

Part II

Medical Diagnostics and Biometrics

Computer-Aided Diagnosis for Breast Tumor Classification Using Microscopic Images of Fine Needle Biopsy

Marek Kowal

Institute of Control & Computation Engineering,
University of Zielona Góra,
Ogrodowa 3b, 65-246 Zielona Góra, Poland
m.kowal@issi.uz.zgora.pl

Abstract. This paper presents an automatic computer system to breast cancer diagnosis. System was designed to distinguish benign from malignant tumors based on fine needle biopsy microscope images. Research is focused on two different problems. The first is segmentation and extraction of morphometric parameters of nuclei present on cytological images. The second concentrates on breast cancer classification using selected features. Analysis of cytologic images is a very difficult task. Especially, nuclei segmentation is extremely challenging. Nuclei often create clusters, overlap each other, their boundaries are not clear and their interiors are not uniform. To cope with this problem, segmentation procedure based on adaptive thresholding and k-means clustering is used to discover nuclei region. Next, conditional erosion is applied to binary image of nuclei to localize nuclei seeds. Finally, nuclei are segmented using seeded watershed (SW) algorithm. A set of 84 features extracted from the nuclei is used in the classification by the k-nearest neighbor (kNN) classifier. The approach was tested on 450 microscopic images of fine needle biopsies. The proposed computer-aided diagnosis (CAD) system is able to classify tumors with reasonable accuracy which reaches 100%.

Keywords: Computer-aided diagnosis, Breast cancer, Image segmentation, Classification.

1 Introduction

Fine needle biopsy (FNB) is an examination that consists in obtaining material directly from the tumor. The collected material is then examined under a microscope to determine the prevalence of cancer cells. This approach requires extensive knowledge and experience of the cytologist responsible for the diagnosis. CAD system can help make the results objective and assist inexperienced specialists. It also allows for screening on a large scale where only difficult and uncertain cases would require further examination by the specialist. Along with the development of advanced vision systems and computer science, quantitative cytopathology has become an important method for detection of diseases [14,29,18,11,12].

J. Korbicz and M. Kowal (eds.), *Intelligent Systems in Technical and Medical Diagnostics*, Advances in Intelligent Systems and Computing 230,
DOI: 10.1007/978-3-642-39881-0_17, © Springer-Verlag Berlin Heidelberg 2014

Currently, there are many researchers dealing with digital cytology in breast cancer diagnosis. Jeleń et al., [18] present an approach based on the level sets segmentation method. Classification efficiency was tested on 110 (44 malignant, 66 benign) images with results reaching up the 82.6%. Niwas et al., [25] show a method based on analysis of nuclei texture using a wavelet transform. Classification with a kNN algorithm on 45 (20 malignant, 25 benign) images gave 93.33% efficiency. Another approach is presented by Malek et al., [22]. They used active contours to segment nuclei. Then, 200 (80 malignant, 120 benign) images were classified with a fuzzy c-means algorithm. The result was 95%. Breast cancer diagnosis was also discussed by Xiong et al., [34]. Partial least squares regression was used to classify 699 (241 malignant, 458 benign) images with the of result 96.57%. However, the authors did not describe the segmentation method used to extract nuclei.

This paper presents a fully automatic breast cancer diagnostic system based on analysis of cytological images of FNB material. The task of the system is to classify a case as benign or malignant. This is done by using morphometric, textural and topological features of nuclei isolated from microscopic images of the tumor. Unfortunately, many authors have reported problems with a segmentation of clumped and overlapped nuclei [19,5]. Segmentation errors introduced by the clustered nuclei give a significant distortion in nuclei features. Eventually, this results in low classification accuracy. To tackle this problem, two stage segmentation procedure was developed. Firstly, adaptive thresholding and k-means clustering were used to background-foreground segmentation. Next, the foreground is conditionally eroded. As the result, centers of the nuclei are determined and used to initialize watershed algorithm. The procedure gives highly satisfactory results for segmentation of clumped and overlapped nuclei.

Finally, 84 features are extracted and the cases are classified as benign or malignant using kNN classifier [4,24]. The entire approach was tested on real medical images obtained from patients of Regional Hospital in Zielona Góra, Poland. As shown later in this paper, the classification accuracy for patients reached 100%. The results demonstrate that a computerized medical diagnosis system based on the proposed method would be effective and can provide valuable, accurate diagnostic information.

The remainder of this paper is organized as follows. Sections 2, describe in detail the process of acquisition of cytological images, preprocessing, segmentation, feature extraction and classification. Section 3 shows experimental results obtained using the proposed approach. The last part of the work includes conclusions and bibliography.

2 Material and Methods

2.1 Study Dataset

The dataset contains 450 microscope images of the cytological material obtained by FNB. The cytological material was collected from 50 patients of the Regional Hospital in Zielona Góra. The set contains 25 benign and 25 malignant cases.

Each patient is represented by 9 images. The number of images was recommended by the specialists from the hospital and allows for correct diagnosis by a pathologist.

Biopsy without aspiration was performed under the control of an ultrasound scanner with a 0.5 mm diameter needle. Smears from the material were fixed in a spray fixative (Cellfix of Shandon company) and dyed with hematoxylin and eosin (h+e). The time between preparation of smears and their preserving in the fixative never exceeded 3 seconds. The images were recorded by a SONY CCD IRIS color video camera mounted an atop Axiophot microscope. The slides were projected into the camera with 160× objective and 2.5× ocular giving together an enlargement of 400×. The images are BMP files, 704x578 pixels, 8 bit/channel RGB. All cancers were histologically confirmed and all patients with a benign disease were either biopsied or followed for a year.

2.2 Image Preprocessing

The first step of the diagnostic procedure is preprocessing. The microscopic images need to be enhanced due to distortions introduced during the acquisition procedure. First, a CCD camera causes the presence of noise. Although barely visible in the original images, the noise is intensified when increasing the contrast later in the preprocessing step and may cause artifacts in the segmentation process. In order to reduce the noise the images are filtered using Gaussian low-pass filer [26]. Then, they are sharpened back with the sharpening filter.

Another image defect requiring removal is a vignette caused by microscope optics. It is removed using a blank slide I_{blank} as a reference. The correction is applied to the images as follows:

$$I' = I + I(1 - I_{mask}), \tag{1}$$

where I_{mask} is a mask representing a decrease in brightness for each pixel and is given as:

$$I_{mask} = \frac{I_{blank}}{max(I_{blank})}. \tag{2}$$

Finally, the histogram is stretched to enhance the contrast and the images are cropped to the size of 697x569 to remove the frame and other artifacts that might be observed on the boarder of the images.

2.3 Cell Nuclei Segmentation

Classification of tumor malignancy requires isolating nuclei from the rest of the image. In literature, many different approaches have been already proposed to extract cells or nuclei from microscope images [1,6,7,21,18,20,23,27]. This task is usually done automatically, using one of the well known methods of image segmentation [28,30,13]. Unfortunately, reliable cell or nuclei segmentation is a challenging task. Very often cells cluster and overlap together and their boundaries are blurred (Fig. 1(a)). Moreover, attempts to generalize segmentation approaches proposed in literature usually fails because such methods work correctly only for specific images. Slides from various sources may vary significantly depending on the method of smear preparation. In order to deal with these problems, an automatic segmentation procedure that integrates results of image segmentation from four different methods is proposed.

The procedure starts from converting the original RGB image into the binary image representing nuclei region. This is done using adaptive thresholding and k-means clustering. Adaptive thresholding is applied to distinguish objects (i.e. nuclei, cytoplasm and red blood cells) from the background. The image is segmented into two subsets, G_{dark} and G_{bright} [28]:

$$\begin{aligned} G_{bright} &= \{g_{i,j} : g_{i,j} > t_{i,j}\}, \\ G_{dark} &= \{g_{i,j} : g_{i,j} \leq t_{i,j}\}, \end{aligned} \tag{3}$$

where $g_{i,j}$ is the pixel luminance value and threshold $t_{i,j}$ is calculated adaptively for subsequent pixels of the image using averaging filter:

$$t_{i,j} = \frac{1}{m^2} \sum_{k=-n}^{n} \sum_{l=-n}^{n} g_{i+k,j+l}, \tag{4}$$

where $n = \frac{m-1}{2}$ and m is the size of the filter window (odd integer). Unknown values outside the bounds of the image are assumed to equal the nearest image boundary value.

In the next step, k-means clustering [16,17] is applied to distinguish nuclei from the rest of the objects. In the considered case, 3 clusters are defined. The clusters correspond to nuclei, red blood cells and cytoplasm. The clustering procedure is carried out in the RGB color space on the subset of pixels provided by adaptive thresholding. The cluster corresponding to the nuclei is determined based on the fact that nuclei are the darkest objects in the image. Next, pixels that belong to the nuclei cluster are used to construct binary image BW. It marks regions in the image where the nuclei are located (Fig. 1(b)).

A key stage of nuclei marker extraction is to correctly detect nuclei centers. The method is based on the concept of conditional erosion [35]. Procedure assumes that the erosion is conducted as long as the size of the processed nucleus is large enough. Two masks for erosion operation are designed. They can be referred as fine and coarse erosion structuring elements. The coarse erosion tend to remain the actual shape but reduces the size of clustered nuclei. This can make the nucleus to disappear because of huge reduction in the size. On the other hand, fine erosion mask is less likely to make the nucleus disappear, but it will lead to the loss of original shape. The erosion operation of the binary image I by the structuring element B is defined by:

$$I \ominus \check{B} = \{x \in \mathbb{R}^2 \mid (B + x) \subset I\}, \tag{5}$$

where $\check{B}$ is a reflection of set B. Conditional erosion is applied to binary image BW obtained in the previous step of segmentation. Threshold T_1 for coarse structuring element B_c and threshold T_2 for fine structuring element B_f are chosen experimentally (T_1=350, T_2=50). Next, nuclei are iteratively eroded using coarse element until the size of all objects is smaller than T_1. Finally, erosion with fine element is applied iteratively to the results obtained during coarse processing. Structuring elements B_c and B_f are designed according to the shape of the nuclei which is similar to an ellipse:

$$B_c = \begin{bmatrix} 0\,0\,0\,1\,0\,0\,0 \\ 0\,0\,1\,1\,1\,0\,0 \\ 0\,1\,1\,1\,1\,1\,0 \\ 0\,1\,1\,1\,1\,1\,0 \\ 0\,1\,1\,1\,1\,1\,0 \\ 0\,0\,1\,1\,1\,0\,0 \\ 0\,0\,0\,1\,0\,0\,0 \end{bmatrix}, \tag{6}$$

$$B_f = \begin{bmatrix} 0\,1\,0 \\ 1\,1\,1 \\ 0\,1\,0 \end{bmatrix}. \tag{7}$$

Objects that have survived the conditional erosion are the markers M used to seed watershed segmentation method (Fig. 1(c)).

The classical watershed algorithm used for nuclei segmentation tend to create many micro-segments [2]. Such oversegmentation makes the results of the watershed method completely useless. To deal with this problem we used SW method, well-known extension of watershed algorithm [35].

Firstly, the topographic surface TS (intensity map) is determined. TS is generated by the Euclidean distance transform of the binary mask of nuclei BW obtained by adaptive thresholding and k-means clustering. Next, the surface is modified accordingly to found markers using morphological reconstruction [32].

The grayscale morphological reconstruction $\rho_I(J)$ can be described as repeated geodesic dilations of an image J, called the marker image, until the contour of the marker image fits under a second image I:

$$\rho_I(J) = \bigvee_{n \geq 1} \delta_I^{(n)}(J). \tag{8}$$

The grayscale geodesic dilation of size n is then given by:

$$\delta_I^{(n)}(J) = \underbrace{\delta_I^{(1)}(J) \circ \delta_I^{(1)}(J) \circ \ldots \circ \delta_I^{(1)}(J)}_{n}, \tag{9}$$

and the elementary geodesic dilation is described by the following relationship:

$$\delta_I^{(1)}(J) = (I \oplus \check{B}) \wedge I, \tag{10}$$

where $\wedge$ stands for the pointwise minimum and $(I \oplus \check{B})$ is the dilation of J by flat structuring element B ($\check{B}$ is a reflection of B).

The algorithm impose the minima of surface TS at the locations specified by the markers M. Modified topographic surface TS_m has regional minima preserved wherever M is nonzero. In this way, the markers are incorporated to

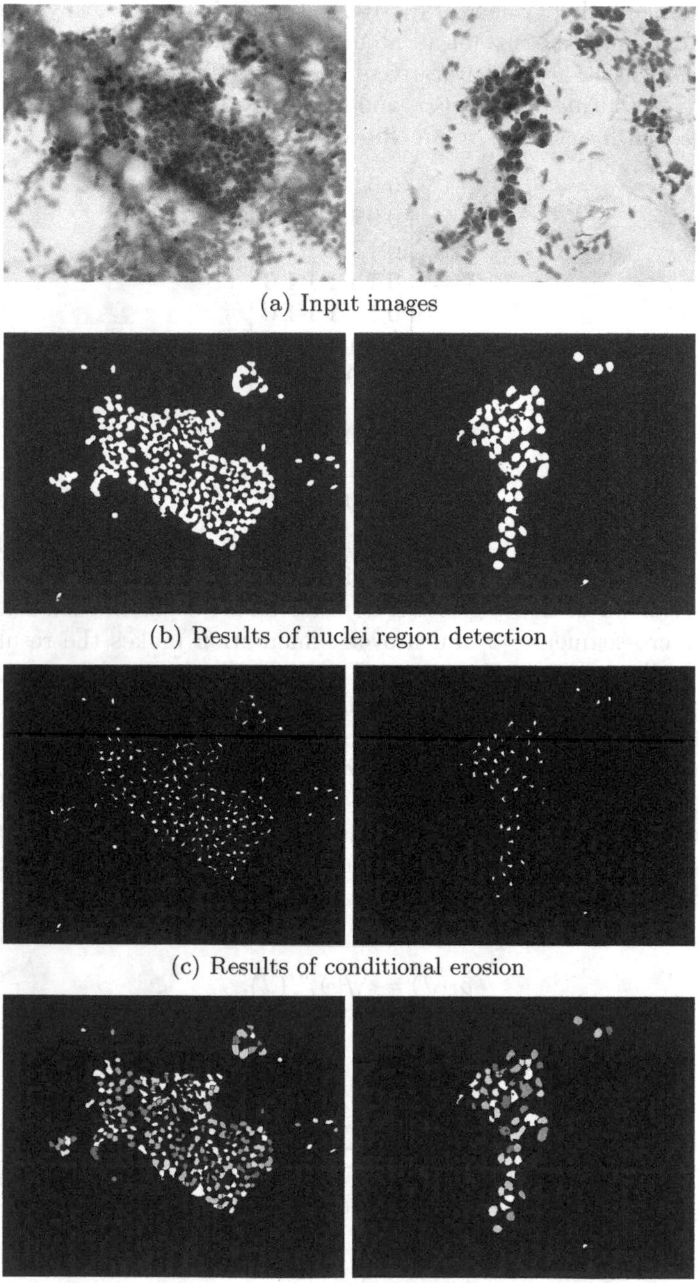

(a) Input images

(b) Results of nuclei region detection

(c) Results of conditional erosion

(d) Results of seeded watershed

Fig. 1. Input images, final and intermediate results of the proposed segmentation procedure original topographic surface. It allows to split the clustered nuclei avoiding the oversegmentation. The results of SW segmentation are stored in the matrix of labels L. The pixels labeled ω_1 make up the first nucleus, the pixels labeled ω_2 make up the second nucleus, and so on (Fig. 1(d)). Matrix L can be directly used to compute the features of nuclei.

2.4 Feature Extraction

For each isolated nucleus 28 features are extracted. Then, for each image, the mean, median and standard deviation are determined giving a total number of 84 features.

The features chosen reflect the observations of cytologists and can be divided into three groups. The first group is related to the size and shape of the nuclei. This is represented by the following features:

- *Area* - the actual number of pixels of the nucleus,
- *Perimeter* - the distance between each adjoining pair of pixels around the border of the nucleus,
- *Eccentricity* (ECC) - the scalar that specifies the ratio of the distance between the foci of the ellipse that has the same second-moments as the segmented nucleus and its major axis length,
- *Major Axis Length* (MjAL)- the length of the major axis of the ellipse that has the same normalized second central moments as the nucleus,
- *Minor Axis Length* (MnAL)- the length of the minor axis of the ellipse that has the same normalized second central moments as the nucleus.

The second group of features is related to the distribution of nuclei in the image. Healthy tissue usually form single-layered structures, while cancerous cells tends to break up which increases the probability of encountering separated nuclei. To express this relation, features representing the distance to centroid of all nuclei, and the distance to k-nearest nuclei are used:

- *Distance to Centroid of All Nuclei* (D2A) - the distance between the geometric center of the nucleus and centroid of all nuclei,
- *Distance to c-Nearest Nuclei* (D2cNN) - sum of distances between the geometric center of the nucleus and geometric centers of c-nearest nuclei; after conducting experiments with different values of c, we decided to set this parameter to 1.

The last group of features is related to the distribution of chromatin in the nuclei. This is represented with texture features based on gray-level co-occurrence matrix (GLCM) [15] and gray-level run-length matrix (GLRLM) [31], as well as the mean and variance of pixel values in each RGB channel.

First four textural features are based on GLCM:

- *Contrast* (CN),
- *Correlation* (CR),
- *Homogeneity* (H),
- *Energy* (EN).

The $N \times N$ matrix P, where N is the number of gray levels, is defined over an image to be the distribution of co-occurring values of pixels at a given offset. In other words, each element of P specifies the number of times a pixel with gray-level value i occurs shifted by a given distance to a pixel with the value j. Here, the mean of four GLCM features determined for offsets corresponding to 0°, 45°, 90° and 135° using eight gray-levels is calculated.

Next eleven textural features are based on gray-level run length matrix (GLRLM):

- *Short Run Emphasis* (SRE),
- *Long Run Emphasis* (LRE),
- *Run Length Nonuniformity* (RLN),
- *Gray-Level Nonuniformity* (GLN),
- *Run Percentage* (RP),
- *Low Gray-Level Run Emphasis* (LGRE),
- *High Gray-Level Run Emphasis* (HGRE),
- *Short Run Low Gray-Level Emphasis* (SRLGE),
- *Short Run High Gray-Level Emphasis* (SRHGE),
- *Long Run Low Gray-Level Emphasis* (LRLGE),
- *Long Run High Gray-Level Emphasis* (LRHGE).

The $N \times M$ matrix p, where N is the number of gray levels and M is the maximum run length, is defined for a given image as the number of runs with pixels of gray level i and run length j. Similar to GLCM, run length matrices for 0°, 45°, 90° and 135° are computed using eight gray-levels.

Finally, the last six features are:

- *Mean R Value, Mean G Value, Mean B Value* - the mean value of pixels of the nucleus in channel R, G and B respectively,
- *Variance of R Value, Variance of G Value, Variance of B Value* - the variance of pixel values of the nucleus in channel R, G and B, respectively.

After the features are extracted and the statistics for the images determined, all input variables are standardized.

2.5 Classification

The images were classified as either represent benign or malignant tumor using kNN classifier with $k = 5$ chosen experimentally [8,3,10]. In the considered approach, images were classified individually. However, the diagnostic decision concerns the patient, and not the single image. The final diagnosis is obtained by a majority voting of the classification of individual images belonging to the same patient (e.g. for given patient, if 5 images were classified as benign and 4 as malignant then the final diagnosis for the patient would be benign). Two classification measures were defined: the percentage ratio of successfully recognized cases to the total number of all cases and the the percentage ratio of successfully classified images to the total number of all images.

The n-fold cross-validation technique was used to measure objectively the performance of the classification algorithm. [9]. The fold was a set of 9 images representing 1 patient. This means the images belonging to the same patient were never at the same time in the training and testing set.

3 Results

In order to verify the effectiveness of the proposed CAD system, entire diagnostic procedure was performed and the classification accuracy was calculated as a result. The system was tested with 450 real medical images. The nuclei were segmented using the method described in Section 2.3. Then, for each image, 84 features were extracted as in Section 2.4. In order to check the discriminant power of the features, classification accuracy (for individual images) was calculated for each single feature using kNN. Table 1 presents obtained results. The

Table 1. Classification accuracy for all 84 individual features

feature	mean	median	STD
Area	66.89	66.89	56.22
Perimeter	59.56	68.89	44.89
ECC	50.89	62.00	48.44
MjAL	55.33	56.89	53.33
MnAL	66.44	66.44	54.22
D2A	77.56	72.89	72.44
D2cNN	72.00	76.89	69.56
CN	76.22	69.11	71.56
CR	75.56	70.22	73.56
H	66.67	63.78	55.56
EN	49.33	52.89	68.89
SRE	74.44	72.00	75.33
LRE	73.56	72.00	46.22
GLN	49.56	42.89	44.22
RLN	53.78	60.22	55.33
PR	50.44	52.89	51.33
LGRE	74.44	69.11	52.44
HGRE	52.00	58.44	53.78
SRLGE	85.11	82.44	79.78
SRHGE	64.44	50.00	73.56
LRLGE	53.33	53.11	61.78
LRHGE	46.22	58.67	50.67
MRV	56.44	62.00	50.44
MGV	84.00	84.22	51.11
MBV	82.89	86.89	52.67
VRV	50.44	51.11	55.56
VGV	70.89	73.33	63.56
VBV	74.89	70.67	79.33

approach was also tested using an suboptimal set of features. The set was found using sequential forward selection algorithm. The suboptimal set of features is as follow: MBV (median), D2CNN (median), SRHGE (mean), MnAL (median), HGRE (median), LRE (mean), LRE (STD), SRHGE (median), LRLGE (mean), LRLGE (STD).

The classification accuracy for images obtained using suboptimal feature set was 95.56%, sensitivity 0.97, specificity 0.94, and Matthews correlation coefficient 0.91. Moreover majority voting gave 100% classification accuracy for patients. This means that all patients were diagnosed correctly.

4 Conclusions

The main objective of the work was to develop an effective computer-aided diagnosis algorithm for breast cancer malignancy classification problem. Proposed approach is based on morphometric examination of cell nuclei. In contrast to normal and benign nuclei, which are typically uniform in appearance, cancerous nuclei are characterized by irregular morphology that is reflected in the parameters described in Section 2.4. Relatively simple methods like adaptive thresholding and k-means clustering were applied successfully to segment nuclei regions. Unfortunately, these methods are not able to properly handle clumped and overlapped nuclei. To overcome the problem, conditional erosion was applied to find the seeds of nuclei. Afterwards, seeds were utilized to initiate the watershed algorithm employed to label the nuclei. Experiments proved that conditional erosion can be a very useful tool for detecting nuclei seeds even when the nuclei are densely clustered. Classification result of 95.56% for individual images and 100% for patients shows that proposed approach can provide valuable and accurate information for a medical specialist.

Another challenge will be applying the whole segmentation and classification system for virtual slides generated by virtual scopes which are able to produce images with extremely high resolutions reaching 9 gigapixels and more. It will be crucial to develop algorithms to find interesting parts of the slides containing valuable information for further processing and diagnosis. Furthermore, it is plan to improve the accuracy of the system by applying more sophisticated methods for feature selection and classification [33].

Acknowledgments. This research was partially supported by the National Science Centre in Poland. The author wish to thank Dr. Roman Monczak from the Regional Hospital in Zielona Góra, Poland for his great help and interesting discussions.

References

1. Al-Kofahi, Y., Lassoued, W., Lee, W., Roysam, B.: Improved automatic detection and segmentation of cell nuclei in histopathology images. IEEE Trans. on Biomedcial Engineering 57(4), 841–852 (2010)

2. Beucher, S., Lantuèjoul, J.: Use of watersheds in contour detection. In: Proceedings of International Workshop on Image Processing, Real-time Edge and Motion Detection/Estimation, Rennes, France, pp. 1–12 (1979)
3. Bishop, C.: Pattern Recognition and Machine Learning. Springer (2006)
4. Breiman, L., Friedman, J., Stone, C.J., Olshen, R.A.: Classification and Regression Trees. Chapman & Hall, Boca Raton (1993)
5. Cheng, J., Rajapakse, J.C.: Segmentation of clustered nuclei with shape markers and marking function. IEEE Transactions on Biomedical Engineering 56(3), 741–748 (2009)
6. Clocksin, W.F.: Automatic segmentation of overlapping nuclei with high background variation using robust estimation and flexible contour models. In: Proc. 12th Int. Conf. Image Analysis and Processing, ICIAP 2003, pp. 682–687 (2003)
7. Cloppet, F., Boucher, A.: Segmentation of overlapping/aggregating nuclei cells in biological images. In: Proc. ICPR, pp. 1–4. IEEE (2008)
8. Cover, T., Hart, P.: Nearest neighbor pattern classification. IEEE Trans. on Information Theory 13(1), 21–27 (1967)
9. Devijver, P., Kittler, J.: Pattern Recognition: A Statistical Approach. Prentice-Hall, London (1982)
10. Duda, R., Hart, P., Stork, D.: Pattern Classification, 2nd edn. Wiley-Interscience (2001)
11. Filipczuk, P., Fevens, T., Krzyak, A., Obuchowicz, A.: Glcm and glrlm based texture features for computer-aided breast cancer diagnosis. Journal of Medical Informatics and Technologies 19, 109–115 (2012)
12. Filipczuk, P., Krawczyk, B., Woźniak, M.: Classifier ensemble for an effective cytological image analysis. Pattern Recognition Letters (2013)
13. Gocławski, J., Sekulska-Nalewajko, J., Kuźniak, E.: Neural network segmentation of images from stained cucurbits leaves with colour symptoms of biotic and abiotic stress. Int. J. Appl. Math and Comp. Sci. 22(3), 669–684 (2012)
14. Gurcan, M.N., Boucheron, L.E., Can, A., Madabhushi, A., Rajpoot, N.M., Yener, B.: Histopathological image analysis: A review. IEEE Reviews in Biomedical Engineering 2, 147–171 (2009)
15. Haralick, R., Shanmugam, K., Dinstein, I.: Textural features for image classification. IEEE Trans. on Systems, Man, and Cybernetics 3(6), 610–621 (1973)
16. Hartigan, J.A.: Clustering Algorithms (Probability & Mathematical Statistics). John Wiley & Sons Inc. (1975)
17. Hartigan, J.A., Wong, M.A.: Algorithm as 136: A k-means clustering algorithm. Journal of the Royal Statistical Society, Series C (Applied Statistics) 28(1), 100–108 (2001)
18. Jeleń, L., Fevens, T., Krzyżak, A.: Classification of breast cancer malignancy using cytological images of fine needle aspiration biopsies. Int. J. Appl. Math and Comp. Sci. 18(1), 75–83 (2010)
19. Jung, C., Kim, C., Chae, S.W., Oh, S.: Unsupervised segmentation of overlapped nuclei using bayesian classification. IEEE Transactions on Biomedical Engineering 57(12), 2825–2832 (2010)
20. Kowal, M., Korbicz, J.: Segmentation of breast cancer fine needle biopsy cytological images using fuzzy clustering. In: Kornacki, J., Raś, Z., Wierzchoń, S., Kacprzyk, J. (eds.) Advances in Machine Learning I, pp. 405–417. Springer, Heidelberg (2010)
21. Krawczyk, B., Filipczuk, P., Woźniak, M.: Adaptive splitting and selection algorithm for classification of breast cytology images. In: Nguyen, N.-T., Hoang, K., Jędrzejowicz, P. (eds.) ICCCI 2012, Part I. LNCS, vol. 7653, pp. 475–484. Springer, Heidelberg (2012)

22. Malek, J., Sebri, A., Mabrouk, S., Torki, K., Tourki, R.: Automated breast cancer diagnosis based on GVF-Snake segmentation, wavelet features extraction and fuzzy classification. J. of Signal Processing Systems 55, 49–66 (2009)
23. Marciniak, A., Obuchowicz, A., Monczak, A., Kołodziński, M.: Cytomorphometry of fine needle biopsy material from the breast cancer. In: Proc. 4th Int. Conf. on Computer Recognition Systems, CORES 2005, pp. 603–609 (2005)
24. Mitchell, T.M.: Machine Learning. McGraw-Hill, Boston (1997)
25. Niwas, I.S., Palanisamy, P., Sujathan, K.: Wavelet based feature extraction method for breast cancer cytology images. In: 2010 IEEE Symp. on Industrial Electronics & Applications (ISIEA), pp. 686–690 (2010)
26. Nixon, M., Aguado, A.: Feature Extraction & Image Processing for Computer Vision, 3rd edn. Academic Press (2012)
27. Obuchowicz, A., Hrebień, M., Nieczkowski, T., Marciniak, A.: Computational intelligence techniques in image segmentation for cytopathology. In: Smoliński, T.G., Milanova, M.G., Hassanien, A.-G. (eds.) Computational Intelligence in Biomedicine and Bioinformatics, pp. 169–199. Springer, Berlin (2008)
28. Sezgin, M., Sankur, B.: Survey over image thresholding techniques and quantitative performance evaluation. Journal of Electronic Imaging 13(1), 146–165 (2003)
29. Śmietański, J., Tadeusiewicz, R., Łuczyńska, E.: Texture analysis in perfusion images of prostate cancer–a case study. Int. J. Appl. Math and Comp. Sci. 20(1), 149–156 (2010)
30. Suri, J.S., Setarhdan, K., Singh, S.: Advanced Algorithmic Approaches to Medical Image Segmentation. Springer, London (2002)
31. Tang, X.: Texture information in run-length matrices. IEEE Trans. on Image Processing 7(11), 1602–1609 (1998)
32. Vincent, L.: Morphological gray scale reconstruction in image analysis: Applications and efficient algorithms. IEEE Transactions on Image Processing 2(2), 176–200 (1993)
33. Woźniak, M., Krawczyk, B.: Combined classifier based on feature space partitioning. Int. J. Appl. Math and Comp. Sci. 22(4), 855–866 (2012)
34. Xiong, X., Kim, Y., Baek, Y., Rhee, D.W., Kim, S.-H.: Analysis of breast cancer using data mining & statistical techniques. In: Proc. 6th Int. Conf. on Software Engineering, Artificial Intelligence, Networking and Parallel/Distributed Computing and 1st ACIS Int. Worksh. on Self-Assembling Wireless Networks, pp. 82–87 (2005)
35. Yang, X., Li, H., Zhou, X.: Nuclei segmentation using marker-controlled watershed, tracking using mean-shift, and Kalman filter in time-lapse microscopy. IEEE Transactions on Circuits and Systems - I 53(11X), 2405–2414 (2012)

The Role and Use of MicroRNA in Cancer Diagnosis

Michał Świerniak[1,2] and Andrzej Świerniak[3]

[1] Department of Nuclear Medicine and Endocrine Oncology,
Maria Sklodowska-Curie Memorial,
Cancer Center and Institute of Oncology, 44-101 Gliwice, Poland
mswierniak@io.gliwice.pl

[2] Department of General, Transplant, and Liver Surgery,
Medical University of Warsaw, 02-097 Warsaw, Poland

[3] Department of Automatic Control,
Silesian University of Technology, 44-101 Gliwice, Poland
andrzej.swierniak@polsl.pl

Abstract. MicroRNAs (miRNAs) are small RNA molecules that post-transcriptionally regulate gene expression. The aim of this paper is to present a survey on published application of these molecules in cancer diagnosis including our own results in this area. As an illustrative example of the use of miRNA in cancer diagnosis we present some main points from our study related to thyroid carcinoma.

Keywords: Medical diagnosis, cancer, MiRNAs, system biology.

1 Introduction

MicroRNAs (miRNAs) are small non-coding regulatory double stranded RNAs that are processed from precursors with a characteristic hairpin secondary structure. Their role as specific regulators of gene expression became one of the most spectacular discoveries in the past decades. These studies were awarded one of the recent Nobel Prizes in Medicine granted to Andrew Fire and Craig Mello [1]. Nevertheless the first study, to our knowledge, related to miRNA was presented in 1993 by Lee et al. [2] who found that among genes of C. Elegans one termed lin-4, did not encode a protein but instead a novel 22-nucleotide small RNA. The second 22-nucleotide small RNA of this type, let-7, a gene also involved in regulation of development of C. elegans was discovered by Reinhart et al. [3]. Homologs of the let-7 gene were identified in other animals including humans [4] that suggested an important and fundamental biological role for this small RNA.

Within the following year, more than 100 additional small regulatory RNAs similar to lin-4 and let-7 were identified in worms, the fruit fly Drosophila, and even in humans (see e.g., [5, 6]). These small non-coding RNAs were named microRNAs (miRNAs). Subsequently, many more short regulatory RNAs were

J. Korbicz and M. Kowal (eds.), *Intelligent Systems in Technical and Medical Diagnostics*, Advances in Intelligent Systems and Computing 230,
DOI: 10.1007/978-3-642-39881-0_18,

identified in almost all multicellular organisms, including flowering plants, worms, flies, fish, frogs, mammals (see[7], and references therein). In invertebrates, miRNAs regulate developmental timing (e.g., lin-4), neuronal differentiation, cell proliferation, growth control, and programmed cell death. In mammals, miRNAs have been found to play a role in embryogenesis and stem cell maintenance, hematopoietic cell differentiation, and brain development. Computational predictions of miRNA targets suggest that up to 30% of human protein coding genes may be regulated by miRNAs [8],[9]. This makes miRNAs one of the most abundant classes of regulatory genes in humans.

MicroRNAs are sequentially processed from longer precursor transcripts that are encoded by miRNA genes. MiRNA genes are referred to by the same name (termed mir) written in italics to distinguish them from the corresponding mature miRNA (termed miR) followed by a number, e.g., mir-1 or miR-1. Sequences of all discovered miRNAs are deposited in a specifically designated public repository miRBase [10]. There are currently about 1600 miRNA genes provided in miRBase. Nevertheless our knowledge of human miRNAs is rather descriptive. MicroRNAs recognize their targets based on sequence complementarity [11]. Mature miRNA is partially complementary to one or more messenger RNAs. In humans, the complementary sites are usually within the 3-untranslated region of the target messenger RNA. To become effective, mature miRNA forms a complex with proteins, termed the RNA-induced silencing complex. The miRNA incorporated into the silencing complex can bind to target messenger RNA by complementary base pairing. This base pairing subsequently causes inhibition of protein translation and/or degradation of the messenger RNA, thus levels of the synthesized protein are consequently reduced, whereas messenger RNA levels may or may not be decreased. Importantly, expression of microRNAs is highly tissue-specific, therefore each cell and tissue type has its own microRNA landscape. This fact is of great importance for the proper understanding of such processes as differentiation, in which cells of the same genotype are developed into phenotypically distinct tissues.

2 MicroRNA and Cancer

Expression of microRNAs has been found to be deregulated in a wide range of human diseases including cancer. However, it remains uncertain whether altered miRNA expression is a cause or consequence of pathological processes. The underlying mechanisms of why and how miRNAs become deregulated are largely unknown. Although bioinformatics approaches can predict thousands of genes that are potentially targeted and regulated by miRNAs based on sequence complementarity, only very few miRNA target genes have been functionally validated. Among hundreds of miRNAs, a fraction have been shown to play a role in a variety of biological processes including development, differentiation, proliferation, and cell death. Since tumorigenesis often involves deregulation of these processes, miRNAs can play roles of oncogenes or tumor suppressors (see e.g. [12]). Emerging data suggest that miRNA are deregulated either by genomic changes or epigenetic mechanisms in certain types of cancer. Many miRNAs are up- or down-regulated in cancers, suggesting that tumorigenesis is associated with changes of expression of numerous miRNAs. There are several reasons why miRNAs could play an important role in human cancer. First of all miRNAs are found to control cell proliferation and apoptosis and their deregulation is known to contribute to cancer development and malignancy. Many miRNA genes are located in fragile sites of genome or regions related with amplification of genes involved in the process of carcinogenesis [13, 14] and in response to anticancer therapy , e.g. drug resistance.

A special role in the studies on relationship between miRNAs and cancer is played by a cluster of six miRNAs, the mir-17-92 cluster. It has been discovered within a region on chromosome 13 known as commonly amplified in human B-cell lymphomas [15]. The same cluster is reported as related to Myc oncogenic pathways and was shown to be overexpressed in lymphoma cell lines carrying this amplification [14]. On the other hand two miRNA genes, mir-15 and mir-16, have been found as located within the 30-kb deletion on the same chromosome 13 which is the most frequent abnormality in chronic lymphocytic leukemia [13]. The same miRNAs may have oncogenic or tumor suppressive activity depending on the context and the cell type they are expressed in. A single miRNA may regulate various unrelated target genes and thereby control opposing activities such as cellular proliferation and apoptosis. The ultimate function of a miRNA may depend on the tissue type it is expressed in on the target genes that are present in this tissue type. For example, some reports demonstrate anti-apoptotic effect of miR-17-92 exerted through various pathways that promote cell proliferation and growth (e.g. [16]) while other reports show its role as a mediator of angiogenesis in tumors induced by the oncogene c-Myc [17]. Interestingly, c-Myc induces expression of mir-17-92 cluster and of E2F1, but expression of the latter is inhibited by microRNAs encoded within mir-17-92 gene.

The question whether global miRNA expression profiles could classify human cancer, posed in the group of Golub [18] is also one of challenges that we are to confront. Due to their high tissue-specificity, microRNA expression profiles clearly differentiate human cancers according to their developmental origin. Many miRNAs have been found to be associated with tumor versus normal breast tissues, invasiveness, molecular subtypes and hormone receptor status, as well as diagnostic and prognostic markers in the breast cancer [19-23]. They may be considered as future diagnostic and predictive tools.

Cancers of epithelial and hematopoietic origin have distinct miRNA profiles. A subgroup of gastrointestinal tumors, which arise from endoderm, was distinguished by miRNA expression patterns. Furthermore, tumors within a single cell lineage such as acute lymphoblastic leukemia were further differentiated according to their underlying genetic abnormality into BCR/ABL-positive tumors, T-cell tumors, and those with MLL gene rearrangement [18]. Finally, the authors applied the miRNA expression profiles they had established to an independent series of 17 poorly differentiated tumors of unknown origin. Based on the differential expression of 217 miRNAs, a correct diagnosis could be established in 12 out of 17 of the tumors. In contrast, gene expression profiling based on 16,000 messenger RNAs did not accurately classify the tumors [18]. This has potential important clinical implications. If miRNAs prove useful for clinical diagnosis, their key advantage might be their high stability. In contrast to most messenger RNAs, they have a long lifespan in vivo and are very stable in vitro, which might allow analysis of paraffin-embedded samples for routine diagnostic applications. The importance of microRNAs has also been shown in the pathogenesis of thyroid cancers. Numerous studies revealed that each type of thyroid cancer exhibits specific signatures of microRNA expression [24, 25]. Deregulated microRNAs not only regulate distinct sets of target genes whose alterations lead to thyroid carcinogenesis, but also allow for distinguishing between the particular cancer subtypes.

Moreover, recent studies based on deep-sequencing technology revealed significant microRNAs heterogeneity, showing that individual miRNA genes may give rise to several mature miRNA products that differ in length. To further clarify the roles of miRNAs in tumorigenesis it seems to be crucial to generate complete datasets of variations in miR sequences and to profile the expression of miR isoforms and their target genes in cancer tissues.
Our studies based on experimental material prepared by biomedical groups from The Laboratory of Genomic Medicine at the Medical University of Warsaw are related to thyroid cancers. In the next section we present some our results in this area as an illustrative example of the issue addressed in this paper.

3 An Example: miRNA in PTC Diagnosis

Papillary thyroid carcinoma (PTC) is the most common form of thyroid malignancies (reaching 80%) and the predisposition to this cancer is heritable, but the factors underlying its pathogenesis remain to be elucidated. Apart from the known BRAF mutations and RET/PTC rearrangements, and from a number of loci that have been proposed as predisposing to PTC, no candidate gene, whose alterations could directly lead to thyroid carcinogenesis, has been identified so far. It seems therefore that genes, whose aberrances underlie thyroid tumorigenesis, may not be of protein-coding type, but include various small, regulatory molecules, such as microRNAs. Indeed, thyroid cancer was the first disease for which a microRNA-related genetic predisposition was found [26].

However, although it is generally acknowledged that microRNAs are important factors implicated in the pathogenesis of thyroid carcinoma, all the previous studies were based on microarray or Real-time PCR microRNA profiling of tumor and control tissue. As such methods allow only for quantification of known and annotated miRNAs, the comprehensive information on the miRNome of thyroid gland and its aberrances in cancer was lacking.

To fully elucidate the role of miRs in PTC we employed the next-generation sequencing to analyze the miRNA profiles of PTC tumors (n=14), unaffected tissue adjacent to tumors (n=14), and normal (control) thyroid tissue (n=14). We received sequences of miRNAs present in the thyroid tissue together with a corresponding number of reads reflecting their expression levels, which were normalized to total reads available for all miRs (RPM, reads per million).

Over 400 miRs deposited in miRbase were found to be significantly expressed in the thyroid tissue. Among them, 124 were significatly deregulated in PTC tumors compared to normal thyroid (FDR<0.05), including 24 strongly upregulated (fold change>2), and 65 strongly downregulated (fold change<0.5). The list of top deregulated miRs comprises miR-146b-5p, miR-221-3p, miR-7-5p, miR-222-3p, confirming the previous microarray profiling. In addition, top-deregulated miRs include several miRs derived from the passenger strand of miRNA precursors, e.g. miR-146b-3p, -7-2-3p, -221-5p,-222-5p [27].

Most mature miRs exist in variable length variants, called isomiRs. Interestingly, our results indicate that 85% of the most abundant miRNA isoforms differ from the standard reference sequence deposited in the miRBase. Moreover, the standard miR sequence was completely absent in 40% of miRs expressed in thyroid gland [27].

Based on the miRNA expression profiles, a diagnostic panel has been proposed to distinguish normal thyroid gland from PTC with the accuracy 93%, sensitivity 86% and specificity 100%.

4 Remarks on Inference Analysis

The bioinformatics analysis is crucial for the whole experiment since even the best results are useless when interpreted incorrectly. Integrative analysis of genome scale data is the most frequently used computational biology approach at the moment. Processing of data from deep sequencing (re-sequencing) experiments creates challenges in many areas of bioinformatics such as massive parallel algorithms required for filtering, matching, alignment, detection, identification, classification, comparison, pattern prediction, discovery of new classes and statistical verification of data of experiments and results of their analysis.

Preprocessing and quality control are the first tasks for the sequencer data outputs. There are several ready to go open software programs so calculations of quality control and filtering out poor quality reads are not as complex as next step of the analysis - alignment. Matching of NGS data to the reference may be performed using existing software for the multiple sequences alignment. There are number of algorithms based on deterministic dynamic programming or discrete-integer programming techniques but the huge number of reads which should be expected makes such algorithms inefficient. Instead, algorithms based on Burrows-Wheeler transform are proposed (such as Bowtie [28] or BWA [29]) as fast and efficient aligners. Those leading programs are gathered in the Galaxy project – a promising platform for comprehensive NGS data analysis. Galaxy contains quality control, alignment and differential analysis tools [30-32]. Additionally, some modifications of these tools have been developed as well [27]. Also alignment techniques based on hidden Markov chains and Vitterbi algorithms seems to be promising and worth to exploring. After alignment normalization step is required followed by filtering of aligned reads based on specific thresholding algorithms. Comparison of obtained patterns between particular groups of samples is generally performed using standard statistical algorithms. Our experience with such methods in class prediction and pattern discovery for data from DNA microarrays experiments suggested that especially techniques based on Support Vector Machines combined with Partial Least Squares and Bootstrapping based validation could lead to efficient algorithms with high accuracy and stability (in the sense of robustness to the choice of experimental material). Similar approach has been used to align and compare selected sequences of miRNAs.

For short RNA sequences data analysis we have proposed an integrated analysis workflow as well. Since the NGS platforms requires a specific length of reads, short RNA fragments in the library were extended with specific adapters, annealed to their $3'$ ends during preparation. Removal of these adapters has been performed on the raw sequence reads using cutadapt software [33] and sequences with the length of 15-28 nucleotides were subject to the further analysis as potential miRNAs. Quality control has been done by FastQC software [34]. To detect all human isomiRs, mature canonical form sequences (based on miRBase) were cut from harpins with 5 nt length flanks from both sides. Reads have been mapped on the such prepared reference using Bowtie [28] with the requirement of perfect matching.

The numbers of mapped reads for each miRNA have been subsequently calculated and provided in two ways : as a number of all sequences mapped to reference sequence (miR expression) and as a number of unique reads mapped to each reference sequence (isomiR expression, followed by normalization data for each sample (RPM normalization) that has allowed us for comparative analysis of miRNA expression. Only significantly expressed miRs (median RPM above 5 within any of the sample types) and isomiRs (expression>1% of the total expression of a particular miRNA, in at least 80% of samples within the sample type) have been passed the filtering step.

Statistical analysis has been based on standard tools. The ratios between number of isomiRs per miRNA among studied groups have been computed using Pearson's chi-squared test. Selection of miRs and isomiRs deregulated between the analyzed types of samples has been performed using Welch t-test. The false discovery rate (FDR) has been used to assess the multiple testing errors, and the maximum allowed proportion of false positive genes was of 5%. Construction of the classifier has been based on extraction Partial Least Squares components that are plugged into Support Vector Machines algorithm. Gene Selection was carried with one-step Recursive Feature Elimination. Performance of the classifier were was estimated by Leaving-One-Out Cross Validation method.

5 Conclusions

The main idea of this paper is to review important aspects of the relationships between human miRNA biology and some features of carcinogenesis. The special attention is focused on possible implications of this relevance for cancer diagnosis and prediction of its development. We mainly give credits to others but we also present some our experience related to the role of miRNA in classification and diagnosis of thyroid carcinomas. Almost all microRNAs exhibit isoforms of variable length and potentially distinct function in thyroid tumorigenesis.

The obtained data in this case is of great importance for understanding of gene regulation networks within the thyroid gland, as it provides information on the miRs that can indeed be important players in this process. As our study has shown, not all the miRs deposited in miRbase are expressed in the thyroid, thus even though in silico analyses reveal potential binding of such miRs to thyroid-expressed genes, such predicted interaction will have no biological relevance. Simultaneously, newly discovered isoforms of miRs constitute a group of previously unknown molecules that regulate a large number of target genes. We do not attempt answering the important question whether the altered miRNA expression observed in cancer is cause or consequence of malignant transformation.

References

1. Fire, A., Xu, S., Montgomery, M.K., Kostas, S.A., Driver, S.E., Mello, C.C.: Potent and specific genetic interference by double-stranded RNA in Caenorhabditis elegans. Nature 391, 806–811 (1998)
2. Lee, R.C., Feinbaum, R.L., Ambros, V.: The C. elegans heterochronic gene lin-4 encodes small RNAs with antisense complementarity to lin-14. Cell 75, 843–854 (1993)
3. Reinhart, B.J., Slack, F.J., Basson, M., Pasquinelli, A.E., Bettinger, J.C., Rougvie, A.E., Horvitz, H.R., Ruvkun, G.: The 21-nucleotide let-7 RNA regulates developmental timing in Caenorhabditis elegans. Nature 403, 901–906 (2000)
4. Pasquinelli, A.E., Reinhart, B.J., Slack, F., Martindale, M.Q., Kuroda, M.I., Maller, B., Hayward, D.C., Ball, E.E., Degnan, B., Muller, P., Spring, J., Srinivasan, A., Fishman, M., Finnerty, J., Corbo, J., Levine, M., Leahy, P., Davidson, E., Ruvkun, G.: Conservation of the sequence and temporal expression of let-7 heterochronic regulatory RNA. Nature 408, 86–89 (2000)
5. Lau, N.C., Lim, L.P., Weinstein, E.G., Bartel, D.P.: An abundant class of tiny RNAs with probable regulatory roles in Caenorhabditis elegans. Science 294, 858–862 (2001)
6. Lee, R.C., Ambros, V.: An extensive class of small RNAs in Caenorhabditis elegans. Science 294, 862–864 (2001)
7. Bartel, D.P.: MicroRNAs: genomics, biogenesis, mechanism, and function. Cell 116, 281–297 (2004)
8. Beitzinger, M., Peters, L., Zhu, J.Y., Kremmer, E., Meister, G.: Identification of Human microRNA targets from isolated Argonaute protein complexes. RNA Biol. 4(2), 76–84 (2007)
9. Lewis, B.P., Burge, C.B., Bartel, D.P.: Conserved seed pairing, often flanked by adenosines, indicates that thousands of human genes are microRNA targets. Cell 120(1), 15–20 (2005)
10. Griffiths-Jones, S.: miRBase: the microRNA sequence database. Methods Mol. Biol. 342, 129–138 (2006)
11. Brennecke, J., Stark, A., Russell, R.B., Cohen, S.M.: Principles of microRNA-target recognition. PLoS Biol. 3, e85 (2005)

12. Calin, G.A., Croce, C.M.: MicroRNA signatures in human cancers. Nat. Rev. Cancer 6, 857–866 (2006)
13. Calin, G.A., Sevignani, C., Dumitru, C.D., Hyslop, T., Noch, E., Yendamuri, S., Shimizu, M., Rattan, S., Bullrich, F., Negrini, M., Croce, C.M.: Human microRNA genes are frequently located at fragile sites and genomic regions involved in cancers. Proc. Natl. Acad. Sci. USA 101, 2999–3004 (2004)
14. He, L., Thomson, J.M., Hemann, M.T., Hernando-Monge, E., Mu, D., Goodson, S., Powers, S., Cordon-Cardo, C., Lowe, S.W., Hannon, G.J., Hammond, S.M.: A microRNA polycistron as a potential human oncogene. Nature 435, 828–833 (2005)
15. Ota, A., Tagawa, H., Karnan, S., Tsuzuki, S., Karpas, A., Kira, S., Yoshida, Y., Seto, M.: Identification and characterization of a novel gene, C13orf25, as a target for 13q31-q32 amplification in malignant lymphoma. Cancer Res. 64, 3087–3095 (2004)
16. Matsubara, H., Takeuchi, T., Nishikawa, E., Yanagisawa, K., Hayashita, Y., Ebi, H., Yamada, H., Suzuki, M., Nagino, M., Nimura, Y., Osada, H., Takahashi, T.: Apoptosis induction by antisense oligonucleotides against miR-17-5p and miR-20a in lung cancers overexpressing miR-17-92. Oncogene 26, 6099–6105 (2007)
17. Chen, C.Z., Li, L., Lodish, H.F., Bartel, D.P.: MicroRNAs modulate hematopoietic lineage differentiation. Science 303, 83–86 (2004)
18. Lu, J., Getz, G., Miska, E.A., Alvarez-Saavedra, E., Lamb, J., Peck, D., Sweet-Cordero, A., Ebert, B.L., Mak, R.H., Ferrando, A.A., Downing, J.R., Jacks, T., Horvitz, H.R., Golub, T.R.: MicroRNA expression profiles classify human cancers. Nature 435, 834–838 (2005)
19. Volinia, S., Galasso, M., Sana, M.E., Wise, T.F., Palatini, J., Huebner, K., Croce, C.M.: Breast cancer signatures for invasiveness and prognosis defined by deep sequencing of microRNA. Proc. Natl. Acad. Sci. USA 109, 3024–3029 (2012)
20. Blenkiron, C., Goldstein, L.D., Thorne, N.P., Spiteri, I., Chin, S.F., Dunning, M.J., Barbosa-Morais, N.L., Teschendorff, A.E., Green, A.R., Ellis, I.O., Tavare, S., Caldas, C., Miska, E.A.: MicroRNA expression profiling of human breast cancer identifies new markers of tumor subtype. Genome. Biol. 8, R214 (2007)
21. Lowery, A.J., Miller, N., Devaney, A., McNeill, R.E., Davoren, P.A., Lemetre, C., Benes, V., Schmidt, S., Blake, J., Ball, G., Kerin, M.J.: MicroRNA signatures predict oestrogen receptor, progesterone receptor and HER2/neu receptor status in breast cancer. Breast Cancer Res. 11, R27 (2009)
22. Foekens, J.A., Sieuwerts, A.M., Smid, M., Look, M.P., de Weerd, V., Boersma, A.W., Klijn, J.G., Wiemer, E.A., Martens, J.W.: Four miRNAs associated with aggressiveness of lymph node negative, estrogen receptor-positive human breast cancer. Proc. Natl. Acad. Sci. USA 105, 13021–13026 (2008)
23. Qian, B., Katsaros, D., Lu, L., Preti, M., Durando, A., Arisio, R., Mu, L., Yu, H.: High miR-21 expression in breast cancer associated with poor disease-free survival in early stage disease and high TGF-beta1. Breast Cancer Res. Treat 117, 131–140 (2009)
24. He, H., Jazdzewski, K., Li, W., Liyanarachchi, S., Nagy, R., Volinia, S., Calin, G.A., Liu, C.G., Franssila, K., Suster, S., Kloos, R.T., Croce, C.M., de la Chapelle, A.: The role of micro-RNA genes in papillary thyroid carcinoma. Proc. Natl. Acad. Sci. USA 102, 19075–19080 (2005)
25. de la Chapelle, A., Jazdzewski, K.: MicroRNAs in thyroid cancer. J. Clin. Endocrinol. Metab. 96, 3326–3336 (2011)

26. Jazdzewski, K., Liyanarachchi, S., Swierniak, M., Pachucki, J., Ringel, M.D., Jarzab, B., de la Chapelle, A.: Polymorphic mature microRNAs from passenger strand of pre-miR-146a contribute to thyroid cancer. Proc. Natl. Acad. Sci. USA 106, 1502–1505 (2009)
27. Swierniak, M., Wojcicka, A., Czetwertynska, M., Stachlewska, E., Maciag, M., Wiechno, W., Gornicka, B., Bogdanska, M., Koperski, L., de la Chapelle, A., Jazdzewski, K.: In-depth characterization of the microRNA transcriptome in normal thyroid and papillary thyroid carcinoma. J. Clin. Endocrinol. Metab. (in print)
28. Langmead, B., Trapnell, C., Pop, M., Salzberg, S.L.: Ultrafast and memory-efficient alignment of short DNA sequences to the human genome. Genome. Biol. 10(3), R25 (2009)
29. Li, H., Durbin, R.: Fast and accurate short read alignment with Burrows-Wheeler Transform. Bioinformatics 25, 1754–1760 (2009)
30. Goecks, J., Nekrutenko, A., Taylor, J.: The Galaxy Team: Galaxy: a comprehensive approach for supporting accessible, reproducible, and transparent computational research in the life sciences. Genome. Biol. 11(8), R86 (2010)
31. Blankenberg, D., Von Kuster, G., Coraor, N., Ananda, G., Lazarus, R., Mangan, M., Nekrutenko, A., Taylor, J.: Galaxy: a web-based genome analysis tool for experimentalists. Curr. Protoc. in Mol. Biol. Chapter 19, Unit 19.10.1-21 (2010)
32. Giardine, B., Riemer, C., Hardison, R.C., Burhans, R., Elnitski, L., Shah, P., Zhang, Y., Blankenberg, D., Albert, I., Taylor, J., Miller, W., Kent, W.J., Nekrutenko, A.: Galaxy: a platform for interactive large-scale genome analysis. Genome Research 15(10), 1451–1455 (2005)
33. Andrews, S.: `http://www.bioinformatics.babraham.ac.uk/projects/fastqc/`
34. Martin, M.M.: Cutadapt removes adaptor sequences from high-throughput sequencing reads. EMBnet. Journal 17.1, 10–12 (2011)

Method of Pattern Detection in Mammographic Images

Jagoda Łazarek, Piotr S. Szczepaniak, and Arkadiusz Tomczyk

Institute of Information Technology
Lodz University of Technology
Wólczańska 215, 90-924 Łódź, Poland
{jagoda.lazarek,arkadiusz.tomczyk}@p.lodz.pl

Abstract. This paper describes a novel method for detecting the boundaries between areas of different brightness. Its goal is to provide a way to detect the boundaries between unclearly separated areas characterised by tonal transition. Boundaries are detected and described by the coordinates of the start and the end point. The proposed method can be used as one of the steps in modern hierarchical image analysis techniques where each step increases semantic knowledge about the image content. In this work it is applied and evaluated using mammograms.

Keywords: medical imaging, mammograms, breast cancer, computer-aided diagnosis, image processing, object detection, active partitions.

1 Introduction

Nowadays, automatic image processing is important because of huge amount of data which is collected continuously. Data analysis without computer support is almost impossible. Medical imaging is one of the domains which provide a lot of data that has to be evaluated. Thus, medical image analysis requires the application of various methods.

Mammography is a commonly used method of medical examination, which helps in breast cancer diagnosis. A mammogram is a grey-scale image created as a result of an X-ray examination of the breast. Mammogram analysis is a very important process which aims at detecting different types of abnormalities [3, 4]. Early and precise diagnosis increases the chances of recovery [5]. There are two main types of abnormalities, namely masses and calcifications. Due to their different characteristics, each of the types requires the application of different detection methods [5].

Mammogram analysis is a challenging area of research, which has been addressed by many authors. Below we outline the existing approaches to breast cancer detection. One of the most common approaches to this problem is founded on the pixel-based methods [5], the main idea of which is to analyse the mammogram pixel by pixel. By doing so, one can determine if the pixel is suspicious, i.e. whether it belongs to the abnormality area (e.g. mass, calcification) or not. Another popular approach makes use of the region-based methods. The latter

J. Korbicz and M. Kowal (eds.), *Intelligent Systems in Technical and Medical Diagnostics*, Advances in Intelligent Systems and Computing 230,
DOI: 10.1007/978-3-642-39881-0_19, © Springer-Verlag Berlin Heidelberg 2014

are founded on the extraction of the regions of interest and their classification. The target regions are classified into two classes suspicious or normal [5]. Recently another innovative approach was proposed in [6–8] where neither pixels nor regions were considered. It was called active partitions and its inspiration comes from active contour methods. It postulates the change of image representation adjusting it to the specific task. A method of automatic preparation of such a representation is presented in this paper.

The paper is organized as follows. Section 2 introduces shortly the active partition method. Next section describes the proposed method of boundary detection between two areas of different brightness characterised by tonal transition. Section 4 presents how the method can be applied to mammogram analysis. This is followed by the discussion of the results. We conclude in Section 5.

2 Active Partitions

As it was mentioned above active partitions were inspired by active contour methods [9–12]. Originally, those methods were developed as a tool for a low-level image segmentation with a possibility of usage of higher-level information (e.g. expert knowledge). The main idea of this approach is to find an optimal contour in the space of considered contours representing a certain image region. The search is performed in an evolution process (optimization) in which the given objective function, called energy, evaluates the quality of the contour. Usually, the energy function consists of two components, namely internal and external, which represent the shape of contour and the features of the ground, respectively. The form of energy should assure that its minima correspond to the sought objects.

The concept of active partitions arose when the relationship between contours and classifiers was noticed and described in [13]. It was revealed that process of contour evolution leads to the optimal labelling (partition) of image pixels set and, since in pattern recognition any objects can be classified, it allowed to formulate a hypothesis that one can use active contour approach even if image is not represented as a set of pixels [14]. That approach in practice requires selection of image representation (set of elements), definition of objective (energy) function able to evaluate the subsets of those elements and choice of the optimization (evolution) method capable of searching for the optimal labelling (partition).

So far two basic alternative image representations were proposed in [6–8] - line segments and circular regions. However, although there was an algorithm allowing to find the second image descriptors automatically, in case of the first one for more demanding images there was no automatic method to generate them. A good exemplification of that problem are mammograms where detection of line segments reflecting the visual expectations of the users appears to be a challenging task. An initial solution of that problem is presented further in this paper.

The choice of line segments as a representation of mammograms has its natural explanation. First of all changing image representation in this way reduces significantly the amount of data that must be analysed. Secondly, the visual aspect that is taken into consideration while analysing of that kind of images are changes in brightness and tonal transitions which can be easily represented by such segments. Finally, that description should allow to detect masses in those images as they have a specific shape (e.g. circumscribed or spicular) which should be easily detectable among line segments as it was depicted in Fig. 1.

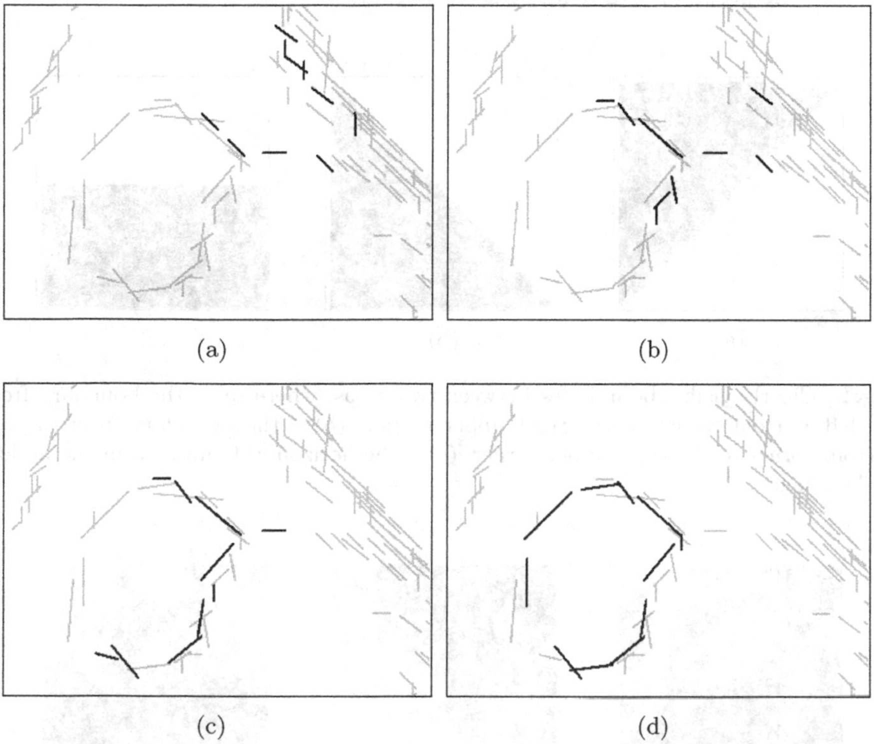

Fig. 1. Example illustrating active partition approach where image content is represented using line segments: (a), (b), (c), (d) - sample successive steps of searching for an optimal labelling (partition)

3 Boundaries Detection

In this section, the idea of the proposed method is described. The method is used to detect the edge between two areas of different brightness, where the boundary between these areas is not clearly visible. The lack of boundary is caused by the tonal transition from one area to another. Our method solves this problem by separating areas of different brightness and gives coordinates of boundary segments in one step - in a contrast to edges which are extracted by filters like Sobel filter or Canny filter. Furthermore, obtained boundaries are very thin, what is important in the next steps of analysis.

Let us analyse the images shown in Figs. 2 and 3. In Fig. 2, the boundaries between two areas are clearly visible, as opposed to the boundaries depicted in Fig. 3. It is easy to notice that there are no straight boundaries which separate those two areas. The edges are ragged and the areas are not homogeneous, but when we look at these images from distance, such details are no more visible. We can see only two areas, a darker one and a brighter one, and a boundary between them. This observation has given rise to the idea of a method which can give the possibility to define this boundary.

(a) (b) (c)

Fig. 2. Clearly visible boundaries between two areas, where (a) – the boundary from the left bottom corner to the right upper corner, (b) – the boundary from the left bottom corner to the right upper corner, (c) – the horizontal boundary, in the middle of the image

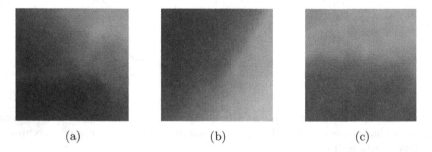

(a) (b) (c)

Fig. 3. Unclearly visible boundaries between two areas, where (a) – the boundary from the left bottom corner to the right upper corner, (b) – the boundary from the left bottom corner to the right upper corner, (c) – the horizontal boundary, in the middle of the image

The above observations have led us to formulate an inverse task - not to find boundaries on the image but to create artificial boundaries and check if they are correct. Let us imagine that the images shown in Figs. 2 and 3 have four artificial, predefined boundaries. These boundaries define the areas. On the basis of calculations and analysis, we decide which boundary is correct.

Let e_1, e_2, e_3 and e_4 be the boundaries, which are defined as follows (where the given coordinates represent the coordinates of the start and the end point of the boundary, respectively) and presented on the image in Fig. 4:

- $e_1 \rightarrow ((x_0, y_0), (x_2, y_2))$,
- $e_2 \rightarrow ((x_0, y_2), (x_2, y_0))$,
- $e_3 \rightarrow ((x_1, y_0), (x_1, y_2))$,
- $e_4 \rightarrow ((x_0, y_1), (x_2, y_1))$.

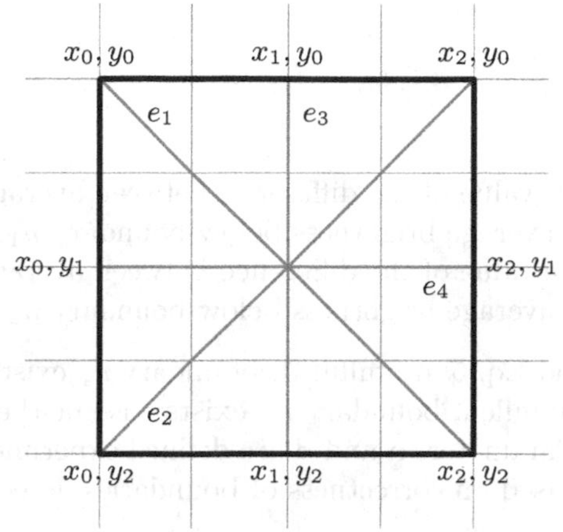

Fig. 4. Mask with artificial boundaries

Boundaries belong to one of the two types - horizontal and vertical boundaries (type 1.) or diagonal boundaries (type 2.). Boundaries of each of the types have to be analysed in pairs – e_1 with e_2 and e_3 with e_4. Such boundaries are opposites to each other, they are mutually exclusive. If one of them exists, second does not. For each pair, the calculation of average brightness for the area above and below the boundaries is done:

- a_{n_1} – average brightness above boundary n_1,
- b_{n_1} – average brightness below boundary n_1,
- a_{n_2} – average brightness above boundary n_2,
- b_{n_2} – average brightness below boundary n_2

where n_1 and n_2 are the analysed pair of boundaries, so there are two cases:

- $n_1 = e_1$, $n_2 = e_2$
- $n_1 = e_3$, $n_2 = e_4$

Values a_{n_1}, b_{n_1}, a_{n_2}, b_{n_2} are used to calculate d_{n_1} and d_{n_2}. These coefficients are needed to decide if and which boundary exists.

$$d_{n_1} = |a_{n_1} - b_{n_1}| \tag{1}$$

$$d_{n_2} = |a_{n_2} - b_{n_2}| \tag{2}$$

$$\frac{d_{n_1}}{d_{n_2}} \geq \alpha, \quad where \quad d_{n_2} > 0 \tag{3}$$

$$\frac{d_{n_2}}{d_{n_1}} \geq \alpha, \quad where \quad d_{n_1} > 0 \tag{4}$$

$$|d_{n_1} - d_{n_2}| > \beta \tag{5}$$

Where:

- d_{n_1} is the absolute value of the difference between average brightness above boundary n_1 and average brightness below boundary n_1,
- d_{n_2} is the absolute value of the difference between average brightness above boundary n_2 and average brightness below boundary n_2.

If conditions Eq. 3 and Eq. 5 are fulfilled, boundary n_1 exists; else if conditions Eq. 4 and Eq. 5 are fulfilled, boundary n_2 exists; else neither boundary n_1 nor boundary n_2 exists. Parameters α and β are defined experimentally – $\alpha = 3$ and $\beta = 15$. Choice was based on correctness of boundaries detection from prepared training set.

4 Application

The method described in Section 3 gives the possibility to detect abnormalities on mammographic images, as it provides a way to detect the boundaries between areas of different brightness. The boundaries are not clearly visible because of the nature of malignant tumours [2]. It is possible to detect relatively big and homogeneous areas like ovals and more complicated shapes like streaks.

The process of detection is performed using a mask with defined dimensions $a \times a$. The mask is moved over the image in step δ, as shown in Fig. 5. For each area below the mask, calculations based on equations from Eq. 3 to Eq. 5 are done and boundaries are created. The above description is just a brief outline of our method. However, to achieve satisfactory results, we need to perform more operations.

The first stage of the process is contrast improvement (the lowest value is mapped to black, the highest value is mapped to white, the rest of the values are spread between them). Thanks to this operation, interesting details are more visible. The detection of edges and streaks is done separately by means of slightly different algorithms described below. The results achieved by means of those algorithms are presented in Figs. 6a and 6b. Based on experiments the chosen parameters are $a = 15$ and $\delta = 5$. The analysed mammograms belong to the MIAS dataset [1].

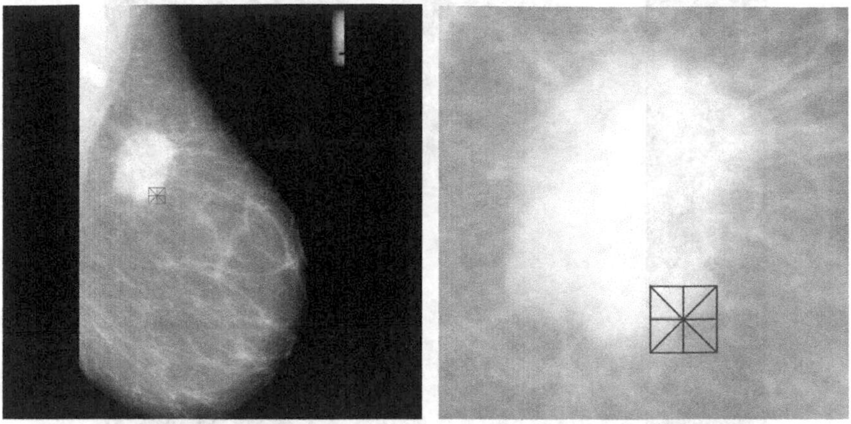

(a) Mammogram with the mask (b) Mammogram with the mask – zoom

Fig. 5. Mammogram with the mask

4.1 Detection of Edges

Detection of edges involves a direct use of the mask with defined dimensions $a \times a$ with defined step δ. As a result of the calculations, we receive an image with detected boundaries. Steps of the algorithm:

1. Choose and open mammogram m.
2. Improve contrast of mammogram m.
3. Move the mask with defined dimensions $a \times a$ with defined step δ over mammogram m and calculate Eqs. 1- 5 (described in Section 3).
4. Apply the post-processing process.
5. Create image with detected boundaries.

4.2 Detection of Streaks

Detection of streaks involves more steps. To improve the quality of the image and visibility of streaks, we use the median filter, which is applied between 2. and 3. step of the algorithm described in Section 4.1. The next step is the same like for edges detection - the mask of defined dimensions $a \times a$ with defined step δ is moved over the image and calculations are performed. As a result, boundaries of streaks are detected.

4.3 Post-Processing Phase

Post-processing eliminates redundant boundaries, which can simplify further analysis of the target regions. The results of preprocessing are presented in Figs. 7a and 7b. Process of elimination is based on simple rules which are presented below (for each type of boundary separately – vertical, horizontal, two types of diagonal; presented in Fig. 4):

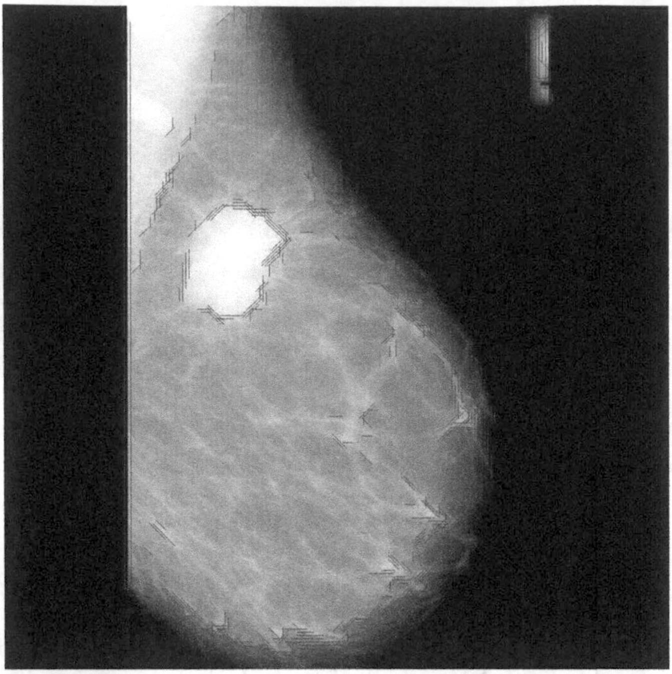

(a) Edges detection (including edges of homogeneous areas).

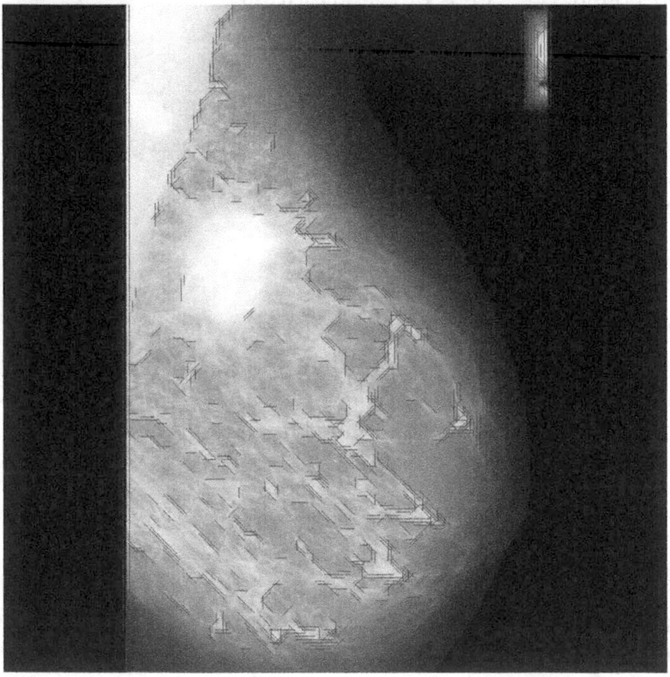

(b) Streaks detection.

Fig. 6. Results of edges and streaks detection

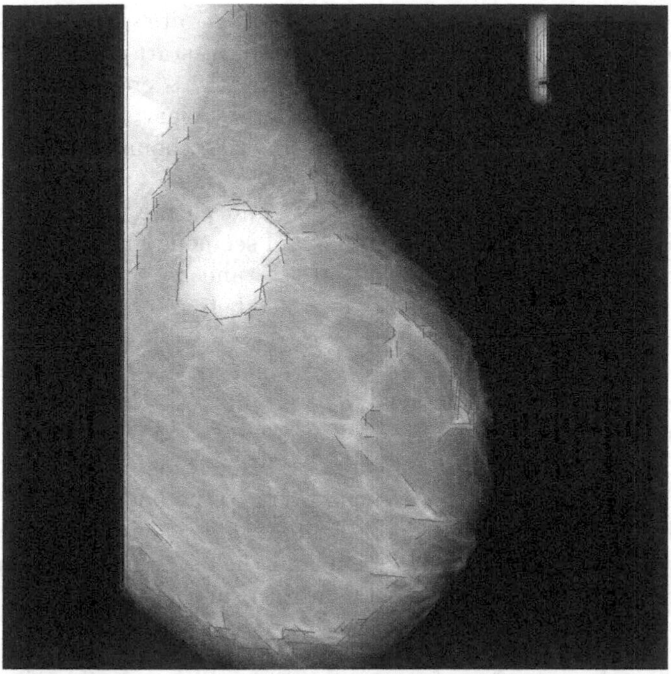

(a) Edges detection (including edges of homogeneous areas).

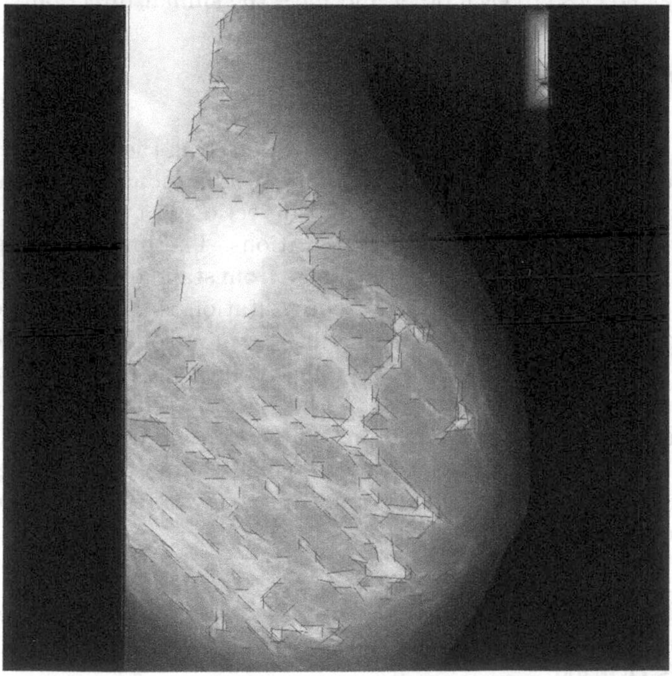

(b) Streaks detection.

Fig. 7. Results of edges and streaks detection after post-processing process

1. If two boundaries of the same type – n_1 and n_2 are collinear and distance between the end point of the boundary n_1 and the start point of the boundary n_2 is less than or equal to δ – the new boundary is created which the start point is the same as the start point of the boundary n_1 and the end point is the same as the end point of the boundary n_2; boundaries n_1 and n_2 are removed.
2. If two boundaries of the same type – n_1 and n_2 are parallel and distance between them is less than or equal to δ and segment which connect the start points of n_1 and n_2 is perpendicular to n_1 and n_2 – the new boundary is created which is placed in a half of distance between n_1 and n_2; boundaries n_1 and n_2 are removed.
3. If two boundaries of the same type – n_1 and n_2 are parallel and distance between them is less than or equal to δ and difference between x coordinates of the start point of the boundary n_2 and the end point of the boundary n_1 is less than or equal to δ and difference between y coordinates of the start point of the boundary n_2 and the end point of the boundary n_1 is less than or equal to δ – the new boundary is created which the start point is the same as the start point of the boundary n_1 and the end point is the same as the end point of the boundary n_2; boundaries n_1 and n_2 are removed.

Further, we plan to connect the boundaries into longer segments and introduce more rules to reduce number of segments. It should be possible if we consider segments that lie close to each other and have the similar direction.

5 Conclusions

In this paper, we have presented a method for detecting the boundaries between areas of different brightness. Our preliminary results indicate that the method can be used as a first step for abnormality detection in mammograms as they approximately correspond with user expectations. The presented approach allows to change the representation of the image from standard description as a set of pixels to set of line segments. This representation is especially useful when proper object localization requires taking into account some external knowledge about expected object's shape. There are many methods that can benefit from this representation such as structural methods, active contours and of course, presented in this paper, active partitions which were an inspiration for this research. The further work will focus on enhancing the post-processing phase to reduce the number of short segments and on using the achieved results as an input for the methods mentioned above.

Acknowledgement. This project has been partly funded with support from National Science Centre, Republic of Poland, decision number DEC-2012/05/D/ST6/03091.

References

1. Mias dataset, `http://peipa.essex.ac.uk/info/mias.html`
2. Guliato, D., Rangayyan, R., Carnielli, W., Zuffo, J., Desautels, J.: Segmentation of breast tumors in mammograms by fuzzy region growing. In: Proceedings of the 20th Annual International Conference of the IEEE Engineering in Medicine and Biology Society, October 30-November 1, vol. 2, pp. 1002–1005 (1998)
3. Huo, Z., Giger, M., Vyborny, C., Metz, C.: Breast cancer: Effectiveness of computer-aided diagnosis – observer study with independent database of mammograms1. Radiology 224(2), 560–568 (2002)
4. Rangayyan, R.M., Ayres, F.J., Desautels, J.L.: A review of computer-aided diagnosis of breast cancer: Toward the detection of subtle signs. Journal of the Franklin Institute 344(34), 312–348 (2007) Special Issue: Medical Applications of Signal Processing, Part I
5. Sampat, M., Markey, M., Bovik, A.: others: Computer-aided detection and diagnosis in mammography. In: Handbook of Image and Video Processing, vol. 2, pp. 1195–1217 (2005)
6. Tomczyk, A., Pryczek, M., Walczak, S., Jojczyk, K., Szczepaniak, P.S.: Spatch Based Active Partitions with Linguistically Formulated Energy. Journal of Applied Computer Science 18(1), 87–115 (2010)
7. Pryczek, M., Tomczyk, A., Szczepaniak, P.S.: Active Partition Based Medical Image Understanding with Self Organized, Competitive Spatch Eduction. Journal of Applied Computer Science 18(2), 67–78 (2010)
8. Tomczyk, A., Szczepaniak, P.S., Pryczek, M.: Cognitive hierarchical active partitions in distributed analysis of medical images. Journal of Ambient Intelligence and Humanized Computing, 1–11 (2012)
9. Kass, M., Witkin, W., Terzopoulos, S.: Snakes: Active Contour Models. Int. Journal of Computer Vision 1(4), 321–333 (1988)
10. Caselles, V., Kimmel, R., Sapiro, G.: Geodesic Active Contours. Int. Journal of Computer Vision 22(1), 61–79 (2000)
11. Cootes, T., Taylor, C.J.: Active shape models - smart snakes. In: Proceedings of 3rd British Machine Vision Conference, pp. 266–275. Springer (1992)
12. Grzeszczuk, R., Levin, D.: Brownian Strings: Segmenting Images with Stochastically Deformable Models. IEEE Trans. on Pattern Analysis and Machine Intelligence 19(10), 100–1013 (1997)
13. Tomczyk, A., Szczepaniak, P.S.: On the Relationship between Active Contours and Contextual Classification. In: Kurzyński, M., et al. (eds.) Proceedings of the 4th Int. Conference on Computer Recognition Systems, CORES 2005, pp. 303–310. Springer, Heidelberg (2005)
14. Tomczyk, A., Szczepaniak, P.S.: Contribution of Active Contour Approach to Image Understanding. In: Proceedings of the 2007 IEEE International Workshop on Imaging Systems and Techniques (IST), CD version only (2007) ISBN: 1-3244-0965-9

An Automatic Method for Cell Membrane Extraction in Histopathology HER2 Images

Martin Tabakov and Szymon Zaręba

Institute of Informatics,
Wroclaw University of Technology, Poland
martin.tabakow@pwr.wroc.pl, 179226@student.pwr.wroc.pl

Abstract. The human epidermal growth factor receptor 2 (HER2) is a biomarker, recognized as a precious prognostic and predictive factor for breast cancer. Nowadays is extremely needed to introduce correct recognition of the HER2 positive breast cancer patients. This can be done by accurate segmentation of the cell membrane of cancer cells that are visualized as HER2 over–expressed on images acquired from corresponding histopathology preparations. To segment this structures automatically, we propose to use fuzzy control system based on Mamadani reasoning and to combined it with Otsu's histogram shape-based image thresholding method. Under this proposal, we have tested different edge detectors.

Keywords: fuzzy control, Mamdani reasoning, image segmentation, image thresholding, Otsu's method, edge detector, histopathology image processing, HER2 breast cancer.

1 Introduction

Nowadays breast cancer is recognized as one of the most common womens cancer. HER2 breast cancer recognition is the problem we have focused on in this paper. HER2 biomarker is recognized as a valuable prognostic and predictive factor for breast cancer. It is over—expressed in approximately 20% of high-grade invasive breast cancers [17]. This type of cancer is also associated with rapid tumor growth, increased risk of recurrence after surgery, weak response to chemotherapy and shortened survival [13]. Over the last few years HER-2/neu found its usefulness to identify likely responders to trastuzumab therapy (Herceptin, Genentech, CA). Studies in HER2 histopathology problematic [18], shown that trastuzumab therapy reduces the risk of recurrence and mortality in patients with early stage breast cancer. On the other hand, this therapy is costly and also information about side effects can be found, thus it is necessary to correctly identify HER2-positive breast cancer patients and at the same time avoid false classification. Immunohistochemistry (IHC) is one of the most widely used method for determining HER2 status. The evaluation of HER2 with IHC involves the visual examination of cell membrane staining with a light microscope and overall slide classification in categories of 0, 1+, 2+, and 3+ corresponding

J. Korbicz and M. Kowal (eds.), *Intelligent Systems in Technical and Medical Diagnostics*, Advances in Intelligent Systems and Computing 230,
DOI: 10.1007/978-3-642-39881-0_20,

to no staining, weak, moderate, and strong staining and membrane compactness/completeness. According to recent research [18], cases scored as 3+ are recommended for trastuzumab therapy, whereas 2+ cases are subject to further testing with fluorescent in situ hybridization (FISH) however, cost of using this test is very high [4].

Therefore, there is great need to develop less complicated and expensive diagnostic process for correct recognition of the corresponding HER-2/neu classes. Solutions of the above problem has been recently proposed by using digital computer systems, in particular image analysis systems. Some review and comparison of subjective and digital image analyses such as local structural segmentation and global scene segmentation, object level, spatially related and multiscale feature extraction, SVM, discrimination and grade-based classificators can be found in [6, 16]. There are different image processing and analysis concepts that are used in the HER-2/neu problem based on feature extraction and analysis, data clustering or other techniques including color pixel classification, nuclei segmentation, cell membrane extracting, measures of cell membrane staining intensity and completeness [8, 1, 3, 12, 7]. In general, the HER2 classification process can be transformed, in terms of digital image processing, into a problem of cell membrane staining and cell membrane connectivity/completeness recognition. Over—expressed cell membrane stain, connectivity and compactness is analyzed and classified, for example strong stained and enough compact cell membrane is very likely to be a 3+ HER-2/neu class cell membrane. However, cell membrane characteristics are rather fuzzy, which creates a major problem for image classification process. Nevertheless to interpret the cell membrane staining and compactness degree, we need first to segment the corresponding cell membrane. Some algorithms for correct cell membrane recognition are proposed [9], but there is still lack of standardized HER2 computed classification procedure, which means that more accurate algorithms should be introduced.

In this paper we propose to use fuzzy control system based on Mamdani reasoning [10], combined with Otsu's thresholding [11], in purpose to extract automatically the target cell membranes (the over–expressed cell membranes) from HER2 histopathology images.

The article is organized as follows: in section 2 some theoretical background, considering the methods used in the proposed research, are described; in section 3, the suggested extraction process is introduced and finally, in section 4 some experiments and results discussion are presented.

2 Theoretical Background

In this section, the preliminaries of fuzzy sets [19] and fuzzy control systems of Mamdani type [10] are presented.

2.1 Fuzzy Sets

Let $X =_{df.} \{x_1, x_2, \cdots, x_n\} \subseteq R$ be some finite set of elements (domain), then we shall call A the fuzzy subset of X, if and only if: $A =_{df.} \{(x, \mu_A(x)) | x \in X\}$, where μ_A is a *function* that maps X onto the real unit interval $[0,1]$, i.e. $\mu_A : X \rightarrow [0,1]$. The function μ_A is also known as the membership *function* of the fuzzy set A, as its values represents the grade of membership of the elements of X to the fuzzy set A. Here the idea is that we can use membership functions, as characteristic functions (any crisp set can be defined by its characteristic function) for fuzzy, imprecisely described sets. Let A and B be two fuzzy subsets of X. Then the basic set operations: *union* and *intersection* of A and B, are defined as follows: $\mu_{A\cup B}(x) =_{df.} \max\{\mu_A(x), \mu_B(x)\}$, $\mu_{A\cap B}(x) =_{df.} \min\{\mu_A(x), \mu_B(x)\}$.

2.2 Fuzzy Control of Mamdani Type

We propose to use fuzzy control concept, as a decision mechanism in our method.

For any fuzzy controller, all input information are fuzzified and then processed with respect to the assumed knowledge base, inference method and the corresponding defuzzification method.

Therefore, a fuzzy controller is composed of the following four elements:

- A rule-base (a set of IF-THEN rules), which contains a fuzzy logic quantification of the experts linguistic description of how to achieve good control.
- An inference mechanism, which emulates the experts decision making in interpreting and applying knowledge.
- A fuzzification interface, which converts controller inputs into information that the inference mechanism can easily use to activate and apply rules.
- A defuzzification interface, which converts the conclusions of the inference mechanism into actual inputs for the process.

Next, we will focus on the concept of Mamdani type fuzzy rules generation.

2.3 Fuzzy Rules

Let $\{X_1, X_2, \cdots, X_n\}$ be a family of finite sets, which defines the corresponding states of primary inputs for a fuzzy control system.

Let $\mathbf{V}^{(X_i)} =_{df.} \{V_1^{(X_i)}, \cdots, V_n^{(X_i)}\}$ be a set of all linguistic variables defined by experts for X_i, i.e. fuzzy subsets of X_i, where $i = 1, 2, \cdots, n$. Then a fuzzy IF-THEN rule antecedent takes the form: (x_1 is $V_{k_1}^{(X_1)}$) o (x_2 is $V_{k_2}^{(X_2)}$) o $\cdots$ $\cdots$ o (x_n is $V_{k_n}^{(X_n)}$), where (x_i *is* $V_{k_i}^{(X_i)}$) means the degree of membership of x_i to $V_j^{(X_i)}$ ($x_i \in X_i; V_j^{(X_i)} \in \mathbf{V}^{(X_i)}$), i.e. $\mu_{V_j^{(i)}}(x_i) \in [0,1]$ ($i = 1, 2, \cdots, n; j = 1, 2, \cdots, k_i$) and o $\in \{\oplus, \otimes\}$, where $\otimes$ is a binary operation over $[0,1]$ ($\otimes : [0,1]^2 \rightarrow [0,1]$) which is commutative, associative, monotonic, and has 1 as unit element. Any such operation is called to be a *t-norm*. The t-norm operator provides the characterization of the AND operator. The dual *t-conorm* $\oplus$ (called also: *s-norm*), characterizing the OR operator, is defined in a similar way having 0 as unit element [2]. Without loss of generality, the Zadehs t-norm (corresponding to the *minimum* operation) is used below.

Let $Y_1, Y_2, \cdots, Y_m$ be a family of fuzzy sets that represents all possible conclusions of a fuzzy control systems, defined over some domain Y — usually a set of real numbers. Then a fuzzy IF-THEN rule conclusion takes the form: "conclusion is Y_p", $p = 1, 2, \cdots, m$.

Thus, the complete form of a classical fuzzy IF-THEN rule is defined as follows:

$$\begin{aligned} Rule =_{df.} \text{ IF } (x_1 \text{ is } V_{k_1}^{(X_1)}) \text{ AND } (x_2 \text{ is } V_{k_2}^{(X_2)}) \text{ AND } \cdots \text{ AND} \\ (x_n \text{ is } V_{k_n}^{(X_n)}) \text{ THEN (conclusion is } Y_p), p \in \{1, 2, \cdots, m\} . \end{aligned} \tag{1}$$

The antecedent value of the rule is interpreted as a weight coefficient that represents the strength of firing the rule.

Next, in our work, as a fuzzy implication function that specifies relation between rule antecedent and corresponding conclusion, we assumed the Larsen implication rule.

2.4 Fuzzy Control Inference Mechanism/Defuzzification Method

In our experiments, we used the Mamadani inference mechanism and the center of gravity defuzzification method. Thus, using the above designations, we can define the system output value (the decision value, in short SO) as follows:

$$\frac{\int \mu_{Y_1' \cup Y_2' \cup \cdots \cup Y_m'}(y) * y}{\int \mu_{Y_1' \cup Y_2' \cup \cdots \cup Y_m'}(y)} . \tag{2}$$

where: $y \in Y$,
$Y_1', Y_2', \cdots Y_m'$ denotes the possible fuzzy rules conclusions modified by the corresponding rule antecedent values, i.e.: let w_r be the antecedent value of the r_{th} fuzzy rule and let denote the conclusion of this rule as Y_r . Then, Y_r' is the corresponding fuzzy set generated from Y_r with respect to the w_r, such that $\mu_{Y_r'}(y) =_{df.} w_r * \mu_{Y_r}(y)$ (assuming the Larsen implication rule).

Typically, referring to the fuzzy control concept applied in recognition process, the corresponding device (or application) is taking action if $(SO \geq T)$, where T is a priori given decision threshold (system parameter).

In our proposal, we intend to use the fuzzy control system as an image transformation, that maps every HER2 image pixel to the set of all gray levels with respect to the SO value (i.e. $SO \in \{0, 1, \cdots, 255\}$). After this transformation is completed, the Otsu's thresholding method [11] is applied over the corresponding gray level image.

2.5 Otsu's Thresholding Method

In computer vision and image processing, Otsu's method is used to automatically perform histogram shape-based image thresholding. The algorithm implies that the image to be thresholded contains two classes of pixels (bi-modal histogram), then computes the optimum threshold by separation of those two classes so that their intra-class variance is minimal.

The formal specification of the Otsu's method is omitted here as it is well known method. Comprehensive description of the algorithm and corresponding calculation examples are provided in [11].

3 Cell Membrane Detection - Method Specification

The main task of this project is the segmentation of the membrane of cancer cells that are HER2 over—expressed from histopathology images. Only having a sufficiently large number of such cells, justifies the use of therapy with monoclonal antibodies. According to recent histology research, we assume that the classification problem of HER2 preparations strictly refers to cancer cell membrane characteristics. As longest we stay in terms of digital image processing, there is a possibility to transform the HER2 classification process, into a problem of cell membrane staining (dark-brown colours) and cell membrane compactness/completeness recognition (see some examples on figure 1, below).

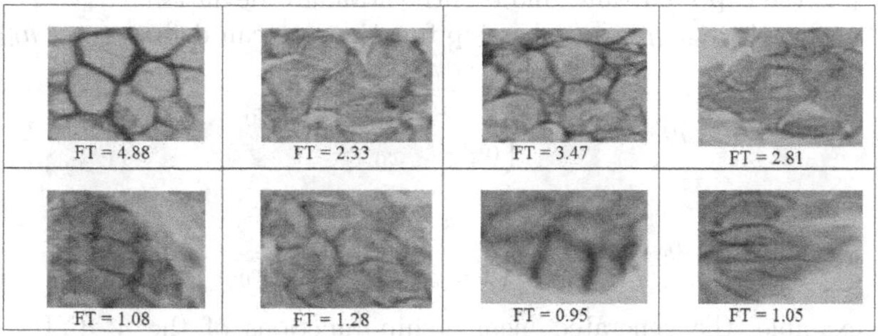

Fig. 1. Example cell membranes with different staining and compactness. The Fish Test (in short FT) examination value is also provided. FT is an objective factor, which gives a clear decision if a histopathology preparation represents suitable case for treatment with monoclonal antibodies. If $(FT \geq 2)$ then a patient should be given the treatment else not (prevent treatment with monoclonal antibodies is not taking into consideration, because of its high cost).

To complete this task, we need first to segment the appropriate cell membranes. This segmentation is realized using the Mamdani inference mechanism and next combined with Otsu's image thresholding.

Next, we precisely explain the construction of the used fuzzification module and knowledge base.

3.1 Fuzzification Module

We assumed in our research that every image pixel is processed. To define the appropriate fuzzification functions, we generated one major image pixel set: Random randomly selected pixels located on the target cell membrane.

The random set was used to introduce common fuzzification over the considered image data. In purpose to fuzzify, generally it is sufficient to introduce three basic linguistic variables: low, medium and high. This fuzzy sets were defined for all the pixels, over the following domains:

- V value of the HSV colour model,
- R, B values of the RGB colour model,
- Edge detectors [5]:
  - $Edge_{Sobel}$: the response of the Sobel edge detector,
  - $Edge_{Kirsh}$: the response of the Kirsch edge detector,
  - $Edge_{Roberts}$: the response of the Roberts edge detector.

Assuming Gaussian data distribution for each feature, we are able to define the fuzzy set medium, by interpreting the corresponding Gaussian probability density function as a membership function:

$$\mu_{medium}(x) =_{df.} e^{\frac{-(x-x_0)^2}{2\sigma^2}} . \tag{3}$$

where x_0 is the expected value and σ is the standard deviation.

Next, using the μ_{medium} membership function, we can define μ_{low} , μ_{high} as well:

$$\mu_{low}(x) =_{df.} \begin{cases} 1 - e^{\frac{-(x-x_0)^2}{2\sigma^2}} & x < x_0 \\ 0 : x \geq x_0 \end{cases} \tag{4}$$

$$\mu_{high}(x) =_{df.} \begin{cases} 0 : x \leq x_0 \\ 1 - e^{\frac{-(x-x_0)^2}{2\sigma^2}} & x > x_0 \end{cases} \tag{5}$$

As an example, the generated membership functions of the V (YUV) and $Edge_{Sobel}$ features are shown below (figure 2).

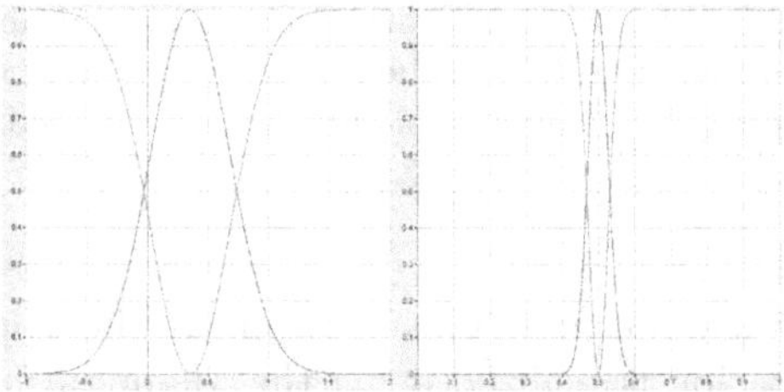

Fig. 2. The generated fuzzification of the V (YUV) and $Edge_{Sobel}$ feature respectively

Then, an appropriate knowledge base was proposed, regarding to the corresponding histopathology knowledge.

3.2 Fuzzy Rule Base

The following fuzzy rules were proposed, in purpose to segment the target cell membrane:

- if BLUE is HIGH or BLUE is MEDIUM or VALUE is HIGH or VALUE is MEDIUM then PIXEL is P_0

- if RED is MEDIUM and (EDGE is MEDIUM or EDGE is HIGH) then PIXEL is P_1
- if RED is MEDIUM and (EDGE is MEDIUM or EDGE is LOW) then PIXEL is P_2
- if (RED is MEDIUM or RED is HIGH)and (EDGE is MEDIUM or EDGE is HIGH) then PIXEL is P_2
- if RED is HIGH and VALUE is HIGH and EDGE is HIGH then PIXEL is P_3
- if RED is HIGH and VALUE is HIGH and EDGE is MEDIUM then PIXEL is P_2,

where $P_1, \cdots, P_3$ (see figure 3, below) are fuzzy sets ($\mu_{P_1}, \cdots, \mu_{P_3} : [0,1] \rightarrow [0,1]$), specified in order to model the histopathology FT value i.e. low domain values (min = 0) correspond to HER2 images with high FT values and high domain values (max = 1) correspond to HER2 images with low FT values.

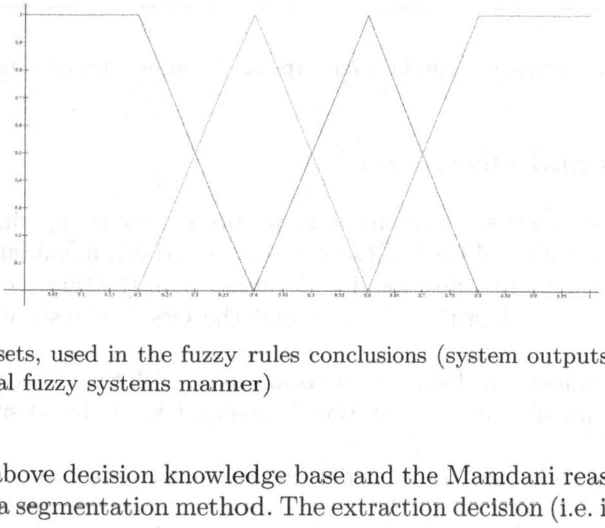

Fig. 3. Fuzzy sets, used in the fuzzy rules conclusions (system outputs gradation defined in classical fuzzy systems manner)

Using the above decision knowledge base and the Mamdani reasoning, we are able to define a segmentation method. The extraction decision (i.e. if a considered image pixel is or is not on the target cancer cell membrane) is strictly related to the final fuzzy system output value. This output value is interpreted as an image gray level, which means that the proposed fuzzy control system transforms every HER2 histopathology image into a grayscale image (dark colours correspond to low SO values and bright colours correspond to high SO values respectively).

Next, the Otsu's method is applied to automatically perform histogram shape-based image thresholding and thus, to provide an automatic target cell membrane extraction. What more, we have examined different segmentation results, separately for the used edge detectors.

Figure 4, below presents the schematic setup (assuming all image pixels) of the proposed automatic cell membrane segmentation process.

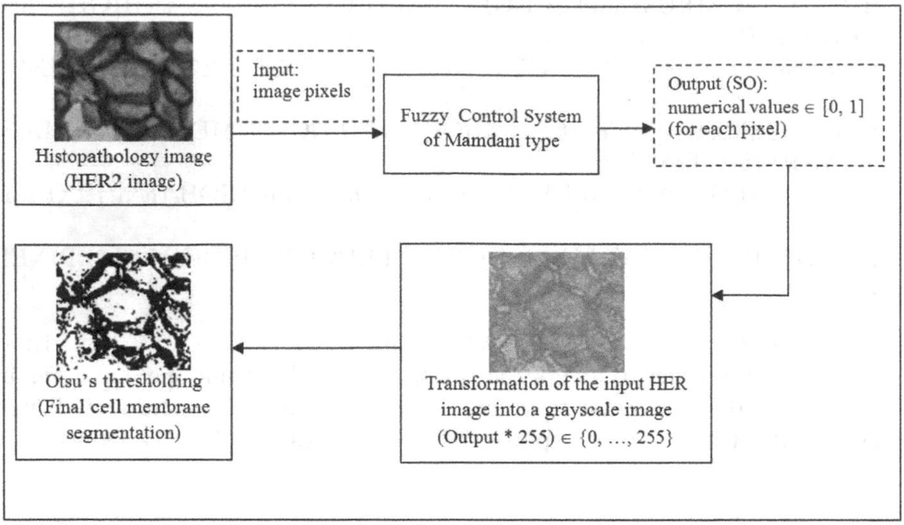

Fig. 4. Schematic setup for over–expressed cell membrane detection

4 Results and Discussion

Considering the above assumptions, next we present some experimental results – sample segmentation of the HER2 over—expressed cell membranes.

The final segmentation decision, i.e. if a pixel is on the target cell membrane or not, is made according the SO value and the Otsu's threshold method (see section 2).

So, for every image pixel a gray level is mapped, with respect to the SO value. Then the cell membrane segmentation is realized with the Otsu's automatic image thresholding.

See some examples, generated under different edge detectors, on figure 5 below.

The generated segmentation results were analyzed by medical experts, who considered the fuzzy system segmentation under Roberts edge detector, as the most appropriate (containing a minimum of artifacts) with respect to the assumed medical problem. Thus below, only the Roberts membrane cell segmentation results were taking into consideration.

The proposed HER2 image processing idea should give segmentation result, that allows to propose a numerical coefficient, that could replace the Fish Test. Over—expressed cancer cells are visible on initial recognition of the HER2 image segmentation result. Shape coefficients, which will affect membrane connections of cancer cells, are proposed to perform classification related to the Fish Test rating.

In our preliminary study, we used the box-counting dimension [14] and a connectivity coefficient (marked below as CC_1) as shape recognition coefficients.

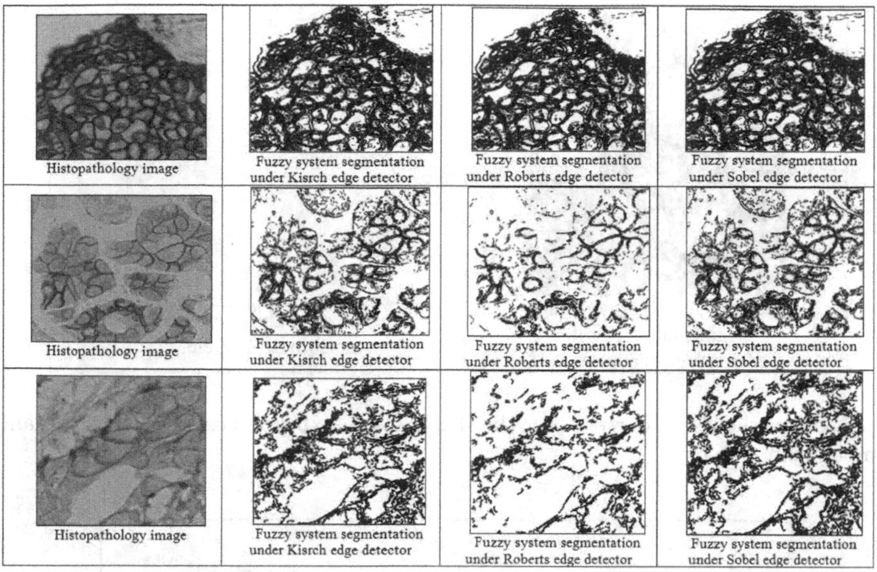

Fig. 5. HER2 over–expressed cancer cell membranes segmentation results

The box counting dimension is a way of determining the fractal dimension of a set, defined in certain metric space. In practice, assuming two dimensional space, it can be used to provide precise numerical characterization of any structure with highly irregular shape.

The CC1 is an often used connectivity coefficient, defined as follows:

$$CC_1 =_{df.} \frac{L^2}{4\pi S} \tag{6}$$

where L is the object's circuit and S denotes the objects area.

Also, to achieve more reliable (more correct) results, we applied post-processing over the generated segmentation results. The proposed post-processing involved morphological operations over the corresponding images, in purpose to segment the so called "objects skeletons". This was done by classical morphological skeleton operator [15] (see some results, on figure 6, below).

Following experiment was provided, to verify thesis, that it is possible to replace the Fish Test with a connectivity coefficient, derived over the cell membrane segmentation results:

- we chose randomly one hundred sub-images of cancer cells from HER2 histopathology images. Fifty were taken from images with $FT < 2$ (1.36; 1.52; 1.75) and the other half from images with $FT \geq 2$ (2.37; 2.81; 4.88),
- the box-counting dimension and CC_1 coefficient were derived for all the cells and presented in the corresponding graph (see figure 7, below).

The achieved values were sorted in purpose to performed better data understanding i.e. values derived from images with $FT < 2$ are sorted in descending

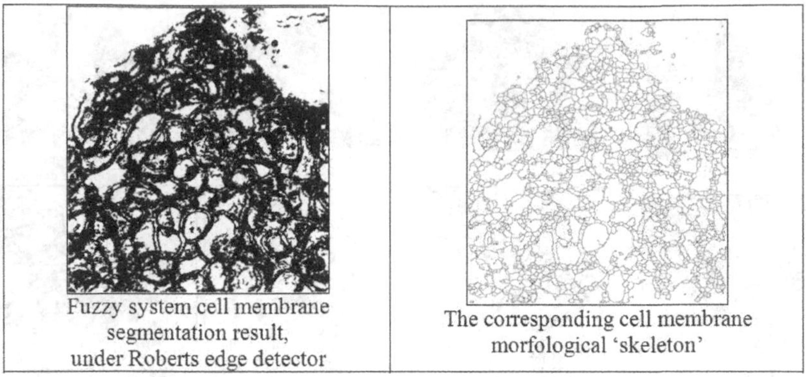

Fuzzy system cell membrane segmentation result, under Roberts edge detector
The corresponding cell membrane morfological 'skeleton'

Fig. 6. Image skeletons, derived over the HER2 over—expressed cancer cell membranes segmentation results

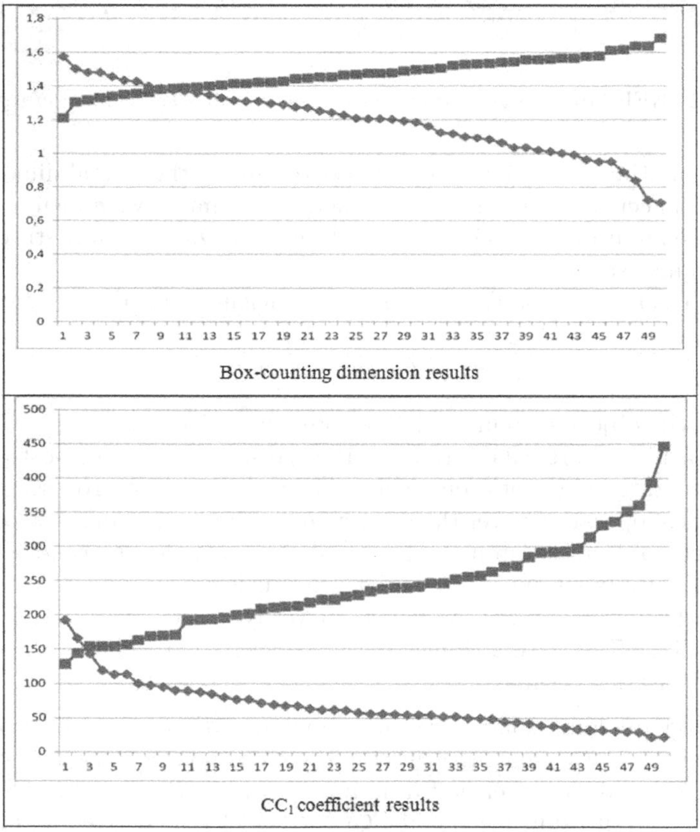

Fig. 7. Blue colour represents cancer cells with $FT < 2$ (not enough HER2 over—expressed) and red colour represents cancer cells with $FT \geq 2$

order, opposite to images with $FT \geq 2$ (sorted in ascending order). This gives a comparison in sense of the "worst case" - the higher shape values of cancer cells derived from histopathology preparations with $FT < 2$ are compared to the lower shape values of cancer cells derived from histopathology preparations with $FT \geq 2$.

In both cases, generally shape coefficient values of cells that are not enough HER2 over—expressed are "under" the values of HER2 over—expressed cells, with better performance of the CC1 shape coefficient. Therefore, this indicates the possibility of replacing the Fish Test with the proposed method, i.e. testing shape coefficients after applying the proposed cell membrane segmentation process.

5 Conclusions

In this article, we propose a fully automated HER2 histopathology image segmentation method of cancer cells membrane. The segmentation process is realized using Mamdani reasoning, combined with Otsu's thresholding, under different edge detectors. This process is very important as its result can be used for classification of the HER2 positive instances, by shape coefficients. This makes possible the replacement of the expensive Fish Test and so facilitate and accelerate the HER2 diagnosis process. As further research, we intend to automate the cell choosing process, required for membrane connectivity analysis as well as we are going to improve the segmentation process, by using type 2 fuzzy control, which will enable better fuzzification of the input image data and therefore better cell membrane recognition.

Acknowledgement. This work was partially supported by Polish Ministry of Science and Higher Education grant No N N518 290540, 2011 — 2014.

References

1. Avoni, M.: Image analysis methods for determining HER2neu Status of breast cancers. BME MSc., DTU Technical University of Denmark, Copenhagen, Denmark, in collaboration with Visiopharm A/S, Hrsholm, MATLAB User Conference, Denmark (2008)
2. Bronstein, I.N., Semendjajew, K.A., Musiol, G., Mhlig, H.: Taschenbuch der Mathematik, p. 1258. Verlag Harri Deutsch (2001)
3. Doyle, S., Agner, S., Madabhushi, A., Feldman, M., Tomaszewski, J.: Automated grading of breast cancer histopathology using spectral clustering with textural and architectural image features. In: IEEE International Symposium on Biomedical Imaging, Paris, France (2008)
4. Gavrielides, M.A., Masmoudi, H., Petrick, N., Myers, K.J., Hewitt, S.M.: Automated evaluation of HER-2/neu immunohistochemical expression in breast cancer using digital microscopy. In: 5th IEEE International Symposium on Biomedical Imaging From Nano to Macro, pp. 808–811 (2008)
5. Gonzalez, R., Woods, R.: Digital Image Processing. Addison-Wesley (2008)

6. Gurcan, N.M., Laura, E., Boucheron, A.C., Madabhushi, A., Rajpoot, N.M., Bulent, Y.: Histopathological Image Analysis - A Review. IEEE Reviews in Biomedical Engeneering 2 (2009)
7. Joshi, A.S., Sharangpani, G.M., Porter, K., Keyhani, S., Morrison, C., Basu, A.S., Gholap, G.A., Gholap, A.S., Barsky, S.H.: Semi-Automated Imaging System to Quantitate Her-2/neu Membrane Receptor Immunoreactivity in Human Breast Cancer. Cytometry A 71(5), 273–285 (2007)
8. Kuo, Y.L., Ko, C.C., Lai, J.Y.: Automated Assessment in HER-2/neu Immunohistochemical Expression of Breast Cancer. In: International Symposium on Computer, Communication, Control and Automation, 3CA 2010, pp. 585–588 (2010)
9. Laurinaviciene, A., Dasevicius, D., Ostapenko, V., Jarmalaite, S., Lazutka, J., Laurinavicius, A.: Membrane connectivity estimated by digital image analysis of HER2 immunohistochemistry is concordant with visual scoring and fluorescence in situ hybridization results: algorithm evaluation on breast cancer tissue microarrays. Diagnostic Pathology 6, 87 (2011)
10. Mamdani, E.H., Assilian, S.: An experiment in linguistic synthesis with a fuzzy logic controller. International Journal of Man-Machine Studies 7(1), 1–13 (1975)
11. Otsu, N.: A threshold selection method from gray-level histograms. IEEE Trans. Sys., Man., Cyber. 9(1), 62–66 (1979)
12. Rexhepaj, E., Brennan, J.D., Holloway, P., Kay, W.E., McCann, H.A., Landberg, G., MDuffy, J.M., Jirstrom, K., Gallagher, M.W.: Novel image analysis approach for quantifying expression of nuclear proteins assessed by immunohistochemistry-application to measurement of oestrogen and progesterone receptor levels in breast cancer. Breast Cancer Research 10 (2008)
13. Ross, J.S., Fletcher, J.A.: The HER-2/neu oncogene in breast cancer: prognostic factor, predictive factor, and target for therapy. Stem Cells 16, 413–428 (1998)
14. Schroeder, M.: Fractals, Chaos, Power Laws: Minutes from an Infinite Paradise, pp. 41–45. W. H. Freeman, New York (1991)
15. Serra, J.: Image Analysis and Mathematical Morphology, vol. 2. Academic Press (1988)
16. Skaland, I., Vestad, I., Janssen, M.A.E., Klos, J., Kjellevold, H.K., Helliesen, T., Baak, A.P.J., Pathol, C.J.: Comparing subjective and digital image analysis HER2neu expression scores with conventional and modified FISH scores in breast cancer. Journal of Clinical Pathology 61(1) (2008)
17. Slamon, D.J., Clark, G.M., Wong, S.G., Levin, W.J., Ullrich, A., McGuire, W.L.: Human breast cancer: correlation of relapse and survival with amplification of the HER2/neu oncogene. Science 235, 177–182 (1987)
18. Wolff, A.C., Hammond, M.E., Schwartz, J.N., Hagerty, K.L., Allred, D.C., Cote, R.J., Dowsett, M., Fitzgibbons, P.L., Hanna, W.M., Langer, A., McShane, L.M., Paik, S., Pegram, M.D., Perez, E.A., Press, M.F., Rhodes, A., Sturgeon, C., Taube, S.E., Tubbs, R., Vance, G.H., van de Vijver, M., Wheeler, T.M., Hayes, D.F.: American Society of Clinical Oncology/College of American Pathologists guideline recommendations for human epidermal growth factor receptor 2 testing in breast cancer. Arch. Path Lab Med. 131, 18–42 (2007)
19. Zadeh, L.: Fuzzy sets. Information and Control 8(3), 338–353 (1965)

Simulation of Influence of the Exciting Sequence Parameters on Results of Spin Echo Method – Virtual Magnetic Resonance Imaging

Przemysław Makiewicz and Krzysztof Penkala

Department of Systems, Signals and Electronics Engineering, Faculty of Electrical Engineering, West Pomeranian University of Technology, Szczecin, Poland
makiewicz@zut.edu.pl

Abstract. In the paper simulation of creating image in the spin echo MRI method is presented. On the basis of three images, with the contrast dependent on proton density and time constants (T1, T2), a model of the original object is created. With the use of this artificial (phantom) object, an image for any arbitrary chosen repetition time (TR) and echo time (TE) sequence parameters can be obtained. This Virtual Magnetic Resonance Imaging (VMRI) technique allows greatly improving medical diagnostic imaging based on the phenomena of magnetic resonance. It is also a useful aid in education.

Keywords: MRI, VMRI, spin echo method, exciting sequence, patient phantom, computer simulation, medical diagnostic imaging.

1 Introduction

The phenomenon of Nuclear Magnetic Resonance (NMR) was discovered in 1945 and described by Felix Bloch and Edward Purcell. A way to exploit this phenomenon in diagnostic imaging was opened firstly in 1973, when Lauterbur invented the NMR tomography [2]. Wide use of this technique as a medical procedure took place over the last several years. For comparison, the X-Rays were successfully used in medicine just a few months after being discovered [8]. It took so long to introduce the magnetic resonance imaging due to the need of improving both the instrument to allow registration of the signal from the test object, and the use of computer technology to obtain the image.

The phenomenon of NMR is related to absorption by the nuclei energy transferred by radio waves (excitation) and emission of energy in the process of relaxation. This phenomenon occurs in nuclei with non-zero spin. Medical imaging using NMR is based mainly on the magnetic resonance of hydrogen nuclei. Isotope 1H is a component of water, whose content in living tissue is 60-80% [11]. For this reason, nuclides of hydrogen became the basis for MRI (Magnetic Resonance Imaging). Under the influence of strong (e.g., 2 T) magnetic field, magnetic moments of hydrogen nuclei can take the position parallel or antiparallel to direction of the magnetic field. Spins of

J. Korbicz and M. Kowal (eds.), *Intelligent Systems in Technical and Medical Diagnostics*, Advances in Intelligent Systems and Computing 230,
DOI: 10.1007/978-3-642-39881-0_21, © Springer-Verlag Berlin Heidelberg 2014

the atoms perform motion called precession, with the speed dependent on the strength of the magnetic field and the type of nuclide. Nuclei can absorb only the energy carried by the RF waves with a characteristic (for a given nuclei and field) frequency, the so called Larmor frequency. For protons, it is equal to 42.6 MHz/T. NMR signal (echo) is obtained by detection and measuring radiation emitted by the nuclei in the process of relaxation.

Contrast in NMR imaging is dependent on three parameters of the test tissue [10]:

- PD - the density of protons (proton density, ρ). This is the amount of nuclei (due to the use of hydrogen nuclei, the term density of protons has been used) that are subject to the phenomenon of resonance, in a unit volume.
- T1 - time of longitudinal relaxation. This is the time constant describing the rate of regrowth of longitudinal magnetization component of the atomic nucleus.
- T2 - transverse relaxation time. This is the time constant describing how rapidly component of transverse magnetization disappears.

There are two parameters of the exciting sequences that determine the influence of tissue parameters on contrast of the image. These are the TR and TE times. TR is the time between successive excitations of imaged tissue, i.e. between the RF pulses. TE describes the time interval between the excitation and measurement of NMR signal (echo). In the literature [1] a mathematical formula can be found describing the influence of the different parameters (TR, TE, T1, T2 and PD - ρ) on the contrast in the spin echo MRI method:

$$S = \rho \cdot \left(1 - e^{-\frac{T_R}{T_1}}\right) \cdot \left(e^{-\frac{T_E}{T_2}}\right) \tag{1}$$

Software integrated with MRI equipment allows some manipulations on acquired images. This includes basic digital image processing procedures. They are useful for a radiologist, who is analyzing, interpreting and describing the images. However, these operations are based on a single picture and thus have limited possibilities. Constructing a model of patient and performing virtual imaging is completely new approach, which gives much better results. Possibility of simulating influence of exciting sequence parameters to improve contrast of images was proposed by the authors in former studies [5]. Presented approach is more flexible then other previously suggested and developed methods [4]. Close collaboration with ophthalmologists and radiologists resulted in first attempts to use VMRI in neuro-ophthalmological diagnosis. Preliminary studies proved benefits of virtual imaging in correct interpretation of difficult to assess pituitary adenoma cases [4,5]. It encouraged the authors to continue research, and to develop VMRI technology. According to the opinions of specialists, this technique can be effectively used as well for diagnosis in orthopedics or tumorous changes in the abdominal cavity, especially small-sized liver tumors.

2 Creating a Phantom Object

Simulation was performed using the Mathcad environment. The basic idea was to create a model of the tested object. Tissue properties affecting the results of imaging are the relaxation times T1, T2 and spin density (protons). The creation of matrices that describe the distribution of these parameters in the tested object allows reconstruction of the result of imaging for the given spin echo sequence. In the simulation experiments, three sample images were used [3].

First, signal corresponding to longitudinal relaxation present in the T1 contrast weighted image was saved to the array aT1. Then similar matrices for the transverse relaxation time T2 (array aT2) and proton density (array PD) were created. For each of the matrices the maximum value of its elements was found and saved in the variables at1max, at2max and apdmax respectively. The parameters for which of the input images were obtained (derived from [3]).

Table 1. Parameters of exciting sequence for the input images

	T1 weighted	T2 weighted	PD weighted
TR	0.3	5	3
TE	0.004	0.1	0.008

In contrast T1 (TE << T2), formula 1 becomes:

$$S = \left(1 - e^{-\frac{T_R}{T_1}}\right) \tag{2}$$

On this basis, the formula for the time T1 in individual pixels of the test object can be derived:

$$T1_{x,y} = TRt1 \cdot \ln\left(2.1 - \frac{aT1_{x,y}}{at1\max}\right) \tag{3}$$

In contrast T2 (TR << T1), formula 1 becomes:

$$S = \left(e^{-\frac{T_E}{T_2}}\right) \tag{4}$$

On this basis, the formula for the time T2 in individual pixels of the test object can be derived:

$$T2_{x,y} = \frac{-TEt2}{\ln\left(2.1 + \frac{aT2_{x,y}}{at2\max}\right)} \tag{5}$$

In contrast PD (TR << T1, TE << T2), formula 1 becomes:

$$S = \rho \tag{6}$$

On this basis, the formula for the proton density in individual pixels of the test object can be derived:

$$PD_{x,y} = 0.95 + \frac{0.05}{apd\,\max} \cdot aPD_{x,y} \tag{7}$$

3 Elimination of Distortions Due to Imperfections in the Input Image

Formulas 2, 4 and 6 are correct if the imaging parameters TR and TE have extreme values. In practice, in the proton density contrast (or any other), the two remaining tissue parameters also affect the image. Figure 1 shows the effect of image (in PD weighted contrast) reconstruction, with the matrices calculated using formulas 3, 5 and 7. The image was created on the basis of the formula 1, taking imaging parameters as for the original image PD (TRpd = 3, TEpd = 0.008). It can be seen that a significant brightening of the brain ventricles occurs. This can be easily explained by the high influence of parameter T2 (for the T2 contrast, the image of cerebrospinal fluid is very bright [1]).

It was necessary to remove the undesired effect, by developing an algorithm that uses the matrices described by formulas 3, 5 and 7. The algorithm involves calculating of values showing the influence of tissue parameters T1, T2 and PD on each of the resultant images. Calculated values allowed removing from the matrices T1, T2 and PD the interference from the two other parameters (e.g., interference from parameters T2 and PD in matrix T1).

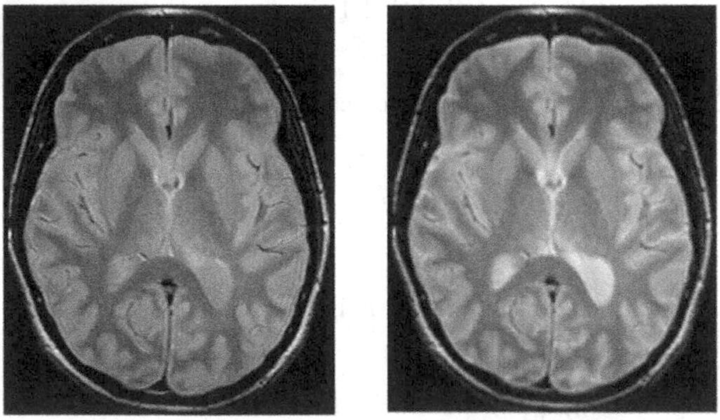

Fig. 1. MRI image of brain in PD contrast, on left input image [3], on right reconstruction (VMRI)

4 MRI Image Reconstruction

The resulting matrices represent a model of an object that was originally illustrated. On the basis of the model, a new image can be created with any chosen parameters (TR, TE). First, a set of three reconstructed images of the sequence parameters TE and TR identical to the original was obtained. The method of image reconstruction is presented below, as an example, for the T1-contrast image. First, formula 1 was implemented:

$$St1_{x,y} = PD_{x,y} \cdot \left(1 - e^{-\frac{TRt1}{T1_{x,y}}}\right) \cdot \left(e^{-\frac{TEt1}{T2_{x,y}}}\right) \tag{8}$$

Then maximal and minimal values were found. Matrix St1 was scaled to fit the full grayscale (values from 0 to 255):

$$St1_{x,y} = \left(St1_{x,y} - st1\min\right) \cdot \left(\frac{255}{st1\max - st1\min}\right) \tag{9}$$

Similarly, images with contrast T2 and PD were reconstructed. Parameters TR and TE (TRt1 and TEt1) occurring in the formula 8 can be freely modified.

5 Results of Simulation

5.1 Verification of VMRI Images

Effects of simulation are presented in three sets of images (Figure 2 – 4). Each set consists of an original image, and the VMRI reconstruction acquired for this same parameters of exciting sequence. There are some slight differences in the brightness of images obtained; however, the characteristics of a particular type of contrast are preserved. This allows to consider presented VMRI technique as valid.

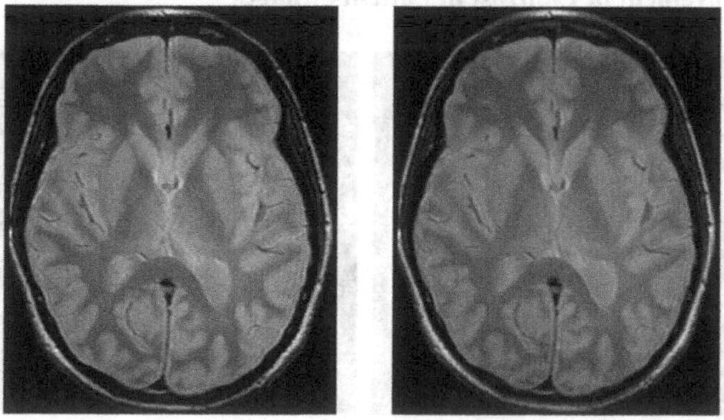

Fig. 2. MRI image of brain in PD contrast, on left input image [3], on right reconstruction (VMRI), both acquired with TR = 3 s and TE = 8 ms

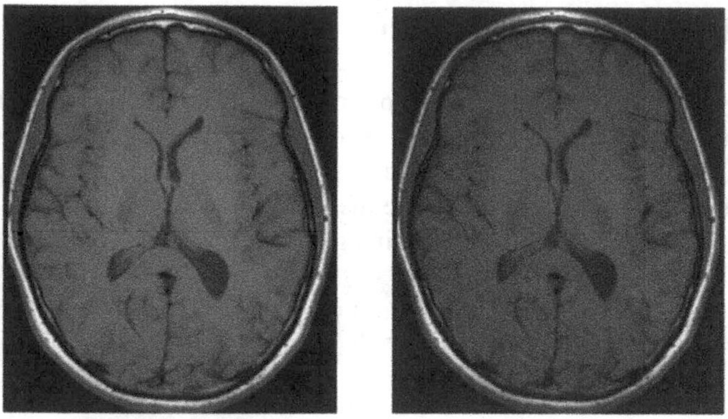

Fig. 3. MRI image of brain in T1 contrast, on left input image [3], on right reconstruction (VMRI), both acquired with TR = 0.3 s and TE = 4 ms

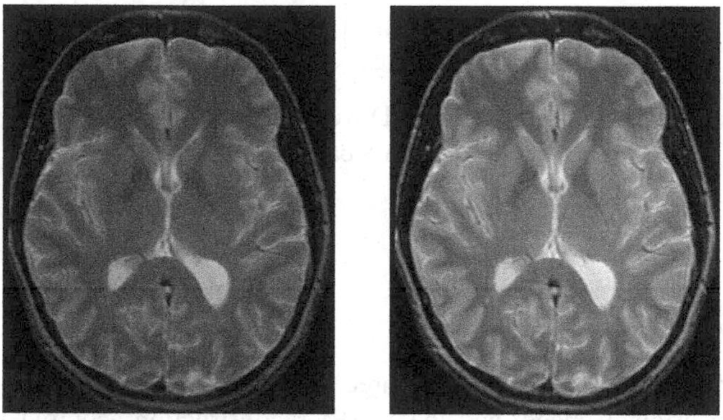

Fig. 4. MRI image of brain in T2 contrast, on left input image [3], on right reconstruction (VMRI), both acquired with TR = 5 s and TE = 100 ms

5.2 Improvement of Contrast in Patients Images

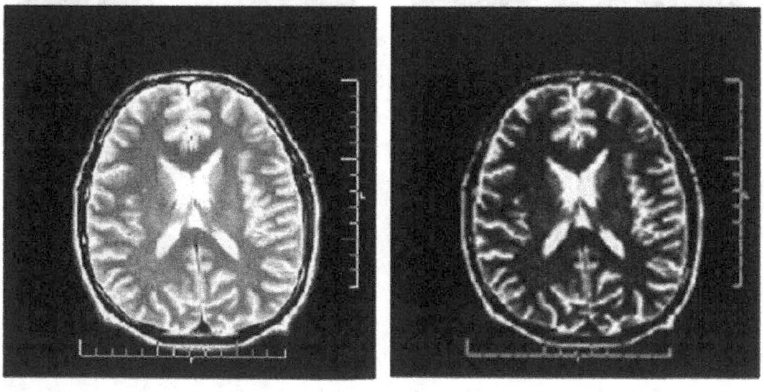

Fig. 5. Improved T2 contrast in MRI image of brain, on left input image (TR = 5 s, TE = 100 ms) [9], on right reconstruction (VMRI: TR = 25 s, TE = 1.25 s)

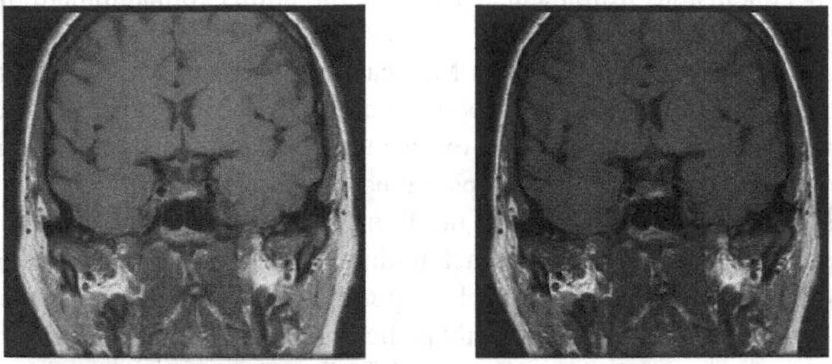

Fig. 6. Improved T1 contrast in MRI image of brain, on left input image (TR = 0.3 s, TE = 4 ms), on right reconstruction (VMRI: TR = 0.12 s, TE = 2 ms)

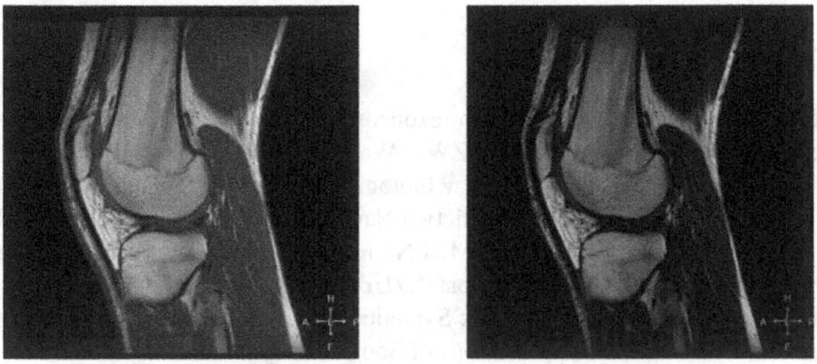

Fig. 7. Improved T2 contrast in MRI image of knee, on left input image (TR = 5.6 s, TE = 98 ms), on right reconstruction (VMRI: TR = 10 s, TE = 250 ms)

6 Conclusions

The purpose of this study was to present the simulation of influence of the exciting sequence parameters on the resultant MRI image. Virtual Magnetic Resonance Imaging allows to create an image for settings that could not be used in any real MRI system. Simulation effects for very long times TR and TE are very interesting. Thanks to the high contrast, as compared to the original results, significant image enhancement can be obtained. Simulations based on creation of phantom test object can be used to manipulate the contrast of images without involvement of the patient and equipment in order to improve the contrast, and to obtain new diagnostic values. This VMRI technique should allow introducing important benefits in medical diagnostic imaging. Despite new diagnostic value, performing virtual imaging allows to save significant amount of money. Sociological benefits for the patient are also important.

In addition, simulation is useful for didactic purposes, mainly for the students of Biomedical Engineering. It provides better way to understand the principles of MRI and contrast dependence for students in technical fields. It is also extremely important for medical students. They are not accustomed to analyzing mathematical formulas and deriving conclusions from them. Therefore, the ability to manipulate the imaging parameters and observe their effect is a very useful teaching aid.

Results of research performed in the Mathcad environment will be used to develop specialized software, designed to support medical MRI diagnostics (after clinical verification) as also for teaching. First experiments using this technique in neurological diagnostics, particularly in neuro-ophthalmology, are very promising. A dedicated project has recently being designed at the Pomeranian Medical University in Szczecin, linking electrophysiological approach to diagnostic problems concerning pituitary adenoma with the radiological one, including the use of VMRI technology. VMRI is also being tested for possible use in orthopedic diagnostics. Other promising application could concern hepatic tumors. Presented findings have recently led to interest of one of the Western Pomerania non-public hospitals in applying the VMRI technology (a license agreement is in preparation).

References

1. Gonet, B.: Obrazowanie magnetyczno-rezonansowe. Zasady fizyczne i możliwości diagnostyczne. Wydawnictwo Lekarskie PZWL, Warszawa (1997) (in Polish)
2. Hausser, K.H., Kalbitzer, H.R.: NMR w biologii i medycynie Badania strukturalne, tomografia, spektroskopia in vivo. Wydawnictwo Naukowe UAM, Poznań (1993) (in Polish)
3. Hess, C.P., Purcell, D.D.: Brain: CT, MRI Noninvasive imaging of central nervous system (2008), `http://knol.google.com/k/brain-ct-mri#`
4. Jara, H.: Patent No. US 7,002,345 B2: Syntetic Images For A Megnetic Resonance Imaging Scanner Using Linear Comibinations of Source Images
5. Makiewicz, P., Penkala, K.: Symulacja wpływu parametrów sekwencji wzbudzającej na wynik obrazowania metodą echa spinowego w rezonansie magnetycznym (in Polish). Inżynieria Biomedyczna – ActaBio-Optica et Informatica Medica 16(2), 76–79 (2010)
6. Makiewicz, P., Penkala, K., Lubiński, W., Walecka, A.: Virtual Magnetic Resonance Imaging (VMRI) as a novel technology supporting diagnostics of the optic nerve diseases. In: 50th ISCEV Symposium Proceedings, Valencia, pp. 58–59 (2012)
7. Makiewicz, P., Penkala, K., Lubiński, W.: Virtual Magnetic Resonance Imaging (VMRI) – innowacyjna technologia wspomagająca diagnostykę schorzeń nerwu wzrokowego. Streszczenia. III Sympozjon Sekcji Neurookulistyki i Elektrofizjologii Klinicznej PTO: 103-105, Międzyzdroje (2011) (in Polish)
8. Pruszyński, B., et al.: Diagnostyka obrazowa: podstawy teoretyczne i metodyka badań. Wydawnictwo Lekarskie PZWL, Warszawa (2000) (in Polish)
9. Tadeusiewicz, R.: Wykład 4 z kursu Techniki Obrazowania Medycznego (2010) (in Polish), `http://upel.agh.edu.pl/msib/course/view.php?id=30`
10. Vlaardingerbroek, M., den Boer, T.,, J.A.: Magnetic Resonance Imaging. Springer, Berlin (1996)
11. Weishaupt, D., Köchli, V.D., Marincek, B.: Wie funktioniert MRI Eine Einführung In Physik Und Funktionsweise der Magnetresonanzbildgebung. Springer, Berlin (2001)

Stateless Q-Learning Algorithm for Training of Radial Basis Function Based Neural Networks in Medical Data Classification

Maciej Kusy and Roman Zajdel

Faculty of Electrical and Computer Engineering, Rzeszow University of Technology, 35-959 Rzeszow, W. Pola 2, Poland
{mkusy,rzajdel}@prz.edu.pl

Abstract. In this article, the stateless Q-learning algorithm is used for the training process of two radial basis function based models: the radial basis function neural network (RBFNN) and the probabilistic neural network (PNN). The training process of considered models consists in the initialization and the adaptation of the smoothing parameter of the networks' activation function in a hidden layer. The main idea of this approach is based on the appropriate computation of the smoothing parameter which relies on its update according to the stateless Q-learning algorithm. The proposed method is tested on six commonly available repository data sets. The prediction ability of the algorithm is assessed by computing the test set error on 10%, 20%, 30%, and 40% of examples drawn randomly from the entire input data. Obtained results are compared with the test errors achieved by PNN trained by means of the conjugate gradient procedure. It is shown that Q-learning method can be applied to the automatic adaptation of the smoothing parameter for both neural networks and provides better prediction ability results.

Keywords: radial basis function neural network, probabilistic neural network, stateless Q-learning algorithm, smoothing parameter, data classification, prediction ability.

1 Introduction

The radial basis function neural network (RBFNN) [1] and the probabilistic neural network (PNN) [2] are the examples of a radial basis function based models commonly used in classification problems. An important field of these models' application is the medical domain [3]–[7].

One needs to realize, that the use of RBFNN and PNN requires the training process which is related to the necessity of the selection of the networks' parameters. This paper introduces the algorithm for an automatic selection of one of the parameters for both models.

The training process of RBFNN is influenced by the choice of four parameters: (a) the number of hidden neurons, (b) the location of radial basis function centres, (c) the value for radial basis function smoothing parameter (denoted

J. Korbicz and M. Kowal (eds.), *Intelligent Systems in Technical and Medical Diagnostics*, Advances in Intelligent Systems and Computing 230,
DOI: 10.1007/978-3-642-39881-0_22,

henceforth as sigma or σ), and (d) the values of weights for output neurons of the model. In this work, we do not focus on the problem (a), (b) and (d). These stages are realized by the learning algorithm provided in [8]. The issue stated in (c) is an open problem, usually solved individually by researchers. For example, in [9], this parameter is set by default to the value of one while in [10] it depends on the number of hidden layer neurons and the distance between the centres.

The training process of PNN pertains solely to the selection of the smoothing parameter of the network's pattern neurons. In the research, this problem has been addressed for some time. For example in [11], Specht suggests the parameters adaptation by perturbing each σ a small amount to find the derivative of the optimization criterion with respect to each sigma. In order to find iteratively the set of sigmas which maximize the optimization criterion, the conjugate gradient descent is proposed. The reference [12] shows numerical procedure for smoothing parameter computation. In successive steps, a rough optimization by the conjugate gradient method is performed and then fine optimization by approximate Newton algorithm takes place. The authors of [13] show a genetic algorithm as the solution. They apply a four-bit decimal coding as the sigma representation. A single parameter is used for the network. The work described in [14] also uses of an evolutionary algorithm based approach. Here a set of smoothing parameters different for each class is adopted as a chromosome where a single value corresponds to a gene. In [15], a gap-based estimation procedure is introduced. The authors provide the constraints on centre values on the basis of the distance between two input points after standardization.

Thus, as one can observe, in both networks the choice of the smoothing parameter constitutes a relevant problem in the training process. In case of RBFNN, σ is one of four parameters which have to be adjusted, while in PNN, it is the only one to be adapted. The simplest strategy of σ selection is a trial and error method which may be tedious and time-consuming. In this work, we propose an alternative algorithm for the smoothing parameter selection of both RBFNN and PNN. It utilizes one of the reinforcement learning algorithms, namely, the Q-learning [16]. This is a directed method of a trial and error approach which aims to maximize the prediction ability of considered networks. The idea of using reinforcement learning as the algorithm for neural network parameter adjustment has already been developed. The authors of [17] apply this algorithm to the computation of the set of optimal weights for multilayer perceptron. The weights are found in the training problem on Iris data base. However, no study has been devoted to the application of the reinforcement learning in the adaptation of the smoothing parameter of radial basis function neural networks.

This paper is structured as follows. In Section 2, the radial basis function based neural networks used in this research are characterized. Section 3 discusses the fundamentals of a stateless Q-learning algorithm. In Section 4, the algorithm of the smoothing parameter selection for a radial basis function neural network and a probabilistic neural network is proposed. Section 5 briefly describes medical data used in this work. The prediction ability of considered networks is here verified. Finally, in Section 6, the conclusions are presented.

2 Neural Networks

In this section, the basis of the neural networks used in this research are discussed. The differences between the models are shown, the training process is highlighted and the way of models' parameter selection and adaptation is explained.

2.1 Radial Basis Function Neural Network

The radial basis function neural network is a type of a feed-forward network which computes some mapping between n-dimensional input and a real-valued target. This mapping is accomplished in two stages: firstly, each of the hidden nodes calculates radial basis function around its center; secondly, the signals determined by all hidden elements are linearly combined to yield the final mapping of the input space. Structurally, RBFNN is composed of two layers: a radial basis hidden layer and a linear output layer. Each of n components of the i-th input vector $\mathbf{x}_i = [x_{i1}, \ldots, x_{in}]$, $i = 1, \ldots, l$, is connected to a hidden layer neurons of radial basis transfer functions

$$y_j^{(1)} = \exp\left(-\frac{\left\|\mathbf{x}_i - \mathbf{w}_j^{(1)}\right\|^2}{2\sigma^2}\right), \tag{1}$$

where $\mathbf{w}_j^{(1)} = [w_{j1}^{(1)}, \ldots, w_{jn}^{(1)}]$ denotes the weight vector assigned to the j-th hidden neuron, σ is the smoothing parameter and $y_j^{(1)}$ is the output of the j-th neuron in a radial basis layer for $j = 1, \ldots, m$. The output of each radial basis neuron depends on the Euclidean distance $\left\|\mathbf{x}_i - \mathbf{w}_j^{(1)}\right\|$ between the input $\mathbf{x}_i$ and the first layer weight vector $\mathbf{w}_j^{(1)}$. The outputs $y_j^{(1)}$ of the hidden neurons are linearly combined with second layer weights $\mathbf{w}_g^{(2)} = [w_{g1}^{(2)}, \ldots, w_{gm}^{(2)}]$ to calculate g-th second layer output

$$y_g^{(2)} = \sum_{j=1}^{m} w_{gj}^{(2)} y_j^{(1)}, \tag{2}$$

where $g = 1, \ldots, G$ and G is the number of classes. In case of the use of RBFNN in classification problem and the output representation in the form of 1 of G, the network output is defined as follows

$$y_{RBFNN} = \arg\max_g \left\{y_g^{(2)}\right\}. \tag{3}$$

The architecture of RBFNN is illustrated in Fig. 1 (a). There are G linear neurons in the output layer of the model. The highest value the g-th output neuron determines the class index to which the classified example belongs. As mentioned in the Section 1, the RBFNN training process consists in the computation of the first and second layer weights ($\mathbf{w}^{(1)}$ and $\mathbf{w}^{(2)}$, respectively), and setting the values for radial basis function smoothing parameter σ. The method of weights computation is not a focus of this work. The algorithm presented in [8] is used for this purpose. Here the attention is paid to the adaptation of the smoothing parameter of the model. In literature, different approaches are proposed. Usually $\sigma = d/\sqrt{2m}$ where d is the maximum distance between the weights $\mathbf{w}_j^{(1)}$ [10]. As shown in exemplary experiments [9], the problem of the smoothing parameter selection plays a crucial role in the generalization ability of this classifier.

2.2 Probabilistic Neural Network

The probabilistic neural network, as RBFNN model is the example of a feed-forward network. Its functionality derives from the Bayes theorem. Assume there exists an input vector $\mathbf{x} \in \mathbb{R}^n$, which is contained in the set of the examples of a given class $g = 1, \ldots, G$. Additionally, let p_g be the probability that the vector $\mathbf{x}$ belongs to the class g, and e_g – the error associated with classifying $\mathbf{x}$ to the class g. Now, if the probability density functions: $y_1(\mathbf{x}), \ldots, y_G(\mathbf{x})$ for all classes are known, then, according to the Bayes theorem, the vector $\mathbf{x}$ is classified to the class g, if

$$p_g e_g y_g(\mathbf{x}) > p_h e_h y_h(\mathbf{x}), \tag{4}$$

for the classes for which $g \neq h$. If we also accept that $p_g = p_h$ and $e_g = e_h$ then the condition $y_g(\mathbf{x}) > y_h(\mathbf{x})$ is sufficient to classify the vector $\mathbf{x}$ to the class g. Usually the Parzen's method [18] is used to determine the probability density function which takes the following form

$$y(\mathbf{x}) = \frac{1}{l \cdot \sigma_1 \cdot \ldots \cdot \sigma_n} \sum_{i=1}^{l} W\left(\frac{x_{i1} - x_1}{\sigma_1}, \ldots, \frac{x_{in} - x_n}{\sigma_n}\right), \tag{5}$$

where $\sigma_1, \ldots, \sigma_n$ denote the standard deviations computed with respect to the average of n variables $x_1, \ldots, x_n$, l is the number of given examples and W is the weighting function. If we assume $\sigma_1 = \ldots = \sigma_n$, regard the Gaussian function as the weighting for W, constrain l to be represented by l_g examples of class g and consider $\mathbf{x}_i$ only from the class g ($i = 1, \ldots, l_g$), then the probability density function in (5) can be formulated as follows

$$y_g(\mathbf{x}) = \frac{1}{l_g (2\pi)^{n/2} \sigma^n} \sum_{i=1}^{l_g} \exp\left(-\frac{\|\mathbf{x}_{i,g} - \mathbf{x}\|^2}{2\sigma^2}\right), \tag{6}$$

where $\mathbf{x}_{i,g}$ is the i-th vector of the class g. The probability density function presented in (6) provides one of G summation neurons in the PNN structure. All the signals defined in (6) are utilized to compute the PNN's output in accordance with the Bayes's decision rule

$$y_{PNN} = \arg\max_g \{y_g(\mathbf{x})\}, \tag{7}$$

where y_{PNN} denotes the predicted class of the pattern $\mathbf{x}$. The architecture of PNN is illustrated in Fig. 1 (b).

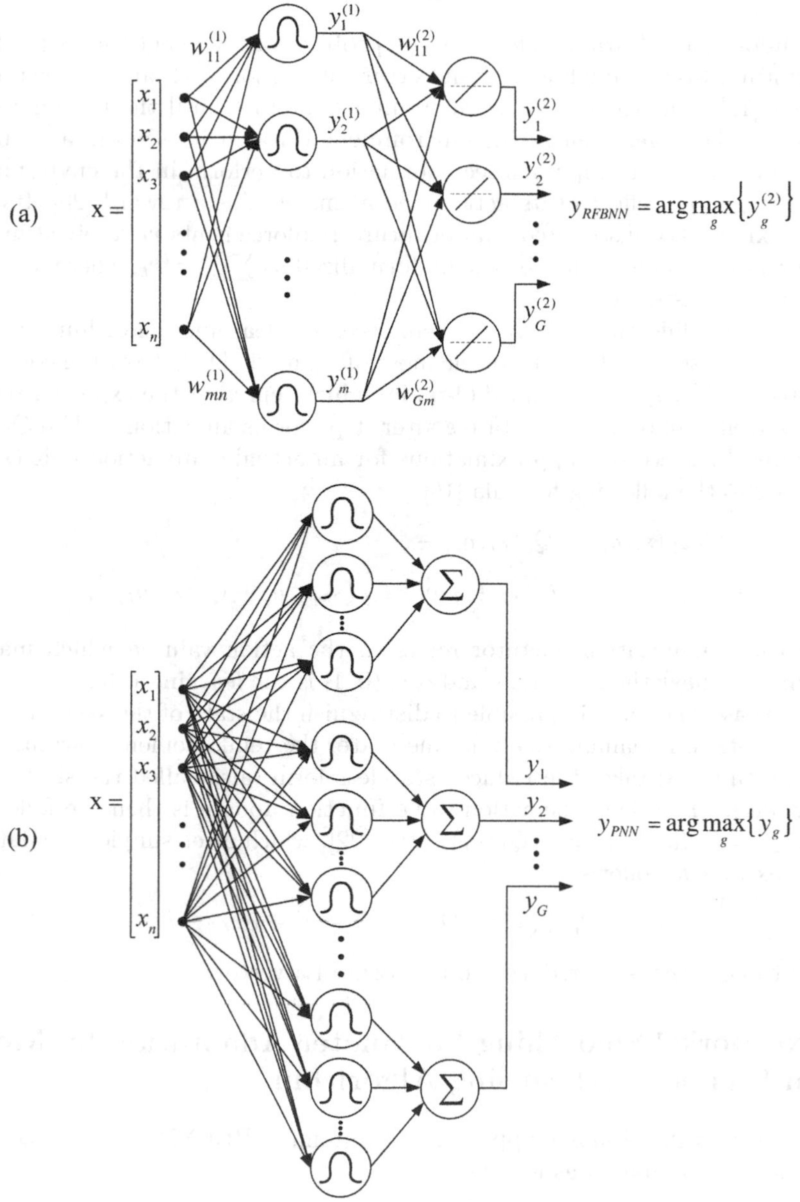

Fig. 1. The architectures of RBFNN – plot (a) and PNN – plot (b). There are two major differences between the structures of the models: 1) PNN requires number of radial basis neurons be equal the number of input examples; 2) In RBFNN, the radial basis neurons are directly connected to the output of the network, while in PNN, these elements, formed in G sub-groups, yield the signals to G summation neurons.

3 Stateless Q-Learning Algorithm

The reinforcement learning addresses the problem of the agent that must learn to perform a task through a trial and error interaction with an unknown environment [19]. The agent and the environment interact until the terminal state is reached. The agent senses the environment through its sensors and, based on its current sensory inputs, selects an action to perform in the environment. Depending on the effect of its action, the agent obtains a reward [20]. Its goal is to maximize the discounted sum of future reinforcements r_t received in long run in any time step t, what is usually formalized as $\sum_{t=0}^{\infty} \gamma^t r_t$, where $\gamma \in [0, 1]$ is the agent's discount rate.

There exist different types of the reinforcement learning algorithms. The Q-learning proposed in [16] is one of the most often used. This algorithm computes the table of $Q(s, a)$ values (called Q-table) which represent the expected reward that an agent can obtain in a state s after it performs an action a. The Q-table is updated by successive approximations for an actual state-action pair (s_t, a_t) according to the following formula [16]

$$Q_{t+1}(s_t, a_t) = Q_t(s_t, a_t) + \\ + \alpha \left(r_t + \gamma \max_a Q_t(s_{t+1}, a) - Q_t(s_t, a_t) \right), \tag{8}$$

where the maximization operator refers to the action value a which may be performed in next the state s_{t+1} and $\alpha \in (0, 1]$ is the learning rate.

In the case when it is impossible to distinguish the state of the system, or the system is static remaining solely in one state, the reinforcement learning algorithm given in (8) takes the reduced stateless form (also called the single-state Q-learning). The rule of the action value function update is then the following: $Q_{t+1}(a_t) = Q_t(a_t) + \alpha (r_t - Q_t(a_t))$ [21], [22], which after simple calculations, can be express as follows

$$Q_{t+1}(a_t) = Q_t(a_t)(1 - \alpha) + \alpha r_t. \tag{9}$$

The Q-table is therefore reduced to the form of a vector.

4 Networks' Smoothing Parameter Adaptation by Means of Stateless Q-Learning Algorithm

The purpose of the proposed approach is to minimize RBFNN and PNN training error measure computed as follows

$$E_{train}(\sigma) = \frac{1}{l_{train}} \sum_{i=1}^{l_{train}} m_i(\sigma), \tag{10}$$

where l_{train} is the cardinality of a training set, and m_i is the indicator of the misclassification defined as follows

$$m_i(\sigma) = \begin{cases} 1 & \text{if} \ y_i(\sigma) \neq o_i \\ 0 & \text{if} \ y_i(\sigma) = o_i \end{cases}, \tag{11}$$

where $y_i(\sigma)$, $i = 1, \ldots, l_{train}$, is the network output defined in (3) or in (7), and o_i is the corresponding target output in the training set for $y_i(\sigma)$. The minimization of $E_{train}(\sigma)$ is achieved by the smoothing parameter update in accordance with the reinforcement learning method. Afterwards, for both trained models, $E_{test}(\sigma)$ error is computed on the independent test sets analogously to (10).

1 Initialize $Q(a)$ arbitrarily for all $a \in A$ and σ_0 according to (12)
2 $E_{train,\min} = E_{train}(\sigma)$ according to (10)
3 Assume the maximum number of training steps $t_{\max}$
4 **for** $t = 1$ **to** $t_{\max}$ **do**
5   Choose action a_t using behaviour policy (e.g. ϵ-greedy)
6     $\sigma_t = \sigma_{t-1} + a_t$
7   Train the network on the training set
8   Calculate $E_{train,t}$
9   **if** $E_{train,t} < E_{train,\min}$ **then**
10         $\sigma_{\min} = \sigma_t$
11         $E_{train,\min} = E_{train,t}$
12   **end**
13     $r_t = E_{train,t-1} - E_{train,t}$
14   Update: $Q_{t+1}(a_t) = Q_t(a_t) \cdot (1 - \alpha) + \alpha \cdot r_t$
15 **end**
16 Calculate $E_{test}(\sigma_{\min})$

Algorithm 1. Stateless Q-learning for RBFNN and PNN σ estimation.

In **Algorithm 1**, we propose the stateless Q-learning method as the solution of the smoothing parameter adaptation for RBFNN and PNN models. The procedure starts with arbitrary initialization of the elements of an action value function. The action a_t is taken from the symmetric set $A = \{-a^{(1)}, -a^{(2)}, \ldots, -a^{(p)}, a^{(p)}, \ldots, a^{(2)}, a^{(1)}\}$. The smoothing parameter σ is initialized according to the following formula

$$\sigma_0 = \underset{\sigma \in [\sigma_{0\_l}, \sigma_{0\_u}]}{\arg\min} \{E_{train}(\sigma)\}, \tag{12}$$

where σ_{0_l} and σ_{0_u} are the lower and upper bounds of the initial value of the smoothing parameter, respectively. After the initialization process, the main loop begins (step **4**). Within this loop, on the basis of current Q values, the choice of an actual action a_t takes place using ϵ-greedy method (step **5**). Then, in step **6**, the smoothing parameter is updated by adding the value of an a_t action. After the training process (step **7**), the training error defined in (10) is computed (step **8**) and compared with its minimum value (step **9**). If an actual error value is smaller than the minimum one, both $\sigma_{\min}$ and $E_{train,\min}$ are updated (steps **10**, **11**). Then, the reinforcement signal is calculated as the difference between the previous and the actual value of the training error (step **13**). The reinforcement signal determined in this manner becomes negative when the training error increases, and it is positive when the training error decreases. Such a form of the reinforcement signal combined with the action value function update (step **14**) strengthens the confidence that the choice of such an action will be beneficial or not. Finally, E_{test} for $\sigma_{\min}$ is computed (step **16**).

5 Empirical Results and Discussion

In the simulations, six UCI machine learning repository [23] medical data sets are used: Wisconsin breast cancer, Pima Indians diabetes, Haberman's survival, cardiotocography, thyroid and dermatology. The cardiotocography and thyroid data sets are the examples of three class databases, while for dermatology data, the number of classes equals 6. In case of the remaining input sets, a binary classification is faced.

After random extraction of 10%, 20%, 30% and 40% of the input data for testing part, the training sets are created using the rest of patterns, i.e. 90%, 80%, 70% and 60% of data, respectively. We intentionally introduce this type of data division since considering all possible training-test subsets is complex from computational point of view – the number of ways of dividing l training patterns into v sets, each of size k, is large, i.e. $l!/(v!(k!)^v)$ [24]. The use of the same training/test sets for RBFNN and PNN makes the obtained results comparable. Both models are trained on the training sets of each database and then validated on the test sets by computing the test set error. For the considered data sets, we aim to find such a value of the smoothing parameter of RBFNN and PNN that ensures the minimum value of the training error. This is realized by means of **Algorithm 1** that is parametrized as follows. The initial values of the action value function Q are set to zero. The six element action set $A_6 = \{-1, -0.1, -0.01, 0.01, 0.1, 1\}$ and ten element action set $A_{10} = \{-10, -1, -0.1, -0.01, -0.001, 0.001, 0.01, 0.1, 1, 10\}$ are used. This allows to modify the smoothing parameter for A_6 by ± 1 or ± 0.1 or ± 0.01 and for A_{10} by ± 10 or ± 1 or ± 0.1 or ± 0.01 or ± 0.001 in the single iteration of the algorithm. The maximum number of the training steps $t_{\max} = 100$. We apply such a value of $t_{\max}$ in order to show that at a relatively small number of epochs it is possible to achieve satisfactory results. The greedy parameter ε is equal to 0.05. The learning rate α of the action value function is set to 0.01. For each network, σ_0 is searched between $\sigma_{0\_l} = 1$ and $\sigma_{0\_u} = 10$ with the step size equal to 1. For the RBFNN model, the number of the hidden neurons is assumed to be $2n + 1$ [25], where n stands for the number of data attributes.

Tables 1 and 2 present E_{test} results for all data partitions, expressed in terms of the percentage of misclassified examples, which are obtained for RBFNN and PNN trained using Q-learning algorithm, respectively. The minimum (*min*), the average (*avr*) and the standard deviation (*sd*) of the test error along with the smoothing parameter $\sigma_{\min}$ for the lowest E_{test} are provided. The results are shown for six (A_6) and ten (A_{10}) element action sets. In Table 3, as the reference results, we present test error values for the PNN model trained using conjugate gradient procedure [26]. This latter model is henceforth denoted by GPNN.

The results presented in the tables lead to the following observations:

Table 1. The test error values (E_{test}, [%]) computed for RBFNN trained using stateless Q-learning algorithm on four considered test subsets

Data set	Action set	Data partitions [%]				*min*	*avr*	*sd*	$\sigma_{\min}$
		60/40	70/30	80/20	90/10				
Breast cancer	A_6	1.8	2.4	4.4	2.9	1.8	2.8	1.1	10.99
	A_{10}	1.8	2.4	4.4	2.9	1.8	2.8	1.1	11.00
Pima Indians diabetes	A_6	33.8	34.4	35.1	32.5	32.5	33.9	1.1	3.31
	A_{10}	27.4	34.3	35.1	35.1	27.4	33.0	3.7	600.41
Haberman's survival	A_6	26.2	21.7	26.2	25.8	21.7	24.9	2.2	8.78
	A_{10}	22.1	21.7	22.9	22.6	21.7	22.3	0.5	20.90
Cardiotocography	A_6	17.3	17.2	16.9	20.1	16.9	17.8	1.5	40.93
	A_{10}	17.3	17.2	17.2	18.2	17.2	17.5	0.5	41.98
Thyroid	A_6	7.1	7.3	7.2	7.1	7.1	7.2	0.1	1.00
	A_{10}	7.1	7.3	7.2	6.8	6.8	7.1	0.2	0.29
Dermatology	A_6	41.5	43.4	44.4	41.7	41.5	42.7	1.4	43.88
	A_{10}	41.5	43.4	44.4	41.6	41.5	42.7	1.4	44.08

Table 2. The test error values (E_{test}, [%]) computed for PNN trained using stateless Q-learning algorithm on four considered test subsets

Data set	Action set	Data partitions [%]				*min*	*avr*	*sd*	$\sigma_{\min}$
		60/40	70/30	80/20	90/10				
Breast cancer	A_6	1.8	2.9	7.3	1.5	1.5	3.4	2.7	1.30
	A_{10}	1.8	2.9	7.3	1.5	1.5	3.4	2.7	1.10
Pima Indians diabetes	A_6	28.9	29.1	31.8	31.2	28.9	30.3	1.5	3.72
	A_{10}	29.3	24.3	28.6	31.2	24.3	28.4	2.9	2.97
Haberman's survival	A_6	23.0	26.1	24.6	23.0	23.0	24.2	1.5	7.57
	A_{10}	23.7	26.1	22.9	22.6	22.6	23.8	1.6	0.07
Cardiotocography	A_6	10.7	11.7	14.4	14.5	10.7	12.8	1.9	0.90
	A_{10}	10.7	11.7	14.4	14.5	10.7	12.8	1.9	0.91
Thyroid	A_6	7.4	7.4	7.4	7.5	7.4	7.4	0.1	1.01
	A_{10}	7.7	8.4	7.1	8.0	7.1	7.8	0.5	0.001
Dermatology	A_6	14.1	12.3	12.5	8.3	8.3	11.8	2.5	1.04
	A_{10}	14.8	12.3	12.5	8.3	8.3	11.9	2.7	1.01

Table 3. The test error values (E_{test}, [%]) computed for probabilistic neural network trained using conjugate gradient method (GPNN) on four considered test subsets

Data set	Data partitions [%]				*min*	*avr*	*sd*	$\sigma_{\min}$
	60/40	70/30	80/20	90/10				
Breast cancer	3.3	2.9	8.1	1.5	1.5	3.9	2.87	0.98
Pima Indians diabetes	33.5	65.6	35.1	29.8	29.8	41.0	16.55	1.01
Haberman's survival	32.7	34.7	39.3	41.9	32.7	37.2	4.20	0.65
Cardiotocography	87.8	18.7	20.5	15.8	15.8	35.7	34.79	0.68
Thyroid	48.2	57.4	52.0	51.6	48.2	52.3	3.80	0.12
Dermatology	45.1	12.3	15.3	11.1	11.1	20.9	16.20	0.97

1. In the classification of breast cancer, Pima Indian diabetes, Haberman's survival, cardiotocography and thyroid data sets, the RBFNN and PNN models with sigma computed by the use of the stateless Q-learning algorithm provide lower test error values than the PNN trained using conjugate gradient method. One should also note that the average test error for GPNN is about seven times higher when thyroid data set is classified.
2. In case of dermatology data set classification, the average test error of PNN trained by means of the proposed approach is almost twice lower in comparison to the GPNN error. Unfortunately, $E_{test} = 42.7\%$ for RBFNN which, in turn, is two times worse result than the test error for GPNN ($E_{test} = 20.9\%$).
3. The increase of the action set from six elements (A_6) to ten elements (A_{10}) generally contributes to the decrease of the test error of both RBFNN and PNN. We need to keep in mind that too many elements in an action set, at relatively small number of epochs (t_{max} is only equal 100), decreases the probability of choosing particular action at the environment exploration stage. This fact may be the reason of worsening test error for PNN after the increase of the number of actions from 6 to 10 in the classification of thyroid and dermatology data sets.
4. For the probabilistic neural network, the stateless Q-learning algorithm is much more efficient training algorithm than the conjugate gradient method since, in each of considered classification cases, PNN in comparison with GPNN model provides significantly lower test error value.
5. It turns out that the constraints on the initial values of the smoothing parameter to the interval [1,...,10] do not limit the final value of $\sigma_{\min}$.

Very poor results obtained by RBFNN model in the classification of dermatology data set (as shown in Table 1) encouraged authors to increase the number of the hidden neurons over initially assumed value recommended in [25]. This solution though did not contribute to any decrease in the test error rate. This particular outcome suggests that: (a) it might be worth to change the smoothing parameter of RBFNN depending on each attribute of the input space or (b) the application of RBFNN to multi-class classification problems (there are six classes in dermatology database) should be thoroughly analysed.

6 Conclusions

In this article, we proposed the application of the stateless Q-learning method to the selection of the smoothing parameter for the radial basis function based neural networks. The results presented in this contribution confirmed that this particular reinforcement learning method can be implemented to the automatic adaptation of this parameter for both models. Due to the use of one hundred iterations in the proposed algorithm we obtained the results, which by applying a trial and error method, might have been achieved after substantially higher number of the experiments. According to the best knowledge of the authors, the stateless Q-learning algorithm has not been proposed so far for the smoothing parameter computation of radial basis function neural networks.

The proposal of an efficient algorithm of the smoothing parameter adaptation for the PNN model is of a particular importance if we regard the value of this parameter different for each variable and class at the same time [12], [15]. In such a case, we face the problem of the choice of the $G \times n$ matrix of the smoothing parameters which is computationally much more complex. The authors plan to develop the algorithm that can handle the adaptation of the smoothing parameters for PNN whose σ is different for each attribute and also a class.

Acknowledgements This work was supported in part by the National Science Centre (Poland) under Grant No. NN 514 705540 and Rzeszow University of Technology Grant No. KPE/DS/2013 and U–8613/DS/2013.

References

1. Broomhead, D.S., Lowe, D.: Multivariable Function Interpolation and Adaptive Networks. Complex Systems 2, 321–355 (1988)
2. Specht, D.F.: Probabilistic neural networks. Neural Networks 3, 109–118 (1990)
3. Maglogiannis, I., Sarimveis, H., Kiranoudis, C.T., Chatziioannou, A.A., Oikonomou, N., Aidinis, V.: Radial basis function neural networks classification for the recognition of idiopathic pulmonary fibrosis in microscopic images. IEEE Trans. Information Technology in Biomedicine 12, 42–54 (2008)
4. Chu, F., Wang, L.: Applying RBF Neural Networks to Cancer Classification Based on Gene Expressions. In: International Joint Conference on Neural Networks, pp. 1930–1934. IEEE Press, Vancouver (2006)
5. Folland, R., Hines, E., Dutta, R., Boilot, P., Morgan, D.: Comparison of neural network predictors in the classification of tracheal-bronchial breath sounds by respiratory auscultation. Artificial Intelligence in Medicine 31, 211–220 (2004)
6. Samanta, B., Bird, G.L., Kuijpers, M., Zimmerman, R.A., Jarvik, G.P., Wernovsky, G., et al.: Prediction of periventricular leukomalacia. Part II: Selection of hemodynamic features using computational intelligence. Artificial Intelligence in Medicine 46, 217–231 (2009)
7. Mantzaris, D., Anastassopoulos, G., Adamopoulos, A.: Genetic algorithm pruning of probabilistic neural networks in medical disease estimation. Neural Networks 24, 831–835 (2011)
8. Chen, S., Cowan, C.F.N., Grant, P.M.: Orthogonal least squares learning algorithm for radial basis function networks. IEEE Trans. Neural Networks 2, 302–309 (1991)
9. Demuth, H., Beale, M.: Neural network toolbox user's guide. The Mathworks, Inc. (1994)
10. Haykin, S.: Neural networks: a comprehensive foundation. Prentice Hall, New Jersey (1999)
11. Specht, D.F., Romsdahl, H.: Experience with adaptive probabilistic neural networks and adaptive general regression neural networks. In: IEEE World Congress on Computational Intelligence 2, pp. 1203–1208. IEEE Press, Orlando (1994)
12. Chtioui, Y., Panigrahi, S., Marsh, R.: Conjugate gradient and approximate Newton methods for an optimal probabilistic neural network for food color classification. Optical Engineering 37, 3015–3023 (1998)

13. Mao, K.Z., Tan, K.-C., Ser, W.: Probabilistic Neural Network Structure Determination for Pattern Classification. IEEE Trans. Neural Networks 11, 1009–1016 (2000)
14. Gorunescu, F., Gorunescu, M., El-Darzi, E., Gorunescu, S.: An evolutionary computational approach to probabilistic neural network with application to hepatic cancer diagnosis. In: IEEE Symposium on Computer-Based Medical Systems, pp. 461–466. IEEE Press, Dublin (2005)
15. Zhong, M., Coggeshall, D., Ghaneie, E., Pope, T., et al.: Gap-Based Estimation: Choosing the Smoothing Parameters for Probabilistic and General Regression Neural Networks. In: IEEE International Joint Conference on Neural Networks, pp. 1870–1877. IEEE Press, Vancouver (2006)
16. Watkins, C.J.C.H.: Learning from delayed Rewards. PhD Thesis, Cambridge University, Cambridge, England (1989)
17. Starzyk, J.A., Liu, Y., Batog, S.: A novel optimization algorithm based on reinforcement learning. In: Tenne, Y., Goh, C.-K. (eds.) Computational Intelligence in Optimization, ALO 7, pp. 27–47 (2010)
18. Parzen, E.: On estimation of a probability density function and mode. Annals of Mathematical Statistics 36, 1065–1076 (1962)
19. Sutton, R.S., Barto, A.G.: Reinforcement learning: An Introduction. MIT Press, Cambridge (1998)
20. Lanzi, P.L.: Adaptive Agents with Reinforcement Learning and Internal Memory. In: Sixth International Conference on Simulation of Adaptive Behavior, pp. 333–342. The MIT Press, Cambridge (2000)
21. Claus, C., Boutilier, C.: The dynamics of reinforcement learning in cooperative multiagent systems. In: Fifteenth National/Tenth Conference on Artificial Intelligence/Innovative Applications of Artificial Intelligence, pp. 746–752. AAAI Press, Madison (1998)
22. McGlohon, M., Sandip, S.: Learning to cooperate in multi-agent systems by combining Q-learning and evolutionary strategy. In: First World Congress on Lateral-Computing, Bangalore, pp. 1–7 (2005)
23. UCI Machine Learning Repository, `archive.ics.uci.edu/ml/datasets.html`
24. Arlot, S.: A survey of cross-validation procedures for model selection. Statistics Surveys 4, 40–79 (2010)
25. Kolmogorov, A.N.: On the representation of continuous functions of many variables by superposition of continuous functions of one variable and addition. Dokl. Akad. Nauk CCCP 114, 953–956 (1957)
26. DTREG predictive modelling software (2013), `http://www.dtreg.com`

Computer-Aided On-line Seizure Detection Using Stockwell Transform

Grzegorz Rutkowski[1], Krzysztof Patan[1], and Paweł Leśniak[2]

[1] Institute of Control and Computer Engineering,
University of Zielona Góra,
ul. Podgórna 50, 65-246 Zielona Góra, Poland
k.patan@issi.uz.zgora.pl
[2] Ward of Neurology and Strokes
Provincial Hospital of Zielona Góra, Poland

Abstract. The paper deals with the designing and implementation of a computer-aided system capable to detect seizures by inspecting EEG records. The system is based on modern signal processing tools, which are able to give a time-frequency representation of a signal. Then using time-frequency representation of EEG data, feature extraction and finally classification of neurological disorders is carried out. The application should be treated as a decision support computer system which was designed to help neurologists in detecting seizures. With the help of such a software the time needed for analysing EEG records can be significantly reduced. The research was carried out using real EEG recordings of epileptic patients as well as healthy subjects prepared with the cooperation of the medical staff of the Ward of Neurology and Strokes of the Provincial Hospital of Zielona Góra, Poland.

Keywords: EEG signal, time-frequency analysis, classification, computer-aided system, medical diagnosis.

1 Introduction

In the literature one can find many techniques for biomedical signal analysis [17,5,7,3]. One of the most popular methods is to measure the activity of neurons on the surface of the cerebral cortex known as ElectroEncephaloGraphy (EEG). This consists in measuring the neural activity on the surface of the cortex by placing electrodes on the scalp. The analysis of these measurements contributes to the subsequent process of diagnosis accuracy and sets the direction for the appropriate treatment for a particular path of dysfunction.

The epilespy is a very important problem. Approximately 1-2% of world's population has epilepsy, about 5% may have at least one seizure during their lifetime, and about 25% of epileptic patients cannot be treated sufficiently by any available therapy [2]. Epileptic seizures can be divided into partial, generalized, unilateral and unclassified [14]. In turn EEG recordings of epileptic patients can be classified into two categories: *inter-ictal* – abnormal activity recorded

J. Korbicz and M. Kowal (eds.), *Intelligent Systems in Technical and Medical Diagnostics*, Advances in Intelligent Systems and Computing 230,
DOI: 10.1007/978-3-642-39881-0_23, © Springer-Verlag Berlin Heidelberg 2014

between epileptic seizures, which has the form of occasional short-term transient waves, and *ictal* – activity recorded during the seizure in the form of long-term polymorphic waves [14].

Generally, the detection of epilepsy is achieved by the so-called visual inspection of EEG records done by an experienced neurophysiologist. However, such an analysis method is time consuming and highly inefficient. Moreover, due to the subjective nature of the analysis, disagreement among experts on the same EEG record is possible. Therefore, there is a plenty of alternative methods of EEG analysis. Among many solutions one can find: parametric methods, spectral analysis methods, artificial neural networks, clustering methods, principal component analysis, independent component analysis, knowlege-based rules, classification algorithms. In the literature one can find large number of automated epileptic seizure detection methods, which reports the achieved seizure detection accuracy from 85.9 to 100%. However, these comparisons were carried out based on data recorded in one EEG channel only. For details the interested reader is referred to [14].

An automatic seizure-detection system can reduce the time required by a neurologist who performs visual inspection EEG data off-line. Such a system can produce the warning signal on-line to produce the information about possible disorders detected in EEG records. The objective of the paper is to design an automated seizure detection system operating on-line on EEG data. The proposed application can be treated as a decision support system able to point out suspicious ranges of a record. The system allows to shorten the time needed for the diagnosis as an expert can focus his attention on areas pointed out by the software. The first version of the application is implemenmted in MATLAB software.

In cooperation with the medical staff of the Ward of Neurology and Strokes of the Provincial Hospital of Zielona Góra, Poland, a database of neurological disorders was prepared. These EEG records were acquired using 16 channels equipment. The data were acquired from both epileptic patients and healthy subjects. With the help of neurologists, 586 seizures from 104 patients were recorded and analysed. Simultaneously, 568 sequences from 61 healthy subjects were derived. Eventually, the database consisted of 1154 EEG sequences. EEG signals were filtered using low-pass filter equal to 35 Hz. The sampling frequency was equal to 500 Hz (sample time equal to 0.002 sec.). An expert in clinical analysis of EEG signals inspected every record visually to score epileptic and normal sequences. Database prepared in this way was used to test the proposed computer-aided diagnostic system.

2 Seizure Detection System

The operation idea of the proposed on-line seizure detection system is based on a classical, quite simple structure, consisting of three main stages (Fig. 1). The first stage is to perform time-frequency analysis on the raw EEG records, acquired using 16 channels equipment. Taking into account the non-stationary

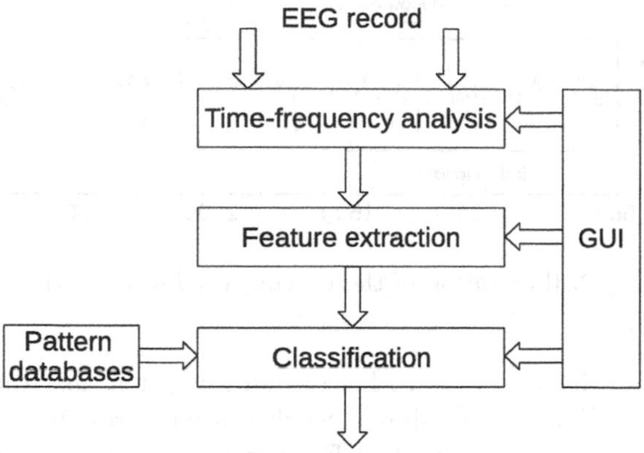

Fig. 1. Operation sequence in the application

character of EEG signals, the classical methods of spectral analysis cannot be applied [13]. To deal with non-stationary signals, a transformation that lead to two-dimentional representation of signals is required. An example of such a method is Short-Time Fourier Transform (STFT), which is defined both in the time and frequency domains. Unfortunately, STFT which uses the fixed width of the window reveals its disadvantages in the case of signals consisting high as well as low frequency components at the same time. These problems can be avoided applying other time-frequency analysis tools such as Wavelet Transform [15] or Stockwell Transform [9]. The latter is a quite new transformation technique and therefore it is described in Section 3. The second stage is called feature extraction. The purpose of feature axtraction process is to select features or attributes and simultaneously reduce the dimension of data [16]. Another problem with feature extraction is to select a proper number of features. The best set of features should be as small as possible but as much informative as possible. Too large set of features leads to the so-called curse of dimensionality, which can significantly decrease the classification quality. Based on the time-frequency representation of the EEG data, features can be derived as [9,12,4]:

- statistical parameters of each component of transformation, e.g. mean value, standard deviation,
- energy of each component of transformation,
- entropy of each component of transformation.

The last stage is classification [1,6]. The feature vector is passed to the classifier which makes a final decision about the neurological condition of a patient. In this study two classes are considered: epileptic patients and healthy subjects. In the literature one can find a plenty of classification methods, among them: K Nearest Neigbours, Naive Bayes, decision trees, Multilayer Perceptron, Suport Vector Machines, etc. At the moment the application uses kNN as the default classifier.

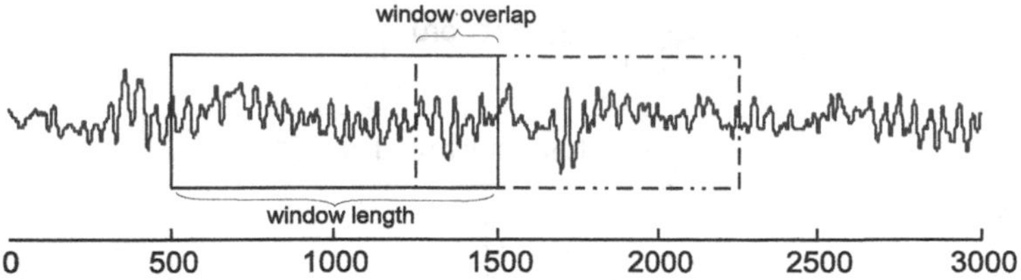

Fig. 2. Illustration of the moving window operation

The objective of the system is to help a neurologist in fast diagnosis of neurological disorders. The system should work on-line, providing an information about the condition of a patient. As EEG records are time sequences its analysis cannot be carried out on a single pattern recorded at time k but on some portion of data. The question is how large portion of data is necessary? Usually, spectral analysis of EEG data is carried out on a short segment of the length 1–4 seconds. Assuming a time window with the length equal to 3 seconds and taking into account the sampling time 0.002 seconds, this leads to the number of samples equal to 1500. This window is moved in time. It is not advisable to make classification every time instant. First of all, assuming that the analysed time sequence consists of hundreds of elements, exchanging one element does not change significantly time sequence characteristics. Secondly, time-frequency analysis of a 16 channels EEG record is time consuming and applying it every time instant can prolong computation burden significantly. Therefore, another parameter introduced here is the window overlap. This parameter controls a frequency of the classification. The window length and window overlap parameters are commonly used when moving windows are applied on time series analysis (Fig. 2).

The considered diagnostic system is based on the mechanisms proposed by the authors in their previous works [8,9]. All implemented methods used for the realization of each block of the system have been tested off-line using the

Algorithm 1. Operating algorithm.

Require: window length, window overlap, analysis method, feature extraction method, type of classifier

```
while stopping criteria are not satisfied do
   set window for analysis
   for each channel do
      perform time-frequency analysis
      determine features
      make a decision about patient state
   end for
   move window
end while
return information (reports) about detected seizures
```

prepared database of neurological disorders. For example, deriving features in the form of the entropy based on the time-frequency representation achieved using the S transform and using kNN classifier, the authors obtained more than 98% of correct classifications [10]. The quality of the classifier has been verified using k-fold cross validation technique [10]. The general operation idea of the application is expressed by Algorithm 1.

3 Stockwell Transform

The S Transform (ST) is a relatively new time-frequency analysis tool developed by Stockwell in 1994 [11]. It can be viewed as a generalization of the Short-Time Fourier Transform (STFT), but instead of a window of the constant size as in the STFT, it uses a scalable Gaussian window. The Stockwell transform of signal x is given as follows:

$$S(t, f) = \int_{-\infty}^{\infty} x(\tau) g(t - \tau) e^{-i2\pi f \tau} d\tau, \quad (1)$$

where $g(t - \tau)$ is the Gaussian function located at $\tau = t$ defined as

$$g(t - \tau) = \frac{|f|}{\sqrt{2\pi}} e^{-\frac{(t-\tau)^2 f^2}{2}}. \quad (2)$$

This makes it possible to use a variable window length. As f increases, the length of the window decreases.

The advantage of S-transform over the short time Fourier transform is that the window width is a function of f rather than a fixed one as in STFT. In contrast to wavelet analysis the S-Transform wavelet can be divided into two parts. The first one is the slowly varying envelope formed by the Gaussian window $g(t - \tau)$ which localizes the time. The second is the oscillatory exponential kernel $e^{-i2\pi f \tau}$ which selects the frequency being localized. Then, it is the time localizing Gaussian that is translated while keeping the oscillatory exponential kernel stationary. As the oscillatory exponential kernel is not translating, it localizes the real and the imaginary com- ponents of the spectrum independently, localizing the phase as well as amplitude spectrum. Thus S-transform retains absolute phase information of the signal which is not provided by wavelet transform. Absolutely referenced phase means that the phase information given by the S-transform is always referenced to time $t = 0$, which is also true for the phase given by the Fourier transform. This is true for each S-transform sample of the time-frequency space. The normalization factor $\frac{|f|}{\sqrt{2\pi}}$ also is very important as it normalizes the time domain localizing window to have unit area. Therefore, the amplitude of the S-transform has the same meaning as the amplitude of the Fourier transform. This provides a frequency invariant amplitude response in contrast to the wavelet transform. The frequency invariant amplitude response means that for a sinusoid with an amplitude M, the S-transform returns an amplitude M regardless of the frequency f.

Figure 3a) presents the ST decomposition of the channel T6-O2. In the case considered, there are coefficients of the high magnitude observed for the small frequencies (about 2 Hz) around the 110th second. This area representes a seizure. Corresponding EEG signals are shown in Fig 3b), were one can observe a significant change of signal characteristics caused by a seizure. In turn, in Fig. 4 one can see the ST decomposition of EEG data of a healty subject. In this case there is no spikes with the high magnitude as observed in Fig. 3. Summarizing, the S transform can be effectively used to represent time-frequency dependencies of

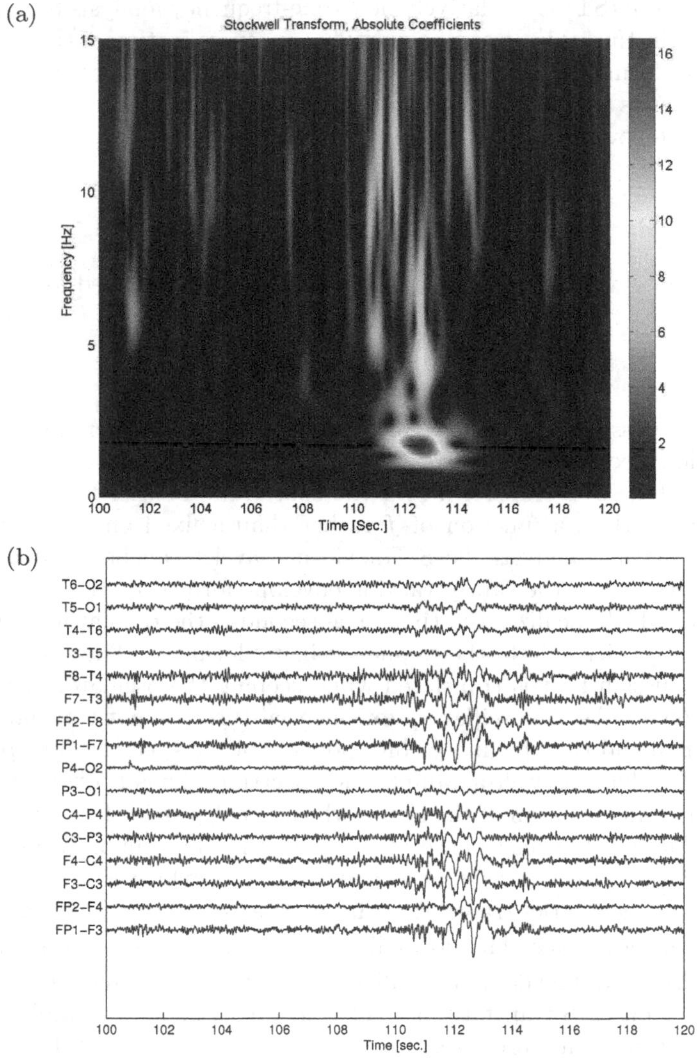

Fig. 3. Time-frequency EEG signal representation using the ST (a), and a EEG record in the time-domain (b)

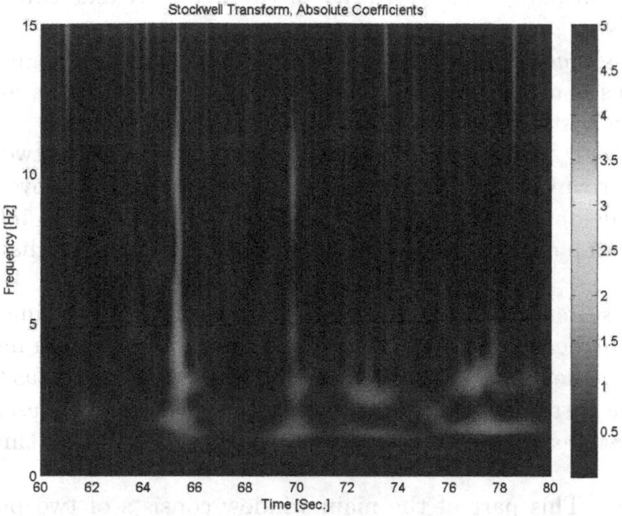

Fig. 4. Stockwell transform of a normal EEG recording

EEG signals. Now the question is how to use it effectively to feature extraction process. The transform (1) is continuous in the both time and frequency. Hopefully, the discrete version of the transform can be easily developed, assuming the sampling time t_s and the sampling frequency f_s. The sampling time $t_s = 0.002$ second is forced by EEG equipment. The sampling frequency was selected taking into account analysis of EEG records. We have decided that the S transform should be performed for the frequency range $(0 - 100)$ Hz with the sampling frequency $f_s = 10$ Hz. The proposed frequency range is quite wide to observe seizures. The chosen sampling frequency is a compromise between a resolution of time-frequency representation of the signal and a computation burden of the procedure. Experiments presented in the previous work of the authors [9] showed that this choise quarantees enough accuracy of time-frequency representation, and simultaneously the computation burden of the transformation is not so high.

The S transform does not have a cross-term problem and yields, a better signal clarity than the Gabor transform. The disadvantage of the S transform is complexity computation and worse clarity than, for example, that of the Wigner or Cohen distribution functions.

4 Automated Detection System

The main window of the diagnosis system is shown in Fig. 5. The window consist of the following parts:

1. *Signal window.* This window shows signals recorded by EEG equipment at the time domain. As we have 16 channels available, values limits on the Y

axis are not displayed due to clarity of the figure. X axis contains current time stamps.
2. *Diagnostic signal window.* This window contains diagnostic signal generated on-line. This is a binary signal, where 1 represents a seizure, and 0 stands for the normal state of a patient.
3. *Moving window parameters information area.* It consists of two edit fields, where a user can set the both window length and windows overlap parameters. The default settings are: the window length equal to 1500, and the window overlap equal to 1400. These parameters cannot be changed during analysis.
4. *Diagnostic systems settings information area.* The window informs about: used time-frequency analysis method, used feature extraction method, used classifier. The default settings are: the Stockwell transformas as the analysis method, the entropy as feature extraction and kNN as the classifier.
5. *Detected seizures information area.* This window displays the time of seizure detection.
6. *Control area.* This part of the main window consists of two push buttons used to control the system. `Start` and `Stop` push buttons starts/restarts and stops/pauses EEG analysis, respectively.

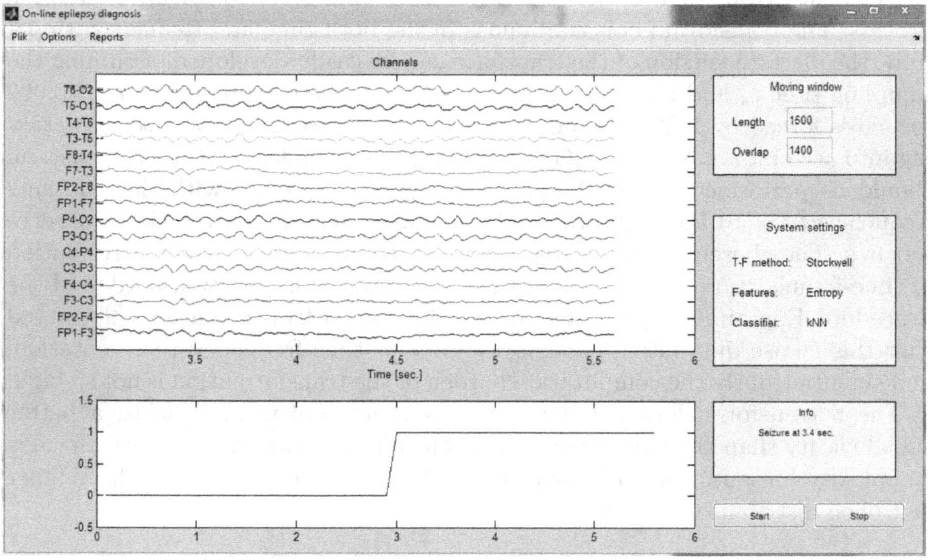

Fig. 5. Computer-aided seizure detection system

The application is also equipped with a menu with the following structure:

- `File`
  - `Load` – connecting to an EEG record,

    - `Save` – save the selected characteristics of the analysis, eg. diagnostic signal, the time of seizure occurrence, a number of seizures during the analysis, etc.,
    - `Exit` – quit the application,
- `Options` – the structure configuration of the diagnostic system,
    - `Analysis` – time-frequency method selection,
        - `ST` – Stockwell transform,
        - `DWT` – discrete wavelet transform,
    - `Features` – attributes method selection,
        - `Statistics` – generate features in the form of basic statistics of the time-frequency components,
        - `Energy` – generate features in the form of the energy of the time-frequency components,
        - `Entropy` – generate features in the form of the entropy of the time-frequency components,
    - `Classifier` – chosee the classifier,
        - `kNN` – k Nearest Neighbour,
        - `MLP` – MultiLayer Perceptron neural network,
        - `Bayes` – Naive Bayes,
        - `SVD` – Support Vector Machines,
- `Reports` – this option allows a user for exporting the report of monitoring/diagnosis to the text file.

Figure 5 shows an example of on-line diagnosis of the EEG signal working with the default settings. The computer system detects a seizure at 3.4 sec., which is signalled by the diagnostic signal plotted in a lower window. Indeed, the data used in this experiment describes an epileptic patient. After analysis, the system will prepare a report for a neurologist with information abount possible seizures. This information can be verified off-line using the time-frequency representation of the data pointed out by the on-line system. Reports and off-line analysis of the data will be included in the application in the forthcomming weeks.

The important problem in the diagnosis is how to avoid false alarms. Let us consider analysis of the EEG signal acquired from a healthy subject. The diagnostic signal (the classifier output) for this case is presented in Fig. 6a). As one can see there, a number of false alarms occurs (change of the signal from 0 to 1). This undesirable effect can be easily eliminated using a time-window. Application of the time-window may prevent the situation when a temporary change of the diagnostic signal raise alert about a seizure. In the present study the time-window with the length of 0.5 sec. was used. If during the period of 0.5 seond the diagnostic signal will keep the value of 1, the system will signal a seizure. Figure 6b) illustrates the situation when the time-window mechanism prevents false decision about a seizure as there are only single, temporary (not permanent), decisions provided by the classifier (cf. Fig. 6a)).

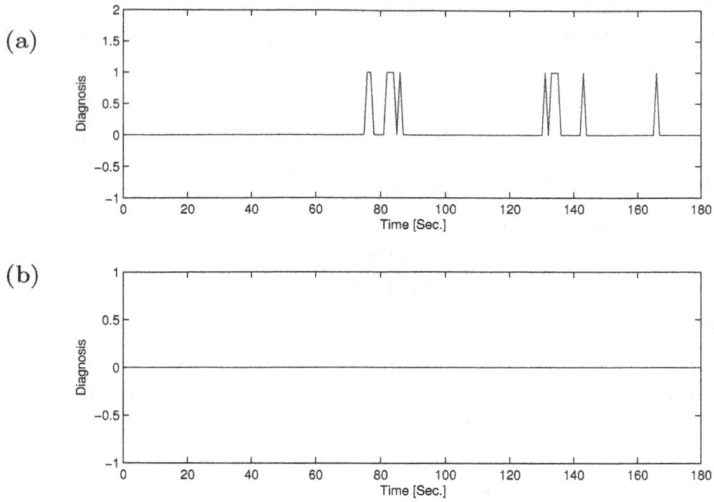

Fig. 6. Diagnostic signal: without the time-window (a), with the time-window (b)

5 Conclusions and Further Development Directions

The paper presents the realization of computer-aided on-line system for seizures detection. The system uses modern signal processing methods as well as outstanding classification algorithms. Close cooperation with specialists in the field of electroencephalography of the Ward of Neurology and Strokes in Zielona Góra, Poland allows further development of the application. The planned extensions include:

- analysis of stored EEG records in off-line mode,
- improving interaction between the application and a user, e.g. selecting important areas of EEG signals for further analysis,
- introducing the diagnosis parameters able to exclude false diagnosis, e.g length of the time-window,
- realization of the analysis reports.

References

1. Duda, R.O., Hart, P.E., Stork, D.H.: Pattern Classification, 2nd edn. Wiley Interscience (2000)
2. Engel, J.: Seizure and Epilepsy. FA, Davis, Philadelphia, PA, USA (1989)
3. Faul, S., Boylan, G., Connolly, S.: An evaluation of automated neonatal seizure detection methods. Clin. Neurophysiol. 116, 1533–1541 (2005)
4. Guo, L., Rivero, D., Pazos, A.: Epileptic seizure detection using multiwavelet transform based approximate entropy and artificial neural networks. J. Neurosci. Methods 193, 156–163 (2010)

5. James, C.J., Jones, R.D., Bones, P.J., Carroll, G.J.: Detection of epileptiform discharges in the eeg by a hybrid system comprising mimetic, self-organized artificial neural network, and fuzzy logic stages. Clin. Neurophysiol. 12, 2049–2063 (1999)
6. Kajdanowicz, T., Kazienko, P.: Multi-label classification using error correcting output codes. J. Applied Mathematics and Computer Science 22(4), 829–840 (2012)
7. McSharry, P., He, T., Smith, L.: Linear and non-linear methods for automatic seizure detection in scalp electro-encephalogram recordings. Med. Biol. Eng. Comput. 40, 447–461 (2002)
8. Patan, K., Rutkowski, G.: Analysis and classification of EEG data: An evaluation of methods. In: Rutkowski, L., Korytkowski, M., Scherer, R., Tadeusiewicz, R., Zadeh, L.A., Zurada, J.M. (eds.) ICAISC 2012, Part II. LNCS, vol. 7268, pp. 310–317. Springer, Heidelberg (2012)
9. Rutkowski, G., Patan, K., Leśniak, P.: Comparison of time-frequency feature extraction methods for EEG signals classification. In: Rutkowski, L., Korytkowski, M., Scherer, R., Tadeusiewicz, R., Zadeh, L.A., Zurada, J.M. (eds.) ICAISC 2013, Part II. LNCS, vol. 7895, pp. 320–329. Springer, Heidelberg (2013)
10. Rutkowski, G., Patan, K.: Extraction of attributes of the EEG signals based on the Stockwell transform. Pomiary, Automatyka, Controla 59(3), 208–211 (2013) (in Polish)
11. Stockwell, R.G., Mansinha, L., Lowe, R.P.: Localization of the complex spectrum: The S transform. IEEE Trans. Signal Processing 44(4), 998–1001 (1996)
12. Subasi, A.: EEG signal classification using wavelet feature extraction and a mixture of expert model. Expert Systems with Applications 32, 1084–1093 (2007)
13. Świercz, E.: Classification in the Gabor time-frequency domain of non-stationary signals embedded in heavy noise with unknown statistical distribution. J. Applied Mathematics and Computer Science 20(1), 135–147 (2010)
14. Tzallas, A.T., Tsipouras, M.G., Tsalikakis, D.G., Karvounis, E.C., Astrakas, L., Konitsiotis, S., Tzaphlidou, M.: Automated epileptic seizure detection methods: A review study. In: Stevanovic, D. (ed.) Epilepsy - Histological, Electroencephalographic and Psychological Aspects. InTech (2012) ISBN: 978-953-51-0082-9
15. Übeyli, E.D.: Combined neural network model employing wavelet coefficients for eeg signals classification. Digital Signal Processing 19, 297–308 (2009)
16. Woźniak, M., Krawczyk, B.: Combined classifier based on feature space partitioning. J. Applied Mathematics and Computer Science 22(4), 855–866 (2012)
17. Yeung, N., Bogacz, R., Holroyd, C.B., Cohen, J.D.: Detection of synchronized oscillations in the electroencephalogram: An evaluation of methods. Psychophysiology 41, 822–832 (2004)

Diagnosis of Obesity with Bioimpedance Method

Monika Cichocka[1] and Józef Tutaj[2]

[1] Faculty of Electrical Engineering, Automatics, Computer Science and Biomedical Engineering, University of Science and Technology AGH, Cracow, Poland
moncich@student.agh.edu.pl

[2] Faculty of Mechanical Engineering, University of Technology, Cracow, Poland
pmtutaj@cyf-kr.edu.pl

Abstract. The article describes the bioimpedance method of obesity diagnosis. Such experiments are very important nowadays, because many people are obese and have problems connected with it. It is essential to check parameters such as a total body water (TBW), fat-free body mass (FFM), fat mass and body fat percentage in order to start a proper treatment and compose a right diet for the patient. Due to that fact, a special meter was constructed. The main aim of the project was to create a cheap, intuitive to operate, easily accessible equipment with high measuring credibility. The projected meter uses laptops and a LabVIEW application dedicated to measurements of bioelectrical impedance. Some results of the meter tests are shown in the article.

Keywords: bioimpedance, dietetics, obesity diagnosis, virtual instrument, LabVIEW.

1 Introduction

Because of the trends which are popular nowadays: a fast lifestyle and carelessness in selecting food, more and more people have problems with being overweight. Therefore, people go to dietetics so that they help them compose a proper diet that will allow them to maintain a slim figure. For this purpose it is necessary to determine the composition of the patient's body by performing non-invasive, yet effective electrical bioimpedance analysis (BIA)[1].

1.1 Problem of Obesity

The obesity is a serious problem with excessive accumulation of body fat. The obesity is diagnosed when the body fat exceeds the physiological needs of organisms. A woman is considered to be obedient when her body fat is higher than 25% of a total body weight and in case of men it is 20%. The thing that matters in not only its amount but also location.

The obesity is a heterogeneous group of conditions with multiple causes. It is determined by an interaction between a variety of factors:

– environmental the westernization of diet and lifestyles,

J. Korbicz and M. Kowal (eds.), *Intelligent Systems in Technical and Medical Diagnostics*, Advances in Intelligent Systems and Computing 230,
DOI: 10.1007/978-3-642-39881-0_24,

- genetic - genetic diseases associated with mutations in genes,
- pharmacological some drugs can cause weight gain, eg. antidepressants and anxiolytics,
- biological caused by autonomic nervous system disorders, eg. damage of the hypothalamus (inflammation or cancer) and ventrolateral nucleus of the medial hypothalamus,
- psychological people who suffers depression often eat more which causes overweight [2].

The main reason is eating too much energetic food and a lack of the physical activity. Thus, it is strongly connected to the problem of overweight.

The obesity causes a variety of health problems. In particular, it is associated with the development of obstructive sleep apnoea, type 2 diabetes mellitus, heart disease, etc. A large obesity leads to disability. Obesity is a social issue in developed countries and in the future there may be an epidemic. It is considered to be one of the hazards of developed societies civilization. Therefore, it is necessary to diagnose this problem early and take action to cure the patient. In order to determine body fat (body composition) and diagnose the obesity, usually BIA measurement is performed or body mass index (BMI) is calculated. The latter is easier and does not require any special device, but the obtained result is approximate and has low reliability of measurement [3].

1.2 Bioimpedance

The bioelectrical bioimpedance is resultant total electrical resistance of tissues. It is the ratio of the applied voltage U to the AC current I flowing through the tissues. It consists of resistance R and reactance X, which is capacitance C in humans.

$$Z(\omega) = \frac{U(\omega)}{I(\omega)} = R + jX(\omega) = R - j\frac{1}{\omega * C} \tag{1}$$

The human bioimpedance is a very important diagnostic parameter. Its value depends on the humans body composition. The electrical properties of the issues changes with the properties of every structural part, eg. intracellular and extracellular substance, cell membrane and all organelles. Resistance depends on electrolyte concentration while capacitance depends on potential differences on cell membranes and varies with the current frequency which was used in examination [4][5]. If DC or low frequency is used, current cannot go through a membrane cell. The membrane behaves like an isolator and the extracellular route is pure resistance. On the other hand, if high frequency is used, a membrane cell behaves like a capacitor, so capacitance becomes associated with it [6].

In connection with the human tissues structure, the human body can be modeled as an electrical circuit which consists of two parallel branches: one branch with resistor and the other with resistor and capacitor connected in series [7]. It is why the body composition (especially water weight and fat mass) can be examined with bioimpedance method.

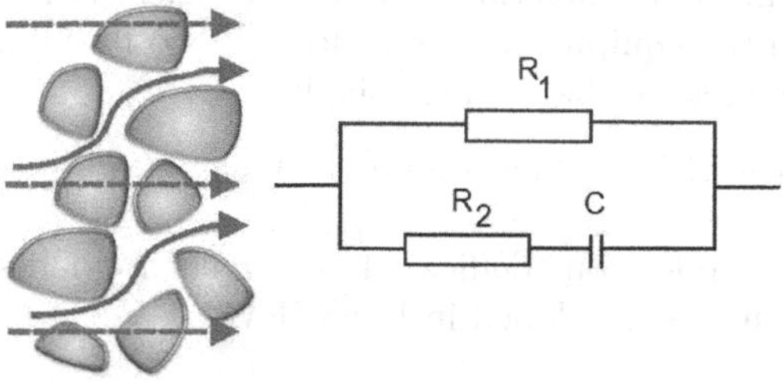

Fig. 1. The low frequency current (represented with a continuous arrow) flow extracellular and the high frequency current (represented with a dotted arrow) flow intracellular which is the reason that the human body can be modeled as resistance and capacitance connected in parallel[6][8].

2 Related Works

Two basic method of bioelectrical impedance analysis and dedicated BIA devices can be distinguished. Single Frequency Bioelectrical Impedance Analysis (SF-BIA) uses only one frequency (usually it is 50 kHz) and current 0,8-1 mA. It is applied to assessment of body composition in healthy subjects [1]. The very precise measurements of patients in postoperative period is performed by Multi Frequancy Bioelectrical Impedance Analysis (MF-BIA) method. A variety of signal frequencies is used: the range is 0-500 kHz (mostly 0, 1, 5, 50, 100, 200 or 500 kHz, but at frequencies below 5 kHz and above 200 kHz poor reproducibility has been noted [7].

There are four standard methods of bioimpedance measurements. Two of them use current source and two use voltage source. The other classification criterion is a number of sensors placed on the patient body. The bipolar methods use only two sensors (eg. hand-to-hand, hand-to-foot or foot-to-foot) while tetrapolar methods use four sensors (hand-to-hand-to-foot-to-foot) [4].

For dietetics diagnosis purpose there is commonly used a single frequency measurement of the whole body with tetrapolar system [1]. Unfortunately, the bioimpedance meters are very expensive so the cost of examination is also quite big and not everybody can afford it.

3 Projected Bioimpedance Meter

The aim of the meter project was to construct a cheap, intuitive and easily accessible equipment with high measuring credibility. Therefore the authors of the paper oryginally invented and constructed a system for bioimpedance measurements using widely accessible laptops, simple electronic circuits and intuitive software (an application in LabVIEW) [10].

Every medical equipment has to be safe for a patient as well as for a physician, so a galvanic separation is needed. Using a laptop running on battery power do not require any additional protection. There is a logical barrier and the battery is loaded only when the equipment is shut down. Two basic functional blocks of the bioimpedance meter can be distinguished:

- hardware (laptop with external sound card, signal adaptor - voltage divider and electrodes),
- software (the application dedicated to measurements of bioelectrical impedance which was developed in LabVIEW).

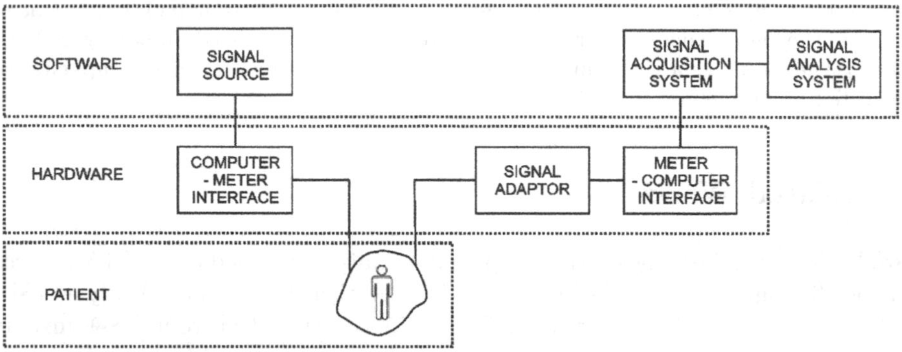

Fig. 2. Block diagram of the projected bioimpedance meter [10]

The above mentioned components are described with details in next subsections.

3.1 Hardware

The meter was created using the bipolar voltage method. In order to decrease cost of the construction, it was decided to use standard inputs and outputs available in every laptop: MIC IN and PHONES OUT. They were examined using HUNG CHANG 5804 40MHz oscilloscope and HAMEG HM8131-2 15MHz generator. Because of their nonlinear amplitude-frequency characteristic, it would be necessary to implement some extra complicated calibration functions in the software. Therefore, external sound card Behringer UCA202 was examined and used (it is not expensive and its inputs/outputs). Unfortunately each laptop has different internal parameters which cause that different signal amplification values are obtained [10]. To minimize artefacts caused by various computer architecture, calibration and parameterization of software have to be performed every time the device is connected to a new laptop for the first time. However, using LINE IN input and LINE OUT output does not require any additional signal preprocessing in software, because these outputs and inputs are linear.

The use of LINE IN and LINE OUT introduced restrictions in generated signal frequency. The minimal signal frequency value which will not be filtered is 100 Hz and the maximum one is 10 kHz regarding to the Nyquist-Shannon sampling theorem (sampling frequency of A/C and C/A converters in sound card is 22

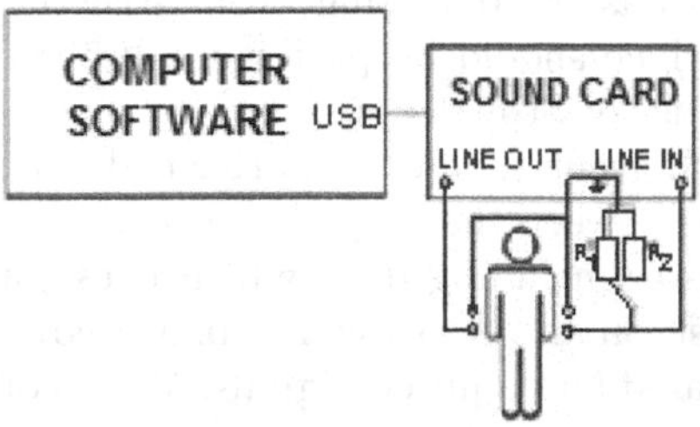

Fig. 3. The projected bioimpedance meter consists of a laptop connected to an external sound card with LINE OUT output and LINE IN input.[10].

kHz). There are also imposed limits of the input signal magnitude which value depends on the laptops type.

A voltage signal is generated and the voltage signal has to be acquired because of the output/input character. Regarding to the figure 2, it was necessary to create an electronic module which will be a kind of diagnostic signal adaptor a voltage divider with an additional resistor. Thus, the voltage value of the LINE IN signal is proportional to the resistance value and the current value which flows through this resistor (it depends on the measured bioimpedance). Such solution enables computing the current value as a ratio of the additional resistors value and the measured voltage value.

The big problem in medical diagnostics and biomedical equipment is caused by noise. It is essential to provide proper equipments and cables shielding and to use good quality sensors. In this case, there were used shielded cables from electrocardiography device and Ag/AgCl single-use electrodes.

3.2 LabVIEW Application

Control and measurement systems developed in LabVIEW are commonly used in engineering. This design platform and development system for a visual programming from National Instruments enables developing applications virtual instruments. When you use them, they are similar to real equipment [11][12]. That is the reason why LabVIEW was chosen to develop an application for bioimpedance measurements.

The author of this paper created the application dedicated to measurements of bioelectrical impedance. The application generates, acquires, processes and analyses the diagnostic signal in bioimpedance measurements.

The main programme is based on the Multiple Loop Architecture using Functional Global Variable.. It enables correct data flow control between programme elements which work in parallel. Three loops have to be performed at the same time:

- Event Handler Loop,
- Generation Loop,
- Acquisition and Analysis Loop.

The Event Handler loop was created using event structure. It defines an event (a code fragment [12]) which should be performed if one of the buttons on the user interface (Start, Pause or Stop) is pushed.

The information which event should be performed and the diagnostic signal data are stored and sent between the loops by specially created global variable. It consists of a data cluster including data which is essential to handling Generation Loop and Acquisition and Analysis Loop. Moreover, there are contained parameters of the generated and aquired signals. Values of the generated signal are to be found there too.

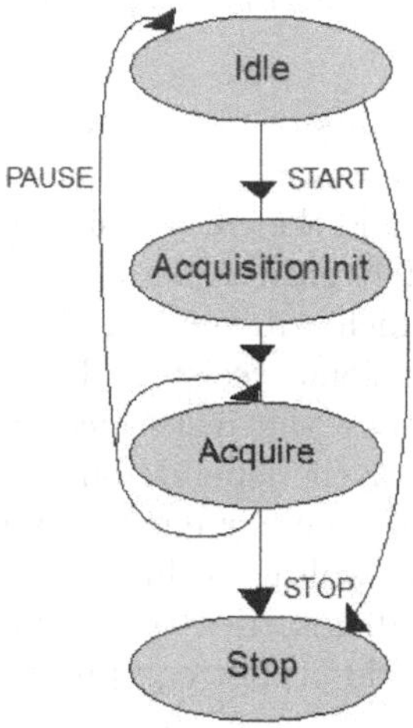

Fig. 4. State diagram of Case Structure in Acquisition and Analysis Loop presents the way how the states can change. The AcquisitionInit state initalizes the whole process, which is performed during Acquire state.

The Generation Loop and The Acquisition and Analysis Loop were created using an architecture which is similar to a state machine architecture as it uses a Case Structure. The difference is in a way of sending information by enum type control between each circuits [11][12]. Instead of a Shift Register, the aforementioned functional global variable was used. Both loops have an idle, initialization, the basic operation and stop states. The basic operation in Generation Loop is to generate a sinusoidal signal with set frequency and magnitude and then to put it in on the computer sound card outputs: PHONES OUT or LINE OUT. The Acquisition and Analysis Loop enables a signal acquisition from the computer sound card inputs: MIC IN or LINE IN, and then its analysis using an implemented algorithm. The acquisition and the analysis parts are combined together. It is caused by laptops memory and calculation capabilities [10]. If these two functions were to be separated, noticeable changes in the program structure would be needed and hardware requirements would increase.

The crucial element of the signal processing block in the application is a calcutation of an instantaneous bioimpedance value. It enables creating a plot of how bioimpedance varies in time (an impedance curve). Moreover, the obtained data is analysed in terms of fat content in the body. The algorithms were implemented to calculate some diagnostic parameters: total body water (TBW), fat-free body mass (FFM), fat mass and body fat percentage [1].

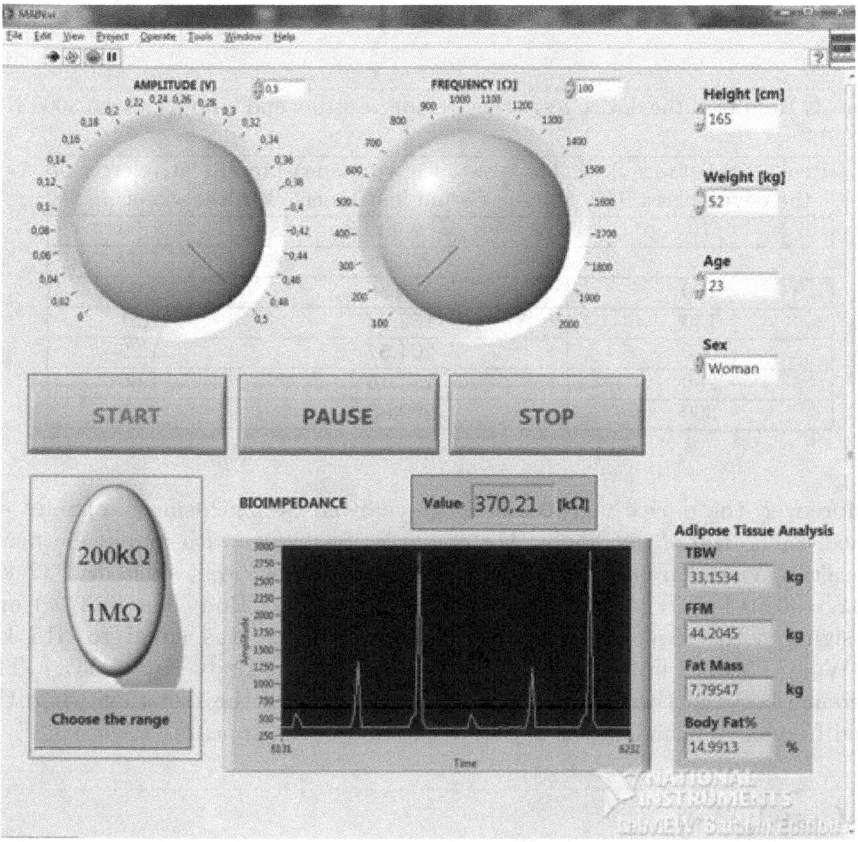

Fig. 5. User interface of the application dedicated to the bioimpedance measurements [10]

The specialized software was prepared in LabVIEW 2011 Student Edition and then compiled in LabVIEW 2010 Professional Development System. The installation file with LabVIEW Run-Time Engine was prepared so that the installation is simple and quick, application costs are low and less computer memory is needed [11].

4 Results

Some tests of the configured device were made. Due to the fact that every laptop has different internal parameters, it was necessary to calibrate the meter every time when a laptop was used for a very first time. Firstly, the meter was examined using reference decade resistance. The measurement error was lower than 2% and it is an acceptable result. No matter what signal frequency was used, the results were the same. It is because the pure resistance does not depend on the frequency.

Table 1. Results of the device test using decade resistors and laptop Lenovo 3000 N20

Real set resistance [kΩ], the device class: 0,05	The resistance value measured by the virtual instrument [kΩ]	Relative error of the meter [in %]
10	9,90	1,00
50	50,81	1,62
100	101,9	1,90
150	152,27	1,51
200	204,57	1,29
250	254,15	1,66
300	299,85	0,05

Moreover, the device was tested on 23-year-old woman using electrodes attached to the top of her hands. For example, for signal with amplitude peak-to-peak 1,5V and frequency 500 Hz the bioimpedance was equal to 332 kΩ (TBW=36,30 kg, FFM=49,71 kg, Fat Mass=2,27 kg, Body Fat=4,37%) and for signal with frequency 10 kHz the bioimpedance was equal to 315 kΩ (TBW=37,95 kg, FFM=51,98 kg, Fat Mass=0,02 kg, Body Fat=0,03%). The difference between these two results is seen if view of the fact that when the signal frequency is higher, the capacitance component appears.

5 Conclusion

In the paper the bioimpedance meter designed to the obesity diagnosis was described. Regarding to the obtained results, it turned out that it is technically and economically possible to construct bioimpedance device using special software in LabVIEW and hardware including laptops and sound cards. It is planned to configure the device with a tablet or a smartphone instead of a laptop. Some changes in the application will also be made. The possibility to connect to the patients database and save the measurements results will be introduce. It will facilitate the physicians work.

There is an issue is with the frequency limits imposed by the sound card input and output. Usually, as it was show in section 2, bioimpedance measurements in dietetics are performed using high frequency signal, but in this solution it was impossible. The device can generate only some of the frequencies used in MF-BIA: 0, 1, 5 and 10 kHz. Due to this fact, the created device can be used in diagnosis of obesity only in a limited way.

Acknowledgments. The work described in this paper was supported by Ministry of Science and Higher Education in Republic of Poland (project no. M-4/84/2013/DS).

References

1. Lewitt, A., Mądro, E., Krupienicz, A.: Podstawy teoretyczne i zastosowania analizy impedancji bioelektrycznej (BIA). Endokrynologia Otyłość Zaburzenia Przemiany Materii 3, 79–84 (2007)
2. Kopelman, P.G.: Obesity as a medical problem. Nature 404, 635–643 (2000)
3. Marra, M., Pasanisi, F., Scalfi, L., Colicchio, P., Chelucci, M., Contaldo, F.: The prediction of basal metabolic rate in young adult, severely obese patients using single-frequency bioimpedance analysis. Acta Diabetologica 40, 139–141 (2003)
4. Grimmes, S., Martinsen, O.G.: Bioimpedance & bioelectricity Basics. Academic Press, Elsevier (2008)
5. Wtorek, J., Nowakowski, A., Pałko, T.: Pomiary bioelektoimpedancyjne. In: Nałęcz, N.: Biocybernetyka i Inżynieria Biomedyczna 2000, Biopomiary, tom. Akademicka Oficyna Wydawnicza EXIT, Warsaw (2001)
6. New imaging methods for the detection of cancer biomarkers, `http://nano.mdx.ac.uk/bioimpedance/`
7. Kyle, U., Bosaeus, I., De Lorenzo, A., Deurenberg, P., Elia, M., Gomez, J.M., et al.: Bioelectrical impedance analysis part I: review of principles and methods. Clinical Nutrition 23, 1226–1243 (2004)
8. International Society for Electrical Bio-impedance, `http://www.isebi.org`
9. Major-Gołuch, A., Miazgowski, T., Krzyanowska-winiarska, B., Safranow, K., Hajduk, A.: Porwnanie pomiarw tuszczu u modych zdrowych kobiet z prawidową masą ciaa za pomocą impedancji bioelektrycznej i densytometrii. Endokr Oty Zab Przem Mat. 6(4), 189–195 (2010)
10. Cichocka, M.: Projekt miernika bioimpedancji z wykorzystaniem środowiska LabVIEW. Thesis on University of Technology in Cracow (supervisor: Tutaj, J.) (2013)
11. National Instruments, `http://www.ni.com`
12. Tłaczała, W.: Środowisko LabVIEW w eksperymencie wspomaganym komputerowo, Wydawnictwo Naukowo-Techniczne, Warsaw (2002)

Swarm Intelligence Algorithms for Multi-level Image Thresholding

Andrzej Marciniak, Marek Kowal, Paweł Filipczuk, and Józef Korbicz

Institute of Control & Computation Engineering,
University of Zielona Góra,
Ogrodowa 3b, 65-246 Zielona Góra, Poland
{a.marciniak,m.kowal,p.filipczuk,j.korbicz}@issi.uz.zgora.pl

Abstract. The multi-level thresholding is one of the most important issues in image segmentation. It is a time consuming problem, i.e. finding appropriate threshold values could take exceptionally long computational time. In this paper we evaluate and compare three meta-heuristic techniques tackling to this problem: ant colony, fireflies and honey bee mating. The proposed methods are validated by illustrative examples. The results show that the investigated methods do not only provide good segmentation results but also their computational effort makes them very efficient approaches.

Keywords: Image segmentation, Multi-level thresholding, Swarm intelligence.

1 Introduction

Image thresholding is the key technique of image segmentation [15]. It is often applied to biomedical image analysis, document processing, inspection of materials, video processing, satellite imagery and many other areas [8,17,18,20]. In literature, a great variety of threshold evaluation methods have been already proposed [5,12,16]. Generally, they can be divided into two classes: non-parametric and parametric. These methods are usually employed to determine single threshold which is used to groups the pixels into two classes: objects and background. However, complex images with many different objects in scene need to be divided into several classes, so multi-level thresholding is required.

Bi-level thresholding can be easily extended to multiple one. Unfortunately, in that case searching for optimal set of thresholds is NP-hard problem, so classical methods are useful only when the number of thresholds is small. To overcome this problem swarm intelligence (SI) algorithms as searching strategies can be applied. Three promising algorithms, i.e., ant colony optimization (ACO), honey bee mating optimization (HBMO) and firefly algorithm (FA) were chosen to solve the problem [1,6,19]. In order to compare the effectiveness of the methods under investigation, set of test experiments was prepared. Results of the experiments indicate that all examined algorithms are efficient enough to the task of image multi-level thresholding.

J. Korbicz and M. Kowal (eds.), *Intelligent Systems in Technical and Medical Diagnostics*, Advances in Intelligent Systems and Computing 230,
DOI: 10.1007/978-3-642-39881-0_25, © Springer-Verlag Berlin Heidelberg 2014

The remainder of this paper is organized as follows. In Section 2, the problem of the multi-level thresholding is formulated and meta-heuristic optimization methods are described. Section 3 gives the description of the experiments and comparative study of the results obtained for implemented meta-heuristics techniques. Concluding remarks are given in Section 4.

2 Algorithms of Swarm Intelligence

According to Collins Dictionary, SI is "an artificial-intelligence approach to problem solving using algorithms based on the self-organized collective behavior of social insects". In recent literature, SI techniques were used in different applications such as clustering, image processing, market segmentation, and benchmark mathematical problems [3,4,7,9,11,13]. Solution strategies presented here, belongs to the class of meta-heuristics, which includes approximate algorithms used to obtain solutions good enough to NP-hard problems in a reasonable amount of computation time. Their task is to tackle the problem of multi-level image thresholding formulated as follows.

Given an image I having pixels with L gray levels, the problem is to find a set of thresholds $T^* = (t_1, t_2, \ldots, t_K)$ which optimizes (maximizes) an objective function

$$F(T) = \frac{\sigma_B^2}{\sigma_T^2} = \frac{\sum_{k=1}^{K+1} \omega_k (\mu_k - \mu_T)^2}{\sum_{i=1}^{L} (i - \mu_T)^2 p_i} \tag{1}$$

where

$$\mu_T = \sum_{i=1}^{L} i p_i, \quad \mu_k = \sum_{i=t_{k-1}}^{t_k} \frac{i p_i}{\omega_k}, \quad \omega_k = \sum_{i=t_{k-1}}^{t_k} p_i, \tag{2}$$

and $p = (p_1, p_2, \ldots, p_L)$ is normalized histogram of image I. This criterion was firstly proposed by Otsu for bi-level thresholding [12] and later extended for multi-level thresholding in [14].

2.1 Ant Colony Optimization

In the early 1990s, ACO was introduced by Marco Dorigo [6] as a novel nature-inspired meta-heuristic, which underlying idea was to use several constructive computational agents simulating the behavior of real ants. The paradigm is based on the observation made by ethologists about the pheromone trails laid by ants to communicate information regarding shortest paths to food. While an isolated ant moves practically at random (exploration force), an ant encountering a previously laid trail can detect it and decide with high probability to follow it and consequently reinforce the trail with its own pheromone (exploitation force). The more the ants follow a trail, the more attractive that trail becomes to be followed what results in a positive feedback loop, during which the probability of choosing a path increases with the number of ants that previously chose the same path. A high-level algorithmic skeleton of ACO meta-heuristic is given in Algorithm 1.

Algorithm 1. ACO algorithm

```
Initialize
while stop condition not met do
  Construct Ant Solutions
  Apply Local Search {optional}
  Update Pheromones
end while
```

After initialization of pheromone values, the main loop consists of three main steps. First, candidate solutions to the problem are constructed by m artificial ants (from scratch). Ants use probabilistic model (pheromone model) to generate solutions by assembling them from the set of solution components, specific to a given application. Once candidate solutions are constructed, these may be improved in an optional local search phase. Finally, the pheromone values are updated in a way that is deemed to bias future sampling toward search space regions containing high quality solutions. In the following, we give a more detailed description of ACO applied to the problem of multi-level image thresholding.

`Initialize`. At the start of algorithm, parameters are set, i.e. all pheromone values are initialized to a value τ_0 and stopping criterion is established.

`Construct Ant Solutions`. A set of m ants constructs candidate solutions to the problem. Each candidate solution is composed of solution components $c_i^j, (i = 1, \ldots, L-1)$, that are instantiations of binary random variables ($j \in \{0,1\}$). A solution component c_i^1 indicates that threshold $t_k, (k = 1, \ldots, K)$ is selected at i-th level of histogram, while c_i^0 indicates that it is not. Each solution component c_i^1 is associated with pheromone trail τ_i. A solution component c_i^1 is chosen with probability

$$p_i\left(c_i^1\right) = \frac{\tau_i}{\sum_j \tau_j}. \tag{3}$$

To meet the cardinality constraint of the number of thresholds (i.e. $\#(c_i^1) = K$), candidate solutions are constructed as follows. First, the probabilities $p_i\left(c_i\right), (i = 1, \ldots, L-1)$ are calculated for all components c_i^1. Then, pseudo-random numbers $Y_i \in (0,1), (i = 1, \ldots, L-1)$ corresponding to solution components are drawn from the standard uniform distribution. Finally, the random variable $Z_i = p_i - Y_i$ is calculated and K components corresponding to maximum realizations of Z_i are chosen.

Once all candidate solutions are constructed, they are evaluated according to the objective function. The value of best solution T_{bs} is stored in memory as best-so-far solution. T_{bs} can be updated in the subsequent iterations with the value of the solution obtained as current iteration best solution, if it has a better objective function value.

`Apply Local Search`. Once the best-so-far solution T_{bs} is established, it is further improved by applying local search based on the single-flip neighborhood. In the single-flip neighborhood a solution $T = (t_1, t_2, \ldots, t_K)$ is a neighbor of T' if they are different in exactly one variable t. In addition, we assumed that the variable realizations in neighboring solutions must differ by a value of 1 . This means that in case of M-level thresholding, $2M$ neighboring solution to T_{bs} is evaluated.

`Update Pheromone Trails`. The pheromone update rule is based on elitist strategy, which is popular ACO-variant and consists of two parts. First, a pheromone evaporation, which uniformly decreases all the pheromone values is performed. Second, the best-so-far solution from the current or earlier iteration is used to increase the values of pheromone deposit on solution components that are parts of best solution. The update rule is given by:

$$\tau_{ij} \leftarrow (1 - rho)\tau_{ij} + wF(T_{bs}) \tag{4}$$

where $\rho \in (0, 1]$ is a parameter called evaporation rate, F stands for quality function and w is weighting parameter.

2.2 Honey Bee Mating Optimization

HBMO algorithm belongs to the general class of SI methods that models the behaviors of social insects. It is inspired by the marriage behavior of honey-bees [1]. In order to define the artificial model of bee mating process it is necessary to present the real mating-flight in honey-bees. Honey bee colony consists of queens, drones and workers. Queens are responsible for colony reproduction so they are specialized in egg-laying. If the colony contains only one queen, it is called a monogynous otherwise a polygynous. In this work only monogynous colonies are considered. Drones are male individuals in the colony. Their task is limited to propagate one of their mothers genome without alteration of their genetic composition. Drones enable females to act genetically as males during the reproduction. Workers bees are responsible for brood care and sometimes lay eggs. Broods arise either from fertilized (potential queens or workers) or unfertilized eggs (drones). A mating process starts with a dance performed by the queen who then flight far from the nest. Drones follow the queen and seven to twenty drones can mates with her. In each mating, selected drones transfer their sperm to queen's spermatheca. When the queen lays eggs, she fertilize some eggs by injecting at random a mixture of the sperms stored in the spermatheca.

In order to develop the algorithm, some simplifications was introduced to the model of mating procedure. Population of drones $D = \{D_1, D_2, \ldots, D_{n_d}\}$ is generated once at the beginning of the algorithm and does not change during the algorithm. In contrast to the original HBMO algorithm it was decided to ignore the brood care role of workers during reproduction procedure. The mating flight starts with the queen Q initialized with some speed $S(0)$. The flight is continued until the spermatheca of capacity n_s is full or the speed drops below given threshold S_{min} and then the queen returns to the nest. After each mating iteration, the queen's speed is reduced according to the following formulae:

$$S(k+1) = \alpha S(k), \tag{5}$$

where $\alpha \in [0, 1]$ is the decreasing factor. Subsequent drones involved in the mating process are chosen randomly with uniform distribution from the set D.

Algorithm 2. HBMO algorithm

```
Initialize randomly drones D and queen Q
while stop condition not met do
  while speed is above the threshold AND spermatheca is not full do
    Select randomly drone D_rand from set D
    Compute the probability of mating for drone D_rand
    if the drone D_rand passes the probabilistic condition then
      Add its sperm to the queens spermatheca
    end if
    Decrease the speed of queen
  end while
  for i = 1 : number of broods do
    Select randomly sperm P_rand from spermatheca
    Generate brood B_i based on P_rand and Q
  end for
  for j = 1 : number of mutations do
    Select randomly brood B_rand
    Apply mutation to B_rand
  end for
  Find the brood B_best with the highest value of objective function
  if objective function of B_best > objective function of queen Q then
    Replace current queen Q with B_best
  end if
end while
```

The probability of adding the sperm of selected drone D_{rand} to the spermatheca of queen Q (successful mating) is given by annealing function:

$$P(Q, D_{rand}) = e^{\frac{-|f(Q)-f(D_{rand})|}{S(k)}}, \tag{6}$$

where $f(Q)$ and $f(D_{rand})$ are the objective function values of the queen and the randomly chosen drone respectively. When the mating flight is over the queen starts laying n_c eggs and fertilize them by randomly selecting with uniform distribution a sperm from the queen's spermatheca. The brood B_j is generated by modifying the queen's genome using the following equation:

$$B_j = Q + \beta(P_{rand} - Q), \tag{7}$$

where $B_j = [B_j^1, B_j^2, \ldots, B_j^c]$ is the j-th brood, P_{rand} is the randomly selected drone's genome from the spermatheca, $\beta \in [0, 1]$ is the randomly generated value with uniform distribution during each breeding process and c is the size of the genome. In the next step, n_m randomly selected broods with uniform distribution are affected by the mutation defined by the following formulae:

$$B_{rand}^k = B_{rand}^k \pm (\delta - \epsilon) B_{rand}^k, \tag{8}$$

where k and $\delta \in [0, 1]$ are randomly chosen with uniform distribution, $\epsilon \in [0, 1]$ is predefined and the $+$ or $-$ sign occurs with equal probability. Finally the brood

B_{best} with the highest value of objective function is selected. It replaces current queen if the objective function of B_{best} is higher then the objective function of the queen Q. The pseudo code of main steps of HBMO is given in Algorithm 2.

2.3 Firefly Algorithm

FA is an optimization technique developed in 2007 by Yang [19]. The algorithm is inspired by the flashing behavior of fireflies. Each firefly interacts with each other with certain strength. The attractiveness depends on the light intensity, which is associated with the objective function, and the distance between the fireflies. It is assumed that there is no sexual distinction between fireflies. Furthermore, less bright firefly will always be attracted by brighter and the brightest one will be moved randomly.

The light intensity $I(r)$ varies according to the inverse square law

$$I(r) = \frac{I_s}{r^2} , \tag{9}$$

where I_s is the intensity at the source and r is the distance between two fireflies. Light is absorbed in the media with an absorbtion coefficient γ. The combination of the inverse square law and the absorbtion can be approximated to avoid singularity at $r = 0$ in I_s/r^2 by

$$I(r) = I_0 e^{-\gamma r^2} , \tag{10}$$

where I_0 is the original light intensity. As the firefly attractiveness is proportional to the brightness seen by its neighbor, the attractiveness function is determined by

$$\beta(r) = \beta_0 e^{-\gamma r^2} , \tag{11}$$

where β_0 is the attractiveness at $r = 0$. To determine the distance affecting the attractiveness between any two fireflies i and j at positions x_i and x_j Euclidean measure is used

$$r_{ij} = \|x_i - x_j\| . \tag{12}$$

In each iteration less attractive fireflies move to the brighter ones. The movement of the firefly i at location x_i to firefly j at location x_j is expressed by

$$x_i(t+1) = x_i(t) + \beta_0 e^{-\gamma r}(x_j - x_i(t)) + \alpha \epsilon_i \tag{13}$$

where α is the randomization parameter, and ϵ_i is a vector of random numbers. In this work the vector ϵ_i is determined using a random number generator uniformly distributed in $[-0.5, 0.5]$.

The pseudo code of main steps of PA is given in Algorithm 3.

Algorithm 3. FA algorithm

```
Initialize with random positions and evaluate light intensities
while t < MaxGeneration do
  for i = 1 : number of fireflies do
    for j = 1 : number of fireflies do
      if I_i < I_j then
        Move firefly i towards j
      end if
      Update light intensities
    end for
  end for
  Find the firefly with the highest light intensity
end while
```

3 Experimental Results

In this section, we will evaluate the performance of the multi-level thresholding methods presented in the previous section. Bearing in mind that all investigated methods are stochastic global optimization algorithms, we applied experimental methodology designed for numerical evaluation of stochastic algorithms proposed in [2]. To fairly compare and present evaluation study, the following experimental procedure was performed.

1. Assemble a set of 5 test images and their histograms.
2. Test selected thresholding algorithms on all test problems.
3. Perform 30 replications for each test histogram and each algorithm.
4. Stop each algorithm after a fixed number of function evaluations for each replication.
5. Record only the function evaluations that obtain improving objective function values (best-so-far).
6. Perform graphical and statistical analysis on the data.

Since presented algorithms associate different meanings and computational effort with iteration, the number of function evaluations is used instead of it. To directly compare the performance of algorithms, we keep the computational effort constant, i.e. the length of each run is set to 1000 function evaluations. The parameters of algorithms have been adjusted in the process of preliminary testing and their values are as follows:

ACO population of ants = 10, $w = 2$, $\rho = 0.01$, $\tau_0 = 0.5$,
HBMO $n_d = 60$, $n_s = 30$, $n_c = 20$, $n_m = 1$, $\alpha = 0.98$, $\epsilon = 0.5$, $S(0) = 1$, $S_{min} = 0.05$,
FA population of fireflies = 10, $\gamma = 0.0001$, $\alpha = 10$.

Objective function values for the best found solutions and their median over 30 runs for all compared algorithms are presented in Table 1.

Table 1. The median of objective function values and the best solution found (in brackets) for 30 runs for all presented algorithms and all testing images.

Image	ACO	HBMO	FA
2 thresholds			
1	0.8708 (0.8708)	0.8708 (0.8708)	0.8708 (0.8708)
2	0.8625 (0.8625)	0.8625 (0.8625)	0.8625 (0.8625)
3	0.9331 (0.9331)	0.9331 (0.9331)	0.9331 (0.9331)
4	0.9067 (0.9067)	0.9067 (0.9067)	0.9067 (0.9067)
5	0.9305 (0.9305)	0.9305 (0.9305)	0.9305 (0.9305)
3 thresholds			
1	0.9298 (0.9298)	0.9298 (0.9298)	0.9298 (0.9298)
2	0.9184 (0.9184)	0.9184 (0.9184)	0.9184 (0.9184)
3	0.9622 (0.9622)	0.9622 (0.9622)	0.9622 (0.9622)
4	0.9483 (0.9483)	0.9483 (0.9483)	0.9483 (0.9483)
5	0.9847 (0.9847)	0.9847 (0.9847)	0.9847 (0.9847)
4 thresholds			
1	0.9561 (0.9561)	0.9560 (0.9561)	0.9561 (0.9561)
2	0.9505 (0.9505)	0.9504 (0.9505)	0.9505 (0.9505)
3	0.9736 (0.9737)	0.9735 (0.9737)	0.9736 (0.9737)
4	0.9672 (0.9672)	0.9672 (0.9672)	0.9672 (0.9672)
5	0.9898 (0.9898)	0.9898 (0.9898)	0.9898 (0.9898)
5 thresholds			
1	0.9695 (0.9695)	0.9691 (0.9695)	0.9694 (0.9695)
2	0.9615 (0.9615)	0.9611 (0.9615)	0.9614 (0.9615)
3	0.9817 (0.9818)	0.9811 (0.9817)	0.9817 (0.9818)
4	0.9773 (0.9773)	0.9773 (0.9773)	0.9773 (0.9773)
5	0.9940 (0.9940)	0.9940 (0.9940)	0.9920 (0.9940)
6 thresholds			
1	0.9765 (0.9768)	0.9758 (0.9767)	0.9766 (0.9767)
2	0.9713 (0.9716)	0.9714 (0.9716)	0.9711 (0.9716)
3	0.9861 (0.9862)	0.9828 (0.9862)	0.9861 (0.9862)
4	0.9832 (0.9834)	0.9834 (0.9834)	0.9833 (0.9834)
5	0.9962 (0.9962)	0.9962 (0.9962)	0.9945 (0.9962)
7 thresholds			
1	0.9817 (0.9822)	0.9813 (0.9820)	0.9819 (0.9821)
2	0.9785 (0.9804)	0.9799 (0.9803)	0.9802 (0.9803)
3	0.9887 (0.9894)	0.9863 (0.9891)	0.9863 (0.9893)
4	0.9870 (0.9873)	0.9873 (0.9873)	0.9872 (0.9873)
5	0.9971 (0.9971)	0.9971 (0.9971)	0.9966 (0.9971)

To capture the progression of the algorithms in terms of the improvements of the objective function values and the variation of these values in all replications, we constructed boxplot diagrams, examples of which are presented in Fig. 1. Note that the scale of the horizontal axis is not equal to length of

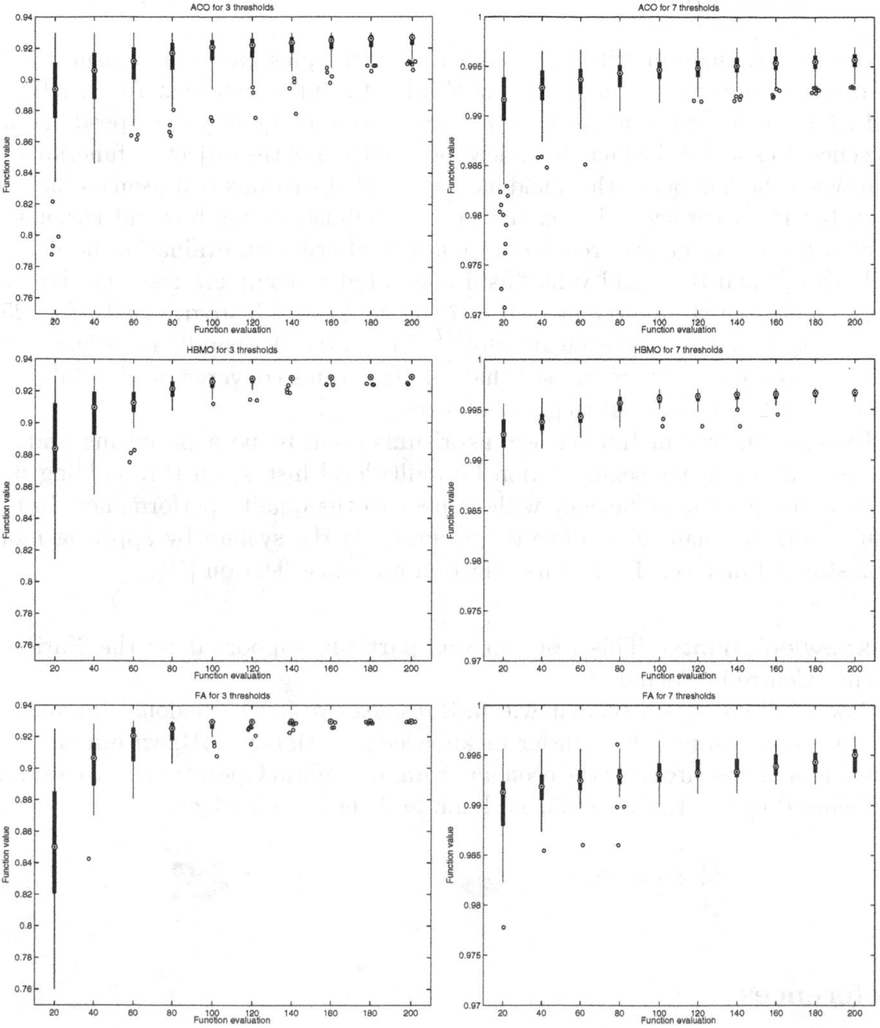

Fig. 1. Performance of investigated algorithms estimated over 30 independent runs. Boxplots show statistics of the best-so-far solution found over evaluations. The x-axis shows evaluations, the y-axis shows the boxplot of the best value of thresholding criterion calculated every 20 evaluations. Black dot inside a white circle denotes median, symbol 'o' denotes outlier.

runs - it is constrained to the first 200 evaluations in order to visualize differences in behavior of algorithms. The edges of the black boxes are the first and third quartiles (respectively q_1 and q_3), while the whiskers extend to the most extreme data points not considered outliers. Points are drawn as outliers if they are larger than $q_3 + 1.5(q_3 - q_1)$ or smaller than $q_1 - 1.5(q_3 - q_1)$.

4 Conclusions

In this paper, three multilevel thresholding techniques based on swarm intelligence algorithms were examined. The obtained results show that ACO, HBMO and FA are comparable algorithms in term of solution quality and speed of convergence. In spite of the fact that a few local optima of the objective function appear when the number of threshold increases, all algorithms still seems to be efficient, both in accuracy and robustness (what is illustrated in box-whisker plots). SI algorithms can greatly reduce the number of criterion evaluation needed to find the optimal threshold values as if compared to complete research. For example, an exhaustive searching of $K = 7$ thresholds in a histogram with $L = 256$ gray levels would require evaluation of $\binom{L-1}{K} \approx 1.2 \cdot 10^{13}$ candidate solutions to find the optimal one. One can see that all algorithms converge quickly (already after about 200 iterations) to good solutions.

To sum up, Swarm Intelligence algorithms seem to be a promising and effective tool for image segmentation by multi-level histogram thresholding due to their computation efficiency with respect to the quality performance. In the future work we plan to improve the accuracy of the system by applying more sophisticated methods for feature selection and classification [10].

Acknowledgments. This research was partially supported by the National Science Centre in Poland.

Paweł Filipczuk is a scholar within Sub-measure 8.2.2: Regional Innovation Strategies, Measure 8.2: Transfer of knowledge, Priority VIII: within the Regional human resources for the economy Human Capital Operational Programme co-financed by the European Social Fund and the state budget.

HUMAN CAPITAL
HUMAN – BEST INVESTMENT!
Lubuskie
Worth your while
EUROPEAN UNION
EUROPEAN
SOCIAL FUND

References

1. Abbasss, H.A.: Marriage in honey bee optimization (hbo): a haplometrosis polygynous swarming approach. In: Proc. Cong. on Evolutionary Computation, pp. 207–214 (2001)
2. Montaz, A.M., Khompatraporn, C., Zabinsky, Z.B.: A numerical evaluation of several stochastic algorithms on selected continuous global optimization test problems. Journal of Global Optimization 31(4), 635–672 (2005)
3. Amiri, B., Fathian, M.: Integration of self organizing feature maps and honey bee mating optimization algorithm for market segmentation. J. of Theoretical and Applied Information Technology, 70–86 (2007)
4. Karaboga, D., Basturk, B.: On the performance of artificial bee colony algorithm. Applied Soft Computing 8, 687–697 (2008)
5. Kapur, J.N., Sahoo, P.K., Wong, A.K.C.: A new method for gray-level picture thresholding using the entropy of the histogram. Computer Vision Graphics Image Processing 29, 273–285 (1985)

6. Dorigo, M., Stuetzle, T.: Ant colony optimization: Overview and recent advances. In: Gendreau, M., Potvin, J.-Y. (eds.) Handbook of Metaheuristics. International Series in Operations Research & Management Science, vol. 146, pp. 227–263. Springer, US (2010)
7. Fathian, M., Amiri, B., Maroosi, A.: Application of honey bee mating optimization algorithm on clustering. Applied Mathematics and Computation 190, 1502–1513 (2007)
8. Sezgin, M., Sankur, B.: Comparison of thresholding methods for non-destructive testing applications. In: Proc. IEEE Int. Conf. Image Process, pp. 764–767 (2001)
9. Soltani, M., Chaari, A., Hmida, F.B.: A novel fuzzy c-regression model algorithm using a new error measure and particle swarm optimization. Int. J. Appl. Math and Comp. Sci. 22(3), 617–628 (2012)
10. Woźniak, M., Krawczyk, B.: Combined Classifier Based on Feature Space Partitioning. Int. J. Appl. Math and Comp. Sci. 22(4), 855–866 (2012)
11. Horng, M.-H.: A multilevel image thresholding using the honey bee mating optimization. Applied Mathematics and Computation 215, 3302–3310 (2010)
12. Otsu, N.: A threshold selection method from gray-level histograms. IEEE Trans. Sys. Man. and Cyber. 9, 62–66 (1979)
13. Filipczuk, P., Wojtak, W., Obuchowicz, A.: Automatic nuclei detection on cytological images using the firefly optimization algorithm. In: Piętka, E., Kawa, J. (eds.) ITIB 2012. LNCS, vol. 7339, pp. 85–92. Springer, Heidelberg (2012)
14. Liao, P.S., Chen, T.S., Chung, P.C.: A fast algorithm for multilevel thresholding. J. Inf. Sci. Eng. 17(5), 713–727 (2001)
15. Gonzalez, R.C., Woods, R.E.: Digital Image Processing. Prentice Hall, New Jersey (2001)
16. Sezgin, M., Sankur, B.: Survey over image thresholding techniques and quantitative performance evaluation. J. Electronic Imaging 13(1), 146–165 (2004)
17. Kryjak, T., Gorgoń, M.: Parallel implementation of local thresholding in mitron-c. Int. J. Appl. Math and Comp. Sci. 20(3), 571–580 (2010)
18. Diyana, W.M., Zaki, W., Faizal, M., Fauzi, A., Besar, R., Munirah, W.S.H., Ahmad, W.: Multi-level segmentation method for serial computed tomography brain images. In: Proc. IEEE Int. Conf. on Signal and Image Processing Applications, pp. 107–112 (2009)
19. Yang, X.S.: Nature-Inspired Metaheuristic Algorithms, 2nd edn. Luniver Press, Frome (2010)
20. Li, Z., Liu, C., Liu, G., Yang, X., Cheng, Y.: Statistical thresholding method for infrared images. Pattern Anal. Applic. 14, 109–126 (2011)

Multimedia Database Techniques for Medical Diagnosis Processes Support

Liliana Byczkowska-Lipińska and Agnieszka Wosiak

Institute of Information Technology, Lodz University of Technology,
ul. Wolczanska 215, 90-924 Lodz, Poland
{liliana.byczkowska-lipinska,agnieszka.wosiak}@p.lodz.pl

Abstract. Technological capabilities in the field of medical imaging contribute to the increasing use of image analysis in diagnostic medical systems. Medical imaging is one of the key sources of information for health professionals. It is largely due to the fact that the accuracy of the conclusions drawn by the physicians from this form of presentation of the data is very large in comparison with the numerical data. The characteristics of multimedia content requires suitable database solutions which are issues of multimedia databases techniques.

This paper presents the methods for managing and analyzing medical images stored in the multimedia database in a standard DICOM format. The research on the capabilities of the system was carried out for the diagnosis of juvenile idiopathic arthritis. The work has shown the usefulness of multimedia databases' techniques in the study area of medical diagnostics.

Keywords: multimedia databases, medical diagnosis, DICOM, medical imaging.

1 Introduction

Technological capabilities in the field of medical imaging contribute to the increasing use of image analysis in diagnostic medical systems. Medical imaging is derived from a number of tomography studies, primary including radiography (RTG), ultrasonography (USG), computed tomography (CT) and magnetic resonance imaging (MRI). The methods of collecting and storing medical images can be performed using almost any database management system. However, the analysis of this type of information is a complex issue and requires advanced information technologies. The appropriate management of medical image is related to the issues concerning database design and specificity of multimedia data [4,6,15].

The process of designing the system for medical diagnostics of imaging data, includes, inter alia, the need of protecting privacy while sharing and accessing the medical images, the protection against data loss and achieving the best system performance, that means the quick access to local, remote and archived medical images. These issues concern nearly every database system, but in the case of

J. Korbicz and M. Kowal (eds.), *Intelligent Systems in Technical and Medical Diagnostics*, Advances in Intelligent Systems and Computing 230,
DOI: 10.1007/978-3-642-39881-0_26,

medical systems combined with the specificity of multimedia data, they are of particular importance.

The capabilities of implementing complex medical systems, that include the specificity of medical imaging, can be significantly increased through the use of multimedia databases. The comparative analysis of medical images improves the patient's diagnostic process by identifying relevant cases or those significantly deviate from the norm [1,9].

Nowadays medical imaging studies are used in a very wide range. They are key elements in the diagnosis of atrophic and degenerative changes of the osteoarticular system and give an accurate picture of the position of a tumor changes. Medical imaging, in particular the study using magnetic resonance imaging, is specifically useful in detection of multiple sclerosis, vascular disorders and Alzheimer's disease [2,17]. Ultrasound and X-rays are the main kinds of techniques for assessing the occurrence of juvenile idiopathic arthritis, which is the subject of this paper.

2 Multimedia Database Systems

2.1 Characteristics of Multimedia Databases

The popularity of multimedia techniques in various areas of life, especially in medical imaging and diagnostic systems, has resulted in the development of multimedia databases. The solutions offered by the multimedia database systems can facilitate the implementation of the module that supports medical imaging. Currently, the abilities to store and manage multimedia data through relational database management systems are precisely defined in the SQL/MM standard (SQL Multimedia and Application Packages) awarded by ISO (International Organization for Standardization) and IEC (International Electrotechnical Commission) [10]. Since 1999 the SQL/MM standard has been expanded and updated. Now it consists of six parts: Framework, Full-Text, Spatial, Still Image, Data Mining and Metadata Registries. The standard does not define methods for handling audio and video data. The SQL/MM standard has not been implemented in most of relational database management systems, except of Oracle Database, which is described later in this paper.

Oracle Database has implemented SQL/MM Still Image type, which is recommended in the standard. Besides it offers its own multimedia types:

- ORDAudio used for audio storage and processing,
- ORDVideo designed to store and manipulate video objects,
- ORDImage used to store and manipulate images,
- ORDDoc for storing and managing heterogeneous multimedia objects.

Storing medical images in a database handled by a particular database management system allows using many other mechanisms provided by the database management system, implementation of which in the external application could be difficult and time consuming. First of all, these are security and audit issues.

Database solutions allow controlling user privileges at many levels, starting at defining the permissions for the entire database down to particular rows and columns of each table. Moreover, these steps allow full monitoring of system's users by generating log files with all the events recorded in the system.

2.2 Metadata for Medical Imaging

The process of searching and comparing multimedia objects requires obtaining information on the content stored in the graphical object. Linking images with their metadata is an effective way to manage these images. The amount and type of metadata depends on the type of graphical objects that are considered. There are the following types of metadata:

- external metadata,
- semantic metadata,
- internal metadata (also called signal metadata).

External metadata include attributes of multimedia object (file type, type of compression, encoding scheme, size) and attributes describing the subject (author, date of creation). They are usually stored as strings of characters. The implementation of a homogeneous format of external metadata provides quick access to basic information on the graphical objects. The standard for medical images has been implemented and its characteristics is presented later in this paper.

Semantic metadata give information about objects, characters, events, that are represented by the multimedia content. These include low-level metadata, which result directly from the construction of the resource and high-level metadata, that are driven through the necessary domain knowledge. The basic and easiest method of obtaining the semantic metadata is implemented by inputting comments added to the images as keywords. These comments are stored in a database, regardless of the place of where the images are stored (inside the database or outside of the database). Semantic metadata are added manually in most of the systems, though there are some specialized systems that make it possible to automatically generate annotations [5,16]. Inputting semantic metadata is time consuming and often difficult. Moreover, in some systems, the comments stored as keywords would not be necessary if there were other possibilities of marking images.

Internal metadata are defined by properties of the signal characteristics, which are extracted from the multimedia content. Generating this type of metadata is realized automatically, but it is the most complex solution and it usually requires pattern recognition techniques and artificial intelligence methods. Moreover, queries on the internal metadata require algorithms of extraction and comparison for signal characteristics. Solutions involving the internal metadata can support (or completely eliminate) the need to use keywords [8,11,12].

The constant increase in the number of devices that generate diagnostic images caused the need to standardize medical format of images. DICOM standard

(Digital Imaging and Communications in Medicine) awarded by the National Electrical Manufacturers Association (NEMA) defines methods for storing bits of data for the picture, as well as about 2,000 markers defining the metadata associated with the image and additional information (e.g. patient data, case study) [13].

Extracting metadata from DICOM objects stored as ORDDicom type is performed by implemented methods. The extracted metadata are available as an XML format. Extraction of metadata may relate to their entire set or only a subset of the most commonly used attributes, resulting in improved search and indexing performance. The most commonly used DICOM metadata include:

- MODALITY - the type of device from which the study is being taken,
- STUDY INSTANCE UID - a unique instance identifier,
- PATIENT ID - the patient's identifier,
- PATIENT'S NAME - the name of the patient,
- PATIENT'S BIRTH DATE - date of birth of the patient,
- BODY PART EXAMINED - studied part of the body,
- VIEW POSITION - the alignment of the body,
- STUDY DATE - date of the study,
- REFERRING PHYSICIAN'S NAME - name of the physician ordering the medical examination,
- PERFORMING PHYSICIAN'S NAME - the name of the attending physician.

The acquisition of these data directly from the imaging study enables binding them automatically with the results of other studies stored in the database.

2.3 Managing Medical Images in Multimedia Databases

Database management system used in diagnostics medical systems based on medical images can significantly improve the process of analysis. Oracle Multimedia tools offers built-in types and methods associated with DICOM objects. The data in DICOM standard are stored in a table whose columns consist of thumbnail image in JPEG format and XML documents for the corresponding metadata. There are five main parts of Oracle Database architecture that enable handling medical images in DICOM format [14]:

- a repository implemented for the DICOM data model that is a part of the Java Virtual Machine (JVM),
- DICOM parser,
- the XML encoder,
- compliance validator,
- and images' converter.

Metadata of XML document are extracted from the DICOM content. Mapping of the DICOM content is based on a defined schema called a mapping document. DICOM content encoded in binary format can be stored directly in the database as an object of BLOB or BFILE type. Data can also be loaded into memory by the parser, and then mapped to XML based on attribute definitions in the mapping document, stored in a repository data model.

The objects in the DICOM standard are stored in suitable columns in tables. These columns are defined as ORDDicom type [7]. The representation of the object of this type consists of the metadata stored in XML format, the original DICOM image content and main and additional attributes of the image. Other Oracle Database Server mechanisms can be used through encoding metadata of DICOM images in the appropriate XML type and columns of attributes. It includes defining of B-trees indexes and XPath queries.

Metadata for graphics files can be saved in binary formats (IPTC - Information Interchange Model and EXIF - Exchangeable Image File Format), or in XML format (XMP - eXtensible Metadata Platform). Oracle Multimedia provides all types of metadata in the XML format based on XML schemas. The schemas define the structure of XML data, and (indirectly) the structure of object - relational type of storage in a database. These schemes are implemented during the installation of Oracle Multimedia tools. The access to metadata is realized by referring to XML elements using XPath format. This makes changes in the schema possible without affecting other elements of an implementation.

Image metadata are extracted by the getMetadata() method. This method returns the XMLSequenceType of available metadata type. The access to the image metadata is always obtained, but other metadata are dependent on the source file format.

The process of extracting metadata from DICOM objects stored in the ORDDicom type is performed by the determined method. Extracted metadata are available in XML format. The following code of PL/SQL stored procedure explains how to obtain the metadata from the DICOM object:

```
select dicom_result into dicom_var
from IMAGING_RESULTS
where id_o = 1;
metadata_var := dicom_var.extractMetadata();
```

Data searching can be performed by the metadata contained in the DICOM object. The various attributes of the object may be obtained using methods getAttributeByName() and getAttributeByTag(). The implementation of the methods is shown as follows:

```
select dicom_result into dicom_var
from IMAGING_RESULTS
where id_o = inID;
result_var := dicom_var.getAttributeByName('Patient ID');
```

Multimedia databases offer the ability to compare internal metadata of images, which is extremely important in the medical analysis of individual cases. While converting the image content contained in the ORDDicom object type to the SI_StillImage object, it is possible to access the internal attributes of the image: the average image color, color histogram and the distribution of colors in the image. The representations of individual features can be described by the following equations:

$$AvgCol = \left(\frac{\sum_{j=1}^{N} R_j}{N}, \frac{\sum_{j=1}^{N} G_j}{N}, \frac{\sum_{j=1}^{N} B_j}{N} \right) \tag{1}$$

$$ColHist = \left\{ \begin{matrix} (I_j, P_j) \, | I_j \in ColorValue, 0 \leq P_j \leq 1, \\ \sum_{0 \leq j \leq 1} P_j = 1, 0 \leq j \leq 1 \end{matrix} \right\} \tag{2}$$

$$PosColor = \{(j, AvgCol_j) | 0 \leq j \leq 1\} \tag{3}$$

where: R - red component value for the j sample, G - green component value for the j sample, B - blue component value for the j sample, $AvgCol$ - the average color for the j sample, N - number of samples.

The similarity measure between the vector F of image features and the query vector Q is a weighted metric D determined as:

$$D = (F - Q)^T \cdot A \cdot (F - Q) \tag{4}$$

where: $(F - Q)^T$ - transpose of a matrix *(F-Q)*, A - the matrix of weight factors.

The feature vector F and the query vector Q may include either color average, color histogram or positional color. It may also include all of the features. The order of the features does not affect the comparison result, as well as various resolutions of the pictures and differences in number for bin in color histograms [3].

The similarity coefficient between the F and Q images for a list of attributes can be written as:

$$D_F = \frac{\sum_{i=1}^{n} f_i SI\_SCORE(Q) \cdot W_i}{\sum_{i=1}^{n} W_i} \tag{5}$$

where: D_F - the value of similarity coefficient between F and Q, f_i - the value of the ith-feature of images,*SI_SCORE* - the name of the method for calculating the similarity value, W_i - weight factor of ith-feature.

The implementation of the method for the list of features is shown as follows:

```
select *
from
IMAGING_RESULTS imgres1, IMAGING_RESULTS imgres2
where
imgres2.nazwa_pliku = 'image01.dcm'
and imgres1.id_o <> imgres2.id_o
and si_featurelist(
si_averagecolor(imgres2.image_m), 0.2,
si_texture(imgres2.image_m), 0.3,
si_positionalcolor(imgres2.image_m),0.5
).si_score(imgres1.image_m) < 10;
```

The value of the similarity factor of the image to the master list of properties is defined by the weighted average for all the attributes. The specificity of medical images requires modification of the weight factors. There are the attributes and methods defined for reading and setting the values of individual properties. The radiological studies are mainly the grayscale images and the best results are obtained with comparing positional color only or increasing weight factor for the component related to the position of the color in the image.

3 Experimental Studies

The module of a diagnostic medical system was implemented for the purpose of this study. The research was conducted based on actual patient data, which were made anonymous before the analysis process. Data collected in the system database concern pediatric rheumatology, which is the field that will be the future application of the system. Database consisted of 40 children with juvenile idiopathic arthritis. Experimental studies were carried out using the Oracle Database for storing objects using ORDDicom and SI_StillImage types. The data obtained in the diagnosis of patients included:

- personal data stored as alphanumeric values,
- laboratory studies which are numeric or nominal values,
- imaging studies stored in DICOM format and descriptions related to these pictures,
- specialist consultations (usually rehabilitation and ophthalmology) stored as descriptions with additional possibility of the assessment of nominal scale.

The purpose of the system was to examine the relationship between the results of all types of patients' studies with particular emphasis on image data. The size of images depended on digital imaging modalities and for computed radiography was 2048 x 2048 pixels (4 megapixels). The data were subjects for statistical analysis, which included the evaluation of descriptive measures and the process of statistical inference. The following signal (internal) metadata were extracted from the medical images and taken into account during the analysis: the average color of the image, the color histogram, texture and color distribution in the image. Due to the characteristics of the study imaging (radiology studies), which were acquired as grayscale images, a special emphasis was placed on the distribution of colors in an image.

Medical support systems and diagnostic decision-making processes are usually built in three-tier architecture as the basic model shown in Fig. 1. The solution of this type implements all the application logic in a separate layer outside the database. The image processing and analysis outside the database requires:

- access to the database each time the query is performed,
- obtain the relevant results,
- send the relevant results to the outer layer,
- perform appropriate operations,
- send the results back to the database.

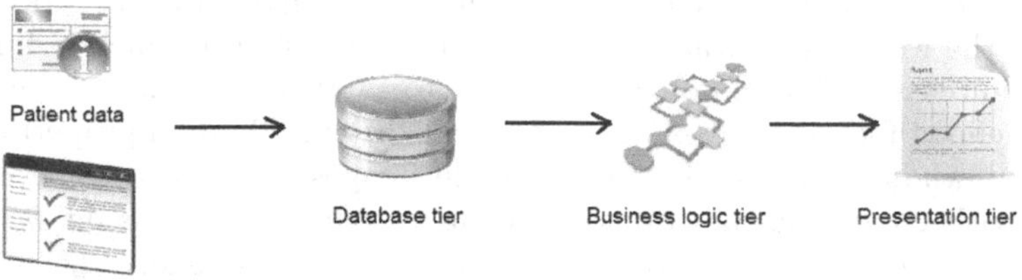

Fig. 1. The schema of a computer system diagnostic module in the three-tier architecture (Source: Own work)

The implementation of multimedia databases takes some logic related to the processing and analysis of images on the database server, as shown in Fig. 2.

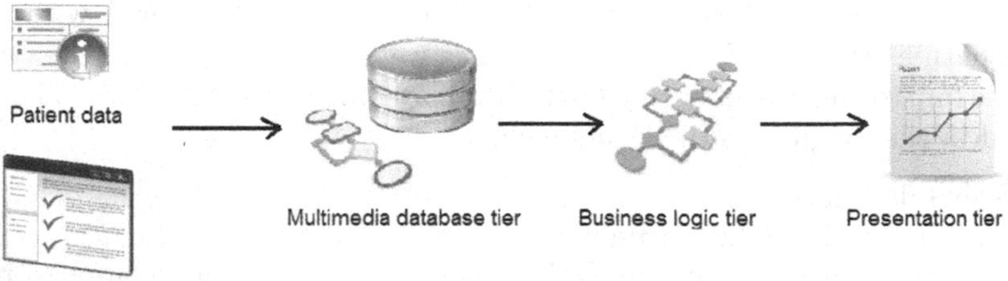

Fig. 2. The schema of a computer system diagnostic module including a multimedia database (Source: Own work)

This approach reduces the amount of time associated with the acquisition of data from the database and related to network bandwidth.

A group of patients with known juvenile idiopathic arthritis was examined. According to the carried analysis, the system returned known cases of the greatest relevance of the image data. The similarity of images was estimated using signal metadata: average color, color histogram and positional color. The results of the average image color and color histogram did not returned expected results. The system returned either most of the cases from the database if the similarity limit was high (score_limit=30 in a range of 0 to 100) or none of the cases if the similarity limit was low (score_limit= 10 in a range of 0 to 100). For positional color and the similarity limit set to 10, the system returned usually 1-10 images depending on the analyzed case. The positional color was estimated based on the equation number (3) using 36 segments (rectangles) of each image (N=36). The number of segments affects the analysis time and to reduce the negative impact of the increased number of segments on the time of analysis, data on the distribution of colors in each segment can be collected during the implementation of the database and stored in a row. These actions make no important charges during the input process and have no negative influence on database capacity and, therefore, they are definite advantages. The result of positional color comparison is shown in Fig. 3.

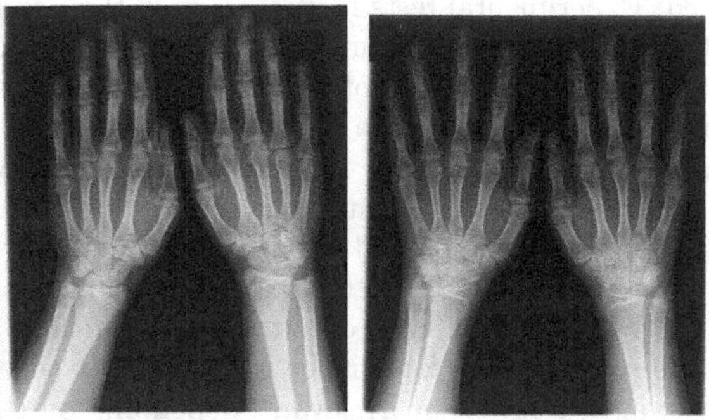

Fig. 3. The best comparison result of the data stored in the database (Source: Own work)

In addition, the research also focused on determining the relationship between the signal metadata extracted from the images' content and other patients' studies. Their relationship allowed to get results of particular interest, including the cases of the special type of juvenile idiopathic arthritis (oligoarticular, polyarticular or systemic) or the cases with ocular complications.

The results of the analysis showed that the inclusion of image signal metadata in the analysis process is warranted, and the method itself gives satisfactory results of the comparison. The solution works well when assessing the relationship between the data signal coming from images, and other data stored in the database, for example, the type of identified disease.

4 Conclusions

Databases that meet the expectations of medical diagnostics and medical imaging, require specific project assumptions and more resources than databases built for other purposes. Multimedia databases offer additional solutions for complex multimedia data and therefore can be widely used in medicine applications.

Modules responsible for data mining play an important role in medical diagnosis systems. The methods of data analysis, in particular statistical analysis, allow detecting regularities and dependencies that have not been noticed before and thus improving the quality of diagnosis. Carrying the process of medical data analysis, first it is necessary to identify the most important attributes of the assessment of the patient. Then, by applying the selected methods, the system should check the relationship between the values of different attributes and relate them to the classification of patients.

Most of the existing diagnostic systems allows you to perform statistical analysis on the basis of numerical data, without examining their relationship with the image data. In this research, we have examined the relationship between test results obtained in different forms. Correlations between the external metadata of DICOM objects and signal metadata extracted from multimedia content have been examined. Experimental tests have confirmed the possibility of using multimedia techniques in database systems for medical diagnosis.

The design of system envisaged use of multimedia databases' technologies, which initially had to be implemented in a single system's module for medical images evaluation. Further work on the project involve the construction of a module that will bind the multimedia databases with pattern recognition techniques implemented in part through the database server, and partly in the outer layer of the logical application tier. Then the evaluation of image data in terms of signal metadata may be the initial stage of automatic medical diagnosis, which allows narrowing the set of results for further processing by separating objects considered to be the most relevant or irrelevant, depending on the needs of the system's users.

References

1. Berner, E.S. (ed.): Clinical Decision Support Systems, Theory and Practice. Springer Science + Business Media, LLC (2007)
2. Byczkowska-Lipinska, L., Wosiak, A.: Image signal metadata analysis in diagnosis of Alzheimer's disease. Przeglad Elektrotechniczny (Electrical Review), R. 88 NR 12b, 148–150 (2012)
3. Chakravarti, R., Meng, X.: Study of Color Histogram Based Image Retrieval. In: 2009 Sixth International Conference on Information Technology: New Generations, ITNG 2009, pp. 1323–1328 (2009)
4. Curtise, K.C., Zhonghua, N., Sun, G.: Development of an online automatic computed radiography dose data mining program: a preliminary study. Computer Methods and Programs in Biomedicine 97, 48–52 (2009)
5. Dacheng, T., Dong, X., Xuelong, L.: Semantic Mining Technologies for Multimedia Databases. Information Science Reference/Idea Group Publishing (2009)
6. Furuie, S., Rebelo, M., Moreno, R., Santos, M., Bertozzo, N., Motta, G., Pires, F., Gutierrez, M.: Managing Medical Images and Clinical Information: InCors Experience. IEEE Transactions on Information Technology in Biomedicine 11(1) (2007)
7. Gettys, B., Mauro, J.: Oracle Database 10g Release 2 DICOM Medical Image Support, An Oracle White Paper (2005)
8. Gutierrez Martinez, J., Nunez Gaona, M.A., Aguirre Meneses, H., Delgado Esquerra, R.E.: Design and Implementation of a Medical Image Viewing System Based on Software Engineering at Instituto Nacional de Rehabilitacion. In: PAHCE Conference, Mexico (2009)
9. Hillestad, R., Bigelow, J., Bower, A., Girosi, F., Meili, R., Scoville, R., Taylor, R.: Can Electronic Medical Record Systems Transform Health Care? Potential Health Benefits, Savings, and Costs, Health Affairs 24(5), 1103–1117 (2005)
10. Information technology - Database languages - SQL Multimedia and Application Packages (2001)
11. Jaime-Castillo, S., Medina, J.M., Sanchez, D.: A System to Perform CBIR on X-Ray Images Using Soft Computing Techniques, FUZZ-IEEE, Korea (2009)
12. Marcosa, E., Acuna, C.J., Vela, B., Caveroa, J.M., Hernandez, J.A.: A database for medical image management. Computer Methods and Programs in Biomedicine 86, 255–269 (2007)
13. National Electrical Manufacturers Association: Digital Imaging and Communications in Medicine (DICOM) (2009)

14. Oracle®Multimedia DICOM Developer's Guide 11g Release 2 (11.2), Oracle Database Documentation Library
15. Prevedello, L., Andirole, K., Hanson, R., Kelly, P., Khorasani, R.: Bussiness Intelligence Tools for Radiology: Creating a Prototype Model Using Open Source Tools. Journal of Digital Imaging 23(2) (2010)
16. Troncy, R., Huet, B., Schenk, S. (eds.): Multimedia Semantics: Metadata, Analysis and Interaction. Wiley-Blackwell (2011)
17. Wosiak, A.: Zastosowanie multimedialnych baz danych w diagnostyce medycznej. In: Kieltyka, L. (ed.) Narzedzia informatyczne w zarzadzaniu. Wybrane zagadnienia, pp. 44–56. Wydawnictwo Politechniki Czestochowskiej, Czestochowa (2011)

Reliability and Error Probability for Multimodal Biometric System

Michał Szczepanik and Ireneusz Jóźwiak

Wrocław University of Technology, Institute of Informatics
Wybrzeże Wyspiańskiego 27, 50-370 Wrocław, Poland
{michal.szczepanik,ireneusz.jozwiak}@pwr.wroc.pl
http://www.ii.pwr.wroc.pl

Abstract. In this paper, the authors compare existing multimodal biometric systems and analyze their reliability. The main focus is placed on a multi sensorial biometric system which recognizes two or more different and independent biometric characteristics. The authors also propose a method to calculate the error probability for this system, treating it as a multiple classifier for a multi class.

Keywords: biometric, error probability, multiple classifier, multimodal biometric system.

1 Introduction

Nowadays biometric systems are more popular than two or three years ago, since readers and scanners are two four times cheaper. Biometric systems for authentication or user identification use physiological characteristic or behavioral traits [10]. Physiological biometrics can identify by face characteristics, iris, DNA, handprints or fingerprints. Behavioral biometrics are related to the behavior of a person, like: typing rhythm, gait, or voice. Most biometrics are considered individually and do not change throughout the whole live. Unfortunately, such a claim is false, because each scar or burn can modify the structure of our body. The likelihood of people with the same fingerprint distribution is $4 * 10^{-7}$, because the distribution does not depend only on genes, but also on the degree of fetal blood flow during the forming of the fingerprint, which is why twins have different fingerprints [6][12]. Immutability of fingerprints depends mainly on damage because the wounds and scars may temporarily disrupt the structure [1][2]. Behavioral traits depend on the degree of fatigue or awareness of the body, caused by taking medication or other drugs. In the case of the eyes iris, which was recognized as one of the most secure biometrics, a team of researchers from the University of Notre Dame under the direction of prof. Kevin Bowyer has shown that it is sensitive to aging [3]. The result is, that over time the system will gradually generate more errors in the case of the same person.

J. Korbicz and M. Kowal (eds.), *Intelligent Systems in Technical and Medical Diagnostics*, Advances in Intelligent Systems and Computing 230,
DOI: 10.1007/978-3-642-39881-0_27, © Springer-Verlag Berlin Heidelberg 2014

Table 1. Comparison of most popular biometric technique

Technique	FAR(%)	FRR(%)
Fingerprints	0.2000	0.0100
Hand geometry	0.2000	0.2000
Retina	10.0000	0.0010
Iris	0.0005	0.0005
Face geometry	1.0000	0.5000

2 Quality Assessment of Biometric Algorithms

In order to adopt biometric technologies, there are evaluated some factors, including the reliability, availability, maintainability and performance of biometric-based verification and identification systems and applications. There are two most important performance metrics for usability and security of biometric systems [16]: False Accept Rate (FAR), also called False Match Rate (FMR), is the probability that the system incorrectly matches the input pattern to a non-matching template from the database. It measures the percent of invalid inputs which are incorrectly accepted. False Reject Rate (FRR), also called False Non-Match Rate (FNMR), is the probability that the system fails to detect a match between the input pattern and a matching template from the database. It measures the percent of valid inputs which are incorrectly rejected. They can be presented mathematically as:

$$FAR(T) = \int_{Th}^{1} p_i(x)dx \tag{1}$$

$$FRR(T) = \int_{0}^{T_h} p_i(x)dx \tag{2}$$

where T_h is the value of a threshold used in the the algorithm. Both FAR and FRR are functions of a threshold T. When T decreases, the system has more tolerance to intraclass variations and noise; however, FAR increases. Similarly, if t is lower, the system is more secure and FRR decreases. From point of view of usability, each system where$FRR < 0.01\%$ can be accepted, but for a biometric system, the most important parameter is FAR, which should be near 0. The best for developers is a compromise when FRR is bigger than FAR, but for most biometrics, system value of this parameter is about 10% when FAR is near 0. The parameter, which defines value where FAR equals FRR is known as Equal Error Rate (EER).

3 Usability and Requirements of a Biometric System

Biometric systems are increasingly used as something new and safe. No one really does check whether the users needs are actually provided. A simple example might be to use a fingerprint identification system in one of the banks. The system had a very good proprietary fingerprint recognition algorithms, but it didn't verify the humanity of the input. Another would be new cash machines which analyze finger vein to identify a user. Everything would be fine, if not for the device operating temperature (0°C – 35°C) which was not acceptable in on our countrys weather and climate.

4 Multimodal Biometric System

Multiple classifier or multimodal biometric systems take input from one or multiple sensors which recognizes one or more different biometric characteristics. For example, a system with fingerprint and finger vein recognition, or a system for face recognition based on two different algorithms for recognizing it. There are five general types of multiple classifier algorithms [14][18]:

- Multi algorithmic biometric system (MABS)
- Multi sensorial biometric system (MSBS)
- Multi biometric characteristic system (MBCS)
- Multi instance biometric system (MIBS)
- Hybrid multi-modal biometric system (HMMB)S

Multi algorithmic biometric system takes data from a single sensor and process it using two or more different algorithms. Multi sensorial biometric system takes data about the same biometric characteristics from two or more different sensors (from one instance). Multi biometric characteristic system takes data about two or more biometric characteristics from two different sensors. Multi instance biometric systems use one sensor, or possibly more sensors, to capture samples of two or more different instances of the same biometric characteristics. Example is capturing images from multiple fingers. Hybrid multi-modal biometric system connects two or more different type of multiple classifier. A Mechanism that can combine the classification results from each biometric channel is called as biometric fusion. Fusions can be made on each step (level) of biometric system process.

4.1 Decision-Making in Multi-module Biometric System

Multi-module biometric systems are also distinguishable because of the way the system of decision-making. There are several systems of decision-making due to dependencies between modules:

- A system of weights
- A system of primary decision-maker

The system of weights makes decisions based on the results of several individual decision-making modules, each of which belongs to a particular biometric module. Each result is expressed as a percentage, determining the degree of compliance with the biometric characteristics of the pattern. Each result is attributed

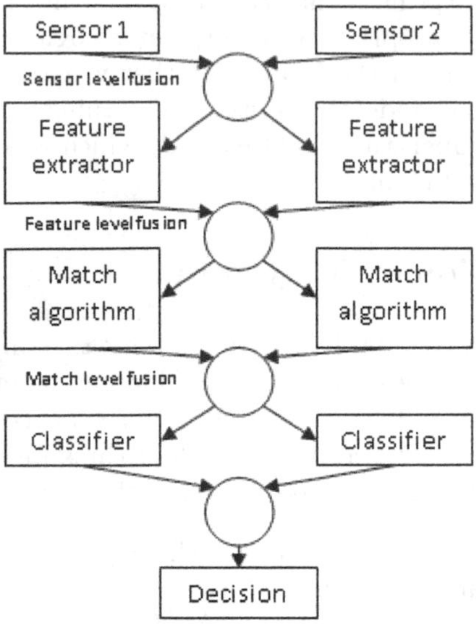

Fig. 1. Fusion in multimodal biometric system

to the weight of the module in the final decision. The case of the dual-system of weights is described by this formula:

$$d = p_A w_A + p_B w_b \tag{3}$$

where: p_A is the degree of compliance with the original features of the module A:$< 0.1 >$, p_B is the degree of compliance with the original features of the module B: $< 0.1 >$, w_A is the weight of module A: $< 0.1 >$, w_B is the weight of module B: $< 0.1 >$. The system of primary decision-maker is used when one or more modules for specific values of the decision taken by the entire system irrespective of the decisions of the other modules. It applies when one of the modules has a very negligible (could be that the system is infallible), one of the FAR or FRR factors assessment compared to the other.

4.2 Reliability of Biometric System

Reliability of any biometric system can be determined basing on the reliability of the sensor, processing algorithms and biometric recognizing. The reliability of the sensors have influence on three major events:

- Physical sensor failure (FP)
- Biometric feature is damaged (BF)
- Read incorrectly, such as modified biometric features due to dirt (RI)

Sensor failure event can be expressed by the formula:

$$SF = FP \cup BF \cup RI \tag{4}$$

On the wrong decision taken as a result of the algorithm being affected by the following events: - Filtering algorithm error / noise correction (FAE) - Classifier error (CE) - Decision algorithm error (DAE) An event of algorithm error can be expressed by the formula:

$$AE = FAE \cup CE \cup DAE \tag{5}$$

Based on these events, you can develop a reliability model for each type of multimodal biometric system.

Multi-algorithm biometric system uses a single sensor to collect data about a feature, and then these data is processed by two independent algorithms. In the case of fingerprint recognition, algorithms can be such an algorithm, which is based on patterns (called Pattern-Based Templates) [7] and flexible matching minutiae (called Elastic minutiae matching) [5].

$$MABSF = SF \cup (AE_1 \cap AE_2) \tag{6}$$

Multi sensorial biometric system is the one which gathers biometric traits using one of two independent sensors, eg. fingerprints are analyzed from two fingers, each by an independent reader. Readers may be identical, but this increases the risk of failure due to defects in the build, or different, for example, one of our second ultrasonic microphones. In the next phase, the comparison algorithm is used for the analysis of one of the biometric features.

$$MSBSF = (SF_1 \cap SF_2) \cup AE \tag{7}$$

Hybrid system is the one that has several sensors and algorithms to identify a biometric feature. Each of the algorithms can analyze data from only one sensor (or just a few). It is currently the most widely used type of biometric systems for multi-module one trait.

$$HBSF = (SF_1 \cup AE_1) \cup (SF_2 \cup AE_2) \tag{8}$$

Multi instance biometric system has two independent sensors. The data from each sensor is processed by independent algorithms. An example of such, is a developed system, presented in the chapter experiment.

$$MIBSF = (SF_1 \cap SF_2) \cup (AE_1 \cap AE_2) \tag{9}$$

A system with one sensor has only one physical sensor for analyzing two biometric features, such as high-resolution hand scanner that allows for the reading of a fingerprint.

$$MBSF = (SF) \cup (AE_1 \cup AE_2) \tag{10}$$

Table 2. Comparison of decision method

Decision method	FAR(%)	FRR(%)
fingerprint only	0.300	0.100
veins only	0.020	0.020
fingerprint and veins	0.010	0.015

5 Reliability and Error Probability for Multimodal Biometric System

Base on the problem of characteristic for a multimodal biometric system which uses different sensors, we can assume that they are independent. We can also assume that each sensor output is a separate classifier or decision maker, which provides information to which class belongs the tested object. In general, this system is multiple classifier for a multi class problem [9] where as a classifier we understand a separate sensor with algorithms dedicated for it, and as a class we understand the human with biometric characteristic specific for him. Base on that, the probability of an error for this system is presented as an expression:

$$P_E = \sum_{k_1=0}^{L} \sum_{k_2=0}^{L} \cdots \sum_{k_K=0}^{L} (P_{MF} \left[H_E + (1 - H_E) \left(1 - \frac{1}{1 + H_D} \right) \right]), \tag{11}$$

where:

$$P_{MF} = f(k_1, k_2, ..., k_K, L, P_1, P_2, ..., P_K), \tag{12}$$

is the probability mass function of the multinomial distribution for non-negative integers $k_1, k_2, ..., k_K$ and H_E is the Heavisides function gives information if the proper class received fewer votes than the wrong class. The number of ties is represented by the formula:

$$H_D = (\sum_{i=2}^{K} \delta(s_1, s_i). \tag{13}$$

6 Experiment

The developed system is a group of dedicated systems and biometric sensors. It consists of a module based on a fingerprint reader and a recognition module in the finger vein - Hitachi Finger Vein Unit H1. The fingerprint reader is equipped with a recognition algorithm based on minutiae groups and the selective attention Sigma-if artificial neural network [8]. The following table presents the results for each module individually, and the entire system. The module is based on a decision of weights, which are respectively 0.4 and 0.6. As you can see hybrid system has a much better FAR and FRR, than any of the systems themselves.

Table 3. Comparison of weight in multimodal biometric system

Fingerprint	Veins	FAR(%)	FRR(%)
0.5	0.5	0.010	0.015
0.7	0.3	0.220	0.050
0.3	**0.7**	**0.010**	**0.012**
0.6	0.4	0.180	0.020
0.4	0.6	0.020	0.013
0.8	0.2	0.280	0.070
0.2	0.8	0.020	0.020

7 Conclusion

Fingerprints are very vulnerable to damage, which reduces the level of security and usability. The proposed multimodal system can operate also in the case of such damages, because the analysis of veins does not have the disadvantages of fingerprints analysis. The system based on two different characteristics can simultaneously identify and authenticate the user. It is planned to expand the system in the future by analyzing human characteristics such as temperature and pH, to make the system more resistant to the trick.

Acknowledgements. This publication is co-financed by the European Union under the European Social Fund.

References

1. Bazen, A.M., Gerez, S.H.: Fingerprint Matching by Thin-plate Spline Modelling of Elastic Deformations. Pattern Recognition 36(8), 1859–1867 (2003)
2. Bebis, G., Deaconu, T., Georgiopoulos, M.: Fingerprint Identification Using Delaunay Triangulation. In: Proc. IEEE International Conference on Intelligence, Information and Systems, pp. 452–459 (1999)
3. Burge, M., Bowyer, K.: Handbook of Iris Recognition. Advances in Computer Vision and Pattern Recognition 11, 103–127 (2013)
4. Cappelli, R., Lumini, A., Maio, D., Maltoni, D.: Fingerprint Classification by Directional Image Partitioning. IEEE Transactions on Pattern Analysis Machine Intelligence 21(5), 402–421 (1999)
5. He, Y., Ou, Z.: Fingerprint matching algorithm based on local minutiae adjacency graph. Journal of Harbin Institute of Technology 10(05), 95–103 (2005)
6. Hicklin, A., Watson, C., Ulery, B.: How many people have fingerprints that are hard to match, NIST Interagency Report 7271 (2005)
7. Hong, L., Wan, Y., Jain, K.: Fingerprint image enhancement: Algorithm and performance evaluation. IEEE Transactions on Pattern Analysis and Machine Intelligence 20, 777–789 (1998)
8. Huk, M.: Backpropagation generalized delta rule for the selective attention Sigma-if artificial neural network. Applied Mathematics and Computer Science 22(2), 449–459 (2012)

9. Huk, M., Szczepanik, M.: Multiple classifier error probability for multi-class problems. Maintenance and Reliability 3, 12–17 (2011)
10. Jain, A.K., Ross, A., Nandakumar, K.: Introducing to biometrics. Spinger (2011)
11. Maltoni, D., Maio, D., Jain, A.K., Prabhakar, S.: Handbook of Fingerprint Recognition, 2nd edn. Springer (2009)
12. Pankanti, S., Prabhakar, S., Jain, A.K.: On the individuality of fingerprints. In: Proceedings of Computer Vision and Pattern Recognition (CVPR), pp. 805–812 (2001)
13. Parziale, G., Niel, A.: A fingerprint matching using minutiae triangulation. In: Zhang, D., Jain, A.K. (eds.) ICBA 2004. LNCS, vol. 3072, pp. 241–248. Springer, Heidelberg (2004)
14. Ratha, N.K., Govindaraju, V.: Advances in Biometrics: Sensors, Algorithms and Systems. Springer (2007)
15. Ross, A., Dass, S.C., Jain, A.K.: A deformable model for fingerprint matching. Pattern Recognition 38(1), 95–103 (2005)
16. Ross, A., Nandakumar, K., Jain, A.K.: Handbook of Multibiometrics. International Series on Biometrics. Springer (2011)
17. Szczepanik, M., Szewczyk, R.: Fingerprint identification algorithm. KNS 1, 131–136 (2008) (in Polish)
18. Wayman, J.L., Jain, A.K., Maltoni, D., Maio, D.: Biometric Systems. Technology, Design and Performance Evaluation, 1st edn. Springer (2005)

A Simulation Model for Emergency Evacuation Time and Low-Cost Improvements of a Hospital Facility Using FlexSim Healthcare: A Case Study

Jerzy Mikulik[1], Witold Aleksander Cempel[1], Stanisław Kracik[2], and Dawid Dąbal[1]

[1] AGH University of Science and Technology, Krakow, Poland
{jmikulik,wcempel,dawid.dabal.030}@zarz.agh.edu.pl
[2] Dr Jozef Babinski Hospital, Krakow, Poland
stanislaw.kracik@babinski.pl

Abstract. Proper preparation for emergency situation like: terrorist threat, flood, fire, etc., in hospital facilities, where people entrust their safeness in hands of medical and nonmedical personnel is important issue. In this paper we present the simulation model of emergency evacuation from the facility of The Voivodship Centre of Addiction Therapy in Krakow, a part of Dr Jozef Babinski Hospital in Krakow (Poland). We used FlexSim Healthcare 3D Simulation Software to find answers for following questions: how much time the evacuation will take, depending on different time frames during the day (we call them scenarios), and what kind of low-cost improvements can be made to shorten evacuation time. The use of low-cost improvements resulted in a nearly 50% reduction of the total evacuation time in each of the five analyzed scenarios. We presented in this paper the aim of the simulation model, methodology, course of research, results, and conclusions.

Keywords: simulation, modeling, evacuation, hospital, FlexSim Healthcare.

1 Introduction

Emergency situation like: terrorist threat, flood, fire, building disaster, etc., fortunately aren't frequent occurrence in Poland, but when it happened may result in costly and sometimes irreversible loss. This is the reason why preparation for such circumstances is necessary, especially in hospital facilities, where people entrust their safeness in hands of medical and nonmedical personnel.

One of the typical solutions in mentioned emergency situation is evacuation people from the facility. Time of evacuation is one of major issues in such a case. The less time people need to leave the building, the better. So, the question is how much time it will take to leave the building and what can be done to shorten this time. Well know method to do this is to make real experiments. It is maybe good solution in office building, but in hospital it is extremely difficult. So, a simulation modeling might be the right solution in such a case.

J. Korbicz and M. Kowal (eds.), *Intelligent Systems in Technical and Medical Diagnostics*, Advances in Intelligent Systems and Computing 230,
DOI: 10.1007/978-3-642-39881-0_28, © Springer-Verlag Berlin Heidelberg 2014

In this paper we present a case study of simulation model the emergency evacuation from the facility of The Voivodship Centre of Addiction Therapy in Krakow as a part of Dr Jozef Babinski Hospital in Krakow (Poland). We used FlexSim Healthcare 3D Simulation Software (FlexSim HC) to find answers: how much time will take evacuation depending on different time frames during the day and what low-cost improvements can be made to shorten evacuation time. In this case we use models as a tools to support and extend the power of thinking [1].

In this paper we present the aim of the simulation model, methodology, course of research, results, and conclusions.

2 The Aim of the Simulation Model

A model is [2] "a symbolic representation of a set of objects, their relationships and their allowable motions". M. Pidd [1] define a model as an external and explicit representation of a part of reality as seen by the people who wish to use the model to understand, to change, to manage and to control that part of reality. Modeling stress or rather the activity of the modeler according to C. Cempel [3] means searching (and probably finding) a system of compelling attributes and relations in regards to the aim. Simulation [4] means the contraction or expansion of a model's space-time in order to recognize a reality better.

The aim of simulation modeling in this case was the determination of the time of evacuation of people from a hospital building and the suggestion of low-cost improvements, which will shorten the evacuation time. Low-cost improvements mean simple changes in the real system, which do not need special investment and which can be done by hospital staff. The facility selected for such improvements in this case study is The Voivodship Centre of Addiction Therapy, a part of Dr Jozef Babinski Hospital in Krakow (Poland).

Evacuation circumstances:

- terrorist threat;
- bioterrorism;
- flood;
- fire;
- building disaster;
- security incidents caused by people.

3 Methodology

Modern simulation and modeling tools are usually characterized as [5] either continuous variable systems (CVS) or discrete event systems (DES). Evacuation is an event, so DES should be good enough to fulfill the aim of modeling and the simulation study previously described.

Taking the above issues into consideration, FlexSim Healthcare 3D Simulation Software (FlexSim HC) has been chosen to model the emergency evacuation from the facility of The Voivodship Centre of Addiction Therapy in Krakow. FlexSim Healthcare was designed for the modeling purposes of hospital facilities.

Each system, also hospitals, has one or more of the following behavioral characteristics [6]:

— components within a system are subject to their own random events;
— random events in the environment impact the system;
— system behavior is time dependent;
— system components have complex interactions, and thus there are many connecting paths within a system.

The above issues enforce well-ordered modeling. Such an approach is necessary, especially when the modeler tries to avoid the "garbage in – garbage out" syndrome.

The general view of a simulation model in a decision support role is illustrated in Figure 1. In any type of system, decision makers need to understand, analyze, design, and manage the system [6]. All of the mentioned activities involve making decisions. Without the use of a model, decision makers will use a variety of methods to gather information from a real system, process it, and then choose an action that causes a change in the system. Simulation can be a powerful tool since real data is not available and costly to obtain [7]. Model-based decision making in general, and simulation-based decision making in particular, provide a means for decision makers to investigate a representation of the system, experiment with alternatives, and predict the effect of proposed changes, all external to the real system [6]. This approach greatly increases the decision space (allows many more alternatives to be evaluated), is not intrusive on the real system, and enables risk of the actions to be assessed.

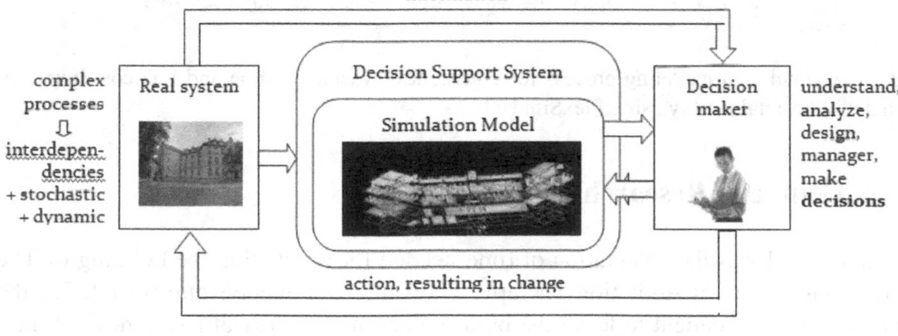

Fig. 1. A model-based decision support system (Source: [6], p. 6)

Taking the above issues into consideration, the specific methodology of a simulation model for the emergency evacuation time and low-cost improvements of a hospital facility using FlexSim Healthcare was formulated (Fig. 2).

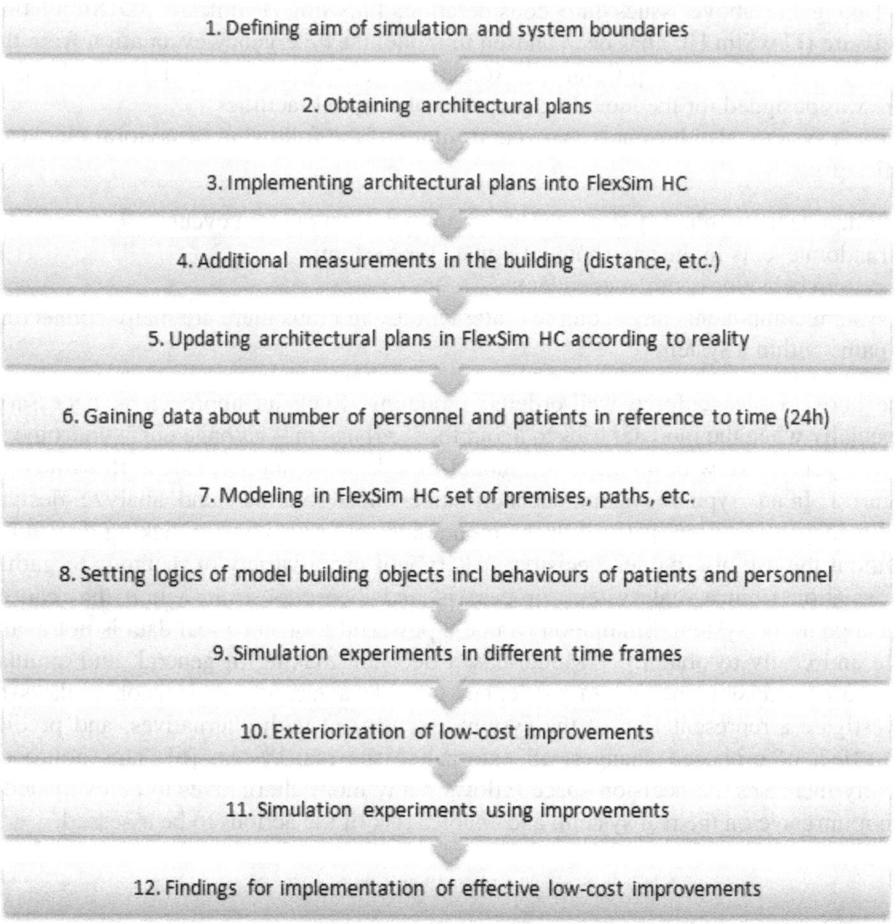

Fig. 2. A simulation modeling process for emergency evacuation time and low-cost improvements of a hospital facility using FlexSim HC

4 Course of Research

The main goal was the calculation of time needed for evacuating the building of The Voivodship Centre of Addiction Therapy. The simulation models aim is to define the time needed for a patient to leave the building and move away at a distance 100 meters from the building.

In order to obtain a proper model of the hospital building, architectural plans were acquired which were then transferred into .dxf files using AutoCAD. The building has five stories, three staircases and three main entrances. Each story has been drawn as individual file to enable the latter recreation of vertical structure of the building in the 3D environment while taking in to consideration the differences of height between specific stories.

Next additional measurements of the building concerning critical parameters for the evacuations were made:

- door width;
- corridor width;
- staircase and railing width;
- location of potential obstacles, their width and the narrowing of the corridors.

All the found differences were included in the updated layouts.

Interviews with the management of specific wards were carried out in order to define the number of patients and medical staff at a given moment in time. Due to limitations of this paper only data for one scenario, concerning the activity of the building in late morning, will be described.

On the basis of the interviews, four types of patients were classified (Table 1):

- type I – patients of twenty-for-hour wards which can leave the building on their own;
- type II – patients of twenty-for-hour wards which require additional transport to leave the building ;
- type III – patients of the day-care wards;
- type IV – outpatients.

Table 1. Number of staff and patients in the building

Patients/Staff	Number
Patients type I	87
Patients type II	8
Patients type III	6
Patients type IV	7
Medical doctors	4
Nurses	12
Therapists	21
Administration	21

The model was built using FlexSim HC basic object classes:

- Patients Arrivals (source) –resource used to generate patients that will travel through the model;
- Patient Exit (exit) – the termination point for patients in the model;
- Patient Processing – resource used to simulate some type of process, or forced delay that patients need to experience;
- Patient Queuing – resource used to simulate the patients general place to wait before continuing on with their track;
- Network Nodes –resource used to designate paths that Task Executors will use when executing a travel activity.

In order to obtain a proper process model of evacuation, the accurate recreation of the paths on which the patients and staff travels is essential. This step is crucial to identify bottlenecks off the system. FlexSim HC uses the following parameters to determine restrictions of object movements on paths:

- speed limit – maximum speed with which object can travel between two given network nodes;
- current/virtual distance – distance between two given network nodes, current or defined by modeler;
- spacing – distance between two travelers moving in the same direction, this parameter is used to determine pace at which the objects can enter paths.

These parameters differ depending on the path type which is being recreated in the model for example in network nodes representing staircases [8] or in network nodes simulating doors [9].

A general view of the a model of The Voivodship Centre of Addiction Therapy build in FlexSim Healthcare is shown in Fig.3.

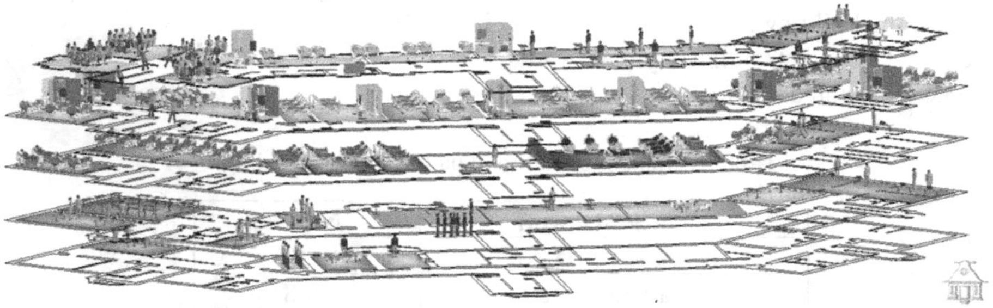

Fig. 3. A model of The Voivodship Centre of Addiction Therapy build in FlexSim HC

In order to recreate the logic behind specific objects FlexSim HC uses PatientTracks for patients and GlobalProcesses for staff. Activities included in tracks are described by:

- activity type – type of activity for example patient transport, patient escort, processes, etc.;
- activity ID – activity numeric identifier;
- predecessors – ensures correct order of performing activities;
- milestone – label of activity used for reporting purposes;
- start time – earliest simulation time that activity can start;
- fixed/variable cost – used for assigning fixed monetary value and rate per hour that this activity will cost;
- process time – allows the setting of process duration (statistical, mathematic expression or static number);
- allocation priority – allows the setting of priority for sharing staff resources ;
- preempt resources – allows an activity to preempt resources;
- patient destination – allows the setting of ending locations for patients travel activity;

- staff/equipment/transportation requirements – allows the use of resources during the activity;
- many other options and user defined advanced functions.

After building the logic behind objects, the evacuation event which start is determined by a global variable was created. The simulation experiments were conducted for every evacuation scenario depending on the time of day. The result obtained include the total evacuation time and the evacuation time of individual patients.

The lower bound total evacuation time during late morning is 16,32 minutes while the upper bound is 16,78 minutes. During the first five minutes of evacuation the building 91% of the patients were evacuated (Fig. 4, 5).

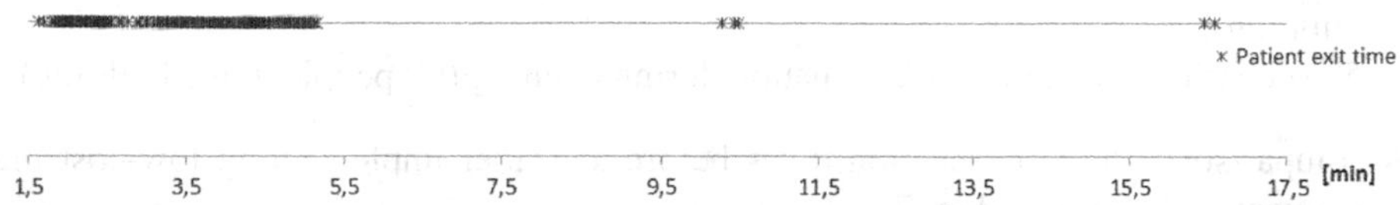

Fig. 4. Patients exit times during the first five minutes of simulation before low cost improvements

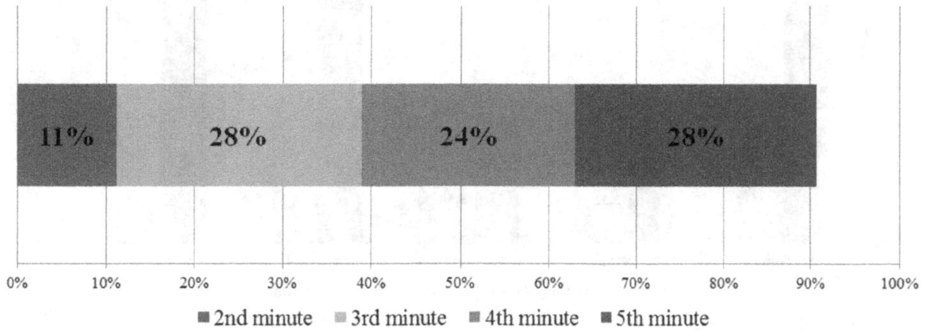

Fig. 5. Percentage of patients evacuated at a given minute

The next step was implementing low-cost improvements such as removing obstacles and changes in the evacuation protocol. After their implementation, the simulation experiments were conducted once again.

Thanks to the implemented improvements the evacuation time was shortened to 5,25 minutes (mean) while 96% of all the patients were successfully evacuated in the first five minutes (Fig. 6).

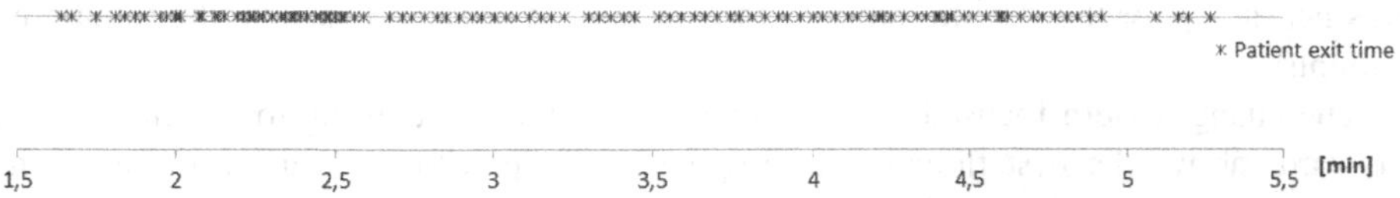

Fig. 6. Patients exit times during the first five minutes of simulation after low-cost improvements

5 Results

The use of low-cost improvements resulted in a nearly 50% reduction of the total evacuation time in each of the five analyzed scenarios:

- 1st scenario – simulation of evacuation during night time (99 people in the building);
- 2nd scenario – simulation of evacuation during early morning (111 people in the building);
- 3rd scenario – simulation of evacuation during late morning and midday (166 people in the building);
- 4th scenario – simulation of evacuation during afternoon (108 people in the building);
- 5th scenario – simulation of evacuation during evening (99 people in the building).

A comparison of total evacuation times before and after implementing low-cost improvements are shown in Fig. 7.

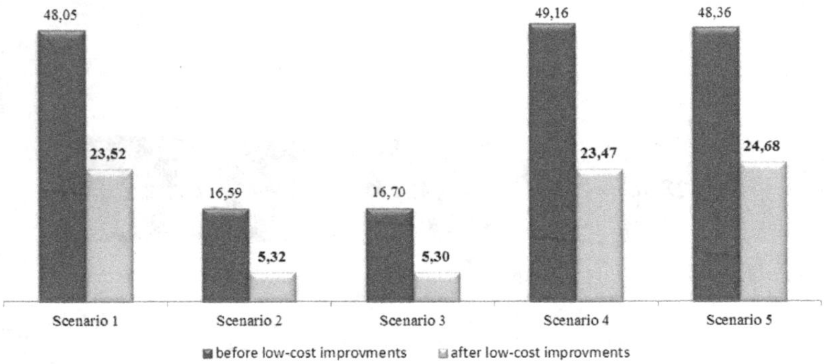

Fig. 7. A comparison of total evacuation times before and after implementing low-cost improvements [min]

The most crucial improvements which enabled a great reduction in the total evacuation time were the changes in the evacuation protocol. Up to this point the medical staff was only responsible for patients residing on their wards. In the case off The Voivodship Centre of Addiction Therapy, patients which require additional transport during evacuation are located on only one ward on the first floor, where the staff number is not sufficient enough to evacuate every patient in one round. That is why it was necessary for the staff to return to the building in order to evacuate the remaining patients.

The changes were focused on implementing protocol according to which the staff of wards above the first floor is also required to help with the transportation of the patients on the first floor while coming down on their way to leave building. Thanks to this modification it is possible to evacuate all the patients in one or maximum two rounds.

Additionally, in the case of a time loop between the occurrence of the evacuation circumstances and the signal to start evacuation, the second wings of the main entrances should try to be opened as they constitute the bottlenecks off the system.

The staff should be informed about the new protocol by way off cyclical trainings which are to take place at least two times a year and be conducted by a health and safety officer. Furthermore, the curriculum of these trainings should include the problem of unloading the congestion near doors and information about crowd control methods.

Another low-cost improvement was removing superfluous obstacles in corridors (flowerpots, unnecessary tables and chairs) as well as from staircases (benches) which reduce flow capacity of paths. Thanks to this it was possible to increase the number of patients who were able to evacuate the building in the first five minutes.

6 Conclusions

The aim of simulation modeling was the determination of the time of evacuation of people from The Voivodship Centre of Addiction Therapy building and the suggestion of low-cost improvements, which will shorten the evacuation time. The aim has been fulfilled.

Results justify the use of FlexSim Healthcare and proposed simulation modeling process to estimate evacuation time from hospital facility, as well as influence on time by low-cost improvements.

Authors are considering organize evacuation event in reality engaging students in this hospital facility. Comparison of the results might enlarge science value of this case study.

References

1. Pidd, M.: Tools for Thinking, Modeling in Management Science. John Wiley & Sons, Ltd., Chichester (2003)
2. Franciois, C. (ed.): International Encyclopedia of Systems and Cybernetics. K.G.Saur, Munchen (1997)
3. Cempel, C.: Systems Theory and Engineering (Teoria i inżynieria systemów). Wydawnictwo Naukowe Instytutu Technologii Eksploatacji, Radom (2008)
4. Cempel, C.: Modern methodological and philosophical issues (Nowoczesne zagadnienia metodologii i filozofii badań). Wydawnictwo Naukowe Instytutu Technologii Eksploatacji, Radom (2003)
5. Petropoulakis, L., Giacomini, L.: A hybrid simulation system for manufacturing processes. Integrated Manufacturing Systems 8(4), 189–194 (1997)
6. Beaverstock, M., Greenwood, A., Lavery, E., Nordgren, W.: Applied Simulation Modeling and Analysis using FlexSim. FlexSim Software Products, Orem (2012)

7. Lin, Y., Fedchenia, I., LaBarre, B., Tomastik, R.: Agent-Based Simulation of Evacuation: An Office Building Case Study. In: Klingsch, W.W.F., Rogsch, C., Schadschneider, A., Schreckenberg, M. (eds.) Pedestrian and Evacuation Dynamics 2008, pp. 347–357. Springer, Heidelberg (2010)
8. Pauls, J.L., Fruin, J.J., Zupan, J.M.: Minimum Stair Width for Evacuation, Overtaking Movement and Counterflow- Technical Bases and Suggestions for the Past, Present, Future. In: Waldau, N., Gattermann, P., Knoflacher, H., Schreckenberg, M. (eds.) Pedestrian and Evacuation Dynamics, pp. 57–69. Springer, Heidelberg (2005)
9. Helbing, D., Farkas, I., Vicsek, T.: Simulating dynamical features of escape panic. Nature 407, 487–490 (2000)

Part III

Robotics and Computer Vision

Diagnostic Requirements in Multi-robot Systems

Cezary Zieliński and Tomasz Kornuta

Institute of Control and Computation Engineering,
Faculty of Electronics and Information Technology
Warsaw University of Technology, Nowowiejska 15/19, 00-665 Warsaw, Poland
{C.Zielinski,T.Kornuta}@elka.pw.edu.pl

Abstract. The paper presents a generic robot control architecture facilitating the collection of data for the purpose of diagnostics. Each robot within the system is treated as an embodied agent composed of effectors, receptors and a control system. Its behaviour is described in terms of transition functions and: initial, terminal and error conditions. Data within such systems can be collected at the control system sampling rate, thus both continuous monitoring and event driven one are possible.

Keywords: robot control architecture, robot programming.

1 Introduction

Diagnostics is defined as the art or science of diagnosis, where the latter is treated in technical sciences as the determination or analysis of the cause or nature of a problem or a situation [1]. Process diagnostics deals with fault: detection, isolation (localization) and identification (assessment of the extent of failure) [5]. Monitoring for the purpose of diagnostics consists in real-time gathering and processing of process (system) variables [9]. Depending on the type of the controlled process diverse approaches to system monitoring and modelling are followed [11]. Some of data obtained by monitoring is used as an input to the system model and some for comparison with the output produced by that model. The difference forms the residue, which is used for fault detection. Residues show the discrepancy between the observed system behaviour and the expected one (obtained from the model). State space, transmittance, neural, fuzzy etc. models can be used for that purpose [7]. Other forms of fault detection exist too, e.g. comparison of data obtained by monitoring with the defined constraints. Fault detection methods fall into two vast groups: those utilising the relationships between process variables and those based on checking the constraints imposed on the parameters associated with process variables [8]. The former require the process model, while the latter need the formulation of constraints.

This paper focuses on the technical means for diagnostic data acquisition and processing in robot control systems. General treatment of this subject is possible only if a generic architecture of a multi-robot system and a generic method of describing its behaviour are proposed. The paper postulates a general control system structure [16], not only facilitating its design, but also enabling the collection and processing of data necessary for the analysis leading to the isolation

J. Korbicz and M. Kowal (eds.), *Intelligent Systems in Technical and Medical Diagnostics*, Advances in Intelligent Systems and Computing 230,
DOI: 10.1007/978-3-642-39881-0_29,

and identification of faults. Thus the paper provides a generic structure of a robot control system and defines a method of specifying its behaviour enabling the description of the necessary actions in both normal and abnormal situations. It focuses on such a creation of the control architecture and means of describing the system behaviour that diagnosis of abnormal functioning can be detected and possibly remedied. Here only the technical means that are introduced into a robot control system enabling the collection of data necessary for the generation of residues, checking of constraints or extraction of symptoms [8] are discussed. This also includes the possibility of introducing specific diagnostic behaviours for testing the current condition of the system, as in e.g. [10].

2 Embodied Agent

The presented specification method relies on three concepts: Agent, Behaviour and Transition Function. The former pertains to the architecture of the system, while the latter two describe its activities. In [13] an Agent is defined simply as an entity that perceives and acts. Further this definition is expanded by adding an imperative for reaching the desired goal, which influences the undertaken actions in order to maximize the agent's expected utility, and hence the term Rational Agent is created. In [15] the concept of Embodiment [3,4] is utilised to define an Embodied Agent – the name underscores that robotic systems possess a corporeal body in the form of physical effectors and receptors.

2.1 General Inner Structure of an Embodied Agent

In general, a j-th agent a_j contains: effectors E_j, influencing the environment, receptors (exteroceptors) R_j collecting data about the environment, and the control system C_j governing the required behaviours. The control system is composed of three types of entities: virtual effectors e_j, virtual receptors r_j and the control subsystem proper c_j. The virtual effectors e_j and virtual receptors r_j present to the control subsystem c_j the effector and the environment respectively in a form that is appropriate for the purposes of control The system can contain many exteroceptors $R_{j,l}$ and many virtual receptors $r_{j,k}$, where l and k are their respective designators. Information aggregation being the responsibility of virtual receptors consists in the composition of readings obtained from several exteroceptors or in the extraction of the required information from a complex sensor. Moreover the readings obtained from a single group of exteroceptors may be processed in several ways, thus many virtual receptors can be formed. Virtual receptor readings are conveyed to the control subsystem c_j which influences the virtual effectors e_j. The Control Subsystem must be able to both reconfigure exteroceptors R_j and adjust the method how the Virtual Receptors r_j aggregate readings. Hence the reverse connections with those subsystems within the agent exist. Usually one agent a_j controls a single effector E_j, but it can be presented to the control subsystem in different forms, hence the multiplicity of virtual effectors $e_{j,n}$. Each virtual effector influences the actuators of the effector

E_j, e.g. assuming the effector to be a manipulator, c_j produces commands with arguments representing the desired end-effector poses expressed in the operational space. In such a case two virtual effectors can be created, one using the inverse kinematics to compute the positions in the configuration space (joint coordinates) and another one using the inverse Jacobian to compute joint position increments. The control subsystem must be able to establish the current state of the effector E_j. Thus a reverse connection between the two must exist. Hence the readings produced by the proprioceptors (e.g. motor encoders) are processed by the virtual effector e_j to produce a representation acceptable by the control subsystem c_j. An agent should also be able to establish a two-way communication with other agents $a_{j'}$, $j \neq j'$. The resulting structure is presented in fig. 1.

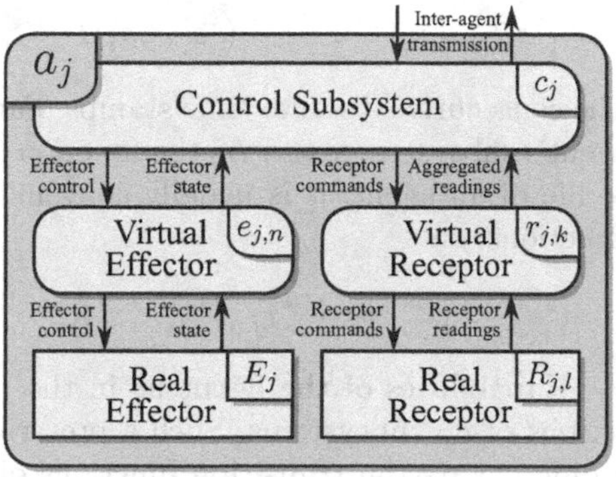

Fig. 1. General structure of an agent a_j

2.2 Notation

Even out of this superficial description of an agent it is evident that the number of different components of the agent is significant and, moreover, each of those components has its own elements (described further on in the paper). To make the description of such a system comprehensible a consistent denotation method is necessary. However to simplify the notation no distinction is being made between the denotation of a component and its state – the context is sufficient in this respect. In the assumed notation a one-letter symbol located in the centre (i.e. E, R, e, r, c designates the component. To reference its subcomponents or to single out the state of this component at a certain instant of time extra indices are placed around this symbol. The left superscript designates the referenced buffer of the component or, in the case of a function, its type. The right superscript designates the time instant at which the state is being considered. The left subscript tells whether this is an input (x) or an output (y) buffer. If there is no left subscript the internal memory of the subsystem is referenced. The right subscript may may contain elements separated by comas. They refer to the ordinal numbers of: the agent, its component and subcomponent, or the ordinal number of the function, e.g. ${}^{e}_{x}c^{i}_{j}$ denotes the contents of the Control Subsystem input buffer of the agent a_j acquired from the Virtual Effector at instant i.

2.3 General Subsystem Behaviour

The general work-cycle of any subsystem s, where $s \in \{c, e, r\}$, of any agent is:

- acquire data from the associated subsystems through the input buffers,
- compute the next subsystem state by evaluating the transition function ${}^{s}f_{j}$,
- dispatch the results via the output buffers to the associated subsystems.

The functioning of a subsystem s of the agent a_j can be described by a single function that processes the data contained in the input buffers ${}_{x}s_{j}$ and internal memory ${}^{s}s_{j}$ to produce the output buffer values ${}_{y}s_{j}$ and update the internal memory ${}^{s}s_{j}$. Hence the transition function ${}^{s}f_{j}$ describing the subsystem behaviour is defined as:

$$\left[{}^{s}s_{j}^{i+1},\, {}_{y}s_{j}^{i+1}\right] := {}^{s}f_{j}({}^{s}s_{j}^{i},\, {}_{x}s_{j}^{i}), \tag{1}$$

where i and $i+1$ are the consecutive discrete time stamps. Function (1) describes the evolution of the agent's subsystem state. As the function (1) should be useful throughout the whole life of an agent, it is usually convenient to decompose it into a set of partial functions:

$$\left[{}^{s}s_{j}^{i+1},\, {}_{y}s_{j}^{i+1}\right] := {}^{s}f_{j,u}({}^{s}s_{j}^{i},\, {}_{x}s_{j}^{i}), \tag{2}$$

where $u = 0, \ldots, {}^{s}n_{f}$. Capabilities of the agent lie in the multiplicity and diversity of partial functions of its subsystems. Such a prescription requires rules of switching between different partial transition functions of a subsystem, thus three additional Boolean valued functions (predicates) are required:

- ${}^{s}f_{j}^{\sigma}$ defining the Initial Condition,
- ${}^{s}f_{j}^{\tau}$ representing the Terminal Condition and
- ${}^{s}f_{j}^{\varepsilon}$ detecting an Error Condition.

The first one selects the transition function for cyclic execution, the second determines when this cyclic execution should terminate, while the third detects an abnormal situation in the execution of the behaviour governed by an associated transition function ${}^{s}f_{j}$. This enables the introduction of the multi-step evolution of the subsystem in a form of a Behaviour $\mathcal{B}_{j,u}$ defined as:

$${}^{s}\mathcal{B}_{j,u} \triangleq {}^{s}\mathcal{B}_{j,u}\left({}^{s}f_{j,u},\, {}^{s}f_{j,u}^{\tau},\, {}^{s}f_{j,u}^{\varepsilon}\right). \tag{3}$$

Additionally, an execution pattern presented in fig. 2 is proposed. The symbol $s_{j^{\bullet}}$, where $j^{\bullet} \in \{j, j'\}$, denotes all subsystems associated with s_j (in the case of Control Subsystem some of them may not belong to the same agent, hence j' appears). Besides that, the Behaviours along with the Initial Conditions enable the presentation of the general behaviour selection mechanism in a form of a state graph as in fig. 3. The ${}^{s}\mathcal{B}_{j,0}$ is the default behaviour, activated when no other Behaviour is required, while ${}^{s}\mathcal{B}_{j,\varepsilon}$ is the behaviour executed when an abnormal situation is detected, hence its initial condition is the logical sum of all error conditions $\bigcup_{u=0}^{{}^{s}n_{f}} {}^{s}f_{j,u}^{\varepsilon}$, where ${}^{s}n_{f}$ is the number of normal behaviours of the

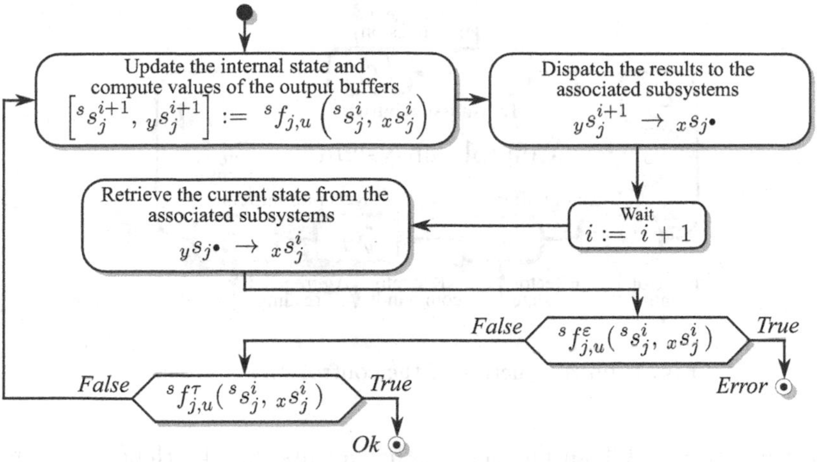

Fig. 2. General flow chart of a subsystem behaviour ${}^{s}\mathcal{B}_{j,u}$

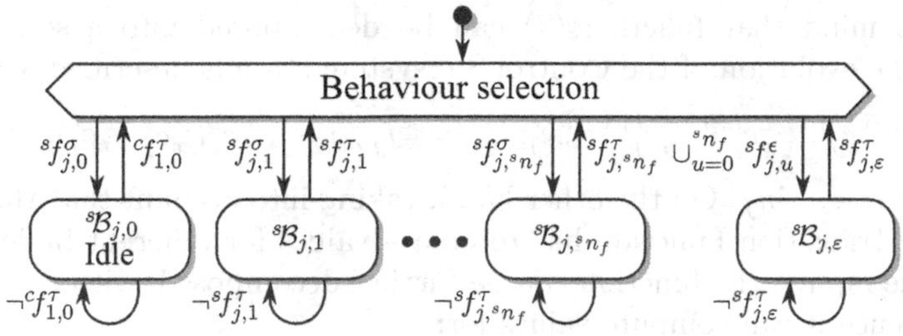

Fig. 3. State graph of the behaviour selection automaton

subsystem s. The collection of data for the purposes of monitoring normal behaviour is the responsibility of the transition functions ${}^{s}f_{j,u}$, while a reaction to the detected abnormal situation is the responsibility of a special ${}^{s}\mathcal{B}_{j,\varepsilon}$. Behaviour selection is executed as a stateless switch.

2.4 Control Subsystem ($s = c$)

The Control Subsystem is primarily responsible for task execution. The data types it operates on form the highest level ontology. It is up to the designer of the system to establish this ontology. Once this is done those data types are used to construct the inner structure of the Control Subsystem buffers. Three types of input and output buffers connecting the Control Subsystem to three different associated subsystems can be distinguished: buffers to the Virtual Effectors,

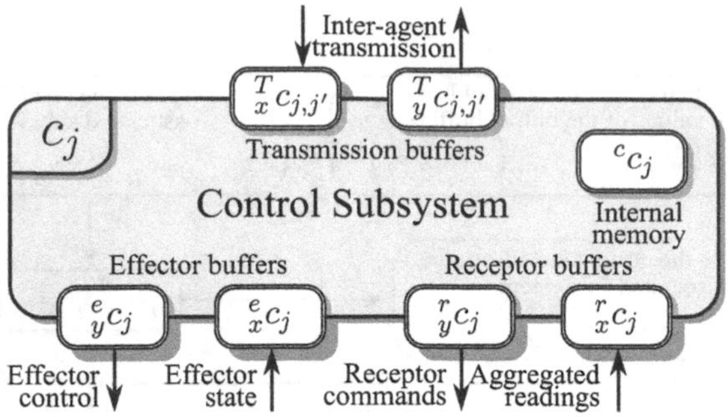

Fig. 4. Inner structure of the control subsystem c_j

Virtual Receptors and Transmitters to other agents (fig. 4). Hence the general Control Subsystem Transition Function can be formulated as:

$$\left[{}^c c_j^{i+1},\, {}^e_y c_j^{i+1},\, {}^r_y c_j^{i+1},\, {}^T_y c_j^{i+1}\right] := {}^c f_j({}^c c_j^i,\, {}^e_x c_j^i,\, {}^r_x c_j^i,\, {}^T_x c_j^i). \tag{4}$$

Having in mind that function (4) can be decomposed into a set of partial functions the evolution of the Control Subsystem state is described as:

$$\left[{}^c c_j^{i+1},\, {}^e_y c_j^{i+1},\, {}^r_y c_j^{i+1},\, {}^T_y c_j^{i+1}\right] := {}^c f_{j,m}({}_x c_j^i,\, {}^e_x c_j^i,\, {}^r_x c_j^i,\, {}^T_x c_j^i), \tag{5}$$

where $m = 1, \ldots, {}^c n_f$. On the other hand, taking into account that the Control Subsystem Transition Function has to create values for different buffers basing on the same inputs, the function can be further decomposed, with respect to its outputs. Hence it can compute values for:

- internal memory ${}^c c_j$ of the Control Subsystem by using ${}^{c,c}f_j$,
- output buffer of the Virtual Effectors ${}^e_y c_j$ using ${}^{c,e}f_j$,
- output buffer of the Virtual Receptors ${}^r_y c_j$ computing ${}^{c,r}f_j$ and
- output buffer of the Transmitter ${}^T_y c_j$ utilising ${}^{c,T}f_j$.

All of them are computed using the data contained in the internal memory ${}^c c_j$ and input buffers ${}^e_x c_j$, ${}^r_x c_j$ and ${}^T_x c_j$. Hence the resulting compound formula for the Control Subsystem state evolution can be defined as follows:

$$\begin{cases} {}^c c_j^{i+1} := {}^{c,c}f_{j,m}({}^c c_j^i,\, {}^e_x c_j^i,\, {}^r_x c_j^i,\, {}^T_x c_j^i) \\ {}^e_y c_j^{i+1} := {}^{c,e}f_{j,m}({}^c c_j^i,\, {}^e_x c_j^i,\, {}^r_x c_j^i,\, {}^T_x c_j^i) \\ {}^r_y c_j^{i+1} := {}^{c,r}f_{j,m}({}^c c_j^i,\, {}^e_x c_j^i,\, {}^r_x c_j^i,\, {}^T_x c_j^i) \\ {}^T_y c_j^{i+1} := {}^{c,T}f_{j,m}({}^c c_j^i,\, {}^e_x c_j^i,\, {}^r_x c_j^i,\, {}^T_x c_j^i) \end{cases}, \tag{6}$$

$m = 1, \ldots, {}^c n_f$, where ${}^c n_f$ is the number of control subsystem transition functions. This decomposition defines the general Control Subsystem Behaviour as:

$${}^c\mathcal{B}_{j,m} \triangleq {}^c\mathcal{B}_{j,m}({}^{c,c}f_{j,m},\, {}^{c,e}f_{j,m},\, {}^{c,r}f_{j,m},\, {}^{c,T}f_{j,m},\, {}^c f^{\tau}_{j,m},\, {}^c f^{\varepsilon}_{j,m}). \tag{7}$$

Behaviour selection and switching rules are identical to the ones shown earlier.

2.5 Virtual Effector ($s = e$)

The responsibility of the Virtual Effector is to present the real effector to the control subsystem in a form that facilitates computation of control values. To do this it transforms the proprioceptive information obtained directly from the effector into values accepted by the control algorithm (e.g. it transforms encoder readings into a homogeneous matrix representing the pose of the end-effector with respect to the world coordinate frame). As the control values will also appear as, for instance, homogeneous matrices, those too must be transformed into values acceptable to the effector control hardware (e.g. PWM ratios). This also is the responsibility of the Virtual Effector. Thus the Virtual Effector operates on values delivered to it through buffers and in the same way it dispatches the values produced by it to the cooperating subsystems. Some computations might need values from previous control cycles, thus internal memory is involved. The mentioned transformations usually change the ontology – the sets of concepts that are used by neighbouring subsystems.

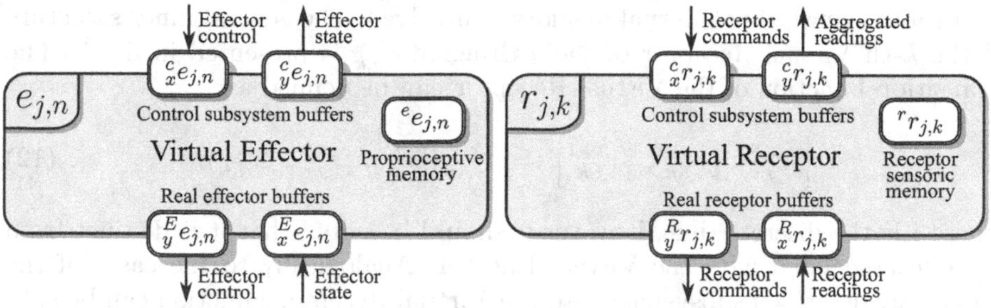

Fig. 5. Inner structures of the virtual effector $e_{j,n}$ and the virtual receptor $r_{j,k}$

The structure of the n-th Virtual Effector of agent a_j is presented in fig. 5. The state evolution of $e_{j,n}$ is described by a general Virtual Effector Function:

$$\left[{}^{e}e^{\iota+1}_{j,n},\ {}^{c}_{x}e^{\iota+1}_{j,n},\ {}^{E}_{x}e^{\iota+1}_{j,n}\right] := {}^{e}f_{j,n}\left({}^{e}e^{\iota}_{j,n},\ {}^{c}_{x}e^{\iota}_{j,n},\ {}^{E}_{x}e^{\iota}_{j,n}\right), \tag{8}$$

where ι denotes the discrete time. We introduce a new symbol due to the fact that the Control Subsystem and Virtual Effector have their own, typically different, frequencies of operations. The function (8) can also be decomposed into partial functions, which results in a set of ${}^{e}n_f$ partial functions:

$$\left[{}^{e}e^{\iota+1}_{j,n},\ {}^{c}_{y}e^{\iota+1}_{j,n},\ {}^{E}_{y}e^{\iota+1}_{j,n}\right] := {}^{e}f_{j,n,p}({}^{e}e^{\iota}_{j,n},\ {}^{c}_{x}e^{\iota}_{j,n},\ {}^{E}_{x}e^{\iota}_{j,n}), \tag{9}$$

where $p = 1, \ldots, {}^{e}n_f$. Considering further the decomposition taking into account the output buffers, the functioning of a virtual effector can be described by using Virtual Effector Memory Function ${}^{e,e}f_{j,n,p}$, Proprioceptive Function ${}^{e,c}f_{j,n,p}$ and Real Effector Control Function ${}^{e,E}f_{j,n,p}$:

$$\begin{cases} {}^{e}e_{j,n}^{\iota+1} := {}^{e,e}f_{j,n,p}({}^{e}e_{j,n}^{\iota}, {}^{c}_{x}e_{j,n}^{\iota}, {}^{E}_{x}e_{j,n}^{\iota}) \\ {}^{c}_{y}e_{j,n}^{\iota+1} := {}^{e,c}f_{j,n,p}({}^{e}e_{j,n}^{\iota}, {}^{c}_{x}e_{j,n}^{\iota}, {}^{E}_{x}e_{j,n}^{\iota}) \\ {}^{E}_{y}e_{j,n}^{\iota+1} := {}^{e,E}f_{j,n,p}({}^{e}e_{j,n}^{\iota}, {}^{c}_{x}e_{j,n}^{\iota}, {}^{E}_{x}e_{j,n}^{\iota}) \end{cases} . \quad (10)$$

The Behaviour ${}^{e}\mathcal{B}_{j,n,p}$ is defined as:

$$ {}^{e}\mathcal{B}_{j,n,p} \triangleq {}^{e}\mathcal{B}_{j,n,p}({}^{e,e}f_{j,n,p}, {}^{e,c}f_{j,n,p}, {}^{e,E}f_{j,n,p}, {}^{e}f_{j,n,p}^{\tau}, {}^{e}f_{j,n,p}^{\varepsilon}). \quad (11)$$

2.6 Virtual Receptor ($s = r$)

The Virtual Receptor primarily aggregates the sensor readings into a form acceptable by the control subsystem. On the other hand the control subsystem might want to configure or change the mode of operation of the sensors. Thus the Virtual Receptor operates on values delivered to it through buffers and through the buffers it dispatches the values produced by it to the cooperating subsystems. Some computations might need values of previous aggregations (e.g. for filtering or averaging), thus internal memory is involved. The general inner structure of the k-th Virtual Receptor of the j-th agent $r_{j,k}$ is presented in fig. 5. The Transition Function of the Virtual Receptor can be defined as:

$$\left[{}^{r}r_{j,k}^{\iota+1}, {}^{c}_{y}r_{j,k}^{\iota+1}, {}^{R}_{y}r_{j,k}^{\iota+1}\right] := {}^{r}f_{j,k}({}^{r}r_{j,k}^{\iota}, {}^{c}_{x}r_{j,k}^{\iota}, {}^{R}_{x}r_{j,k}^{\iota}), \quad (12)$$

where ι is the discrete time, however it should be noted that it is distinct from the one associated with the Virtual Effector. Analogically to the cases of the previously described subsystems also the Virtual Receptor Function can be subsequently decomposed into a set of n_{f_r} partial functions, each defined as:

$$\left[{}^{r}r_{j,k}^{\iota+1}, {}^{c}_{y}r_{j,k}^{\iota+1}, {}^{R}_{y}r_{j,k}^{\iota+1}\right] := {}^{r}f_{j,k,t}({}^{r}r_{j,k}^{\iota}, {}^{c}_{x}r_{j,k}^{\iota}, {}^{R}_{x}r_{j,k}^{\iota}), \quad (13)$$

where $t = 1, \ldots, n_{f_r}$. The following further decomposition of each partial function takes into account the outputs, and thus Sensoric Memory Functions ${}^{r,r}f_{j,k,t}$, Receptor Reading Aggregation Functions ${}^{r,c}f_{j,k,t}$ and Sensoric Memory Functions ${}^{r,R}f_{j,k,t}$ are distinguished, what results in:

$$\begin{cases} {}^{r}r_{j,k}^{\iota+1} := {}^{r,r}f_{j,k,t}({}^{r}r_{j,k}^{\iota}, {}^{R}_{x}r_{j,k}^{\iota}, {}^{c}_{x}r_{j,k}^{\iota}) \\ {}^{c}_{y}r_{j,k}^{\iota+1} := {}^{r,c}f_{j,k,t}({}^{r}r_{j,k}^{\iota}, {}^{R}_{x}r_{j,k}^{\iota}, {}^{c}_{x}r_{j,k}^{\iota}) \\ {}^{R}_{y}r_{j,k}^{\iota+1} := {}^{r,R}f_{j,k,t}({}^{r}r_{j,k}^{\iota}, {}^{R}_{x}r_{j,k}^{\iota}, {}^{c}_{x}r_{j,k}^{\iota}) \end{cases} , \quad (14)$$

where $t = 1, \ldots, {}^{r}n_{f}$. Having specified both the inner structure as well as the functions describing the state evolution for the Virtual Receptor we can define the Behaviour ${}^{r}\mathcal{B}_{j,k,t}$ as:

$$ {}^{r}\mathcal{B}_{j,k,t} \triangleq {}^{r}\mathcal{B}_{j,k,t}({}^{r,r}f_{j,k,t}, {}^{r,c}f_{j,k,t}, {}^{r,R}f_{j,k,t}, {}^{r}f_{j,k,t}^{\tau}, {}^{r}f_{j,k,t}^{\varepsilon}). \quad (15)$$

2.7 Diagnostic Needs of an Embodied Agent

Having defined the actions of all subsystems the behaviour of the whole Embodied Agent stems from the interoperation of its subsystems. Virtual effectors and receptors can carry out their own diagnostics locally, by using the ${}^{e,e}f_{j,u}$ and ${}^{r,r}f_{j,u}$ transition functions, but usually this will be just the initial preprocessing of data that has to be transferred to the control subsystem due to low computational capabilities of those subsystems. Anyway the results of the thus performed computations need to be conveyed to the control subsystem, so that it can decide as to the further actions. Interoperation means that the subsystems of the embodied agent work together to achieve a goal. The main goal is known only to the Control Subsystem, which coordinates the work of associated Virtual Effectors and Virtual Receptors. This also singles out the control subsystem as the one that should be responsible for catering to the diagnostic needs of the agent. If no abnormal situation is detected the transition functions ${}^{e,c}f_{j,n,p}$ and ${}^{r,c}f_{j,k,t}$ of the virtual effector $e_{j,n}$ and virtual receptor $r_{j,n}$ respectively should log the diagnostic data into the output buffers ${}^{c}_{y}e_{j,n}$ and ${}^{c}_{y}r_{j,k}$. From there, through the input buffers ${}^{e}_{x}c_{j,n}$ and ${}^{r}_{x}c_{j,k}$, the control subsystem c_j will obtain this data. The control subsystem c_j can either process that data internally, storing both this data and the results of its processing in ${}^{c}c_j$, or dispatch this data to another agent $a_{j'}$ for processing. In both cases processing is carried out for the purpose of detection, isolation and identification of faults. In the former case the transition functions ${}^{c,c}f_{j,m}$ will be involved, while in the latter ${}^{c,T}f_{j,m}$ will be used.

In the case of abnormal functioning of the system the situation becomes more difficult. Hence monitoring the system performance before the situation deteriorates excessively is a reasonable engineering practice that should be taken into account during its design. If the above mentioned monitoring through an appropriate definition of all subsystem transition functions fails to detect a problem before it occurs the last resort is the detection of an error condition through testing ${}^{s}f^{\varepsilon}_{j,u}$. In that case the subsystem, that has detected an error condition, should respond depending on the class of the error that occurred. In [14] three classes of errors were singled out: non-fatal, fatal and system errors. Non-fatal errors are due to inappropriate values of arguments or results of computations. Those can usually be remedied by a certain action of the control subsystem. Fatal errors are usually due to malfunction of the hardware. If the hardware is not redundant then this type of error usually leads to the termination of the currently realised task. In rare situations the system can pursue some goals that do not require the use of the hardware that failed. The system errors are due to the malfunction of the control hardware or software. If there is no redundancy in this hardware/software a catastrophe is highly probable. In all situations the system should be left in a safe state, if possible. Thus all those errors should cause the invocation of appropriate ${}^{s}\mathcal{B}_{j,\varepsilon}$ behaviours. It should be noted that diagnostic processing conducted by transition functions should lead to an early detection of abnormal functioning of the system. This in turn should lead to triggering of a non-fatal error which is much easier to remedy as an appropriate definition of ${}^{s}\mathcal{B}_{j,\varepsilon}$ behaviours for this class of errors is usually fairly simple.

3 Diagnostics in Robot Programming Frameworks

As currently robot programming frameworks are the basis for the creation of robot system controllers we analyse the diagnostic mechanisms offered by one of the most widespread robot programming frameworks – ROS (Robot Operating System) [12]. A programming framework is a library of software modules (functions, procedures, objects or components) supplemented by general use patterns [6]. ROS offers a variety of tools enabling storing, processing, analysis and visualisation of diagnostic data. However, first the reader has to be introduced to two basic concepts: nodes and topics. A ROS-based controller typically forms a network of cooperating processes called nodes, that can be located both on the same machine or can be distributed over a network of machines. Those nodes communicate with each other via a mechanism called topic, which is an implementation of the publisher-subscriber design pattern. The advantage of ROS is that due to topics those nodes do not depend on each other, and in particular they do not need to know whether someone reads from or sends data to their topics – ROS takes care of data delivery to subscribers as soon as the publisher decides to write something. Those mechanisms also form the basis for the diagnostic functions offered by ROS. ROS offers the so-called bags. Bags are typically created by subscribing to one or more ROS topics and storing every received message in a file. These files can subsequently be played back in ROS to the same topics they were recorded from, or even remapped to new topics. As the result, by using bags one can simulate once again the conditions leading to the occurrence of an error. Aside of bags, ROS also offers a specialized diagnostic toolchain, designed to collect information from hardware drivers for operators for analysis, troubleshooting and logging. It is named ROS diagnostics (after the name of a ROS package containing it) and it contains tools for collecting, publishing, analysing and viewing diagnostics data. Generally, the whole idea of the ROS diagnostics is built around a specialized *diagnostics* topic, in which hardware drivers and devices can publish special messages with their names, statuses and specific information. ROS diagnostics offers several tools for data collection, aggregation, analysis and viewing, e.g. for categorization of information collected at runtime or display of the processed data in a graphical form. Typically, those categories group diagnostic information related to a given part of the system, e.g. wheels, stereopair, lidar. This aggregated information is used by the control system to monitor the state of the associated hardware, for example, the ${}^{e}f^{\varepsilon}$ condition of the Virtual Effector monitors the state of its effectors, while ${}^{r}f^{\varepsilon}$ monitors the status of the Virtual Receptor. Nevertheless it should be stressed that the mentioned diagnostics is associated with finding the cause of a hardware malfunction, whereas it is obvious that in a real, unstructured and uncertain environment the majority of problems will be caused by the improper realization of a given high-level task. As a result the error condition ${}^{c}f^{\varepsilon}$ of the Control Subsystem not only processes the information about errors detected by the Virtual Effectors and Receptors, but also takes into account the status of the executed task, thus provides a higher level view of the situation. Hence the mechanisms provided by ROS deliver the low-level information that must be put into the proper context by the Control Subsystem. How to establish the right relationship between the low-level (ROS) monitoring and the high level context and subsequent reaction to malfunction is still an open research question.

4 Conclusions

The paper presents a general robot control system structure. Its operation is based on the definition of behaviours that in turn rely on transition functions. Monitoring the system state for the purposes of diagnostics has to be carried out by those transition functions. The behaviour of the system triggered due to the detection of an error should try to remedy the error situation or at least leave the system in a safe mode of operation. This behaviour is again defined in terms of transition functions, so no new tools are needed both for the purpose of specification and implementation of such behaviours. As there are three classes of errors at least three types of error remedying behaviours usually appear. System design is usually carried out in two stages. First the transition functions governing normal behaviour of the system are defined. In the second stage those functions are extended to take care of system monitoring data transfer and processing. Such an incremental design by extension of the specification of the transition functions is fairly simple. The definition of behaviours reacting to the detected errors requires new transition functions, however their definition is more difficult as especially in the case of fatal and system errors the necessary system behaviour is not obvious.

An interesting concept worth investigation is the inclusion of special diagnostic behaviours (e.g. [10]), which could be invoked during normal functioning of the system pursuing its goals. Obviously those behaviours collecting diagnostic data would again have to be specified in terms of transition functions. Another interesting possibility is mutual observation of robots and the effects of their actions on the environment (stigmergy [2,17]). In such a case one robot can detect an erroneous actions of another one. In effect it can execute a remedying behaviour, which of course would be defined in terms of transition functions.

Both monitoring for the purpose of diagnostics as well as error recovery from the detected errors has to be executed by the agent itself, even in multi-agent systems. If a multi-agent system is fully distributed with the agents acting independently, then nothing more can be done. In the case of multi-agent systems having a coordinator some of the diagnostic processing can be relegated to the coordinator. Obviously the behaviours of the coordinator again are defined in terms of transition functions.

Acknowledgement. This project was funded by the National Science Centre according to the decision number DEC-2012/05/D/ST6/03097.

References

1. Webster's Encyclopedic Unabridged Dictionary of the English Language. Portland House, New York (1989)
2. Bonabeau, E., Dorigo, M., Theraulaz, G.: Swarm Intelligence: From Natural to Artificial Systems. Oxford University Press, New York (1999)
3. Brooks, R.A.: A robust layered control system for a mobile robot. IEEE Journal of Robotics and Automation 2(1), 14–23 (1986)
4. Brooks, R.A.: Intelligence without reason. Artificial Intelligence: Critical Concepts 3, 107–163 (1991)
5. Cholewa, W., Kościelny, J.: Introduction. In: Korbicz, J., Kościelny, J., Kowalczuk, Z., Cholewa, W. (eds.) Fault Diagnosis. Models, Artificial Intelligence, Applications, pp. 3–28. Springer, Heidelberg (2004)
6. Kaisler, S.: Software Paradigms. Wiley Interscience (2005)
7. Kościelny, J.: Models in the diagnostics of processes. In: Korbicz, J., Kościelny, J., Kowalczuk, Z., Cholewa, W. (eds.) Fault Diagnosis. Models, Artificial Intelligence, Applications, pp. 29–58. Springer, Heidelberg (2004)
8. Kościelny, J.: Process diagnostics methodology. In: Korbicz, J., Kościelny, J., Kowalczuk, Z., Cholewa, W. (eds.) Fault Diagnosis. Models, Artificial Intelligence, Applications, pp. 59–118. Springer, Heidelberg (2004)
9. Kościelny, J.: State monitoring algorithms for complex dynamic systems. In: Korbicz, J., Kościelny, J., Kowalczuk, Z., Cholewa, W. (eds.) Fault Diagnosis. Models, Artificial Intelligence, Applications, pp. 721–762. Springer, Heidelberg (2004)
10. Kościelny, J., Bartyś, M., Syfert, M., Pawlak, M.: Models in the diagnostics of processes. In: Korbicz, J., Kościelny, J., Kowalczuk, Z., Cholewa, W. (eds.) Fault Diagnosis. Models, Artificial Intelligence, Applications, pp. 865–902. Springer, Heidelberg (2004)
11. Korbicz, J., Kościelny, J., Kowalczuk, Z., Cholewa, W. (eds.): Fault Diagnosis. Models, Artificial Intelligence, Applications. Springer, Heidelberg (2004)
12. Quigley, M., Gerkey, B., Conley, K., Faust, J., Foote, T., Leibs, J., Berger, E., Wheeler, R., Ng, A.: ROS: an open-source Robot Operating System. In: Proceedings of the Open-Source Software Workshop at the International Conference on Robotics and Automation, ICRA (2009)
13. Russell, S., Norvig, P.: Artificial Intelligence: A Modern Approach. Prentice Hall, Upper Saddle River (1995)
14. Zieliński, C.: Reaction to errors in robot systems. In: Third International Workshop on Robot Motion and Control, RoMoCo 2002, Bukowy Dworek, Poland, November 9-11, pp. 201–208 (2002)
15. Zieliński, C.: Specification of behavioural embodied agents. In: Kozłowski, K. (ed.) Fourth International Workshop on Robot Motion and Control (RoMoCo 2004), June 17-20, pp. 79–84 (2004)
16. Zieliński, C., Kornuta, T., Boryń, M.: Specification of robotic systems on an example of visual servoing. In: 10th International IFAC Symposium on Robot Control (SYROCO 2012), vol. 10, pp. 45–50 (2012)
17. Zieliński, C., Trojanek, P.: Stigmergic cooperation of autonomous robots. Journal of Mechanism and Machine Theory 44, 656–670 (2009)

Spatial Uncertainty Assessment in Visual Terrain Perception for a Mobile Robot

Przemysław Łabęcki and Piotr Skrzypczyński

Institute of Control and Information Engineering,
Poznan University of Technology, ul. Piotrowo 3A, 60-965 Poznań, Poland
{pl,ps}@cie.put.poznan.pl

Abstract. This paper addresses the issue of reliable terrain modelling from passive stereo vision data. An analytical uncertainty model for a dense stereo vision system is proposed. This model propagates the uncertainty from calibration through image processing, allowing for calculation of the uncertainty of measured 3D point coordinates. The use of this model in terrain mapping is shown with quantitative results.

Keywords: Uncertainty, stereo vision, mobile robot, terrain model.

1 Introduction

A high level of autonomy in a mobile robot can be achieved only if the robot perceives the surroundings and plans the motion on the basis of a terrain map. The classic 2D maps are insufficient for robots operating in natural or unstructured terrain, such as search and rescue robots. Such a map can be constructed from the sensory data in the form of an elevation grid, which is a 2.5D tessellated representation of the surrounding terrain. Elevation grids were conceived for early outdoor robots with 3D laser sensors [9]. In these maps each cell contains a value that represents the height of the terrain at that cell. Elevation maps can be built from point clouds yielded by 3D laser scanners, however, some search and rescue robots, such like legged machines, cannot carry bulky and heavy 3D laser scanners. Therefore, other sensing modalities are considered for these robots, such as passive stereo vision (Fig. 1A).

Despite high computational cost of image processing, passive stereo vision was successfully used in terrain mapping for walking robots [7,15]. However, very few works published so far consider stereo vision measurement uncertainty, and propagation of this uncertainty to the terrain map. Particularly, if the terrain map has to be used for local motion planning, as shown in our earlier work [1], the calculated elevation values should be accompanied by their spatial uncertainty estimates. Assessment of the spatial uncertainty allows to adapt the motion planning procedures to the terrain map, e.g. by raising the feet of a walking robot to the soecific height, depending on the known elevation uncertainty in the map. Moreover, uncertainty assessment is required for efficient multisensor data integration in the map. When the stereo camera is used together with

J. Korbicz and M. Kowal (eds.), *Intelligent Systems in Technical and Medical Diagnostics*, Advances in Intelligent Systems and Computing 230,
DOI: 10.1007/978-3-642-39881-0_30,

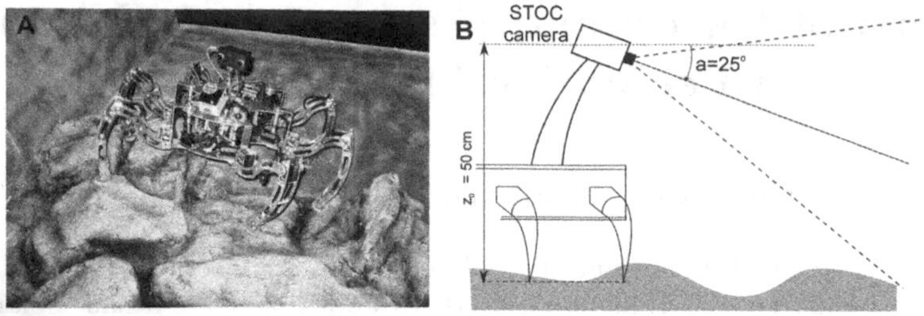

Fig. 1. Walking robot with the STOC camera (A), and configuration of the visual terrain perception system (B)

another exteroceptive sensor, such as laser scanner, the whole system should be properly calibrated [12], and the uncertainty can be taken into account when weighting the elevation measurements.

In our previous work we presented the elevation map updating algorithm, and the spatial uncertainty model for the laser scanner used to perceive the terrain immediately in front of the robot [2]. The elevation map is used for foothold planning [1]. However, to efficiently plan the robot's body path while perceiving the terrain, a sensory system with much longer range than the pitched down 2D laser scanner is needed. We employ the Videre Design STOC (STereo On a Chip) passive stereo vision camera (Fig. 1B), which yields a dense depth map in real-time, owing to the hardware-embedded image processing. Unfortunately, the STOC camera range data are sparse and corrupted by various artefacts. In [10] we presented methods for artefacts removal, and a simple, *ad hoc* measurement uncertainty measure based on the distance of the pixel from the image centre. The uncertainty measure was determined as the resolution of the camera in z_c axis, weighted by the distance w of the pixel from the image centre:

$$w = 1 - \frac{\sqrt{(u - C_x)^2 + (v - C_y)^2}}{\frac{1}{2} I_w}, \tag{1}$$

where (u, v) are the coordinates of the given pixel, (C_x, C_y) is the location of the principal point of the camera and I_w is the width of the image. While this model is simple in computation, it has an important drawback – the obtained values do not have any physical interpretation. They are merely weights, that can be used to decide if a new measurement should replace an old one in the elevation map. However, these values cannot be used to tell the robot's control algorithm the height at which it should raise a foot to "safely" move the leg from one point to another, avoiding a collision with the terrain.

Therefore, a need emerged for replacing it with a model based on systematical uncertainty analysis [11]. In this paper, an uncertainty model of the stereo vision system creating a dense depth map is presented. The model is derived analytically, based on the uncertainty propagation through the distortion and rectification algorithms. Also the uncertainty of elevation measurements was determined, taking into account the specific orientation of the camera in the terrain mapping task.

2 Related Work

Some of the grid-based algorithms for rough terrain mapping use *ad hoc* formulations for sensor fusion, e.g. Ye and Borenstein [17] apply some heuristics when they fuse consecutive laser range measurements into an elevation grid — their certainty measure is a simple hit/miss counter. On the other hand, there are 2.5D mapping methods, which are based on probabilistic foundations. The *locus* method of [9] applies simple geometric approximation in order to obtain elevation uncertainty from the sensor error model. This approximation, however, does not account for the uncertainty of measurements. Recently, a method for efficient volumetric mapping was presented, which employs the notion of positive and negative occupancy mass to represent the environment [5].

Stereo vision was used for terrain modelling with walking robots in few projects, mostly because typical stereo systems impose high costs of 3D points computation. In [7] and [15] walking robots with stereo-based perception are shown to traverse autonomously rugged terrain, but in both cases the computations are done off-board, and there is no explicit propagation of the spatial uncertainty from the stereo data to the elevation map.

The existing uncertainty models in stereo vision mainly consider systems, in which the stereo matching is applied to discrete features, and no dense depth map is created [4,14]. In [13], three-dimensional Gaussian distributions were applied to model the measurement error caused by the discrete nature of digital images used in stereo vision, and it was shown, that presenting the uncertainty in a form of a covariance matrix of the coordinates allows for reduction of the localization error. In [16] an optimal configuration of a stereo vision system, that allowed to minimize the depth measurement error, was shown. In [6] an analysis of uncertainties in a system that creates a dense depth map was made, however the optical distortion of the camera lenses was not taken into consideration and the pixel location error was calculated using only the reprojection error.

3 The Stereo Vision System

We consider the Videre Design STOC stereo vision camera as an example of the efficient visual sensor for terrain perception. The aforementioned camera utilizes the stereo vision algorithms of the Small Vision System (SVS) [8], which allows for creation of a dense depth map of the scene. The data processing is performed in the camera (using a FPGA structure), which allows to reduce the workload of the robot's on-board computer.

In the SVS system, matching of the points between the left and right image is performed on previously rectified images. Therefore, only corresponding image rows need to be analysed in order to find a match. The matching itself is performed using a correlation-based method, described in [8], which allows to calculate the disparity value (i.e. the displacement of the point on the left image with relation to the point on the right image, (Fig. 2A) that can be easily used to determine the depth of the point. The STOC camera, aside from grey scale images, can also produce disparity images. In such images, each pixel holds the calculated disparity value. Figure 2B shows an example disparity image, in which the disparity values are coded with color. Because [8] does not define which rectification algorithm is used in SVS, we assume that a method similar to the one proposed by J.-Y. Bouguet [3] is used.

Fig. 2. Disparity as the difference in the location of the matching points on two images (A), and a disparity image (B)

3.1 Image Rectification and the Stereovision System Model

The matching of pixels from the left and right image is performed on rectified images. The rectification process is such a transformation of a stereo image pair, that makes them correspond to images taken in an ideal stereo vision system, in which both images are co-planar and their corresponding rows lay along the same lines. In such situation, the epipolar lines are located along image rows, and the stereo matching is limited to corresponding rows.

In order to determine the parameters of such transformation, one should use intrinsic parameters of both cameras, the parameters of the optical distortion of the lenses and the extrinsic stereo vision system parameters. Therefore, calibration of the stereo vision system is required. The camera model used in the rectification algorithm is based on a pinhole camera model and takes radial and tangential distortion into consideration. If the coordinates of the point in 3D space $\mathbf{p_c} = [x_c \; y_c \; z_c]^T$ are known, then the coordinates x, y after the projection are as follows:

$$[x \; y]^T = \left[\tfrac{x_c}{z_c} \; \tfrac{y_c}{z_c}\right]^T \tag{2}$$

These coordinates are then distorted:

$$\begin{bmatrix} x_d \\ y_d \end{bmatrix} = \begin{bmatrix} (1 + k_1 r^2 + k_2 r^4 + k_3 r^6)x + 2p_1 xy + p_2(r^2 + 2x^2) \\ (1 + k_1 r^2 + k_2 r^4 + k_3 r^6)y + 2p_2 xy + p_1(r^2 + 2y^2) \end{bmatrix}, \tag{3}$$

where k_1, k_2, k_3, p_1, p_2 are distortion coefficients, determined in the calibration procedure, and r is the distance from the point to the image centre $r = \sqrt{x^2 + y^2}$. Finally, the coordinates on the distorted image are calculated as follows:

$$\begin{bmatrix} u_d \\ v_d \end{bmatrix} = \begin{bmatrix} x_d f_x + c_x \\ y_d f_y + c_y \end{bmatrix} \tag{4}$$

where f_x and f_y determine the focal length in pixels (in the vertical and horizontal direction, respectively), and c_x and c_y are the coordinates of the principal point of the camera.

All the coefficients used in the equations (2) – (4) are determined in the camera calibration process. In this process, also the location of the right camera in the left camera reference frame is computed. This location is determined by six extrinsic parameters: T_x, T_y, T_z, θ_x, θ_y, θ_z. The first three coefficients comprise the position vector and the remaining three determine the orientation in an axis-angle form and can be transformed to a rotation matrix using Rodrigues' rotation formula. Based on these 6 coefficients one can calculate rotation matrices that allow for image rectification. At first, each image is rotated toward the other image by half of the rotation given by θ_x, θ_y, θ_z. Therefore, if Rod() represents the Rodrigues' rotation formula, then the left and right rotation matrices are, respectively:

$$\mathbf{R}_{cL} = \text{Rod}([\theta_x \ \theta_y \ \theta_z]^T/2), \mathbf{R}_{cR} = \text{Rod}(-[\theta_x \ \theta_y \ \theta_z]^T/2). \tag{5}$$

Application of these rotation matrices will bring the images into one plane. Next, a transformation that ensures the correspondence of rows is performed. This transformation rotates the images in such a way, that their rows are parallel to the line connecting the centres of both images. The rotation axis can be thus calculated as the cross product of a vector along an image row and the vector $\mathbf{t}$, that connects the centres of the images (both vectors given in the coordinates of the left coplanar image):

$$\mathbf{l} = \mathbf{t} \times [-1 \ 0 \ 0]^T, \tag{6}$$

where $\mathbf{t} = \mathbf{R}_{cL}^{-1} \cdot [T_x \ T_y \ T_z]^T$. The rotation angle can be calculated from the definition of the dot product of the two vectors:

$$\omega = \arccos\left(\frac{\mathbf{t} \cdot [-1 \ 0 \ 0]^T}{||\mathbf{t}||}\right). \tag{7}$$

Finally, the matrix that ensures the correspondence of rows is calculated using Rodrigues' rotation formula as:

$$\mathbf{R}_{rect} = \text{Rod}\left(\frac{\mathbf{l}}{||\mathbf{l}||} \cdot \omega\right). \tag{8}$$

The matrix that joins both rotations is calculated as the product of the two following matrices:

$$\mathbf{R}_L = \mathbf{R}_{rect}\mathbf{R}_{cL}, \mathbf{R}_R = \mathbf{R}_{rect}\mathbf{R}_{cR}. \tag{9}$$

A point in the undistorted camera coordinates can be transformed using the above matrices to the rectified coordinates x_{rp}, y_{rp}, z_{rp}. For the left and right images, the $\mathbf{R}_L$ and $\mathbf{R}_R$ matrices are used, respectively:

$$[x_{rp}\ y_{rp}\ z_{rp}]^T = \mathbf{R}_L \cdot [x_d\ y_d\ 1]^T, \qquad [x_{rp}\ y_{rp}\ z_{rp}]^T = \mathbf{R}_R \cdot [x_d\ y_d\ 1]^T. \tag{10}$$

This point is then projected onto the imaging plane, and the projected coordinates are transformed to the rectified image coordinates:

$$\begin{bmatrix} u_r \\ v_r \end{bmatrix} = \begin{bmatrix} x_r f_{rx} + c_{rx} \\ y_r f_{ry} + c_{ry} \end{bmatrix}, \tag{11}$$

where $x_r = \frac{x_{rp}}{z_{rp}}$ and $y_r = \frac{y_{rp}}{z_{rp}}$. The intrinsic parameters of the rectified image: f_{rx}, f_{ry}, c_{rx}, c_{ry} are selected arbitrarily to maximize the common field of view of the left and right rectified image. Most commonly $f_{rx} = f_{ry}$ is selected.

Theoretically, the rectification process should comprise of distortion removal and then the rotation of the image to a rectified form. In practice, the order of the operations is opposite. This is due to the complicated form of the equations (3) that describe the optical distortion, which makes it difficult to determine the relationship that would allow to remove the distortion. Therefore, instead of transforming the pixels from the distorted to the rectified coordinates, for each pixel of the rectified image, its corresponding distorted coordinates are calculated. However, for the use of uncertainty analysis, an equation that removes the distortion had to be determined. To this end, linear approximation of the equations (3) was used. After bringing (3) to the implicit form:

$$\begin{array}{l} 0 = (1 + k_1 r^2 + k_2 r^4 + k_3 r^6)x + 2p_1 xy + p_2(r^2 + 2x^2) - x_d = f_x \\ 0 = (1 + k_1 r^2 + k_2 r^4 + k_3 r^6)y + 2p_2 xy + p_1(r^2 + 2y^2) - y_d = f_y \end{array}, \tag{12}$$

linear approximation in the point $\mathbf{p} = [k_{1p}\ k_{2p}\ k_{3p}\ p_{1p}\ p_{2p}\ x_{dp}\ y_{dp}\ x_{up}\ y_{up}]^T$ was performed:

$$\begin{aligned} 0 = {} & \left.\frac{\partial f_x}{\partial k_1}\right|_{\mathbf{p}} (k_1 - k_{1p}) + \left.\frac{\partial f_x}{\partial k_2}\right|_{\mathbf{p}} (k_2 - k_{2p}) + \left.\frac{\partial f_x}{\partial k_3}\right|_{\mathbf{p}} (k_3 - k_{3p}) \\ & + \left.\frac{\partial f_x}{\partial p_1}\right|_{\mathbf{p}} (p_1 - p_{1p}) + \left.\frac{\partial f_x}{\partial p_2}\right|_{\mathbf{p}} (p_2 - p_{2p}) + \left.\frac{\partial f_x}{\partial x_d}\right|_{\mathbf{p}} (x_d - x_{dp}) \\ & + \left.\frac{\partial f_x}{\partial y_d}\right|_{\mathbf{p}} (y_d - y_{dp}) + \left.\frac{\partial f_x}{\partial x_u}\right|_{\mathbf{p}} (x_u - x_{up}) + \left.\frac{\partial f_x}{\partial y_u}\right|_{\mathbf{p}} (y_u - y_{up}) \\ 0 = {} & \left.\frac{\partial f_y}{\partial k_1}\right|_{\mathbf{p}} (k_1 - k_{1p}) + \left.\frac{\partial f_y}{\partial k_2}\right|_{\mathbf{p}} (k_2 - k_{2p}) + \left.\frac{\partial f_y}{\partial k_3}\right|_{\mathbf{p}} (k_3 - k_{3p}) \\ & + \left.\frac{\partial f_y}{\partial p_1}\right|_{\mathbf{p}} (p_1 - p_{1p}) + \left.\frac{\partial f_y}{\partial p_2}\right|_{\mathbf{p}} (p_2 - p_{2p}) + \left.\frac{\partial f_y}{\partial x_d}\right|_{\mathbf{p}} (x_d - x_{dp}) \\ & + \left.\frac{\partial f_y}{\partial y_d}\right|_{\mathbf{p}} (y_d - y_{dp}) + \left.\frac{\partial f_y}{\partial x_u}\right|_{\mathbf{p}} (x_u - x_{up}) + \left.\frac{\partial f_y}{\partial y_u}\right|_{\mathbf{p}} (y_u - y_{up}). \end{aligned} \tag{13}$$

In the above equation, the vector $\mathbf{p}$ holds the values of the distortion coefficients, as well as the coordinates of the point in the distorted and undistorted image. This means, that for every pixel, the linearisation must be performed independently. If we assume:

$$
\begin{aligned}
A = {} & \left.\frac{\partial f_x}{\partial k_1}\right|_{\mathbf{p}} (k_1 - k_{1p}) + \left.\frac{\partial f_x}{\partial k_2}\right|_{\mathbf{p}} (k_2 - k_{2p}) + \left.\frac{\partial f_x}{\partial k_3}\right|_{\mathbf{p}} (k_3 - k_{3p}) \\
& + \left.\frac{\partial f_x}{\partial p_1}\right|_{\mathbf{p}} (p_1 - p_{1p}) + \left.\frac{\partial f_x}{\partial p_2}\right|_{\mathbf{p}} (p_2 - p_{2p}), \\
B = {} & \left.\frac{\partial f_y}{\partial k_1}\right|_{\mathbf{p}} (k_1 - k_{1p}) + \left.\frac{\partial f_y}{\partial k_2}\right|_{\mathbf{p}} (k_2 - k_{2p}) + \left.\frac{\partial f_y}{\partial k_3}\right|_{\mathbf{p}} (k_3 - k_{3p}) \\
& + \left.\frac{\partial f_y}{\partial p_1}\right|_{\mathbf{p}} (p_1 - p_{1p}) + \left.\frac{\partial f_y}{\partial p_2}\right|_{\mathbf{p}} (p_2 - p_{2p}),
\end{aligned}
\tag{14}
$$

then the system of equations (13) can be solved to calculate the undistorted pixel coordinates as a function of the distorted coordinates:

$$
\begin{bmatrix} x_u \\ y_u \end{bmatrix} = \begin{bmatrix}
\dfrac{\left.\frac{\partial f_y}{\partial y_u}\right|_{\mathbf{p}} A - \left.\frac{\partial f_x}{\partial y_u}\right|_{\mathbf{p}} B + \left(\frac{\partial f_y}{\partial y_u}\frac{\partial f_x}{\partial x_d} - \frac{\partial f_y}{\partial x_d}\frac{\partial f_x}{\partial y_u}\right)\Big|_{\mathbf{p}} x_d + \left(\frac{\partial f_y}{\partial y_u}\frac{\partial f f_x}{\partial y_d} - \frac{\partial f_y}{\partial y_d}\frac{\partial f_x}{\partial y_u}\right)\Big|_{\mathbf{p}} y_d}{\left(\frac{\partial f_y}{\partial x_u}\frac{\partial f_x}{\partial y_u} - \frac{\partial f_x}{\partial x_u}\frac{\partial f_y}{\partial y_u}\right)\Big|_{\mathbf{p}}} \\
-\dfrac{\left.\frac{\partial f_y}{\partial x_u}\right|_{\mathbf{p}} A - \left.\frac{\partial f_x}{\partial x_u}\right|_{\mathbf{p}} B + \left(\frac{\partial f_y}{\partial x_u}\frac{\partial f_x}{\partial x_d} - \frac{\partial f_x}{\partial x_u}\frac{\partial f_y}{\partial x_d}\right)\Big|_{\mathbf{p}} x_d + \left(\frac{\partial f_y}{\partial x_u}\frac{\partial f_x}{\partial y_d} - \frac{\partial f_x}{\partial x_u}\frac{\partial f_y}{\partial y_d}\right)\Big|_{\mathbf{p}} y_d}{\left(\frac{\partial f_y}{\partial x_u}\frac{\partial f_x}{\partial y_u} - \frac{\partial f_x}{\partial x_u}\frac{\partial f_y}{\partial y_u}\right)\Big|_{\mathbf{p}}}
\end{bmatrix}.
\tag{15}
$$

3.2 Calculating the 3D Point Coordinates

The location of the point in 3D space depends upon its location on the rectified image, the calculated disparity value and the extrinsic parameters of the stereo vision system. The z coordinate is calculated as the first one:

$$
z_c = \frac{T_{rx} \cdot f_r}{d}. \tag{16}
$$

In the above equation f_r is the focal length selected for the rectified image, d is the disparity value for the given pixel and T_{rx} is the distance between the coordinate frames of both rectified images. The disparity is calculated as the difference of the location in the u axis on the left and right rectified image:

$$
d = |u_{rR} - u_{rL}|, \tag{17}
$$

while T_{rx} is the mutual position of the cameras after being rectified:

$$
\begin{bmatrix} T_{rx} \\ T_{ry} \\ T_{rz} \end{bmatrix} = \mathbf{R_r} \begin{bmatrix} T_x \\ T_y \\ T_z \end{bmatrix}. \tag{18}
$$

In the equation (18) T_x, T_y, T_z are the parameters acquired from the calibration of the stereo vision system and $\mathbf{R_r}$ is given by (9). After the coordinates in z_c axis have been determined, one can calculate the remaining coordinates:

$$
x_c = \frac{(v_r - c_{xr})}{f_r} z, \; y_c = \frac{(u_r - c_{yr})}{f_r} z. \tag{19}
$$

In the above equation v_r and u_r are the pixel coordinates in the left rectified image and f_{xr}, f_{yr}, c_{xr}, c_{yr} are the intrinsic parameters of the left rectified image.

4 Measurement Uncertainty in a Stereo Vision System

The most obvious causes of measurement uncertainty in a stereo vision system are limited image resolution and incorrect matching. Because matching is performed on images that were rectified using parameters estimated in the calibration procedure, also the uncertainty of calibration influences the overall measurement uncertainty. While the incorrect matching is difficult to investigate due to its random nature, the two remaining causes can be analysed, because the rectification algorithm as well as the 3D coordinates calculation are given by analytical equations.

The calculations leading to the determination of measurement uncertainty are performed in several stages. First, separately for the left and right image, the uncertainty of pixel location in the undistorted image is determined, followed by the calculation of the uncertainty of pixel location in the rectified image. On this basis, the uncertainty of the computed disparity can be determined, as well as its influence on the uncertainty of 3D measurements, and in the following step, the uncertainty of elevation values in the grid map. To conserve space in the remainder of this paper only calculations for the left image are presented for the uncertainty of the pixel location in the undistorted and rectified images. However, before the uncertainty of the disparity can be determined, one must perform the same calculations also for the right image.

After expressing the uncertainty of the intrinsic camera parameters and the uncertainty of the distortion coefficient as a covariance matrix, and using the Jacobian matrix of the equation (15) with respect to the aforementioned parameters, one can calculate the covariance matrix for the pixel location on the undistorted image.

$$\boldsymbol{\Sigma}_{uL} = \mathbf{J}_u \cdot \boldsymbol{\Sigma}_{dL} \cdot \mathbf{J}_u^T. \tag{20}$$

In (20) $\mathbf{J}_u$ is the Jacobian matrix of the equation (15) with respect to the distortion coefficients and intrinsic camera parameters (k_1, k_2, k_3, p_1, p_2, f_x, f_y, c_x, c_y), while $\boldsymbol{\Sigma}_{dL}$ is the covariance matrix for these parameters and $\boldsymbol{\Sigma}_{uL}$ is the calculated covariance matrix for the pixel location in the undistorted image.

The propagation of the uncertainty through the rectification algorithm can be performed in a similar way:

$$\boldsymbol{\Sigma}_{rL} = \mathbf{J}_{rL} \cdot \boldsymbol{\Sigma}_{ueL} \cdot \mathbf{J}_{rL}^T, \tag{21}$$

where $\mathbf{J}_{rL}$ is the Jacobian matrix of the equation (11) with respect to the pixel location in the undistorted image and the extrinsic camera parameters (x_u, y_u, T_x, T_y, T_z, θ_x, θ_y, θ_z), $\boldsymbol{\Sigma}_{ueL}$ is the covariance matrix for the pixel coordinates in the undistorted image, and $\boldsymbol{\Sigma}_{rL}$ is the covariance matrix for the pixel coordinates in the rectified image.

Using the uncertainty of the location of the corresponding pixels in the rectified images, one can determine the disparity uncertainty. The disparity standard deviation is calculated as the sum of the standard deviations in the u axis in the left and right rectified images.

$$\sigma_d = \sigma_{u_{rL}} + \sigma_{u_{rR}}. \tag{22}$$

When calculating the uncertainty in the u axis of the rectified images, one must remember that the pixel coordinates in the left image are the coordinates for which the disparity is calculated, while the coordinates in the right image are the coordinates of the corresponding pixel (i.e. shifted by the disparity value with respect to the left image).

At this point, one should also take into account the limited image resolution. The SVS algorithm determines the disparity with the resolution of 1/16 pixel, which determines the lower bound of the disparity error. It is therefore assumed that the disparity error is 3 times the standard deviation of the disparity. Thus, the value of the disparity standard deviation cannot be lesser than $\sigma_{d_{min}} = 1/48$ pixel. After the disparity standard deviation has been calculated, one can compute the uncertainty of the location of the point in 3D space. Using equations (16)–(19) we define the location vector:

$$\mathbf{p} = [x_c \; y_c \; z_c]^T, \tag{23}$$

then the covariance matrix of the location of the point in 3D space can be calculated as:

$$\boldsymbol{\Sigma}_p = \mathbf{J}_p \cdot \boldsymbol{\Sigma}_{rp} \cdot \mathbf{J}_p^T, \tag{24}$$

where $\mathbf{J}_p$ is the Jacobian matrix of the fucntion $\mathbf{p}$ with respect to d, u_r, v_r and the extrinsic parameters of the stereo vision system. and $\boldsymbol{\Sigma}_{rp}$ is the covariance matrix for the disparity, the pixel location in the rectified image and the extrinsic parameters of the stereo vision system, while $\boldsymbol{\Sigma}_p$ is the covariance matrix of the location of the measured point in 3D space.

In the task of terrain map building for a walking robot, the knowledge about the elevation uncertainty is needed. Assuming that the camera is positioned at the elevation z_0=50cm above the ground, and is tilted down by the angle θ_t=25°, the elevation can be calculated as:

$$e = z_c \cdot \sin(\theta_t) - y_c \cdot \cos(\theta_t) + z_0. \tag{25}$$

The variance of the elevation σ_e^2 can be then computed as:

$$\sigma_e^2 = \mathbf{J}_e \cdot \boldsymbol{\Sigma}_{yz} \cdot \mathbf{J}_e^T, \tag{26}$$

where $\mathbf{J}_e$ is the Jacobian matrix of the equation (25) with respect to y_c, z_c and $\boldsymbol{\Sigma}_{yz}$ is the covariance matrix for the coordinates y_c and z_c.

5 Experimental Results

The Videre Design STOC camera was calibrated using the algorithm implemented in [3]. This method also estimates the error values of the calculated parameters. Because the errors are assumed to be 3 times the standard deviations of the determined values, the diagonal elements (variances) of the input covariance matrix can be determined using these errors. As the calibration procedure does not provide any covariances, it is assumed, that the parameters

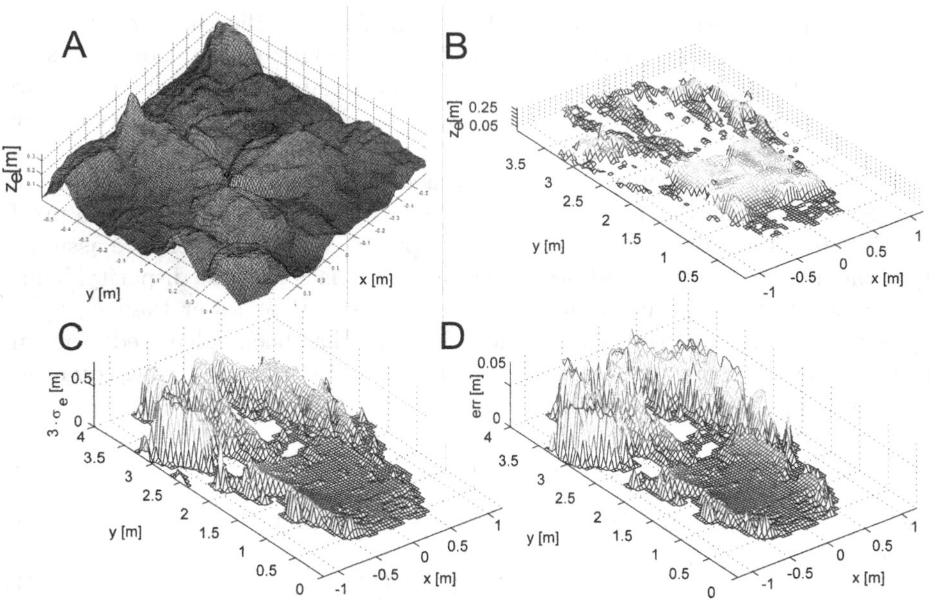

Fig. 3. Ground truth map of the rocky terrain mockup (A), elevation map created using the analytical uncertainty model (B), uncertainty map created using the analytical uncertainty model (C), uncertainty map created using the simple uncertainty model from [10] (D)

calculated by the calibration algorithm are not correlated, thus the matrix $\boldsymbol{\Sigma}_d$ remains diagonal: $\boldsymbol{\Sigma}_d = \mathrm{diag}\left(\sigma_{k1}^2, \sigma_{k2}^2, \sigma_{k3}^2, \sigma_{p1}^2, \sigma_{p2}^2, \sigma_{fx}^2, \sigma_{fy}^2, \sigma_{cx}^2, \sigma_{cy}^2\right)$.

Using the equations from section 4, the uncertainty of pixel location in the left and right rectified image was calculated. Then, uncertainty of elevation measurements was computed.

The uncertainty model was used in the task of terrain mapping with the point clouds obtained from the STOC camera. The map updating algorithm described in details in [2] was used. At the same time, a corresponding uncertainty grid map was created. The experiment was conducted using a rocky terrain mockup, for which a ground truth map is available (Fig. 3A). This map was acquired by scanning the mockup using a Hokuyo URG-04LX laser scanner, moved precisely by an industrial robot arm. When moving the camera towards the mockup, measurements were taken. The measurement uncertainty was used to determine whether the elevation value from the new measurement should replace the old elevation value. It was also used as a weighting factor when calculating the weighted average elevation in a map cell, based on all the points in that cell. Figure 3B shows a terrain map created using the analytical uncertainty model proposed here. Using the same measurements, a similar map was created with the simple uncertainty model presented in [10], and given by (1).

Both elevation maps didn't show significant differences, which can be explained by a similar spatial distribution of the uncertainty in both the simple

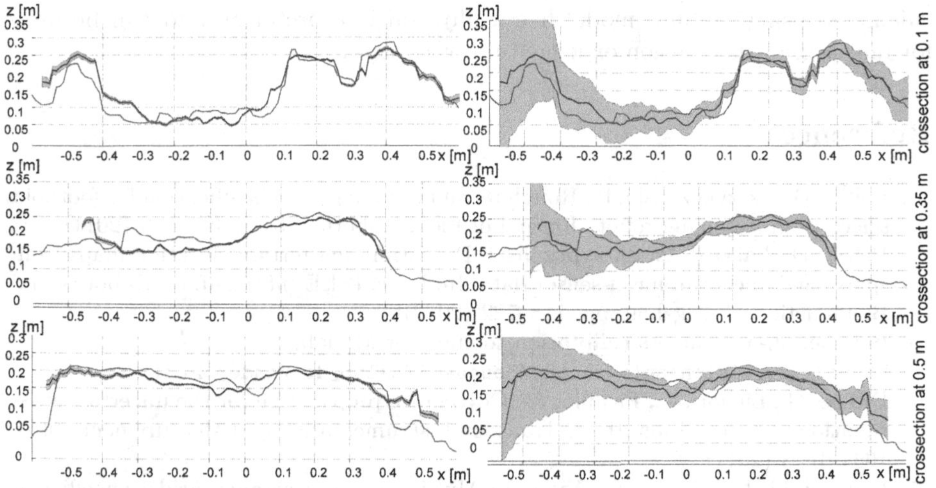

Fig. 4. Horizontal crossections through the elevation map of the mockup, at three different distances from the first measurement with envelopes representing the spatial uncertainty of elevation: left column – simple uncertainty model, right column – analytical uncertainty model

and the analytical model. This is confirmed by the similarity of the shape of uncertainty maps, created by both models (Fig. 3C and 3D). This means, that each time the same measurements were used to replace the old measurements, whether using the simple or the proposed model. It is however important to notice, that opposed to the uncertainty model (1), the results acquired from the proposed model have an easily explainable physical interpretation.

Figure 4 shows a horizontal crossection through the elevation map shown in Fig. 3B (thick line). Also, envelopes representing the uncertainty value for the measured elevation are marked for both the proposed analytical uncertainty model and the model from [10]. The uncertainty acquired from the proposed model (right column) can be interpreted as the error of elevation measurement (taken as three times the standard deviation of elevation), which can be used in legged robot control [1]. Figure 4 also shows (thin lines) a horizontal section through the ground truth map. When using the proposed new uncertainty model, the envelopes encompass the real terrain profile, which is not the case for the simple model.

6 Conclusions

In this paper, an analytical uncertainty model was introduced for the terrain elevation measurements using a stereo vision system that creates a dense depth map. Benefits due to using this uncertainty model in terrain mapping for a walking robot were demonstrated. Although no substantial improvement in map building over the previously used simple model was noticed, the uncertainty calculated using the new model has a physical interpretation and can be used for multisensor data fusion or legged robot control.

References

1. Belter, D., Skrzypczyński, P.: Rough terrain mapping and classification for foothold selection in a walking robot. Journal of Field Robotics 28(4), 497–528 (2011)
2. Belter, D., Łabęcki, P., Skrzypczyński, P.: Estimating terrain elevation maps from sparse and uncertain multi-sensor data. In: Proc. IEEE Int. Conf. on Robotics and Biomimetics, Guangzhou, pp. 715–722 (2012)
3. Bouguet, J.-Y.: Camera calibration toolbox for Matlab, `http://www.vision.caltech.edu/bouguetj/calib_doc/index.html`
4. Di Leo, G., Liguori, C., Paolillo, A.: Covariance propagation for the uncertainty estimation in stereo vision. IEEE Trans. on Instrumentation and Measurement 60(5), 1664–1673 (2011)
5. Dryanovski, I., Morris, W., Xiao, J.: Multi-volume occupancy grids: an efficient probabilistic 3D mapping model for micro aerial vehicles. In: Proc. IEEE/RSJ Int. Conf. on Intell. Robots and Systems, Taipei, pp. 1553–1559 (2010)
6. Kim, W., et al.: Performance analysis and validation of a stereo vision system. In: IEEE Int. Conf. Systems, Man and Cybernetics, pp. 1409–1416 (2005)
7. Kolter, J.Z., Kim, Y., Ng, A.Y.: Stereo vision and terrain modeling for quadruped robots. In: IEEE Int. Conf. on Robotics and Automation, Kobe, pp. 1557–1564 (2009)
8. Konolige, K.: Small vision systems: Hardware and implementation. In: Proc. Eighth Int. Symp. on Robotics Research, Hayama, pp. 111–116 (1997)
9. Kweon, I.S., Kanade, T.: High-resolution terrain map from multiple sensor data. IEEE Trans. Pattern Analysis and Machine Intell. 14(2), 278–292 (1992)
10. Łabęcki, P.: Improved data processing for an embedded stereo vision system of an inspection robot. In: Březina, T., Jabłoński, R. (eds.) Mechatronics: Recent Technological and Scientific Advances, pp. 749–757. Springer (2011)
11. Łabęcki, P.: Measurement uncertainty in a terrain map building task using sterovision. Postępy Robotyki 2, 583–594 (2012) (in Polish)
12. Łabęcki, P., Belter, D.: Sensory system calibration method for a walking robot. Journal of Automation, Mobile Robotics and Intelligent Systems 7(2), 39–45 (2013)
13. Matthies, L., Shafer, S.A.: Error modeling in stereo navigation. IEEE Journal Robot. Autom. 3(3) (1987)
14. Miura, J., Shirai, Y.: An uncertainty model of stereo vision and its applications to vision-motion planning of robot. In: 13th Int. Joint Conf. on Artificial Intelligence, Chambery, pp. 1618–1623 (1993)
15. Rusu, R.B., et al.: Leaving flatland: Efficient real-time 3D perception and motion planning. Journal of Field Robotics 26(10), 841–862 (2009)
16. Sahabi, H., Basu, A.: Analysis of error in depth perception with vergence and spatially varying sensing. Computer Vision and Image Understanding 63, 447–461 (1996)
17. Ye, C., Borenstein, J.: A novel filter for terrain mapping with laser rangefinders. IEEE Trans. on Robotics and Automation 20(5), 913–921 (2004)

Vision-Based Convoy Forming for Mobile Robots

Damian Kowalów and Maciej Patan

Institute of Control and Computation Engineering
University of Zielona Góra
ul. Podgórna 50, 65-246 Zielona Góra, Poland
d.kowalow@gmail.com,
m.patan@issi.uz.zgora.pl

Abstract. An approach is proposed to form a convoy of vehicles autonomously following a given path. Particularly, the part of the complex mobile robot guidance system related to leader-follower control process is presented in detail. The ultimate objective is to use the local vision system of mobile robotic platform to follow the moving goal as accurate as possible simultaneously keeping the constant distance from the leading robot. Then, a proper controller design is proposed together with the implementation for the Amigobot mobile robotic platform from Adept Mobile Robots Inc. A verification of the guidance system performance via navigation experiments is also presented.

Keywords: mobile robots; convoy forming; controller design.

1 Introduction

The importance of autonomous robot control in unknown environment has already been recognized in numerous engineering domains. It is also well known, that many inspirations for practical applications is taken from self-organized behaviors in biological groups. The formation of the convoy or group is a natural phenomenon in nature and occurs in different ways such as fish shoals or keys of birds. All elements of a group cooperate in the herding behavior, that can enhance its effectiveness in detecting predators, finding food and also decrease energy consumption when group is moving [6].

In engineering applications, a similar benefits can be obtained when solving navigation or route planning problems for swarms of mobile robots. Therefore, an extensive research related to the tasks of convoying is conducted, especially in the context of a network of autonomous mobile robots [2,7]. An example here might be the *Safe Road Trains for the Environment* (SARTRE) project, which aims to create a convoy of cars autonomously following the leading truck [9]. Apart from the transport security issues, elimination of the uncontrolled situations on the road generated by the other drivers leads also to significant fuel savings, less motion resistance and environmental benefits. Another motivating example are convoys of *High Mobility Multipurpose Wheeled Vehicles* (HMMWV) used by the US Army during operations in Iraq [1].

J. Korbicz and M. Kowal (eds.), *Intelligent Systems in Technical and Medical Diagnostics*, Advances in Intelligent Systems and Computing 230,
DOI: 10.1007/978-3-642-39881-0_31,

New technology strives to reduce the participation of the people in remote control and take advantage of the artificial intelligence methods in combination with modern sensor technology. Because of easy access to the video equipment and cheap vision subsystems, cameras are one of the very promising ways to support control tasks. Through the use of the vision it is possible to significantly enhance control design related to the navigation of robotic vehicles [5,14]. Simultaneously, vision system forms a base for the application of algorithms of advanced control and navigation [19].

Organized group navigation (including convoying) is a fundamental problem of mobile robotics and constitutes the area of scientific investigation since the late 80s [16,17]. Convoying is a very complex algorithmic task, and although many years of intense research, there is still a lack of cheap and universal systems for its effective solution [4,20]. This fact stands for a strong motivation for the development of new methods and further improvement of existing techniques for different mobile platforms.

The main purpose of the research undertaken here was to establish a practical approach to autonomous steering of mobile robotic platform with the use of simple vision system in order to follow a leader, simultaneously satisfying the constraints related to proper positioning and velocity control. We also wish to demonstrate how the existing control techniques can be combined to build a reliable navigation subsystem and solve the underlying control problems. More precisely, a proper controller design is provided and algorithms are implemented on the Amigobot mobile robotic platform from Adept Mobile Robots Inc. being a popular drive-differential robot.

2 Problem Description

2.1 Robot Control Paradigm

The robot can be controlled via an appropriate algorithm and control can take place with the support of information gathered from internal or external factors. Taking into account the potential ways of controlling the mobile robot four main classes of robot control paradigm can be distinguished [13,10]. One of the first attempts to structured control was the *deliberative control* method which was used to develop an action plan based on data gathered from sensors and internal knowledge or symbolic representations of the world. To be able to assess the results, a complex model is required, which entails large computational burden and demands high amount of memory. In effect, this approach provides the opportunity to determine the optimal control. The paradigm was inspired by the research on artificial intelligence in the late 60s and was successfully implemented by Nilsson on the famous *Shakey* robot [18].

A completely different approach is represented by the so-called *reactive control*. This method uses only the data from the sensors and effectors to build the behaviour in response to the external stimulus. In general, this approach is biologically inspired by the mechanism of reflexes. A first system which successfully exploited this control law was proposed in the mid 80s with the so-called *subsumption architecture* by Brooks [3]. This architecture is still popular and widely used in practice because it leads to relatively simple and low resource demanding algorithms.

The third approach is the *hybrid method* which constitutes a combination of deliberative and reactive methods forming the two subsystems. Hybrid systems combine the advantages of the planning and reactive paradigms and simultaneously try to avoid the disadvantages of these architectures.

As a last technique the *behavioral approach* being a modification of reactive control can be distinguished. This architecture consists in an assignment of specific behaviors to particular goals to achieve. The database of system history is created allowing for implementation of learning algorithms. Thus, the previous errors can be used to maximize current performance, what is very advantageous in highly varying environments. Building a navigation subsystems for the robot which follows the leader, the reactive technique seems to be the most suitable, allowing for immediate adaptation to external conditions such as dynamic change of leader pose.

2.2 Robot Motion Model

With no loss of generality, we assume that the motions of each following vehicle is described by

$$\dot{\xi}(t) = f(\xi(t), u(t)) \quad \text{a.e. on } T, \quad \xi(0) = \xi_0 \tag{1}$$

where $\xi \in \mathbb{R}^d$ is a robot pose, $T = (0, t_f]$ a bounded time interval, a given function $f : \mathbb{R}^d \times \mathbb{R}^\ell \to \mathbb{R}^d$ is required to be continuously differentiable, $\xi_0 \in \mathbb{R}^d$ defines an initial robot configuration, and $u : T \to \mathbb{R}^\ell$ is a measurable control function which satisfies

$$u_{\min} \leq u(t) \leq u_{\max} \quad \text{a.e. on } T \tag{2}$$

for some constant vectors $u_{\min}$ and $u_{\max}$.

Given any initial position ξ_0 and any control function u, there is a unique absolutely continuous function $\xi : T \to \mathbb{R}^n$ which satisfies (1) a.e. on T. In what follows, we will call it the sensor trajectory corresponding to ξ_0 and u.

2.3 Induced State Constraints

In reality, some restrictions on the motions are inevitably induced. First of all, robot should stay within the admissible region Ω_{ad} where there are no obstacles. We assume that it is a compact set defined as follows:

$$\Omega_{\text{ad}} = \{\xi \in \Omega \cup \Gamma \mid b_i(\xi) \leq 0, \, i = 1, \dots, I\} \tag{3}$$

where b_i's are given continuously differentiable functions. Accordingly, the conditions

$$b_i(\xi(t)) \leq 0, \quad \forall t \in T \tag{4}$$

must be fulfilled, where $1 \leq i \leq I$.

2.4 Leader Path Planning

As mentioned above, motion capabilities of the robot may be severely limited. Denoting as $\chi(t)$ the given reference path for leader robot, we shall be primarily concerned with restrictions imposed on accuracy of following this trajectory. The total error over the interval $[0, t_f]$ can be formally defined by analogy to the problems in [23,22,21] as

$$e(t_f) = \int_0^{t_f} \|\chi(t) - \xi(t)\| \,\mathrm{d}t \tag{5}$$

where $\|\cdot\|$ signifies the Euclidean norm.

The goal in the optimal measurement problem is to determine the forces (controls) applied to each vehicle conveying a sensor, which minimize a design criterion $e(t_f)$ under the constraints on the magnitude of the controls and induced state constraints. In order to increase the degree of optimality, in our approach we will regard ξ_0 as a control parameter vector to be chosen in addition to the control function u.

The above control problem can be interpreted as an optimization problem over the set of feasible pairs

$$\mathcal{P} = \big\{(\xi_0, u) \mid \xi_0^j \in \Omega_{\mathrm{ad}}, u : T \to \mathbb{R}^\ell \text{ is measurable,} \\ u_{\min} \le u(t) \le u_{\max} \text{ a.e. on } T\big\}. \tag{6}$$

This leads to the following formulation:

Problem 1. Find the pair $(\xi_0, u) \in \mathcal{P}$ *which maximizes*

$$J(\xi_0, u) = e(t_f) \tag{7}$$

subject to the constraints (1), (2) *and* (4).

2.5 Follower Path Planning

The follower robot should not only properly position itself behind the leader but also adapt its translational velocity in order to keep a proper distance. Therefore the control problem here is a multipurpose one. As a solution to this, a common control structure is the nested control loop [5]. The outer loop is responsible for maintaining an appropriate follower robot pose and determines the velocity required to keep the desired distance from the leader. This is based on the image positioning error generated by the vision system of the follower. The inner loop is responsible for maintaining this velocity as demanded by outer loop. The structure is depicted on the Fig. 1. This is based on the image scaling error of the vision system. The situation when follower is behind the leader is shown on the Fig. 2(b), where error for controller loop is computed from difference between desired zero and actual position of tracked area (part of image representing the leader robot). In such a way, the planning task consists in appropriate tuning of the PID controllers for the inner and outer loops.

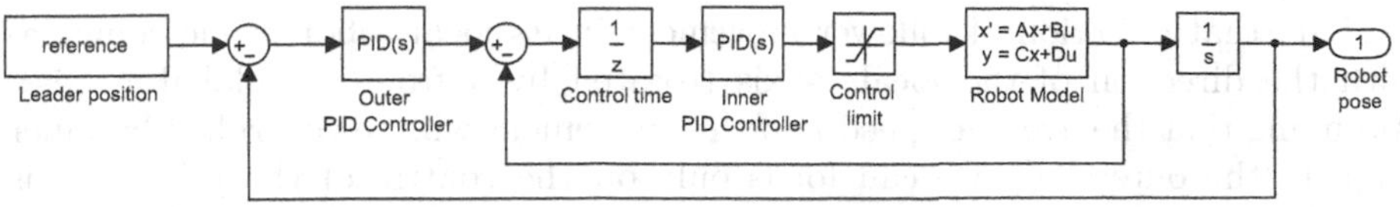

Fig. 1. Position-velocity nested control for the follower path planner

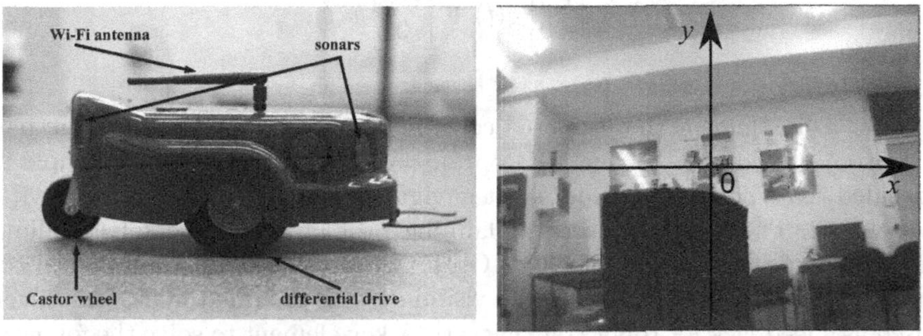

(a) (b)

Fig. 2. Amigobot differential drive robot (a) and view from vision system (b)

3 Illustrative Example

3.1 Modelling and Controller Design

All experiments were carried out using the AmigoBOT mobile robot manufactured by Adept Mobilerobots Inc. This classic differential drive robotic platform developed for the educational and research purposes is depicted in Fig. 2(a). The robot can be controlled by velocity setting for each wheel driven by two independent DC motors. Let us denote as v_l and v_r the velocities of the left and right wheel, respectively.

When the velocities of left and right wheel are constants and $v_l \neq v_r$, robot's center of gravity moves at a constant speed $v = \frac{v_l + v_r}{2}$ along the circumference of a circle and its center is located on the axis of the wheels. Introducing the radiuses of movements r_l and r_r for left and right wheels, respectively, we have the angular velocity

$$\dot{\theta} = \frac{v_r - v_l}{r_r - r_l}, \tag{8}$$

where θ is an azimuth of robot motion. Further, denoting by $u = \frac{v_l - v_r}{2}$ the change of the robot pose in the global (inertial) frame takes the following form

$$\dot{\theta} = -\tfrac{2}{b}u, \tag{9}$$

$$\dot{x} = v\cos\theta, \tag{10}$$

$$\dot{y} = v\sin\theta. \tag{11}$$

where b is the wheelbase of a mobile robot.

Note, that direction of follower movement in its local robot frame coincides with the direction of the local x-axis (rotated by θ from the global x-axis). Assuming that the average speed v of this movement will be controlled by inner loop, in the outer loop we can focus only on the control of the robot y-axis position and azimuth. Consequently, under small θ values, $\sin\theta \approx \theta$, and the robot model can be represented by

$$\dot{\xi}(t) = \begin{bmatrix} 1 & 0 \\ v & 0 \end{bmatrix} \xi(t) + \begin{bmatrix} -\frac{2}{b} \\ 0 \end{bmatrix} u(t), \tag{12}$$

where $\xi = [\theta, y]^\top$.

In order to implement the proposed control scheme a MATLAB program has been written providing GUI and PID optimization features. In this example it was decided that the convoy leader must navigate the pre-determined path with random speed. The leader trajectory planning problem has been solved offline using **fmincon** function from MATLAB Optimization Toolbox. The application has been also integrated with an ACTS vision system provided by the producer and used as a feedback for robot tracking [11]. A key element to solve the follower control task is the ability to suitable choose a PID controllers for a mobile robot [15]. As for the inner loop the proportional architecture was implemented to simplify problem. As for the outer loop the full PID controller was implemented.

The optimal PID parameters were determined using the MATLAB Control System Toolbox and the unconstrained derivative-free minimizer **fminsearch** [8] with the penalty function defined using robot state-space model (12). Graphical

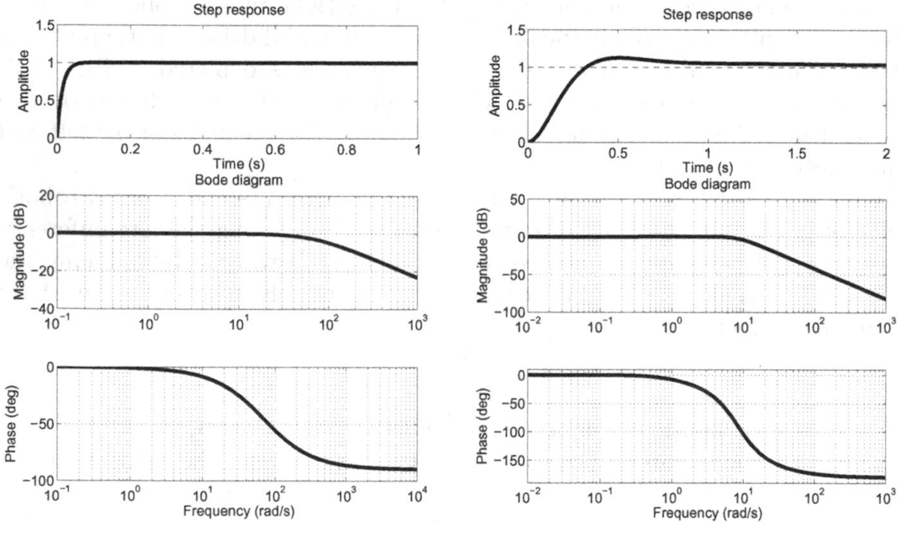

(a) (b)

Fig. 3. Result of using penalty function for controller design (a) and result of using compensator to control mobile robot result of iterative search of controller gains is shown on Fig. 3(a) and the final controller transmittance was

$$C(s) = 12.8 + \frac{0.0004}{s} + 69.5\frac{s}{0.1s+1}. \tag{13}$$

Alternatively, for comparison, the selection of PID parameters was made with the use of MATLAB–build in tool Control and Estimation Manager. As a result the compensator with transmittance

$$C(s) = 3.04 + \frac{0.137}{s} + 6.09\frac{s}{0.1s+1}, \tag{14}$$

has been determined and its simulation on the robot model is shown in Fig. 3(b).

3.2 Experiments with Mobile Robot

The validation of the developed controllers was tested by verification of the follower tracking response in comparison with changes in the position of the

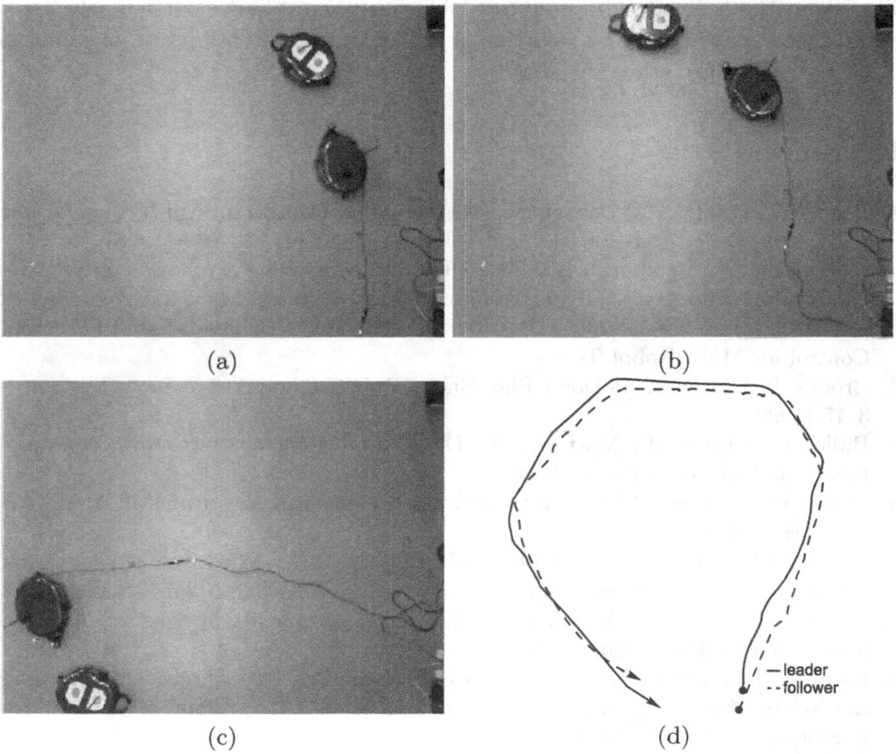

Fig. 4. Realization of an exemplary testing passage for the leader and follower robots (a–c), and the final paths (d)

leader robot. Tests were made on several pre-defined routes, where the robot leader had implemented deliberative control architecture. However, the tracking robot has reactive control architecture on board. The study was initiated by passing to robot a trajectory consisting of one arc passes then the straight lines, route passage of both robots is shown in Fig.4.

Another attempt was crossing on the set of random shapes points, which resembles an oval.

4 Conclusions

The results presented in this work presents the part of the navigation system for convoy creation. It is shown that adequate knowledge of the object model, contributes to the quality of the design of the controller. The ability to design for the controller is especially needed in the control system that may cause danger.

The proposed approach shows the path in the design of a mathematical model in the state space for a mobile robot and a practical choice of regulator for the vision system. Tests consisted in the verification of developed control algorithms with the real robotic platform.

The project is ongoing and future research will be focused on the application of different methods for determining available control methods, particularly based on the state–space control, in particular, the Kalman filtration and the control with feedback of the state variable.

References

1. Balch, T., Arkin, R.C.: Behavior-based Formation Control for Multi-robot Teams. IEEE Transactions on Robotics and Automation 14(6), 926–939 (1998)
2. Belkhouche, F., Belkhouche, B.: Modeling and controlling a robotic convoy using guidance laws strategies. IEEE Transactions on Systems, Man, and Cybernetics, Part B 35(4), 813–825 (2005); Balch, T., Arkin, R.C.: Behavior-based Formation Control for Multi-Robot Teams
3. Brooks, R.A.: Elephants Don't Play Chess. Robotics and Autonomous Systems 6, 3–15 (1990)
4. Bullo, F., Cortés, J., Martínez, S.: Distributed Control of Robotic Networks. Princeton University Press (2009)
5. Corke, P.: Robotics, Vision and Control: Fundamental algorithms in MATLAB. Springer (2011)
6. Couzin, I.D., Krause, J., Franks, N.R., Levin, S.A.: Effective leadership and decision-making in animal groups on the move. Nature 433, 513–515 (2005)
7. Dudek, G.,Jenkin, M., Milios, E., Wilkes, D.: Experiments in Sensing and Communication for Robot Convoy Navigation (1995)
8. Lagarias, J.C., Reeds, J.A., Wright, M.H., Wright, P.E.: Convergence properties of the Nelder-Mead simplex method in low dimensions. SIAM Journal of Optimization 9(1), 112–147 (1998)
9. Coelingh, E., Solyom, S.: IEEE Spectrum IEEE 49(11), 34–39 (2012)
10. Siegwart, R., Nourbakhsh, I., Illah, R.: Introduction to Autonomous Mobile Robots. MIT Press, Boston (2004)

11. Whitbrook, A.: Programming Mobile Robots with Aria and Player. A Guide to C++ Object-Oriented Control. Springer (2010)
12. Aström, K.J., Murray, R.M.: Feedback Systems: An Introduction for Scientists and Engineers. Princeton University Press, Princeton (2008)
13. Murphy, R.: Introduction to AI Robotics. MIT Press, Cambridge (2000)
14. Horn, B.: Robot vision. MIT electrical engineering and computer science series (1986)
15. Ledin, J.: Embedded Control System in C/C++. CMP Media (2004)
16. Smith, R.C., Cheeseman, P.: On the representation and estimation of spatial uncertainty. The International Journal of Robotics Research 5(4), 56–68 (1986)
17. Durrant-Whyte, H.F.: Uncertain geometry in robotics. IEEE Journal of Robotics and Automation 4(1), 23–31 (1988)
18. Nilsson, N.J.: Shakey The Robot, Technical Note 323. AI Center, SRI International, Menlo Park (1984)
19. Santana, A.M., Adelardo, A.M.: Straight-lines modelling using planar information for monocular SLAM. International Journal of Applied Mathematics and Computer Science 22(2), 409–421 (2012)
20. Orjuela, R., Marx, B., Ragot, J., Maquin, D.: Predictive control for trajectory tracking and decentralized navigation of multi-agent formations. International Journal of Applied Mathematics and Computer Science 23(1), 103–115 (2013)
21. Patan, M.: Optimal Sensor Networks Scheduling in Identification of Distributed Parameter Systems. LNCIS, vol. 425. Springer, Heidelberg (2012)
22. Uciński, D., Patan, M.: Sensor network design for the estimation of spatially distributed processes. International Journal of Applied Mathematics and Computer Science 20(3), 459–481 (2010)
23. Tricaud, C., Patan, M., Uciński, D., Chen, Y.: D-optimal trajectory design of heterogeneous mobile sensors for parameter estimation of distributed systems. In: Proc. American Control Conference, ACC 2008, Seattle, USA, Jun 11-13, pp. 663–668 (2008)

Cognitive Memory for Intelligent Systems of Decision-Making, Based on Human Psychology

Zdzisław Kowalczuk and Michał Czubenko

Faculty of Electronics, Telecommunications and Informatics,
Gdańsk University of Technology, Narutowicza 11/12 Gdańsk, Poland
kova@pg.gda.pl, m.czubenko@gmail.com

Abstract. A model of memory, which allows to expand the functionality of autonomous decision-making systems for robotic and human-system-interface purposes, is discussed. The model provides functions and features which make the prospective robotic and interfacing systems more human-like as well as more efficient in solving interactive issues.

Keywords: cognitive system, recognition, decision making, robots, humanoids.

1 Introduction

Nowadays, humanoid robots are designed for many applications (mostly due to their mobile abilities) like fire-fighting, scouting, etc. [3]. There are also several projects that concern modelling of the human brain. They focus on various aspects, like computational models of emotions [13], behaviour-based robotics [4], motivation theories [40], agents systems [37], and many others. New technologies are launched, based on human psychology and artificial intelligence. Certainly, they also contribute to new methods of machine learning.

Expectations for humanoid robots suggest a behaviour similar to human. Thus most of new discoveries have analogies in the nature, according to the cybernetic approach. Similarly, a memory system of an autonomous robot can also be inspired by the human memory. This paper presents a cognitive type of memory, designed for the Intelligent System of Decision-making (ISD) [23,25], based on human psychology. In particular, the ISD project demonstrates how to integrate the models of cognitive and personality psychology into an effective decision-making system.

There are several works which treat the subject of cognitive memory modelling. The multi-agent approach [9] appears to be useful for the concept of this work. It sets classes of ontologies linked to a reasoning engine for the purpose of modelling the system's environment. This allows an agent to infer in a human-like manner. Another approach [49] models the human ability to forget, and an architecture of a long-time memory for a virtual human. An alternative concept proposes a model of attentional learning, for applications in cognitive agents [14]

J. Korbicz and M. Kowal (eds.), *Intelligent Systems in Technical and Medical Diagnostics*, Advances in Intelligent Systems and Computing 230,
DOI: 10.1007/978-3-642-39881-0_32, © Springer-Verlag Berlin Heidelberg 2014

using many types of memories. An important psychological & neurological point of view is described in [41], with limitation to an episodic memory.

2 Human Memory – A Psychological Viewpoint

The human memory can be treated as a process and as an ability to store items. Thus memory can be treated as a capability to provide the service of storing concrete and abstract information. It is a component of intelligence, and it is results from genetic predispositions of humans. Its specific functions are associated with the storage of information that can be progressively and sequentially improved. On the other hand, the memory process is cyclic and consists of universal phases, identical for all people, and the specific phases (mainly, encoding and reconstructing), which can be subjected to improvement [31].

The simplified (classical) memory process can be divided into three basic phases: memorization, storage, and reproduction. Certainly, more phases can be identified, which are responsible for encoding information and adding new aspects to the knowledge (*e.g.* updating the context). Figure 1 shows our view on phases of the memory process based on the work of Tulving [43].

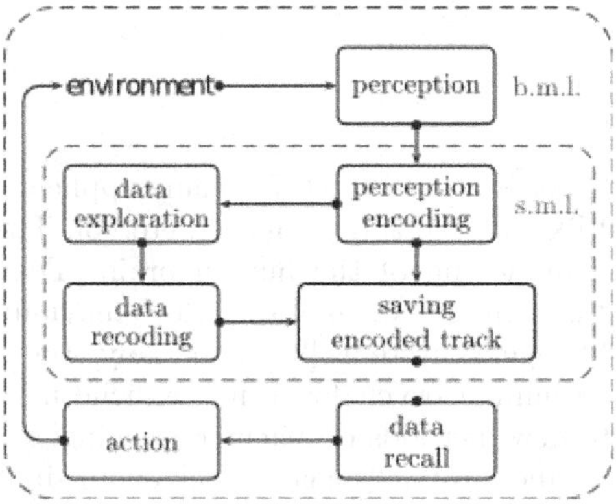

Fig. 1. Phases of a memory process

Within the memory process (of the Tulving type), we can distinguish two loops: a small memory loop (s.m.l.) and a big memory loop (b.m.l.). The small loop is based on the perception and memorising of information, as well as on the analysis of recorded data (in accordance with the knowledge). It concerns receiving multiple information. Whereas the large memory loop (feedback) performs *analysis* of available information and inner *perception* of new (discovered) contexts (deducted facts).

Human memory (as a capability) can be categorized due to the time of storage, elements which are stored, and the types of remembered features. According to the time of storage, memory can be catalogued as [31,34]:

- sensory memory (USTM, ultra-short-time memory) - used for latching and keeping stimuli (rough data or raw information). The average storage time in this kind of memory is about 0.5 seconds. It can be divided with respect to senses [33] (the types of remembered elements):
  - iconic memory
  - echoic memory
  - smell memory
  - touch memory
  - taste memory
- short-time memory (STM), which is a kind of a limited container of data (abstract items needing immediate attention). It can contain about 7 ± 2 abstract elements [32] or about 80 bits of information [15]. The average storage time is 15-18 seconds (what can be extend by a kind of inner repetition). From the viewpoint of the format of writing information, it can be divided into:
  - low-level STM (Low-STM) keeping noticed impressions
  - high-level STM (High-STM) containing identified discoveries (objects)
- long-time memory (LTM), being in line with the common sense of an abstract memory, and storing a wider set of objects and system reactions (its capacity is practically unlimited just as its storage time).

This kind of partition, being used in various scientific projects, e.g. [42,16,45], is also applied in the ISD project (in its cognitive model form shown in Fig. 3).

An even more important division is, however, based on the content (not type!) of stored information. We can thus identify semantic, episodic and procedural memories (Fig. 2). Semantic memory stores knowledge about perceived (not only!) objects, relations between them, and how one should understand facts [44]. Episodic memory indicates the momentum (in time) of appearance of certain situation (objects are emplaced in a certain space/map and in a sequence of relations/interactions between them). Procedural memory contains direct experiences, possible (simple) reactions and their descriptions. The most commonly used reactions became subconscious (as one experiences, for instance, when driving a car).

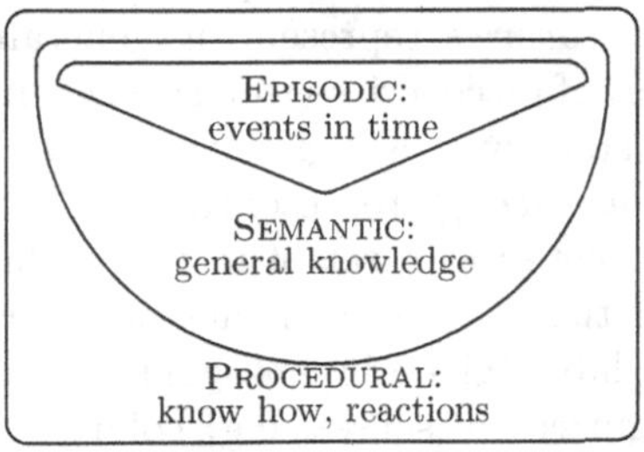

Fig. 2. Structure of memory based on [39]

The latter partition is based on a type of stored information. As mentioned above, the number of memory types matches the number of senses (it is clear that agents cannot store information about the taste in a form of an image). The fundamental types are iconic and echoic, since those two senses are most commonly used by human. Echoic memory can store information about the other senses, only if the corresponding impressions are well described (in a language).

It can be noted that inadvertently our own independently-developed model of cognitive psychology used in the ISD project (Fig. 3), reflects quite close the Tulving processes [25]. Stimuli that are shown in a perception block, are encoded to impressions (the primitives, or basic features of an object) and than recoded into a discovery form (representation of a single object). During this process the discoveries from the LTM (long-time memory) are explored (in the context of currently recoded objects), what refers to the small memory loop. The big memory loop is encapsulated in the thinking process, where the agent selects (through the attention) an action to be performed (actions with respect to both the external <physical> environment and the internal <mind> environment). This approves the coherency of the introduced model of cognitive psychology (in the memory context).

3 Model of Memory for Intelligent Agents

Note that the functional memory structure of Fig 3 coincides with the time of duration, or storing the data. Such a division is useful for all systemic and electronic approaches. It contains a sensory buffer (ultra-short time memory), an operational memory (RAM - short time memory) and a storage memory (hard drive - long-time memory). The memories can be further categorized into to types: semantic and sequential (Fig. 4).

3.1 Semantic Memory

Semantic memory has two kinds of declarations, and two interconnected semantic networks. Both networks consist of discoveries. A discovery (Fig. 5) contains a description of a known (at least in part) object. We could differentiate abstract discoveries (*a-discoveries* representing generalized objects, e.g. horse, flower, etc.) and instances of discoveries (*i-discoveries* meaning specific objects, e.g. the Silver Star, or a sunflower that grows in my garden).

Abstract discovery (*a-discovery*), representing the layer of T-boxes (classical-terminology elements), generalizes a set of instances/facts (*i-discoveries*). Any abstract discovery (other than a sheer abstract, which is a parent to all discoveries), has its own name/label (given by an agent), a list of impressions (simple features which are recognized by senses; e.g. color, texture, shape, etc.), and a list of relations. It may have a list of sub-emotional contexts (emotions associated with certain objects; e.g. when the agent likes flowers <a-discovery>, it would associate with a positive sub-emotion context like 'joy'), and a list of

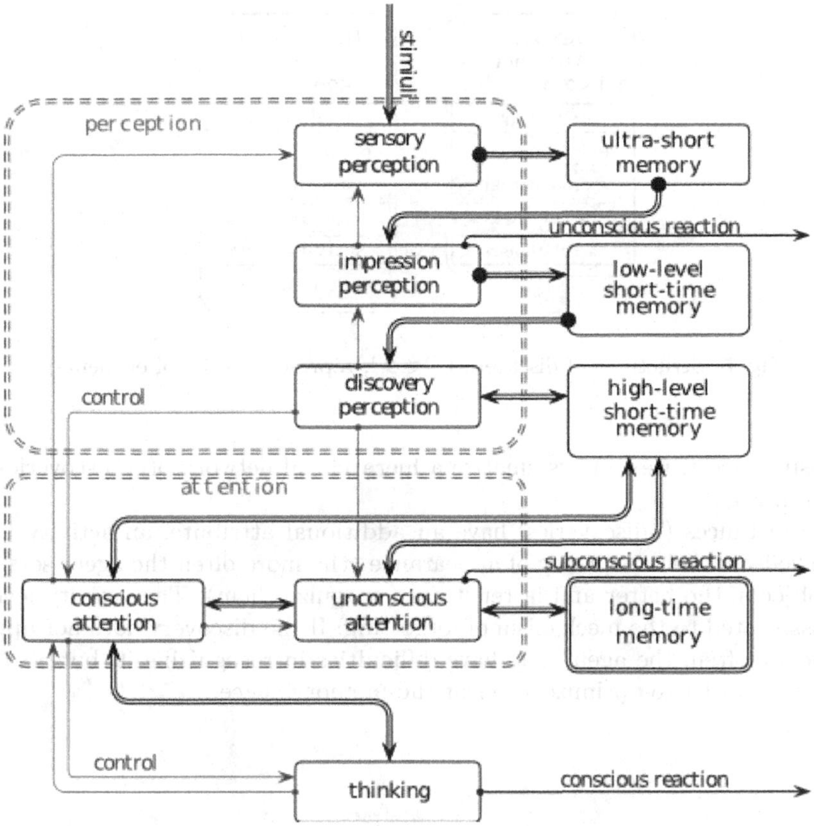

Fig. 3. Model of cognitive psychology [25]

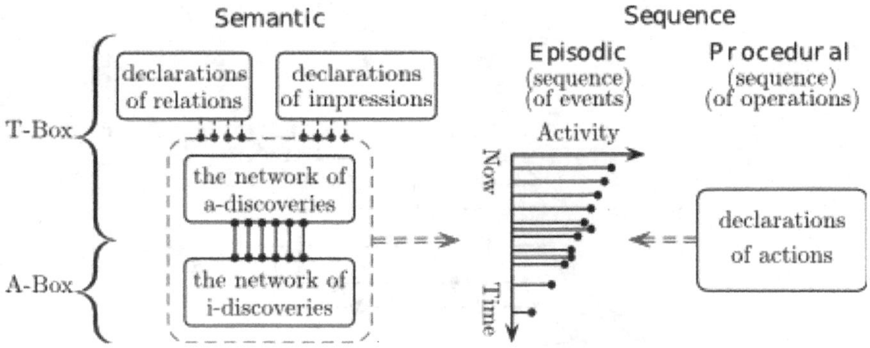

Fig. 4. Long-time memory in the context of the stored content

contextual needs. The need context shows how a particular instance of a given object (i-discovery) influences of the needs system of the agent. An important element of the a-discovery is a list of relations. The relations define a structure of the semantic memory. Assuming only one inheritance relation (parent-children)

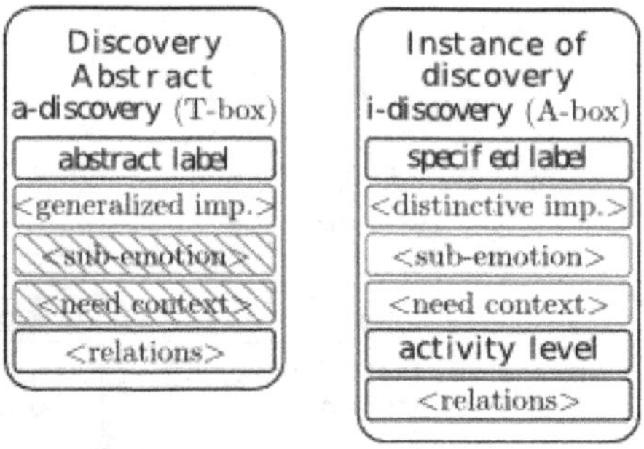

Fig. 5. Structures of discoveries ('< >', represents a list of elements)

and using layers, we can systematize a hierarchical network of a-discoveries, as shown in Fig. 6.

The instances (i-discoveries) have an additional attribute, an activity level, which reflects their frequency of appearance (the more often the agent sees certain objects, the better and faster it can recognize them). The activity level is also associated to the mechanism of forgetting: If the discovery does not appear sufficiency often, the agent may have difficulties in recognizing it, but also can forget it in order to optimize, or clear the memory space.

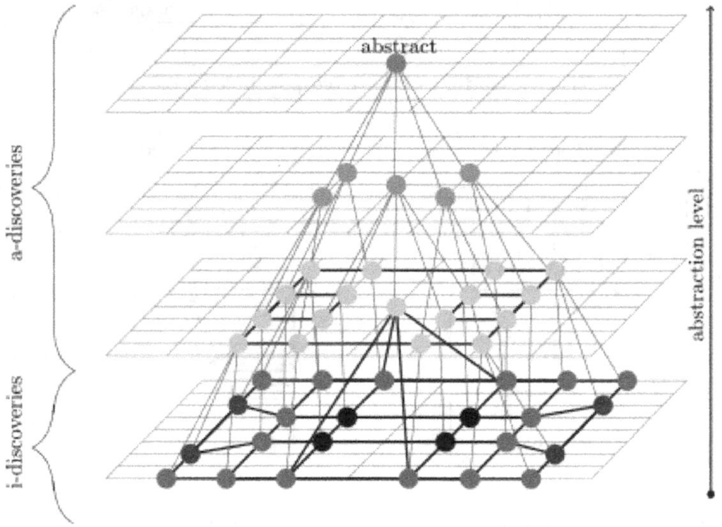

Fig. 6. Model of semantic memory represented as a tree of a-discoveries and i-discoveries. The red lines represent the inheritance relations among the a-discoveries; the black lines denotes arbitrary (networking) relations defined among the T-boxes (the greyness of i-discoveries depicts their activity level.

3.2 Sequential Memory

Sequential memory merges the episodic and procedural memories. Procedural memory is used to store sequences of action primitives. Each action can also be represented by another sequence of simple moves. Moreover, the actions are categorized as inner (mental) and outer (environmental). An agent can perform its own actions. It can also imitate and learn actions identified in the environment. Own actions have different origins (another feature), as: the reaction may be conscious (selected in the thinking process), or sub-conscious (learned by often repetition, without the thinking process: e.g. driving a car, does not require thinking, if the driver is experienced).

Sequences of different complex objects, called *semantic events*, are stored in the episodic memory. A semantic event joins different i-discoveries from the semantic memory and action patterns into one sequence. The sequence has its own location (a space context), momentum of occurrence (in time) and an activity parameter, which allows to filter (for the better performance), forget, and remind semantic events. The parameter is computed as a superposition of curves of forgetting and remembering (activity). The activity curve (A) reflects how many times the remembrance/recollection has been recalled. The forgetting curve (R – memory retention) shown in Fig. 7, is described as:

$$R = e^{-\frac{t}{S}} \tag{1}$$

where S is a relative strength of remembrance (also connected to emotions), and t is the time parameter [11].

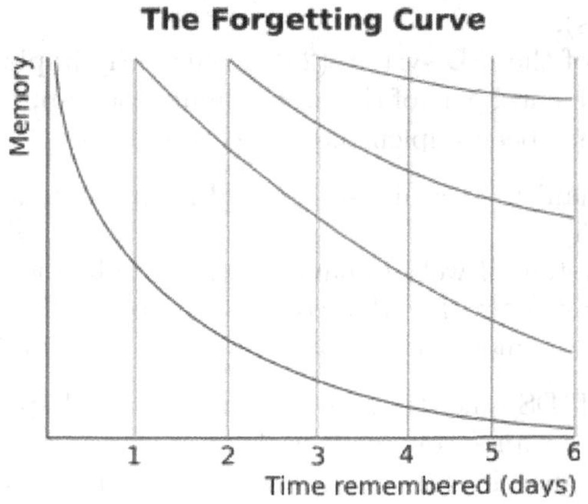

Fig. 7. The exponential nature of forgetting

4 Memory Features

The above model describes only the structure of the memory to be used in autonomous agents, including the most characteristic features of human, such as:

- refreshing remembrances (which are recalled from memory, and modify their activity parameter),
- forgetting remembrances (operation opposite to the above),
- restructuring remembrances (which add new information to known remembrances).

The first two features refers to the semantic network, the episodic memory and the part of the procedural memory that stores the agent's actions (conscious and unconscious). Restructuring remembrances concern all the elements (that can evolve)in the memory. New elements can be added, both primitive and complex. The restructuring process can not only modify items according to the perceived information, but also by imaginary information (facts that have not happened, but that the agent actually believe).

5 Conclusions

The presented model of memory allows us to expand the functionality of the ISD system designed for the purpose of autonomous robotics and decisions making. Moreover, the approach can be utilized in any human-robot interactions, nowadays also referred to as human-system interfaces. It provides functions and features making robots, or other software applications more human-like (friendly and intelligent). Nevertheless, and first of all, such an organization improves the performance of autonomous systems (through an improved access to 'optimally' memorized elements).

The full version of the ISD system (2.0) is currently implemented on a NAO robot, so the experimental part of this work should soon appear. Earlier versions of the ISD system has been implemented and tested, as:

- a decision mechanism of a mini-robot - qFix, presented in [20,26], and described in [22,25],
- an intelligent system of website navigation through voice control [28],
- LinguaBot, an avatar with verbal communication [6,21],
- xDriver - an autonomous driver based on the ISD brain [8].

The ultimate goal of IDS is not to create another, and perhaps more efficient DSS (in some respects), but an autonomous decision-making system based on human psychology and designed for robots and bots or avatars. Experience shows that more or less complex sub-systems, optimising precise compound mathematical criteria, do not contribute to effective systems (being objects of design). On the other hand, full-scale systems designed using complex multi-objective evolutionary computations, have both practical limitations and theoretical, dimensionality (Pareto) concerns [29,17,18,19].

Therefore, appreciating the bulk of developed artificial intelligence elements and methods, we rather try to follow the discoveries concerning the behaviour and performance of human brain - with a hope for new directions in the field of AI. Thus our approaches presents a completely new approach towards the study of both the human brain and the artificial intelligence. In particular, we are trying to preselect and formalise as much as possible of the findings in psychology and philosophy, presented in many pertinent fundamental books [34,33,31,12,2,5], taking into account such features as needs, emotions, mood, memory, etc., which appear most essential for automatic control, robotics, as well as diagnosis and HSI purposes, including intelligent autonomous mobile robots.

Certainly, there are sparse works in artificial intelligence, which develop systems and tools similar to our ideas, see [10], for instance. Some faint forms of cognitivity, representing to a certain extend a similar branch of research, are applied in the context of specific image interpretations (referred to as "understanding") and used in medical diagnostics applications [36,35]. One of the main differences is that the AI usually treats psychological processes in a bottom-up convention, whereas, our approach is clearly top-down and purely cybernetic, since we intend to model an effective functional/operational structure of human brain, instead of the biological one.

References

1. Babu, G.S., Suresh, S.: Meta-cognitive RBF network and its projection based learning algorithm for classification problems. Applied Soft Computing 13(1), 654–666 (2013)
2. Barlow, J.S.: The cerebellum and adaptive control. Cambridge University Press (2005)
3. Breland, S., McKinney, D., Parry, D., Peachey, C.: NRL designs robot for shipboard firefighting. Naval Research Laboratory. SPECTRA (1) (2012)
4. Brooks, R.A.: A robot that walks; Emergent behaviors from a carefully evolved network. Tech. Rep. 1091 (1989)
5. Bruner, J.S., Anglin, J.M.: Beyond the information given: studies in the psychology of knowing. Routledge (2010)
6. Chybiński, A.: LinguaBot – Prezentacyjny awatar z komunikacją... werbalną.... Master's thesis, (Promotor: prof. Z. Kowalczuk) Gdańsk University of Technology, Gdańsk, Poland (2012)
7. Conway, M.A.: Cognitive Models of Memory. MIT Press (1997)
8. Czubenko, M., Ordys, A., Kowalczuk, Z.: Autonomous driver based on intelligent system of decision-making. Cognitive Computation (in preparation, 2013)
9. Dourlens, S., Ramdane-Cherif, A.: Cognitive memory for semantic agents in Robotic interaction. In: 9th IEEE International Conference on Cognitive Informatics (ICCI), pp. 511–517 (July 2010)
10. Duch, W., Setiono, R., Zurada, J.M.: Computational intelligence methods for rule-based data understanding. Proceedings of the IEEE 92(5), 771–805 (2004)
11. Ebbinghaus, H.: Memory: A contribution to experimental psychology. Teachers College, New York (1885)
12. Eckman, P.: Kłamstwo i jego Wykrywanie w Biznesie, Polityce i Małżeństwie. PWN (2012)

13. El-Nasr, M.S., Yen, J., Ioerger, T.R.: Flame - fuzzy logic adaptive model of emotions. Autonomous Agents and Multi-agent Systems 3(3), 219–257 (2000)
14. Faghihi, U., McCall, R., Franklin, S.: A computational model of attentional learning in a cognitive agent. Biologically Inspired Cognitive Architectures 2(0), 25–36 (2012)
15. Fisher, R.P., Craik, F.I.: Interaction between encoding and retrieval operations in cued recall. Journal of experimental Psychology: Human Learning and Memory (3), 701–711 (1977)
16. Ho, W.C., Dautenhahn, K., Lim, M.Y., Vargas, P.A., Aylett, R., Enz, S.: An initial memory model for virtual and robot companions supporting migration and long-term interaction. In: The 18th IEEE International Symposium on Robot and Human Interactive Communication (RO-MAN 2009), pp. 277–284 (2009)
17. Kowalczuk, Z., Białaszewski, T.: Genetic algorithms in multi-objective optimisation of detection observers. In: Korbicz, J., Kościelny, J.M., Kowalczuk, Z., Cholewa, W. (eds.) Fault Diagnosis. Models, Artificial Intelligence, Applications, ch. 13, pp. 511–556. Springer, Berlin (2004)
18. Kowalczuk, Z., Białaszewski, T.: Improving evolutionary multi-objective optimisation by niching. Int. Journal of Information Technology and Intelligent Computing 1(2), 245–257 (2006)
19. Kowalczuk, Z., Białaszewski, T.: Niching mechanisms in evolutionary computations. Int. Journal of Applied Mathematics and Computer Science 16(1), 59–84 (2006)
20. Kowalczuk, Z., Czubenko, M.: Intelligent decision system of an autonomous robot. Seminar on Robot Autonomy (PAN:Poznań and IEEE:RAS) (November 2009) (in Polish)
21. Kowalczuk, Z., Czubenko, M.: DICTOBOT an autonomous agent with the ability to communicate. Zeszyty Naukowe Wydziału ETI Politechniki Gdańskiej. Technologie Informacyjne, pp. 87–92 (2010)
22. Kowalczuk, Z., Czubenko, M.: Interactive cognitive-behavioral decision making system. In: Rutkowski, L., Scherer, R., Tadeusiewicz, R., Zadeh, L.A., Zurada, J.M. (eds.) ICAISC 2010, Part II. LNCS, vol. 6114, pp. 516–523. Springer, Heidelberg (2010)
23. Kowalczuk, Z., Czubenko, M.: Model of human psychology for controlling autonomous robots. In: 15th International Conference on Methods and Models in Automation and Robotics (MMAR), Miedzyzdroje, Poland, pp. 31–36 (August 2010)
24. Kowalczuk, Z., Czubenko, M.: Diagnostyka antropidalnego systemu decyzyjnego. Pomiary Automatyka Kontrola 57(9), 1011–1015 (2011)
25. Kowalczuk, Z., Czubenko, M.: Intelligent decision-making system for autonomous robots. International Journal of Applied Mathematics and Computer Science 21(4), 621–635 (2011)
26. Kowalczuk, Z., Czubenko, M.: Systemic approach to modelling the human brain. The 1st International Kracow Conference Cognitive Science: Consciousness and Volition (Jagiellonian University) (September 2012) (in Polish)
27. Kowalczuk, Z., Czubenko, M.: xEmotion – obliczeniowy model emocji dedykowany dla inteligentnych systemów decyzyjnyjnych. Pomiary, Automatyka, Robotyka (2) (2013)
28. Kowalczuk, Z., Klimczak, J.: System inteligentnej nawigacji sterowanej głosem po serwisie internetowym. Pomiary, Automatyka, Kontrola (in print, 2013)
29. Kowalczuk, Z., Suchomski, P., Białaszewski, T.: Evolutionary multi-objective Pareto optimisation of diagnostic state observers. Int. Journal of Applied Mathematics and Computer Science 9(3), 689–709 (1999)

30. Krishnamurthy, S., Thamilarasu, G., Bauckhage, C.: MALADY: A machine learning-based autonomous decision-making system for sensor networks. In: International Conference on Computational Science and Engineering (CSE), vol. 2, pp. 93–100 (August 2009)
31. Maruszewski, T.: Cognitive Psychology. Gdańskie Wydawnictwo Psychologiczne, Gdańsk, polish Title: Psychologia Poznania (2001)
32. Miller, G.A.: The magical number seven, plus or minus two: Some limits on our capacity for processing informations. Psychological Review (63), 81–97 (1956)
33. Neisser, U.: Cognitive psychology. Appleton-Century-Crofts, New York (1967)
34. Nęcka, E., Orzechowski, J., Szymura, B.: Psychologia Poznawcza. PWN, Warszawa (2008)
35. Ogiela, L., Tadeusiewicz, R., Ogiela, M.R.: Cognitive techniques in medical information systems. Computers in Biology and Medicine 38(4), 501–507 (2008)
36. Ogiela, M.R., Tadeusiewicz, R.: Modern Computational Intelligence Methods for the Interpretation of Medical Images. SCI, vol. 84. Springer, Heidelberg (2008)
37. Ota, J.: Multi-agent robot systems as distributed autonomous systems. Advanced Engineering Informatics 20(1), 59–70 (2006)
38. Papageorgiou, E.I., Kannappan, A.: Fuzzy cognitive map ensemble learning paradigm to solve classification problems: Application to autism identification. Applied Soft Computing 12(12), 3798–3809 (2012)
39. Pettifor, E.: Endel Tulving's monohierarchical multimemory systems model. Tech. rep., Simon Fraser University (2000)
40. Rodríguez, M., Iglesias, R., Regueiro, C.V., Correa, J., Barro, S.: Autonomous and fast robot learning through motivation. Robotics and Autonomous Systems 55(9), 735–740 (2007)
41. Stachowicz, D., Kruijff, G.M.: Episodic-Like memory for cognitive robots. IEEE Transactions on Autonomous Mental Development 4(1), 1–16 (2012)
42. Sun, Z., Lu, Z., Jin, H.: Image retrieval with long-term memory learning and short-time relevance feedback. In: Third International Conference on Intelligent Information Hiding and Multimedia Signal Processing (IIHMSP 2007), vol. 1, pp. 173–177 (2007)
43. Tulving, E.: Ecphoric processes in recall and recognition. In: Brown, J. (ed.) Recall and Recognition, pp. 37–73. Wiley, London (1976)
44. Tulving, E.: How many memory systems are there? American Psychologist 40, 385–398 (1985)
45. Wang, W., Subagdja, B., Tan, A., Starzyk, J.A.: Neural modeling of episodic memory: Encoding, retrieval, and forgetting. IEEE Transactions on Neural Networks and Learning Systems 23(10), 1574–1586 (2012)
46. Widrow, B., Aragon, J.C.: Cognitive memory. Neural Networks (in print, 2013)
47. Widrow, B., Etemadi, M.: Cognitive memory: Human and machine. In: International Joint Conference on Neural Networks (IJCNN), pp. 3365–3372 (June 2009)
48. Xue, M., Zhu, C.: A study and application on machine learning of artificial intelligence. In: International Joint Conference on Artificial Intelligence (JCAI), pp. 272–274 (April 2009)
49. Yang, Z., Fan, Y., Zhang, B., Cheng, X.: A computation memory model with human memory features for autonomous virtual humans. In: International Conference on Computer Application and System Modeling (ICCASM), vol. 3, V3-246–V3-250 (October 2010)
50. Zhang, B.: Cognitive learning and the multimodal memory game: Toward human-level machine learning. In: IEEE International Joint Conference on Neural Networks (IJCNN), pp. 3261–3267 (June 2008)

A Computer Vision System for Evaluation of High Temperature Corrosion Damages in Steam Boilers

Ewaryst Rafajłowicz, Jerzy Wietrzych, and Wojciech Rafajłowicz

Institute of Computer Engineering, Control and Robotics,
Wroclaw University of Technology,
50-370 Wroclaw, Wyb. Wyspianskiego 27, Poland
{ewaryst.rafajlowicz,jerzy.wietrzych,wojciech.rafajlowicz}@pwr.wroc.pl

Abstract. Our aim is to describe a prototype of computer vision system for evaluation changes on rough surfaces. The system can be used for processing images of various kinds of surfaces. However, the version described here is dedicated for evaluating high temperature corrosion damages in the water-walls tubes of steam boilers. Firstly, a general idea of the system is presented and then we concentrate on on implemented methods of evaluating damages of water walls tubes caused by high temperature corrosion. These methods are based on image processing techniques that are tuned to our purposes.

Keywords: image processing, computer vision system, hot corrosion, steam boiler.

1 Introduction

The problem of evaluating roughness of surfaces from images has been of interest for many years. In particular, also corroded surfaces in nano-scale [5] as well as evaluating the rusted parts of bridges [8] have been analyzed by such techniques.

In this paper we concentrate on evaluating high temperature corrosion damages in water-walls tubes of pulverized boilers from high resolution images. Our motivation comes from the practical needs of deciding whether a group of water-walls tubes must be replaced or whether they can be safely used during the next period of hot steam and water supply.

The surface image analysis system has the task of collecting and storing data on the surface of objects exposed to corrosion. The images of the examined surface together with their descriptions and classifications are subject of archiving. The system consists of two programs. A program for archiving images and their analysis results, and a program called `ImageAnalysis`, aimed at analyzing images. The archiving program allows us to run external image analysis programs, to monitor their functioning and archiving the results of the analysis. The functionality of `ImageAnalysis` includes:

J. Korbicz and M. Kowal (eds.), *Intelligent Systems in Technical and Medical Diagnostics*, Advances in Intelligent Systems and Computing 230,
DOI: 10.1007/978-3-642-39881-0_33, © Springer-Verlag Berlin Heidelberg 2014

– calculating the variance of the signal value of the entire image;
– searching local minima and maxima of signal values in the image;
– performing segmentation of the image;
– calculating the variance of the signal in each of the segments found;
– displaying the general list of local minima and maxima of signal values by image segments found;
– calculating the fractal dimension in image segments found.

The latter of the above-mentioned functions is performed by two methods: `CubeCounting` and `Triangulation` (see [1], [2] for their theoretical justifications). `ImageAnalysis` performs different functions based on parameters set at display. These include the name of the image file, the defining of a set of functions to be performed and parameters affecting the segmentation of an image.

The paper is organized as follows. In Section 2 we describe the image acquisition system. Then, selected image processing modules are briefly described and the results of applying them to a corroded pipe are shown. The data base for image management is described in Section 3, in which also a discussion on final decisions is provided. In Section 4 we briefly describe directions of further development of the system functionalities.

2 Image Acquisition and Processing for Corrosion Evaluation

2.1 Image Acquisition

A general scheme of the image acquisition system for corroded surfaces is shown in Fig. 1 (left panel). It consists of
– a box containing two laser range-finders that allow us to obtain precise coordinates of an acquired image in order to locate it on a wall,
– a camera with LEDs and a one-touch control panel (see right panel for its prototype[1]). The system has been designed in such a way that
– images of corroded surfaces are taken easily (one touch mode), – in the same lighting conditions (own strong LEDs),
– at precisely known places on a steam boiler wall.
The precise positioning is provided by two laser rangefinders (see the left panel in Fig. 1) and each image is stored together with its positioning coordinates on a steam boiler wall. A proper focusing and selection of the region of interest are obtained by projecting a laser frame on a wall. The operator just moves the camera toward the wall and he can see which part of the wall will be captured. He can also see two frames and their coincidence indicates that the focusing is proper.

When a sufficiently long sequence of images of steam boiler water-walls tubes is stored at a hard disc of a tablet, then they are moved and registered to the data

[1] The prototype of the system was produced by OPTOSOFT S.A., Wrocław for the Wrocław University of Technology.

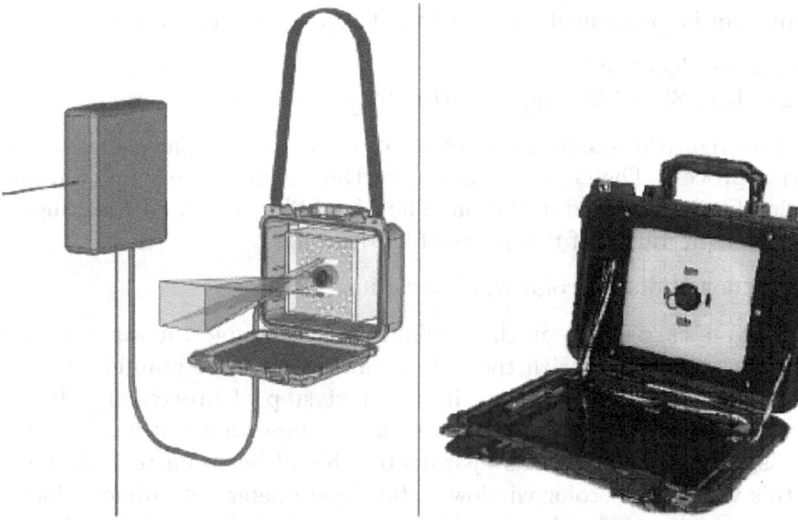

Fig. 1. Image acquisition system

Fig. 2. High temperature corrosion of water-wall tubes – a selected part

base on a desktop computer. The user interface of this data base allows us to display them as a map of original images or as a map of images after processing. Available tools for image processing are described in the next subsections, while in Section 3 details concerning the data base are described.

2.2 Image Filtering

Image filtering is aimed at smoothing the image, so that the slight irregularities became invisible. This allows us to find later, during the segmentation, the fragments of the image that differ significantly from each other. At the same time, minor variations are ignored, helping to avoid the creation of a large number of small segments.

Filtering can be performed by one of the two built-in algorithms:

- GaussianBlur algorithm;
- Pyramid Mean Shift Filtering algorithm from the OpenCV library.

Each of these algorithms require the setting of the respective parameters defining the filtering process. The GaussianBlur algorithm requires the sigma parameter or Gaussian kernel standard deviation. The Mean Shift Pyramid Filtering algorithm requires the values of two parameters defining:

- spatial window radius, – color window radius.

During the tests carried out on the available image samples, it was established that the images processed with the "Mean Shift Filtering Pyramid" algorithm could be better split into segments in the next step of processing. "Pyramid Mean Shift" filtering algorithm parameters are of importance. It has been found that the "spatial radius windows" parameter should be equal to 7, or be very close to this value. The "color window radius" parameter determines which differences in the value of the signal will be considered negligible. It was found by experiments that a good value of the RGB image coding scheme for images of the surface of the water-walls tubes used in the steam boilers is 8.

2.3 Image Segmentation

The segmentation of the image undergoing analysis using the "Graph-Based Image Segmentation" is executed using the algorithm published in [3]. This algorithm was chosen because of the fact that it is entirely oriented to be applied to images in which there are fragments of a certain irregularity of the signal value, such as grass, corroded surfaces, etc. This allows us to select uniform texture fragments, which will be used to create segments, and to perform further analysis on these segments.

The algorithm uses two parameters:

- k – a parameter for the threshold function,
- min – a minimum component size enforced by post-processing.

These values must be chosen experimentally. Typical parameters are k = 500, min = 20. Larger values for k result in larger connected components in the result. The action of this algorithm results in an array, corresponding to the division of the image into pixels. In this array, each element has a value which is a tag identifying the membership to one of the image segments. The content of this array is used later to select pixels in further calculations of fractal dimensions and the position of minima and maxima, etc. Another result is the number of segments found. It can be one of the indicators of the degree of roughness of the surface.

2.4 Estimating Fractal Dimensions

In this case, calculating fractal dimension is performed using the `CubeCounting` and `Triangulation` methods. The original detailed solutions applied here are

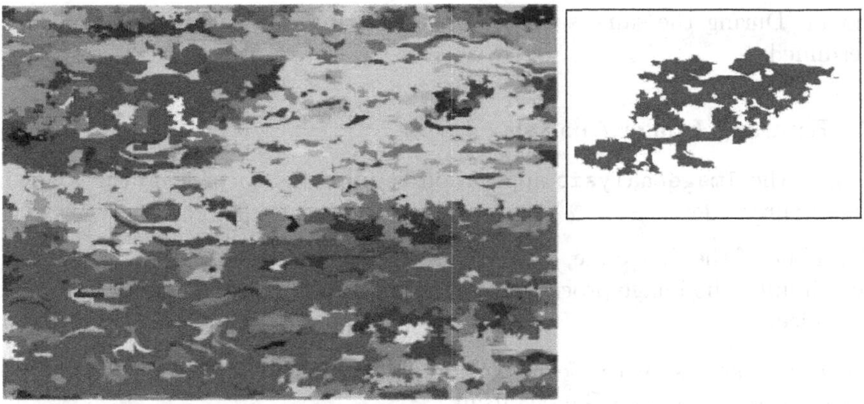

Fig. 3. Hot corroded segment after segmentation (left panel) and mostly damaged part selected (right panel)

algorithms similar to that used in the Gwyddion program [4]. Gwyddion was created by a team headed by David Neas and Petr Klapetek. This module allows automatic calculations of fractal dimensions of all segments or to segments selectively chosen by the operator. Based on the tags assigned during segmentation, is then proceed to the selection of pixels belonging to a segment and to the finding of the minimum and maximum coordinates of the shape representing the segment. We then proceed to the calculation of the width and height of the rectangle that covers the segment. In the next step, the smaller of the two values is selected and accepted as one side of the abstract square covering part of the segment. Then, the square is moved near the geometric center of the segment and the coordinates related to the longer side of the earlier found rectangle are adjusted. Signal values at given points of the designated square are used to calculate fractal dimension. The resulting set of points obtained is considered as an abstract area to be analyzed. It is assumed that the roughness of the examined surface of the metal is higher, the higher the fractal dimension calculated is. If the test surface is completely smooth, which results in the uniform value of the signal, the fractal dimension is two.

Sometimes, the obtained fractal dimension values are lower than the theoretical minimum equal to 2, which may be due to a low number of pixels that compose the analyzed segment. Further analysis of this case is left to the operator. The module includes mechanisms for selection of segments for which the calculated fractal dimensions are large and which allows us for a closer look at each of these segments.

2.5 Local Variance

A widely used measure of the degree of roughness of the surface is its empirically estimated variance of the signal value. The calculation of the variance of the signal value is performed on the entire image and on-demand for each picture segment. During the same sequence of calculations the average signal value is determined.

2.6 Running Image Analysis

Running the `ImageAnalysis` application requires us to provide the following parameters:

– the name of the image file,
– determining the image processing algorithm to be executed and, depending on this choice,

- two parameters for the segmentation algorithm,
- a parameter for the "GaussianBlur",
- two parameters for "Pyramid mean Shift Filtering" algorithm,
- segment identifier - this parameter is needed for selective segment processing.

The results of image processing by a selected method are kept in the data bases that is described in the next section.

3 Data Base of Corroded Surfaces

The archived data are collected in a manner that allows unambiguous identification of the company, in which the study was conducted, and of each of the studied objects, steam boilers and their water-walls tubes.

The tool providing a basis for this system is the SQL Server Compact Edition, a database managing software from the Microsoft company. In databases which are being created, all the necessary data is collected, except for files containing the surface images itself. These images are kept in the file system which is external to the database. The database keeps only the names of these external files. This split was decided due to:

– the size of image collections,
– the variety of formats in which the image files will be provided, that makes it difficult the use them in the database n a coherent way,
– the possible access to the files by using the external programs written in different programming languages and systems without using additional mechanisms to allow this access.

The PhotoArchive module is used to manage a database of images. It provides mechanisms for inserting images into the database and those aimed at facilitating the analysis of the degree of surface corrosion in water-walls tubes. The operator has a choice of two levels of image analysis progress.

3.1 Running Image Processing

The basic level of analysis includes simple image transformation mechanisms such as changes in brightness, contrast, calculation of histograms and image filtering using the Pyramid Segmentation algorithm. Using this method, the final assessment of the pipe corrosion degree is performed by the operator. This assessment is a numerical value and is important when choosing which parts of the steam -boiler water-walls tubes must be replaced. Advanced level of image analysis includes calculations by external applications and the assessment of the degree of water-walls tubes corrosion based on the calculated indicators. With this method, the operator has still some influence on setting parameters of applied image processing modules. It also has the ability to perform sample analysis, in which the results are not yet roughness indicators, but only a converted original image. This allows the selection of parameter values, so that further analysis automatically gives the results which are most relevant to the final assessment of the degree of water-walls tubes corrosion.

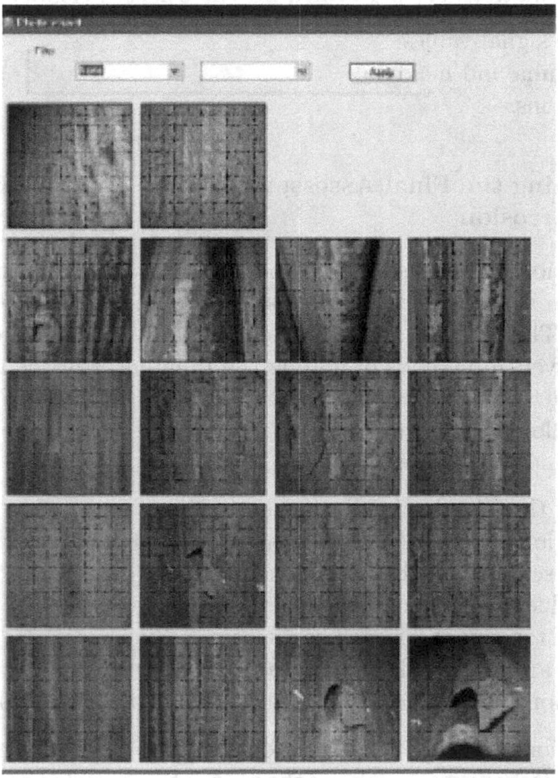

Fig. 4. Map of hot corrosion

3.2 Collection of Data Bases

The image data archiving program can be used by internal company services dealing with objects which are subject to corrosion as well as by specialized units dealing professionally with evaluation of the degree of surface corrosion. For this reason, it was found that the program will allow us to select a database from which to work. Each of the supported databases can collect data on all objects in the company or given objects individually. The choice of the contents of the database belongs to the operator who will use the application. Therefore, working with the application should always start by creating a new database or opening a database created earlier. Creating a new database starts by clicking the New option in the menu as described in Section 3.2.3. Once the database is created, it is possible to save detailed data on the company including all objects under examination. This identification is optional and useful if multiple databases on different companies are stored in a single workstation.

Activating the "Segments analysis" button will result in displaying the splitting of the image into segments as above, and then the selected segment, but this time, all calculations are performed for the selected segment:

- variance of the signal values;
- number of minima and maxima;
- fractal dimensions.

3.3 Determining the Final Assessment of the Degree of Water-Walls Tubes Corrosion

The current version of the system does not contain any uniform mechanism of automatic determination of the final assessment of the degree of water-walls tubes corrosion. The application performs a splitting of the image into segments based on the universal methods of finding the differences in the appearance of a given area, and further attempts to analyze segments as found. There are partial indicators that allow the determination of the degree of surface roughness of each segment. These indicators are:

- the variance of the signal value for the entire image;
- the number of local minima and maxima of the entire image;
- the number of segments to be distinguished in the image;
- the average value of the signal for each segment;
- the variance of the signal for each segment;
- number of local minima and maxima for each segment;
- fractal dimensions calculated for each segment as described above.

The above-mentioned values assists the operator in his subjective assessment. The alternative mechanisms, involving the search for corroded areas based on the analysis of the absolute values describing their color are also taken into account. Such a method, however, would require the prior establishment of a library of patterns of corrosion.

4 Further Development

The computer vision system described in the previous section is a working prototype. We plan to develop it further in in order to:

- enhance its ease of use and ergonomics,
- extend its functionalities in the area of image processing and decisions support.

Below, we discuss briefly only the latter topics with the emphasis on the possibility of implementing methods that are based on our earlier ideas.

We consider extending image filtering techniques by adding filters based on moving averages by extending [6] and edge preserving filters, based on a median calculated along a space-filling curve [7] or resulting from the approach that uses vertical weighting [9], [14]. It is worth considering also a filtering technique that uses a partial differential equation as a filter (see [17] and the bibliography therein).

The estimation of fractal dimensions is crucial for applications of our system. We plan to implement two more techniques, namely a 2D version of the correlation method that was extensively tested in [16] and a multivariate version of box-counting proposed in [18].

Among techniques of decision support we consider implementing the following classes of methods.

Local techniques. Local techniques take into account only small areas around a given pixel and their decisions are also local, i.e., to mark (or not) a given pixel as defect (see [20] and [15]).

Semi-local techniques. Semi-local methods take into account local dependencies between pixels but provide decisions concerning larger areas, e.g., by finding edges (see, e.g., [12] for a fast algorithm and the bibliography therein for classical approaches).

Global decisions. Global decisions are made after inspecting larger parts of surfaces. Potentially useful tools come from two classes of methods. The first one is change-detectors, see [13] for the one based on vertical weighting and the bibliography therein for classical approaches. The second group is based on pattern recognition techniques that are able to take an external context into account [10] and/or put different losses to ordered decisions [10]. In applications to corroded water-walls tubes the external context that influences decisions can come from a history of replacing water-walls tubes. Classifiers proposed in [10] take into account that we should put less penalty when a low corroded surface is classified as a medium corroded and higher penalty when classified as a highly corroded one.

In the above cited papers of our authorship, the problem of corroded surfaces was either not considered at all or a corroded surface served as an illustrative example.

Appendix

The image archiving application was written in C # language using Microsoft Dotnet 3.5. environment. The application also uses Microsoft SQL Server Compact Edition software. The application interface is built using the standard controls available in the Dotnet environment. Wherever possible, efforts were made to provide the access to the lists of records with information using a grid. A complementary way to access the individual fields of records stored in the database were added where required to operate in the text fields of considerable length. Entering the contents of such information through the grid fields is very inconvenient and that is why it has been decided to use another solution.

Several online-published libraries and code elements that support various special features were used to build the internal structure, including but not limited to:

– Exif Utils library, whose author is Stephen M. McKamey. This library is used to read the supplementary information in the Exif format of how the image is created, stored in the image files.
– Methods for grid functionality extension, contained in various libraries published on the Internet without any mention of copyright.
– Methods for cooperation with external programs which work asynchronously, published on the Internet without any mention of copyright.
According to the construction rules of such applications, the application is divided into modules which include:
forms-composing resource definition,
– code created and supported by the so-called Designer which is integrated in the Visual Studio system,
– code written by the author of the program.

The application uses access to the database using ADO Dotnet methodology. This means that while working on RAM managed by the application, a copy of the records retrieved from the disk database is created. Synchronization of RAM status and the disk file is performed when necessary. Sometimes it is done automatically, at changing application statuses, and sometimes at the explicit request of the operator. This obviously forced synchronization is the result of activating the buttons Save or Refresh. It is assumed that only one dataset with tables is created and managed from the main form level. Other forms are available through DataBindingSource mechanisms and publicly available main form methods, forcing the synchronization of the local copy of information with the database on to disk.

The `ImageAnalysis` program was written in C ++ language. Its internal structure uses snippets of code made available under license GNU by the following authors:

– Pedro Felzenszwalb - Image Segmentation
– C. Evans - modules FastHessian, IntegralImage, iPoint, Utils
– David Neas and Petr Klapetek - fractal dimension calculations

The application structure retains largely the original code split on modules.

Acknowledgments. The publication was founded by EIT Knowledge & Innovation Community KIC InnoEnergy. Sole responsibility for the content of this publication lies with the authors and EIT is not responsible for any use that may be made of the information it contains.

The authors express thanks to Professor H. Pawlak-Kruczek and to Dr. T. Hardy for their explanations and suggestions concerning the interpretation of surfaces damaged by hot corrosion.

References

1. Barnsley, M.F.: Fractals Everywhere. Academic Press (1988)
2. Clarke, K.C.: Computation of the fractal dimension of surfaces the triangular prism surface area method. Comput. Geosci. 12, 713–722 (1986)
3. Felzenszwalb, P.F., Huttenlocher, D.F.: Efficient Graph-Based Image Segmentation. International Journal of Computer Vision 59(2), 167–181 (2004)
4. Published under the GNU license, `http://www.gwyddion.net`
5. Livens, S., et al.: A Texture Analysis Approach to Corrosion Image Classification, Microscopy, Microanalysis, Microstructures 7(2), 1–10 (1996)
6. Krzyzak, A., Rafajlowicz, E., Pawlak, M.: Moving average restoration of bandlimited signals from noisy observations. IEEE Transactions on Signal Processing 45(12), 2967–2976 (1997)
7. Krzyzak, A., Rafajlowicz, E., Skubalska-Rafajlowicz, E.: Clipped median and space-filling curves in image filtering. Nonlinear Analysis-Theory Methods and Applications 47(1), 303–314 (2001)
8. Lee, S., Chang, L.-M., Skibniewski, M.: Automated recognition of surface defects using digital color image processing. Automation in Construction 15, 540–549 (2006)
9. Pawlak, M., Rafajlowicz, E.: Jump preserving signal reconstruction using vertical weighting. Nonlinear Analysis-Theory Methods and Applications 47(1), 327–338 (2001)
10. Rafajłowicz, E.: Classifiers sensitive to external context - theory and applications to video sequences. Expert Systems 29, 84–104 (2012)
11. Rafajłowicz, E., Krzyżak, A.: Pattern recognition with ordered labels. Nonlinear Analysis, Theory, Methods & Applications. Series A, Theory and Methods. 71, 1437–1441 (2009)
12. Rafajlowicz, E.: SUSAN edge detector reinterpreted, simplified and modified. In: International Workshop on Multidimensional (nD) Systems, Aveiro, Portugal, pp. 69–74 (2007)
13. Rafajlowicz, E., Pawlak, M., Steland, A.: Nonparametric sequential change-point detection by a vertically trimmed box method. IEEE Transactions on Information Theory 56(7), 3621–3634 (2010)

14. Rafajlowicz, E., Pawlak, M., Steland, A.: Nonlinear image processing and filtering: a unified approach based on vertically weighted regression. International Journal of Applied Mathematics and Computer Science 18, 49–61 (2008)
15. Rafajlowicz, E., Wnuk, M., Rafajlowicz, W.: Local detection of defects from image sequences. International Journal of Applied Mathematics and Computer Science 18(4), 581–592 (2008)
16. Rafajlowicz, E.: Testing (non-) existence of input-output relationships by estimating fractal dimensions. IEEE Transactions on Signal Processing 52(111), 3151–3159 (2004)
17. Rafajlowicz, E., Rafajlowicz, W.: Testing (non-)linearity of distributed-parameter systems from a video sequence. Asian Journal of Control 12(2), 146–158 (2010)
18. Skubalska-Rafajlowicz, E.: A new method of estimation of the box-counting dimension of multivariate objects using space-filling curves 53(5-7), e1281–e1287
19. Skubalska-Rafajłowicz, E.: Clustering of data and nearest neighbors search for pattern recognition with dimensionality reduction using random projections. In: Rutkowski, L., Scherer, R., Tadeusiewicz, R., Zadeh, L.A., Zurada, J.M. (eds.) ICAISC 2010, Part I. LNCS (LNAI), vol. 6113, pp. 462–470. Springer, Heidelberg (2010)
20. Skubalska-Rafajowicz, E.: Local correlation and entropy maps as tools for detecting defects in industrial images. International Journal of Applied Mathematics and Computer Science 18, 41–47 (2008)

Hand Gesture Recognition Using Time-of-Flight Camera and Viewpoint Feature Histogram

Tomasz Kapuściński, Mariusz Oszust, and Marian Wysocki

Rzeszow University of Technology
Department of Computer and Control Engineering
W. Pola 2, 35-959 Rzeszow, Poland
{tomekkap,marosz,mwysocki}@kia.prz.edu.pl

Abstract. Time-of-flight (ToF) cameras acquire 3D information about observed scenes. They are increasingly used for hand gesture recognition. This paper is also related to this problem. In contrast to other works which try to segment the hands we propose using point cloud processing and the Viewpoint Feature Histogram (VFH) as the global descriptor of the scene. To empower the distinctiveness of the descriptor a modification is proposed which consists in dividing the work space into smaller cells and calculating the VFH for each of them. The method is applied to five sample static gestures which are relatively difficult to recognise because hands are not the objects nearest the camera and/or touch each other, touch the head or appear in the background of the face. Results of ten-fold cross validation that justify the proposed approach are given.

Keywords: hand gesture recognition, time-of-flight camera, point cloud, viewpoint feature histogram.

1 Introduction

Human-computer interaction based on hand gestures has several important advantages. It is simple, natural, intuitive contactless, non-invasive, works in a noisy environment and is friendly to people with disabilities and elderly people. However despite of decades of research in the domain, the hand operated devices are not commonly used in our daily live. This is because the developed systems offer reasonable reliability only in controlled laboratory environment. The number and significance of constraints imposed on the developed systems is still important.

Nowadays, the new 3D imaging devices as ToF cameras offer new possibilities but due to their specific nature they require new algorithms and processing schemes. ToF cameras acquire 3D information about observed scenes. They are increasingly used for hand gesture recognition. This paper is also related to this problem. In contrast to other works which try to segment the hands we propose using point cloud processing and VFH as the global descriptor of the scene. To empower the distinctiveness of the descriptor a modification is proposed which

J. Korbicz and M. Kowal (eds.), *Intelligent Systems in Technical and Medical Diagnostics*, Advances in Intelligent Systems and Computing 230,
DOI: 10.1007/978-3-642-39881-0_34, © Springer-Verlag Berlin Heidelberg 2014

consists in dividing the work space into smaller cells and calculating the VFH for each of them. The method is applied to five sample static gestures which are relatively difficult to recognise because hands are not the objects nearest the camera and/or touch each other, touch the head or appear in the background of the face. Results of ten-fold cross validation that justify the proposed approach are given.

The remaining part of this article is organized as follows. Section 2 contains a brief overview of related publications. Section 3 gives basic information on ToF cameras. Section 4 discusses point cloud processing. VFH are considered in Section 5. Section 6 covers the recognition results and Section 7 concludes the paper and indicates directions of future research.

2 Related Work

ToF cameras are now increasingly used for hand gestures recognition, e.g. [3,5,6,13,14,15]. They facilitate the segmentation of the hand and provide additional 3D information used in feature vectors.

Most often, it is assumed that the hands are the objects nearest to the camera [3,5,6,13,15]. In [5,13,15] hands are segmented by thresholding the depth data. This technique works well in the central region of the scene. However, when the gesture is executed in the outermost areas, the same threshold value leads to imprecise segmentation. In such a case some parts of the forearm are also identified as a hand. In [6] the region growing method is used. It starts from the point nearest to the ToF camera and expands the region taking into account similar values. The condition determining when to stop the growing needs to be defined. Moreover, due to non uniform spatial density of the captured points, the region expansion may stop too early resulting in partially segmented hands. The distance between two adjacent points on the hand can vary significantly depending on the hand orientation. Furthermore the assumption that the hands are the objects nearest the camera is not always met. For certain gestures when the hand touches the head, ear, nose, mouth, throat or chest, it is so that the wrist, forearm or elbow becomes the object nearest the camera. In [11] the hand is located based on the fact that it is an end point of human body. This method will fail when the hand touches other parts of the body. The interesting approach is proposed in [14] where the depth and colour cameras are used. Obtained images are undistorted and the ToF image is projected onto the RGB image coordinates. The face is detected, its distance from the camera is measured and used to remove the background by thresholding the depth image. The hands are identified by applying skin colour detection to the remaining pixels. Additional processing to remove the arms is required when the user is wearing short sleeves. In the cited work this is achieved by the assumption that the arm is underneath the hand. This is obviously not fulfilled for the gestures performed in the area of stomach and waist.

The ToF camera provides accurate and complex information about the spatial shape of the visible surface. In [6,15] the 3D information is not added to the feature vectors. In [5] only the coarse information in the form of the hand geodesic centre and the minimum depth point is used. The depth images containing a lot of information about the hand posture are used in [14,13]. The pixel-wise depth distance between the registered templates and the observed gesture is used in classification. Because of the low-resolution of the ToF camera, this approach requires that the user's distance from the camera is not greater than 1 m.

Recently VFH - a new descriptor for 3D point cloud data has been introduced [7]. It encodes the geometry and viewpoint of the surface and is robust to large noise and missing depth information. Moreover, it has been shown experimentally, that the VFH can detect subtle variations in the geometry of objects even for untextured surfaces [7].

Taking into account the above-mentioned experiences we propose a quite different approach. Since accurate and reliable segmentation of the hand is problematic, we propose that the recognition takes place on the basis of the whole body (from the waist up) using the robust descriptor with good distinctive features. This approach can be justified by the fact that the shoulders, elbows and forearms are also involved in the formation of a gesture. Their configuration provides additional information that may be useful in the recognition. To empower the distinctiveness of the descriptor, especially for the gestures with subtle differences in their shape, we propose its modified version which consists in dividing work space into 9 cells.

3 ToF Cameras

In ToF camera 3D information is acquired by measuring the time taken for light to travel from an active illumination source to the objects in the field of view and then back to the sensor [17]. The scene is illuminated by the infrared light pulse, and the imaging sensor measures the time based on the phase shift between the modulated signal sent and reflected. Because this time is proportional to the distance, for each point of the image the depth information about corresponding object point is given. The principle is equivalent to the laser scanning, however, in this case, the entire scene is acquired at once.

Time-of-flight camera products for civil applications began to emerge around 2000 and now they are becoming more and more affordable in consumer electronics. The ToF cameras can be used in the range of 1 to 60 m. The distance resolution is about 1 cm. The depth image resolution of time-of-flight cameras is generally low compared to standard 2D video cameras. Images have a size of about 200 × 200 pixels. TOF cameras operate very quickly, providing up to 100 images per second. In contrast to stereo vision or triangulation systems the camera is very compact. The illumination is placed just next to the lens. Comparing to the other systems the ToF camera does not need a certain minimum base line. In contrast to laser scanning systems no mechanical moving parts are needed. The distance measurement is very efficient and reliable comparing to stereo vision, where complex correlation algorithms have to be implemented. The method is also less dependent on the texture of the object.

ToF cameras deliver a range image and an amplitude image with infrared modulation intensities at video frame rates: the range image (or depth image) contains for each pixel the radial measured distance between the considered pixel and its projection on the observed object, while the amplitude image contains for each pixel the strength of the reflected signal by the object [2].

Considering all the above advantages, ToF cameras seem to be very promising tools in a domain of human - computer interaction. They should reduce restrictions on lighting, backgrounds and clothing, currently imposed on vision-based gesture recognition systems.

In this work the ToF MESA Swiss Ranger 4000 camera has been used [16]. It has the range from 30 cm to 5 m, the depth image resolution 176 × 144, the distance resolution 1 cm and the frame rate up to 50 images/s.

4 Point Cloud Processing

The feature vectors used in this paper for the gesture recognition, were obtained by processing the so called point clouds. Point cloud is a data structure representing a multidimensional set of points [10,7]. In a 3D cloud each point is described by its coordinates x, y, z. In a 4D cloud an additional information about the brightness is available. Point clouds can be obtained using imaging devices, such as stereo cameras, 3D scanners or ToF cameras.

Points were recorded in a clockwise coordinate system whose centre is located at the intersection of the optical axis of the camera with the plane containing its front wall, the x-axis is horizontal and is directed to the left, the y-axis runs vertically and faces up, and the z-axis coincides with the optical axis and the camera is turned towards the object. Sample amplitude image and the projection of the point cloud on the xy plane is shown in Fig. 1.

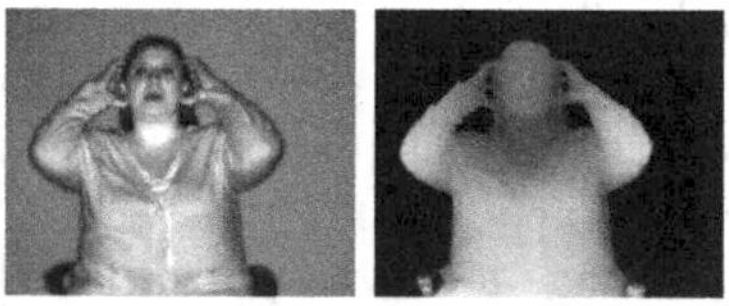

(a) (b)

Fig. 1. Sample amplitude image (a) and input cloud projected on the xy plane (b)

In this work the Point Cloud Library (PCL) was used [10]. First, the so called pass through filter was applied. The input point cloud was filtered along the z axis. Points whose z coordinate is greater than the assumed threshold were rejected. The threshold was selected to remove the background objects and extract the foreground person performing the gestures. For each point in the cloud the ToF camera provide the amplitude value. These values form the amplitude image, that can be acquired together with the cloud. The person's

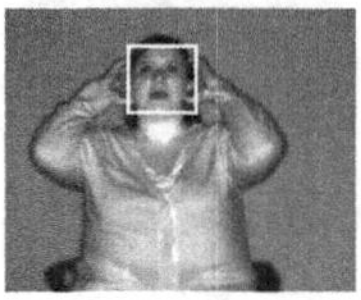

Fig. 2. Head segmentation in the amplitude image

head position in the amplitude image was found using the Haar detector (Fig. 2) [1]. The z coordinate of the point corresponding to the head center, assumed to be person distance from the camera, was increased by 20 cm and chosen as the pass through filter threshold.

Then, the statistical outlier removal filter was used in order to reject sparse, noisy measurements, which may corrupt the results leading to erroneous values of local point cloud characteristics such as surface normals [10]. For each point of the cloud the average distance from all points within its neighbourhood is determined. The neighbourhood size is the parameter of the method. The distribution of the average distances obtained in this way may be described by the global mean μ and standard deviation σ. All the points for which the average is not within the range $< \mu - n\sigma, \mu + n\sigma >$, where n is the filter parameter, are rejected.

The obtained cloud is then downsampled in order to reduce the number of points and speed up further calculations. The so called voxel grid filter is used [10]. This filter creates a 3D voxel grid over the input point cloud data. The physical dimensions of the voxel are the filter parameters. Then, in each voxel (i.e. 3D box), all the points present are approximated with their centroid.

5 Viewpoint Feature Histogram

VFH is the global descriptor of the point cloud. It has been successfully used for object recognition and 6DOF pose estimation [7]. The descriptor consists of two components: (i) a surface shape component, (ii) viewpoint direction component.

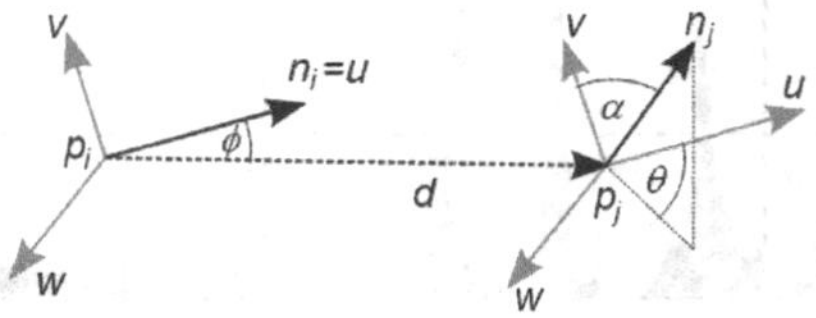

Fig. 3. Calculation of the surface shape component of the VFH

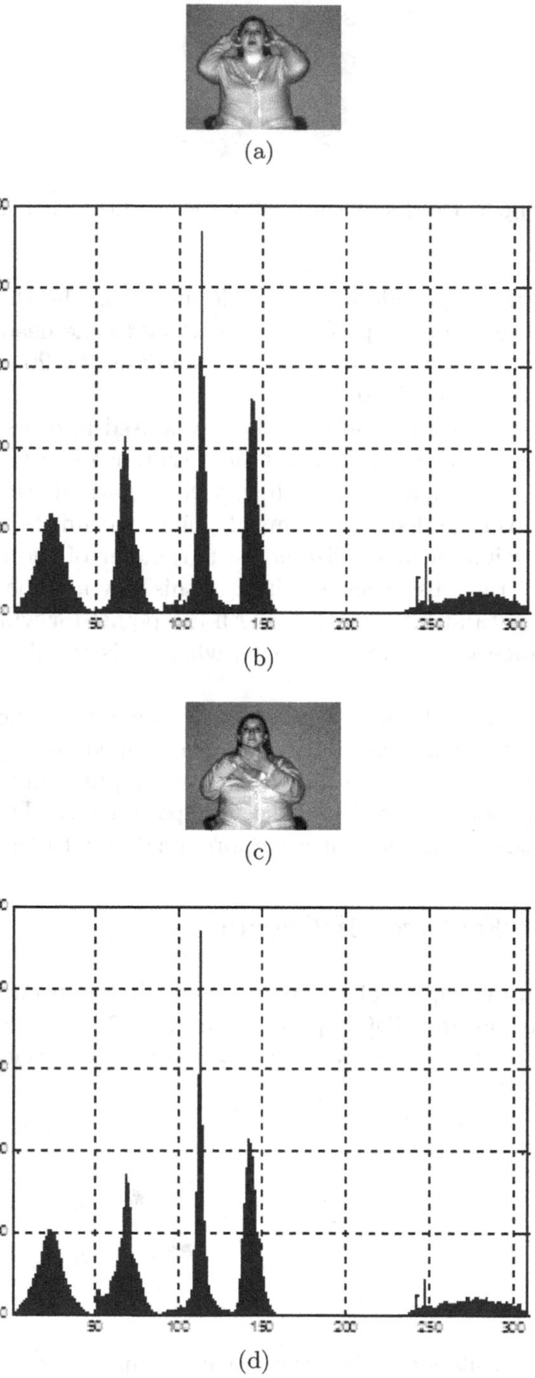

Fig. 4. Amplitude images (a), (c) and corresponding VFH signatures (b), (d) for two static gestures

The first component collects pairwise pan α, tilt ϕ, yaw θ angles between pair of normals (n_i, n_j) calculated for every pair of the points (p_i, p_j) on a surface patch and the distance d between these points, as shown in Fig. 3, where

$$u = n_i \tag{1}$$

$$v = u \times \frac{d}{|d|} \tag{2}$$

$$w = u \times v \tag{3}$$

$$\alpha = v \cdot n_j \tag{4}$$

$$\phi = u \cdot \frac{d}{|d|} \tag{5}$$

$$\theta = arctg\left(\frac{w \cdot n_j}{u \cdot n_j}\right) \tag{6}$$

The viewpoint component is computed by collecting a histogram of the angles that the viewpoint direction makes with each normal. The details of the calculation are given in [8,9,7]. Default descriptor consists of 308 values, 45 for each feature α, ϕ, θ, d and 128 for the viewpoint component.

Sample VFH signatures and the amplitude images for two static gestures are shown in Fig. 4. The obtained VFH signatures look very similar. Therefore to empower the distinctiveness of the descriptor the modified version of VFH is proposed. The modification consists in dividing the user's work space into smaller cells and calculating the VFH for each cell (Fig. 5). The resulting descriptor is the concatenation of the individual VFH.

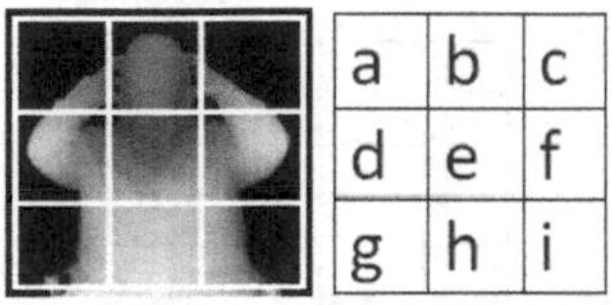

Fig. 5. Dividing the user's work space into smaller cells labeled from a to i

For the gestures shown in Fig. 4 (a) and 4 (c), the VFH signatures were calculated for each cell and compared (see Fig. 6 (a) - (i)). The upper part of each comparison refers to the first gesture, lower to the second. As one can see, VFHs obtained for cells are easier to distinguish. It is evident in the cells a and c, where hands are absent in the second gesture and in the cell e due to visible difference between point cloud shapes in this area.

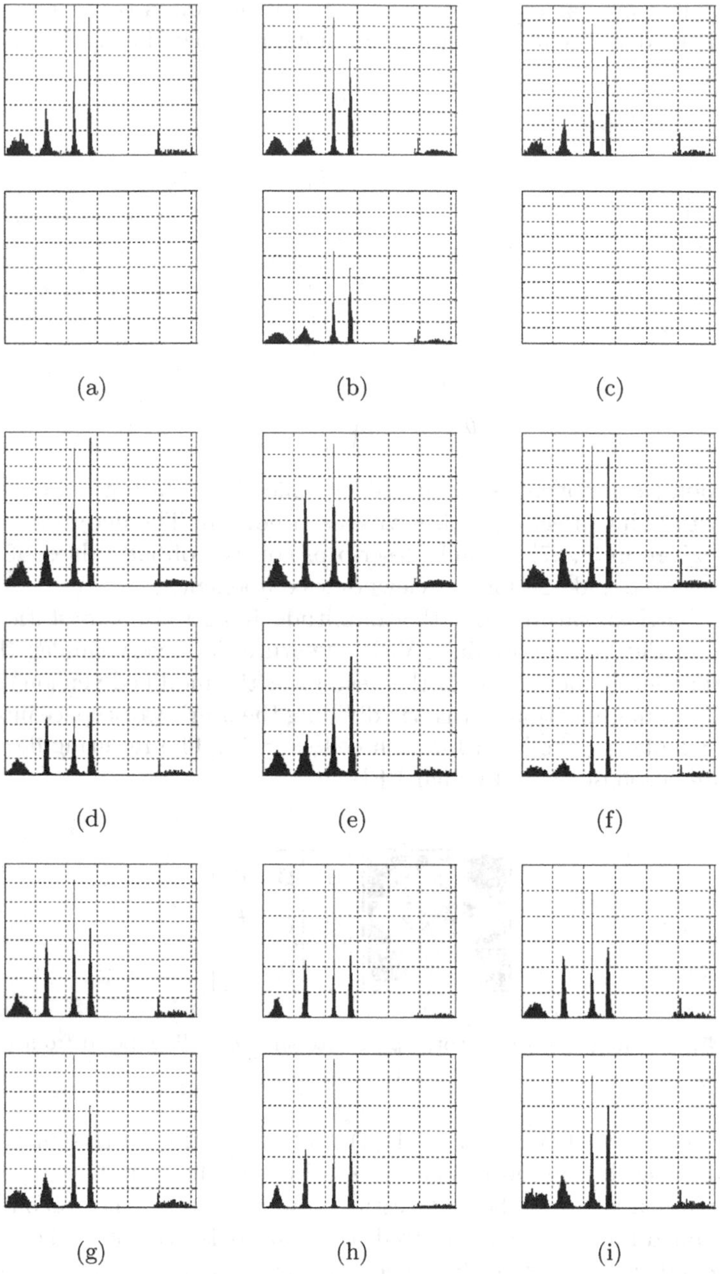

Fig. 6. Comparison of the VFH signatures of two gestures calculated for each cell, letters (a) - (i) correspond to the cells' labels. Horizontal axis contains 308 VFH histogram's bins as it can be seen on Fig. 4

6 Experimental Results

For recognition we used five sample static gestures which are relatively difficult to recognise because hands are not the objects nearest the camera and/or touch each other, touch the head or appear in the background of the face (Fig. 7). Each gesture was performed 20 times.

(a) (b) (c) (d) (e)

Fig. 7. Recognised gestures

We obtained point cloud data descriptors for two cases. The fist case consists in calculating VFH for the entire cloud. In the second case we use nine 3D cells as shown in Fig. 5. For recognition gestures' VHFs need to be compared, and it involves comparison of histograms. There are many different techniques for this task in the literature [1], [4] but we applied a very simple approach leading to good results. Since VHF is composed of five histograms we represented each of them by its mean and standard deviation. In the first case (entire cloud) each gesture is represented by the feature vector with length of 10, whereas for the second case (nine cells) the feature vector has length of 90.

Gestures data were divided into ten disjoint subsets in order to perform cross-validation tests for the nearest neighbour classifier [12] with Euclidean distance. We performed ten experiments in two variants: in variant A we used nine subsets as the training set and the remaining subset as the test set, in variant B the training and the test set were swapped. Results are given in Table 1.

Results obtained by the nearest neighbour classifier are satisfactory, but as one can see, dividing the work space into cells makes the classifier more robust in a case when only a few learning examples are shown. We also tried the k - nearest neighbour classifier with different k, but $k = 1$ yielded best results and only these results were considered.

Because each cell is characterised by five pairs of feature values it is worth examining which of them carry most discriminative information. Therefore we performed further ten-fold cross validation experiments using all combinations of this five pairs. It gives 32 combinations but for the clarity of presentation we show only obtained mean values with single descriptors (see Table 2).

Results presented in Table 2 lead to the conclusion that using combination of features based on histogram of ϕ and on histogram of d (comp. Fig. 3) should give the most interesting result. This turned out to be the best combination in the one cell case (100% in variant A, 95% in variant B). For the nine cell case such a combination gave 100% in variant A, 97.3% in variant B.

Table 1. Cross–validation tests with used feature vectors. Results of recognition are given in %.

	One cell		Nine cells	
Test \Variant	A	B	A	B
1	100.00	93.33	100.00	97.78
2	100.00	91.11	100.00	98.89
3	100.00	98.89	100.00	98.89
4	90.00	88.89	100.00	93.33
5	100.00	92.22	100.00	98.89
6	100.00	97.78	100.00	98.89
7	100.00	92.22	100.00	96.67
8	100.00	93.33	100.00	98.89
9	100.00	91.11	100.00	96.67
10	100.00	93.33	100.00	98.89
Mean	99.00	93.22	100.00	97.78
Minimum	90.00	88.89	100.00	93.33
Maximum	100.00	98.89	100.00	98.89
StDev	3.16	3.03	0.00	1.81

Table 2. Cross–validation tests using only chosen descriptors. Results of recognition are given in %.

	One cell		Nine cells	
Active descriptor \Variant	A	B	A	B
α	50.00	44.33	93.00	89.88
ϕ	96.00	95.44	97.00	94.11
θ	63.00	60.55	91.00	82.88
d	69.00	59.55	99.00	94.11
Viewpoint component	46.00	44.44	88.00	76.77

7 Conclusions and Future Work

In this paper we propose using point cloud processing and the Viewpoint Feature Histogram as the global descriptor of the scene. To empower the distinctiveness of the descriptor a modification is proposed which consists in dividing the working space into smaller cells and calculating the VFH for each of each of them. The method is applied to five sample static gestures which are relatively difficult to recognise because hands are not the objects nearest the camera and/or touch each other, touch the head or appear in the background of the face. Results of ten-fold cross validation and the nearest neighbour classifier are given. The results look promising. However, the approach requires further research. We are aware that our gesture data set should be extended by providing additional gesture classes and adding gestures performed by different people. It will allow to examine discriminative properties of VFHs more thoroughly. As another future direction we plan to use it for recognition of dynamic signed expressions of the Polish sign language. Dividing the work space into cells, considering various classifiers as well as the processing time are other interesting problems. Recognition of fairly complex hand gestures is one of the possible applications of developed approach and allows us to extrapolate the results to different areas. For example it could be used to improve discriminatory properties of any computer vision based decision system with ToF camera, especially working in various lighting conditions. In many manufacturing processes a diagnostic system using an automatic recognition of faulty objects plays important role. It is worth mentioning that the first use of VFH descriptor was involved in providing accurate information about surrounding objects in a robot arm's work area.

Acknowledgment. The ToF MESA Swiss Ranger 4000 camera used in the research presented in this paper has been purchased within the Project POPW.01.03.00−18−012/0 cosponsored by UE within the Operational Program Development of East Poland 2007-20013, Priority I, Modern Economy, Action 1.3 Innovation Support.

References

1. Bradski, G., Kaehler, A.: Learning OpenCV: Computer Vision with the OpenCV Library. O'Reilly Media, Cambridge (2008)
2. Chiabrandoa, F., Piatti, D., Rinaudo, F.: SR-4000 ToF Camera: Further Eexperimental Tests and First Application to Metric Surveys. International Archives of Photogrammetry, Remote Sensing and Spatial Information Sciences XXXVIII, 149–154 (2010)
3. Haubner, N., Schwanecke, U., Drner, R., Lehmann, S., Luderschmidt, J.: Recognition of Dynamic Hand Gestures with Time-of-Flight Cameras. In: Proceedings of ITG/GI Workshop on Self-Integrating Systems for Better Living Environments (Sensyble Workshop), pp. 33–39 (2010)
4. Mason, M., Duric, Z.: Using Histograms to Detect and Track Objects in Color Video. In: Proceedings of the 30th on Applied Imagery Pattern Recognition Workshop, pp. 154–159 (2001)
5. Molina, J., Escudero-Vinolo, M., Signoriello, A., Pardas, M., Ferran, C., Bescos, J., Marques, F., Martinez, J.M.: Real-time user independent hand gesture recognition from time-of-flight camera video using static and dynamic models. Machine Vision and Applications 24, 187–204 (2013)
6. Oprisescu, S., Rasche, C., Su, B.: Automatic static hand gesture recognition using ToF cameras. In: Proceedings of the 20th European Signal Processing Conference (EUSIPCO), pp. 2748–2751 (2012)
7. Rusu, R.B., Bradski, G., Thibaux, R., Hsu, J.: Fast 3D Recognition and Pose Using the Viewpoint Feature Histogram. In: Proceedings of the 23rd IEEE/RSJ International Conference on Intelligent Robots and Systems (IROS), pp. 2155–2162 (2010)
8. Rusu, R.B., Marton, Z.C., Blodow, N., Beetz, M.: Learning informative point classes for the acquisition of object model maps. In: 10th International Conference on Control, Automation, Robotics and Vision (ICARCV), pp. 643–650 (2008)

9. Rusu, R.B., Blodow, N., Beetz, M.: Fast Point Feature Histograms (FPFH) for 3D registration. In: IEEE International Conference on Robotics and Automation (ICRA), pp. 3212–3217 (2009)
10. Rusu, R.B., Cousins, S.: 3D is here: Point Cloud Library (PCL). In: IEEE International Conference on Robotics and Automation, ICRA (2011)
11. Song, L., Hu, R., Xiao, Y., Gong, L.: Real-Time 3D Hand Tracking from Depth Images. In: Proceedings of the 2nd International Conference On Systems Engineering and Modeling (ICSEM-13), pp. 1119–1122 (2013)
12. Theodoridis, S., Koutroumbas, K.: Pattern Recognition, 4th edn. Academic Press (2008)
13. Uebersax, D., Gall, J., Van den Bergh, M., Van Gool, L.: Real-time sign language letter and word recognition from depth data. In: IEEE International Conference on Computer Vision Workshops (ICCV Workshops), pp. 383–390 (2011)
14. Van den Bergh, M., Van Gool, L.: Combining RGB and ToF cameras for real-time 3D hand gesture interaction. In: IEEE Workshop on Applications of Computer Vision (WACV), pp. 66–72 (2011)
15. Zahedi, M., Manashty, A.R.: Robust Sign Language Recognition System Using ToF Depth Cameras. World of Computer Science and Information Technology Journal (WCSIT) 1, 50–55 (2011)
16. Mesa Imaging AG SwissRanger4000, `http://www.mesa-imaging.ch/`
17. Time of Flight Cameras, `http://www.metrilus.de/time-of-flight-cameras/`

Part IV

Various Problems of Technical Diagnostics

Robust Residual Generator Design for Systems with Distributed Time Delay

Dušan Krokavec and Anna Filasová

Technical University of Košice, Faculty of Electrical Engineering and Informatics,
Department of Cybernetics and Artificial Intelligence,
Letná 9, 042 00 Košice, Slovakia
{dusan.krokavec,anna.filasova}@tuke.sk
http://web.tuke.sk/kkui

Abstract. The paper is engaged with the framework of designing robust residuals for continuous-time systems with distributed time delay. The Lyapunov-Krasovskii functional principle is enforced by imposing the integral partitioning method and the new delay-dependent design conditions are established in terms of linear matrix inequalities. The stability conditions are derived with respect to the incidence of structured matrix variables in the linear matrix inequality formulation.

Keywords: linear matrix inequality, systems with distributed time delays, Lyapunov-Krasovskii functional, robust residual generators.

1 Introduction

The master aspect for designing a fault-tolerant control (FTC) system is based on the diagnostics operations that solve the fault detection and isolation (FDI) problems. This procedure most commonly used residual signals, generated by fault detection filters (FDF), followed by their evaluation within decision functions. Objective is to create residuals that are as a rule zero in the fault free case, maximally sensitive to faults, as well as robust to disturbances and model uncertainties. Like this, faults are generally detected by setting a threshold on the residual signal. The core of the FTC strategy is so an FDI system and a control parameter reconfiguration procedure, working in pursuance of the information delivered by the FDI. Research in FTC has attracted many investigators, and is now the subject of wide range of book publications (see, e.g., [2], [3]).

The full decoupling of faults and unknown system inputs cannot be, in general, realized completely, and so residual fault sensitivity and robustness has to be analyzed. In order to quantify sensitivity to faults as well as robustness to unknown system inputs, the fault detection filter design task becomes, in general, an multiple-objective optimization problem [10]. Other of the most commonly eaten way to design robust FDF is H_∞ filtering, introducing a reference model and formulating the design of residual generator as an H_∞ optimization (see [11] and the reference therein). The useful adaptation the reference residual model approach in the given field can be found in [1], [13].

J. Korbicz and M. Kowal (eds.), *Intelligent Systems in Technical and Medical Diagnostics*, Advances in Intelligent Systems and Computing 230,
DOI: 10.1007/978-3-642-39881-0_35,

Models with the distributed time-delay (DsTD) representation are exploited, e.g., in the modeling of combustion in the combustion chamber of a rocket motor. [5], [12]. Reflecting the fact that the discrete time-delay delineations are not complete related to systems with DsTD, alternative stability conditions for systems with DsTD were derived (e.g., in [8], [9]). The readers are referred to [4], and the reference therein, for the current status report about the delay-dependent stability conditions and stability analysis of such systems.

The presented approach to robust FDF design, destined for systems with DsTD and unknown disturbances, sets out to obtain the delay-dependent stability conditions. The Lyapunov-Krasovskii functional is enforced by imposing the integral partitioning and the conditions are formulated with respect to requisite of structured matrix variables. Exploiting the duality principle to [4], [6], [7], the fontal design criteria are derived for the FDF design, intended for systems with constant DsTD. The robustness is achieved through reference residual model, adapting the technique presented in [1]. To the best of authors knowledge, this approach has not yet been investigated for this kind of systems.

The outline of this paper is as follows: Section 2 introduces the model with DsTD and Section 3 places the preliminary results. In Section 4, the estimator structure is introduced and in Section 5 the reference residual model design conditions are derived. The robust FDF design conditions are proven in Section 5 and Section 6 gives a illustrative example. Section 7 presents some conclusions.

Throughout the paper, the following notations are used: $\boldsymbol{x}^T$, $\boldsymbol{X}^T$ denotes the transpose of the vector $\boldsymbol{x}$ and matrix $\boldsymbol{X}$, respectively, $diag[\cdot]$ denotes a block diagonal matrix, for a square matrix $\boldsymbol{X} < 0$ means that $\boldsymbol{X}$ is a symmetric negative definite matrix, the symbol $\boldsymbol{I}_n$ indicates the n-th order unit matrix, $I\!R$ denotes the set of real numbers, $I\!R^{n\times r}$ refers to the set of $n\times r$ real matrices and $L_2\langle 0, +\infty)$ is the space of square-integrable vector functions over $\langle 0, +\infty)$.

2 System Model

The systems under consideration are MIMO dynamic systems with DsTD. The considered class is represented by the next state-space equations with the initial condition

$$\dot{\boldsymbol{q}}(t) = \boldsymbol{A}\boldsymbol{q}(t) + \boldsymbol{A}_h \int_{t-h}^{t} \boldsymbol{q}(x)\mathrm{d}x + \boldsymbol{B}\boldsymbol{u}(t) + \boldsymbol{B}_f\boldsymbol{f}(t) + \boldsymbol{B}_d\boldsymbol{d}(t)\,, \tag{1}$$

$$\boldsymbol{y}(t) = \boldsymbol{C}\boldsymbol{q}(t)\,, \tag{2}$$

$$\boldsymbol{q}(\theta) = \boldsymbol{\varphi}(\theta), \quad \forall\,\theta \in \langle -(h+\Delta h), 0\rangle, \quad m\Delta h = h\,, \tag{3}$$

where $h \in (0, h_m\rangle$ is the magnitude of time-delay, m is a partitioning factor, Δh is maximum deviation of h, $\boldsymbol{q}(t) \in I\!R^n$, $\boldsymbol{u}(t) \in I\!R^r$, and $\boldsymbol{y}(t) \in I\!R^p$ are vectors of the state, input and output variables, respectively, matrices $\boldsymbol{A} \in I\!R^{n\times n}$, $\boldsymbol{A}_h \in I\!R^{n\times n}$, $\boldsymbol{B} \in I\!R^{n\times r}$, $\boldsymbol{C} \in I\!R^{p\times n}$, $\boldsymbol{B}_f \in I\!R^{n\times r_f}$, $\boldsymbol{B}_d \in I\!R^{n\times r_d}$ are real matrices and $\boldsymbol{d}(t) \in I\!R^{r_d}$ is the disturbance input that belongs to $L_2\langle 0, +\infty)$. It is considered that $h \in I\!R$ and $m \in I\!N$ are known, the fault is bounded, the value of $\boldsymbol{f}(t)$ is set to zero until a fault is occurred and the couple $(\boldsymbol{A}, \boldsymbol{C})$ is observable.

3 Preliminary Results

Proposition 1. [7] *If* $\boldsymbol{N}$ *is a positive definite symmetric matrix, and* $\boldsymbol{M}$ *is a square matrix of the same dimension then*

$$\boldsymbol{M}^{-T}\boldsymbol{N}\boldsymbol{M}^{-1} \geq \boldsymbol{M}^{-1} + \boldsymbol{M}^{-T} - \boldsymbol{N}^{-1} . \tag{4}$$

Proposition 2. [6] *(Jensen's inequalities) Let* $f(\boldsymbol{x}(p))$, $\boldsymbol{x}(p) \in I\!R^n$, $\boldsymbol{X} = \boldsymbol{X}^T > 0$, $\boldsymbol{X} \in I\!R^{n\times n}$ *is a real integrable vector function of the form*

$$f(\boldsymbol{x}(p)) = \boldsymbol{x}^T(p)\boldsymbol{X}\boldsymbol{x}(p) \tag{5}$$

such that there exist well defined integrations

$$\int_{-c}^{0}\int_{t+r}^{t} f(\boldsymbol{x}(p))\,dp\,dr > 0, \qquad \int_{t-c}^{t} f(\boldsymbol{x}(p))\,dp > 0 , \tag{6}$$

for $c > 0$, $c \in I\!R$, $t \in \langle 0, \infty)$. *Then*

$$\int_{-c}^{0}\int_{t+r}^{t} \boldsymbol{x}^T(p)\boldsymbol{X}\boldsymbol{x}(p)\,dp\,dr \geq \tfrac{2}{c^2}\int_{-c}^{0}\int_{t+r}^{t} \boldsymbol{x}^T(p)\,dp\,dr\,\boldsymbol{X}\int_{-c}^{0}\int_{t+r}^{t} \boldsymbol{x}(p)\,dp\,dr , \tag{7}$$

$$\int_{t-c}^{t} \boldsymbol{x}^T(p)\boldsymbol{X}\boldsymbol{x}(p)\,dp \geq \tfrac{1}{c}\int_{t-c}^{t} \boldsymbol{x}^T(p)\,dp\,\boldsymbol{X}\int_{t-c}^{t} \boldsymbol{x}(p)\,dp . \tag{8}$$

4 Estimator Structure

To construct FDF, the following state estimator is considered

$$\dot{\boldsymbol{q}}_e(t) = \boldsymbol{A}\boldsymbol{q}_e(t) + \boldsymbol{A}_h\int_{t-h}^{t} \boldsymbol{q}_e(x)\mathrm{d}x + \boldsymbol{B}\boldsymbol{u}(t) + \boldsymbol{J}(\boldsymbol{y}(t) - \boldsymbol{y}_e(t)) , \tag{9}$$

$$\boldsymbol{r}(t) = \boldsymbol{V}\boldsymbol{C}(\boldsymbol{q}(t) - \boldsymbol{q}_e(t)), \quad \boldsymbol{y}_e(t) = \boldsymbol{C}\boldsymbol{q}_e(t) , \tag{10}$$

where $\boldsymbol{q}_e(t) \in I\!R^n$ is the estimator state vector, $\boldsymbol{y}_e(t) \in I\!R^p$ is observed output vector, $\boldsymbol{r}(t) \in I\!R^{p_r}$ is residual signal, $\boldsymbol{J} \in I\!R^{n\times p}$ is the observer gain matrix and $\boldsymbol{V} \in I\!R^{p_r\times p}$ is the residual weighting matrix.

Defining the estimation error as

$$\boldsymbol{e}(t) = \boldsymbol{q}(t) - \boldsymbol{q}_e(t) , \tag{11}$$

then, from (1), (2) and (9), (10), the differential equation for the error is

$$\dot{\boldsymbol{e}}(t) = (\boldsymbol{A} - \boldsymbol{J}\boldsymbol{C})\boldsymbol{e}(t) + \boldsymbol{A}_h\int_{t-h}^{t} \boldsymbol{e}(x)\mathrm{d}x + \boldsymbol{B}_f\boldsymbol{f}(t) + \boldsymbol{B}_d\boldsymbol{d}(t) , \tag{12}$$

respectively, and the residual signal equation takes the form

$$\boldsymbol{r}(t) = \boldsymbol{V}\boldsymbol{C}\boldsymbol{e}(t) = \boldsymbol{H}\boldsymbol{e}(t), \qquad \boldsymbol{H} = \boldsymbol{V}\boldsymbol{C},\ \boldsymbol{H} \in I\!R^{p_r\times n} . \tag{13}$$

Since a time delay occurrence in (12), adapting the approach given in [1], a reference residual model is proposed to be used. Finding the reference residual model, the additive constraint in FDF design task is to minimize the worst case H_∞ norm between the generated residual and the reference one.

5 Reference Residual Model

5.1 Choice of Reference Residual Model

Generally, (12) can be rewritten as

$$\dot{e}(t) = (\boldsymbol{A} - \boldsymbol{J}\boldsymbol{C})e(t) + \boldsymbol{A}_h \int_{t-h}^{t} e(x)\mathrm{d}x + \begin{bmatrix} \boldsymbol{B}_d & -\boldsymbol{B}_f \end{bmatrix} \begin{bmatrix} \boldsymbol{d}(t) \\ -\boldsymbol{f}(t) \end{bmatrix} . \tag{14}$$

Considering, for the sake of simplicity, $r_f = r_d = r_g$, and using the equivalent observer structure with the same cross-bonds between $\boldsymbol{d}(t)$ and $\boldsymbol{f}(t)$, the observer equation of the reference residual model can be chosen in the form

$$\dot{\boldsymbol{e}}^\diamond(t) = (\boldsymbol{A} - \boldsymbol{J}^\diamond\boldsymbol{C})\boldsymbol{e}^\diamond(t) + \boldsymbol{A}_h \int_{t-h}^{t} \boldsymbol{e}^\diamond(x)\mathrm{d}x + \begin{bmatrix} \boldsymbol{B}_d & -\boldsymbol{B}_f \end{bmatrix} \boldsymbol{T}^\diamond \begin{bmatrix} \boldsymbol{d}(t) \\ -\boldsymbol{f}(t) \end{bmatrix} , \tag{15}$$

where the cross-bonds matrix $\boldsymbol{T}^\diamond$ is selected as

$$\boldsymbol{T}^\diamond = \begin{bmatrix} \boldsymbol{I}_{r_g} & \boldsymbol{I}_{r_g} \\ \boldsymbol{I}_{r_g} & \boldsymbol{I}_{r_g} \end{bmatrix} = \begin{bmatrix} \boldsymbol{I}_{r_g} \\ \boldsymbol{I}_{r_g} \end{bmatrix} \begin{bmatrix} \boldsymbol{I}_{r_g} & \boldsymbol{I}_{r_g} \end{bmatrix} = \boldsymbol{N}^\diamond \boldsymbol{N}^{\diamond T}, \qquad \boldsymbol{N}^{\diamond T} = \begin{bmatrix} \boldsymbol{I}_{r_g} & \boldsymbol{I}_{r_g} \end{bmatrix} . \tag{16}$$

Applying (16), the residual reference model takes the form

$$\dot{\boldsymbol{e}}^\diamond(t) = (\boldsymbol{A} - \boldsymbol{J}^\diamond\boldsymbol{C})\boldsymbol{e}^\diamond(t) + \boldsymbol{A}_h \int_{t-h}^{t} \boldsymbol{e}^\diamond(x)\mathrm{d}x + \boldsymbol{G}^\diamond\boldsymbol{g}^\diamond(t) , \tag{17}$$

$$\boldsymbol{r}^\diamond(t) = \boldsymbol{H}^\diamond\boldsymbol{e}^\diamond(t) , \tag{18}$$

where, with $\boldsymbol{G}^\diamond \in \mathbb{R}^{n \times r_g}$, $\boldsymbol{g}^\diamond(t) \in \mathbb{R}^{r_g}$,

$$\boldsymbol{G}^\diamond = \begin{bmatrix} \boldsymbol{B}_d & -\boldsymbol{B}_f \end{bmatrix} \boldsymbol{N}^\diamond, \qquad \boldsymbol{g}^\diamond(t) = \boldsymbol{d}(t) - \boldsymbol{f}(t) . \tag{19}$$

Note, for a delay-free model ($\boldsymbol{A}_h = \boldsymbol{0}$), (17), (18) are linear and

$$\boldsymbol{R}^\diamond(s) = \boldsymbol{H}^\diamond(s\boldsymbol{I} - (\boldsymbol{A} - \boldsymbol{J}^\diamond\boldsymbol{C}))^{-1}(\boldsymbol{B}_d - \boldsymbol{B}_f) = \boldsymbol{R}_d^\diamond(s) - \boldsymbol{R}_f^\diamond(s) . \tag{20}$$

Therefore, if

$$\sqrt{\delta} \geq \|\boldsymbol{R}^\diamond(s)\|_\infty \geq \|\boldsymbol{R}_d^\diamond(s)\|_\infty - \|\boldsymbol{R}_f^\diamond(s)\|_\infty , \tag{21}$$

where $\delta > 0$, $\delta \in \mathbb{R}$, and $\sqrt{\delta}$ is an upper bound of H_∞ norm of (21), then

$$\int_{0}^{\infty} \left(\boldsymbol{r}^{\diamond T}(x)\boldsymbol{r}^\diamond(x) - \delta\, \boldsymbol{g}^{\diamond T}(x)\boldsymbol{g}^\diamond(x)\right)\mathrm{d}x > 0 \tag{22}$$

and the minimization of δ can be interpreted as minimization of $\|\boldsymbol{R}_d^\diamond(s)\|_\infty$ while as maximization of $\|\boldsymbol{R}_f^\diamond(s)\|_\infty$.

5.2 Residual Reference Model Design

Theorem 1. *The residual reference model (17), (18) is stable with quadratic performance* $\|\boldsymbol{R}^{\diamond}(s)\|_{\infty} < \sqrt{\delta}$ *if there exist a symmetric positive definite matrix* $\boldsymbol{P}^{\diamond} \in I\!R^{n\times n}$*, matrices* $\boldsymbol{Y}^{\diamond} \in I\!R^{n\times p}$, $\boldsymbol{H}^{\diamond} \in I\!R^{p_r\times n}$ *and a positive scalars* $\delta, \gamma \in I\!R$ *such that*

$$\boldsymbol{P}^{\diamond} = \boldsymbol{P}^{\diamond T} > 0, \quad \delta > 0, \quad \gamma > 0\,, \tag{23}$$

$$\begin{bmatrix} \boldsymbol{P}^{\diamond}\boldsymbol{A} + \boldsymbol{A}^T\boldsymbol{P}^{\diamond} - \boldsymbol{Y}^{\diamond}\boldsymbol{C} - \boldsymbol{C}^T\boldsymbol{Y}^{\diamond T} & * & * & * \\ \boldsymbol{G}^{\diamond T}\boldsymbol{P}^{\diamond} & -\delta\boldsymbol{I}_{r_g} & * & * \\ \boldsymbol{A}_h^T\boldsymbol{P}^{\diamond} & \boldsymbol{0} & -\gamma\boldsymbol{I}_n & * \\ \boldsymbol{H}^{\diamond} & \boldsymbol{0} & \boldsymbol{0} & -\boldsymbol{I}_{p_r} \end{bmatrix} < 0\,. \tag{24}$$

When the above conditions hold, the residual reference model gain matrices are

$$\boldsymbol{J}^{\diamond} = \boldsymbol{P}^{\diamond -1}\boldsymbol{Y}^{\diamond}, \quad \boldsymbol{H}^{\diamond}\,. \tag{25}$$

Hereafter, $*$ *denotes the symmetric item in a symmetric matrix.*

Proof. To obtain delay independent design conditions, Lyapunov function is introduced as follows

$$v(\boldsymbol{e}^{\diamond}(t)) = \\ = \boldsymbol{e}^{\diamond T}(t)\boldsymbol{P}^{\diamond}\boldsymbol{e}^{\diamond}(t) + \int_0^t \big(\boldsymbol{r}^{\diamond T}(x)\boldsymbol{r}^{\diamond}(x) - \delta\,\boldsymbol{g}^{\diamond T}(x)\boldsymbol{g}^{\diamond}(x) - \gamma\,\boldsymbol{e}_h^{\diamond T}(x)\boldsymbol{e}_h^{\diamond}(x)\big)\mathrm{d}x\,, \tag{26}$$

where $\boldsymbol{P}^{\diamond} = \boldsymbol{P}^{\diamond T} > 0$, $\delta > 0$, $\gamma > 0$ and

$$\boldsymbol{e}_h^{\diamond T}(x) = \int_{x-h}^{x} \boldsymbol{e}^{\diamond}(\vartheta)\mathrm{d}\vartheta\,. \tag{27}$$

The derivative of (26) with respect to t is

$$\dot{v}(\boldsymbol{e}^{\diamond}(t)) = \boldsymbol{e}^{\diamond T}(t)\boldsymbol{P}^{\diamond}\dot{\boldsymbol{e}}^{\diamond}(t) + \dot{\boldsymbol{e}}^{\diamond T}(t)\boldsymbol{P}^{\diamond}\boldsymbol{e}^{\diamond}(t) + \\ + \boldsymbol{r}^{\diamond T}(t)\boldsymbol{r}^{\diamond}(t) - \delta\,\boldsymbol{g}^{\diamond T}(t)\boldsymbol{g}^{\diamond}(t) - \gamma\,\boldsymbol{e}_h^{\diamond T}(t)\boldsymbol{e}_h^{\diamond}(t) \tag{28}$$

and substituting (17), (18), then (28) takes the form

$$\dot{v}(\boldsymbol{e}^{\diamond}(t)) = \boldsymbol{e}^{\diamond T}(t)\big((\boldsymbol{A} - \boldsymbol{J}^{\diamond}\boldsymbol{C})^T\boldsymbol{P}^{\diamond} + \boldsymbol{P}^{\diamond}(\boldsymbol{A} - \boldsymbol{J}^{\diamond}\boldsymbol{C}) + \boldsymbol{H}^{\diamond T}\boldsymbol{H}^{\diamond}\big)\boldsymbol{e}^{\diamond}(t) + \\ + \boldsymbol{e}^{\diamond T}(t)\boldsymbol{P}^{\diamond}\boldsymbol{A}_h\boldsymbol{e}_h^{\diamond}(t) + \boldsymbol{e}_h^{\diamond T}(t)\boldsymbol{A}_h^T\boldsymbol{P}^{\diamond}\boldsymbol{e}^{\diamond}(t) - \gamma\,\boldsymbol{e}_h^{\diamond T}(t)\boldsymbol{e}_h^{\diamond}(t) + \\ + \boldsymbol{e}^{\diamond T}(t)\boldsymbol{P}^{\diamond}\boldsymbol{G}^{\diamond}\boldsymbol{g}^{\diamond}(t) + \boldsymbol{g}^{\diamond T}(t)\boldsymbol{G}^{\diamond T}\boldsymbol{P}^{\diamond}\boldsymbol{e}^{\diamond}(t) - \delta\,\boldsymbol{g}^{\diamond T}(t)\boldsymbol{g}^{\diamond}(t)\,. \tag{29}$$

Thus, defining the composite vector

$$\boldsymbol{e}_c^{\diamond T}(t) = \big[\boldsymbol{e}^{\diamond T}(t)\ \boldsymbol{g}^{\diamond T}(t)\ \boldsymbol{e}_h^{\diamond T}(t)\big]\,, \tag{30}$$

(29) can be rewritten as

$$\dot{v}(\boldsymbol{e}^{\diamond}(t)) = \boldsymbol{e}_c^{\diamond T}(t)\boldsymbol{P}_c^{\diamond}\boldsymbol{e}_c^{\diamond}(t) < 0\,, \tag{31}$$

$$P_c^{\diamond} = \begin{bmatrix} (A - J^{\diamond}C)^T P^{\diamond} + P^{\diamond}(A - J^{\diamond}C) + H^{\diamond T}H^{\diamond} & * & * \\ G^{\diamond T}P^{\diamond} & -\delta I_{r_g} & * \\ A_h^T P^{\diamond} & 0 & -\gamma I_n \end{bmatrix} < 0 \,. \quad (32)$$

With the notation

$$Y^{\diamond} = P^{\diamond}J^{\diamond} \,, \quad (33)$$

applying the Schur complement property, (32) implies (24). ■

6 Residual Filter Design Conditions

Changing back $T^{\diamond} = I_{2r_g}$, the reference residual model (17)-(18) takes the form

$$\dot{e}^{\diamond}(t) = (A - J^{\diamond}C)e^{\diamond}(t) + A_h \int_{t-h}^{t} e^{\diamond}(x)\mathrm{d}x + B_f f(t) + B_d d(t) \,, \quad (34)$$

$$r^{\diamond}(t) = H^{\diamond}e^{\diamond}(t) \,, \quad (35)$$

where $J^{\diamond}$, $H^{\diamond}$ are the parameters outlined using conditions given in the previous sections. The overall model, incorporating the state estimate error (12), (13) and the reference residual model (34), (35) is suggested to be expressed as

$$\dot{e}^{\bullet}(t) = (A^{\bullet} - J^{\bullet}C^{\bullet})e^{\bullet}(t) + A_h^{\bullet} \int_{t-h}^{t} e^{\bullet}(x)\mathrm{d}x + G^{\bullet}g^{\bullet}(t) \,, \quad (36)$$

$$r^{\bullet}(t) = (H^{\bullet} - H^{\star})e^{\bullet}(t) \,, \quad (37)$$

where

$$e^{\bullet}(t) = \begin{bmatrix} e(t) \\ e^{\diamond}(t) \end{bmatrix}, \; g^{\bullet}(t) = \begin{bmatrix} f(t) \\ d(t) \end{bmatrix}, \; G^{\bullet} = \begin{bmatrix} B_f \; B_d \\ B_f \; B_d \end{bmatrix}, \quad (38)$$

$$A^{\bullet} = \mathrm{diag}\begin{bmatrix} A \; A \end{bmatrix}, \; C^{\bullet} = \mathrm{diag}\begin{bmatrix} C \;\; J^{\diamond}C \end{bmatrix}, \; J^{\bullet} = \mathrm{diag}\begin{bmatrix} J \; I_n \end{bmatrix}, \quad (39)$$

$$A_h^{\bullet} = \mathrm{diag}\begin{bmatrix} A_h \; A_h \end{bmatrix}, \; H^{\bullet} = \begin{bmatrix} H \; 0 \end{bmatrix}, \; H^{\star} = \begin{bmatrix} 0 \; H^{\diamond} \end{bmatrix}, \quad (40)$$

$e^{\bullet}(t) \in \mathbb{R}^{2n}$, $g^{\bullet}(t) \in \mathbb{R}^{r_f + r_d}$, $A^{\bullet}, A_h^{\bullet} \in \mathbb{R}^{2n \times 2n}$, $G^{\bullet} \in \mathbb{R}^{2n \times (r_f + r_d)}$, $C^{\bullet} \in \mathbb{R}^{(p+n) \times 2n}$, $J^{\bullet} \in \mathbb{R}^{2n \times (p+n)}$, $H^{\bullet}, H^{\star} \in \mathbb{R}^{p_r \times 2n}$.

These matrix structures have to be defined for this reason, in order to use structured LMI matrix variables.

Theorem 2. *The residual filter (9), (10), associated with the reference model (17), (18), is stable with quadratic performance* $\|R^{\bullet}(s)\|_{\infty} < \sqrt{\delta}$ *if there exist symmetric positive definite matrices* $P^{\bullet}, U, V \in \mathbb{R}^{2n \times 2n}$, $W \in \mathbb{R}^{2mn \times 2mn}$, *matrices* $H \in \mathbb{R}^{p_r \times n}$, $Y \in \mathbb{R}^{n \times p}$ *and a positive scalar* $\delta \in \mathbb{R}$ *such that*

$$P^{\bullet} = P^{\bullet T} > 0, \; U = U^T > 0, \; V = V^T > 0, \; W = W^T > 0, \; \delta > 0 \,, \quad (41)$$

$$\begin{bmatrix} \Delta_{11} & * & * & * \\ Y^{\circ}T_A & -b^{-2}(2P^{\bullet} - V) & * & * \\ G^{\bullet T}(P^{\bullet}T_I + a^4 Y^{\circ}T_A) & 0 & -\delta I_{r_f + r_d} + b^2 G^{\bullet T}VG^{\bullet} & * \\ (H^{\bullet} - H^{\star})T_I & 0 & 0 & -I_{p_r} \end{bmatrix} < 0 \,, \quad (42)$$

$$\boldsymbol{\Delta}_{11} = \boldsymbol{T}_I^T \boldsymbol{Y}^\circ \boldsymbol{T}_A + \boldsymbol{T}_A^T \boldsymbol{Y}^\circ \boldsymbol{T}_I - \boldsymbol{T}_V^T \boldsymbol{V} \boldsymbol{T}_V + \boldsymbol{T}_U^T \boldsymbol{U}^\circ \boldsymbol{T}_U + \boldsymbol{T}_W^T \boldsymbol{W}^\circ \boldsymbol{T}_W \,, \tag{43}$$

$$\boldsymbol{T}_I = \begin{bmatrix} \boldsymbol{I}_{2n} \; \boldsymbol{0} \; [\boldsymbol{0} \cdots \boldsymbol{0}] \; \boldsymbol{0} \end{bmatrix}, \quad \boldsymbol{T}_V = b^{-1} \begin{bmatrix} a^2 \boldsymbol{I}_{2n} \; -\boldsymbol{I}_{2n} \; [\boldsymbol{0} \cdots \boldsymbol{0}] \; \boldsymbol{0} \end{bmatrix}, \tag{44}$$

$$\boldsymbol{T}_U = \begin{bmatrix} a\boldsymbol{I}_{2n} & \boldsymbol{0} \\ \boldsymbol{0} & a^{-1}\boldsymbol{I}_{2n} \end{bmatrix} \begin{bmatrix} \boldsymbol{I}_{2n} & \boldsymbol{0} & [\boldsymbol{0} \cdots \boldsymbol{0}] & \boldsymbol{0} \\ \boldsymbol{0} & \boldsymbol{I}_{2n} & [\boldsymbol{0} \cdots \boldsymbol{0}] & \boldsymbol{0} \end{bmatrix}, \; \boldsymbol{T}_W = \begin{bmatrix} \boldsymbol{0}_w & [\boldsymbol{I}_{2mn} \; \boldsymbol{0}_w] \\ \boldsymbol{0}_w & [\boldsymbol{0}_w \; \boldsymbol{I}_{2mn}] \end{bmatrix}, \tag{45}$$

$$a = \sqrt{\frac{h}{m}}, \quad b = \frac{h}{\sqrt{2}m}, \quad \boldsymbol{U}^\circ = \begin{bmatrix} \boldsymbol{U} & \boldsymbol{0} \\ \boldsymbol{0} & -\boldsymbol{U} \end{bmatrix}, \quad \boldsymbol{W}^\circ = \begin{bmatrix} \boldsymbol{W} & \boldsymbol{0} \\ \boldsymbol{0} & -\boldsymbol{W} \end{bmatrix}, \tag{46}$$

$$\boldsymbol{Y}^\circ = \begin{bmatrix} [\boldsymbol{P}^\bullet \; \boldsymbol{Y}^\bullet] \; \boldsymbol{P}^\bullet \; [\boldsymbol{P}^\bullet \cdots \boldsymbol{P}^\bullet] \; \boldsymbol{P}^\bullet \end{bmatrix}, \tag{47}$$

$$\boldsymbol{T}_A = diag \left[\begin{bmatrix} \boldsymbol{A}^\bullet \\ -\boldsymbol{C}^\bullet \end{bmatrix} \; \boldsymbol{A}_h^\bullet \; diag \begin{bmatrix} \boldsymbol{A}_h^\bullet \cdots \boldsymbol{A}_h^\bullet \end{bmatrix} \; \boldsymbol{0} \right], \tag{48}$$

$$\boldsymbol{P}^\bullet = diag \begin{bmatrix} \boldsymbol{P}_1^\bullet \; \boldsymbol{P}_2^\bullet \end{bmatrix}, \; \boldsymbol{H}^\bullet = \begin{bmatrix} \boldsymbol{H} \; \boldsymbol{0} \end{bmatrix}, \; \boldsymbol{Y}^\bullet = diag \begin{bmatrix} \boldsymbol{Y} \; \boldsymbol{P}_2^\bullet \end{bmatrix}, \tag{49}$$

where the matrices $\boldsymbol{Y}^\circ \in \mathbb{R}^{2n\times((m+2)2n+p+n)}$, $\boldsymbol{W}^\circ \in \mathbb{R}^{4mn\times 4mn}$, $\boldsymbol{U}^\circ \in \mathbb{R}^{4n\times 4n}$, $\boldsymbol{P}^\bullet \in \mathbb{R}^{2n\times 2n}$, $\boldsymbol{Y}^\bullet \in \mathbb{R}^{2n\times(n+p)}$, $\boldsymbol{H}^\bullet \in \mathbb{R}^{p_r\times 2n}$ *are structured matrix variables and* $\boldsymbol{T}_U \in \mathbb{R}^{4n\times 2(m+2)n}$, $\boldsymbol{T}_I, \boldsymbol{T}_V \in \mathbb{R}^{2n\times 2(m+2)n}$, $\boldsymbol{T}_W \in \mathbb{R}^{4mn\times 2(m+2)n}$, $\boldsymbol{T}_A \in \mathbb{R}^{(2(m+2)n+p+n)\times 2(m+2)n}$.

When the above conditions hold, the residual filter gain matrices are given by

$$\boldsymbol{J} = \boldsymbol{P}_1^{\bullet -1}\boldsymbol{Y}, \quad \boldsymbol{H} \,. \tag{50}$$

Proof. The Lyapunov-Krasovskii functional is now defined as follows

$$v(\boldsymbol{e}^\bullet(t)) = v_0(\boldsymbol{e}^\bullet(t)) + v_1(\boldsymbol{e}^\bullet(t)) + v_2(\boldsymbol{e}^\bullet(t)) + v_3(\boldsymbol{e}^\bullet(t)) \,, \tag{51}$$

where, with $\boldsymbol{P}^\bullet = \boldsymbol{P}^{\bullet T} > 0$, $\boldsymbol{W} = \boldsymbol{W}^T > 0$, $\boldsymbol{U} = \boldsymbol{U}^T > 0$, $\boldsymbol{V} = \boldsymbol{V}^T > 0$,

$$v_0(\boldsymbol{e}^\bullet(t)) = \boldsymbol{e}^{\bullet T}(t)\boldsymbol{P}^\bullet\boldsymbol{e}^\bullet(t) + \int_0^t \left(\boldsymbol{r}^{\bullet T}(x)\boldsymbol{r}^\bullet(x) - \delta\, \boldsymbol{g}^{\bullet T}(x)\boldsymbol{g}^\bullet(x)\right)\mathrm{d}x \,, \tag{52}$$

$$v_1(\boldsymbol{e}^\bullet(t)) = \int_{t-\frac{h}{m}}^t \boldsymbol{e}_h^{\bullet T}(x)\boldsymbol{W}\boldsymbol{e}_h^\bullet(x)\mathrm{d}x, \; v_2(\boldsymbol{e}^\bullet(t)) = \int_{-\frac{h}{m}}^0 \int_{t+\vartheta}^t \boldsymbol{e}^{\bullet T}(x)\boldsymbol{U}\boldsymbol{e}^\bullet(x)\mathrm{d}x\mathrm{d}\vartheta \,, \tag{53}$$

$$v_3(\boldsymbol{e}^\bullet(t)) = \int_{-\frac{h}{m}}^0 \int_{\vartheta}^0 \int_{t+\lambda}^t \dot{\boldsymbol{e}}^{\bullet T}(x)\boldsymbol{V}\dot{\boldsymbol{e}}^\bullet(x)\mathrm{d}x\mathrm{d}\lambda\mathrm{d}\vartheta \tag{54}$$

and the integral partition is considered as follows

$$\boldsymbol{e}_h^\bullet(t) = \int_{t-h}^t \boldsymbol{e}^\bullet(x)\mathrm{d}x \sim \begin{bmatrix} \boldsymbol{e}_{h1}^\bullet(t) \; \boldsymbol{e}_{h2}^\bullet(t) \end{bmatrix} \,, \tag{55}$$

$$\boldsymbol{e}_{h1}^\bullet(t) = \int_{t-\frac{h}{m}}^t \boldsymbol{e}^\bullet(x)\mathrm{d}x, \quad \boldsymbol{e}_{h2}^\bullet(t) = \begin{bmatrix} \int_{t-\frac{2h}{m}}^{t-\frac{h}{m}} \boldsymbol{e}^\bullet(x)\mathrm{d}x & \cdots & \int_{t-h}^{t-(m-1)\frac{h}{m}} \boldsymbol{e}^\bullet(x)\mathrm{d}x \end{bmatrix}, \tag{56}$$

$$\boldsymbol{e}_{h3}^\bullet(t) = \int_{t-h-\frac{h}{m}}^{t-h} \boldsymbol{e}_h^\bullet(x)\mathrm{d}x, \quad \boldsymbol{e}^\bullet(t-\tfrac{h}{m}) \sim \begin{bmatrix} \boldsymbol{e}_{h2}^\bullet(t) \; \boldsymbol{e}_{h3}^\bullet(t) \end{bmatrix} . \tag{57}$$

Hence, the derivative of $v(\boldsymbol{e}^{\bullet}(t))$ with respect to t is given by

$$\dot{v}(\boldsymbol{e}^{\bullet}(t)) = \dot{v}_0(\boldsymbol{e}^{\bullet}(t)) + \dot{v}_1(\boldsymbol{e}^{\bullet}(t)) + \dot{v}_2(\boldsymbol{e}^{\bullet}(t)) + \dot{v}_3(\boldsymbol{e}^{\bullet}(t)) . \tag{58}$$

Using the time-derivative of the first item, with (36), (37) it follows

$$\begin{aligned}\dot{v}_0(\boldsymbol{e}^{\bullet}(t)) &= \boldsymbol{e}^{\bullet T}(t)\boldsymbol{P}^{\bullet}\dot{\boldsymbol{e}}^{\bullet}(t) + \dot{\boldsymbol{e}}^{\bullet T}(t)\boldsymbol{P}^{\bullet}\boldsymbol{e}^{\bullet}(t) + \boldsymbol{r}^{\bullet T}(t)\boldsymbol{r}^{\bullet}(t) - \delta\, \boldsymbol{g}^{\bullet T}(t)\boldsymbol{g}^{\bullet}(t) = \\ &= \left[(\boldsymbol{A}^{\bullet} - \boldsymbol{J}^{\bullet}\boldsymbol{C}^{\bullet})\boldsymbol{e}^{\bullet}(t) + \boldsymbol{A}_h^{\bullet}\boldsymbol{e}_h^{\bullet}(t) + \boldsymbol{G}^{\bullet}\boldsymbol{g}^{\bullet}(t)\right]^T \boldsymbol{P}^{\bullet}\boldsymbol{e}^{\bullet}(t) + \\ &\quad + \boldsymbol{e}^{\bullet T}(t)\boldsymbol{P}^{\bullet}\left[(\boldsymbol{A}^{\bullet} - \boldsymbol{J}^{\bullet}\boldsymbol{C}^{\bullet})\boldsymbol{e}^{\bullet}(t) + \boldsymbol{A}_h^{\bullet}\boldsymbol{e}_h^{\bullet}(t) + \boldsymbol{G}^{\bullet}\boldsymbol{g}^{\bullet}(t)\right] + \\ &\quad + \boldsymbol{r}^{\bullet T}(t)\boldsymbol{r}^{\bullet}(t) - \delta\, \boldsymbol{g}^{\bullet T}(t)\boldsymbol{g}^{\bullet}(t)\end{aligned} \tag{59}$$

and, evidently, the second item time-derivative is

$$\dot{v}_1(\boldsymbol{e}^{\bullet}(t)) = \boldsymbol{e}_h^{\bullet T}(t)\boldsymbol{W}\boldsymbol{e}_h^{\bullet}(t) - \boldsymbol{e}_h^{\bullet T}(t - \frac{h}{m})\boldsymbol{W}\boldsymbol{e}_h^{\bullet}(t - \frac{h}{m}) . \tag{60}$$

Taking into consideration (8) and (56), it yields

$$\begin{aligned}\dot{v}_2(\boldsymbol{e}^{\bullet}(t)) &= \int_{-\frac{h}{m}}^{0} \boldsymbol{e}^{\bullet T}(t)\boldsymbol{U}\boldsymbol{e}^{\bullet}(t)\mathrm{d}\vartheta - \int_{-\frac{h}{m}}^{0} \boldsymbol{e}^{\bullet T}(t+\vartheta)\boldsymbol{U}\boldsymbol{e}^{\bullet}(t+\vartheta)\mathrm{d}\vartheta = \\ &= \tfrac{h}{m}\boldsymbol{e}^{\bullet T}(t)\boldsymbol{U}\boldsymbol{e}^{\bullet}(t) - \int_{t-\frac{h}{m}}^{t} \boldsymbol{e}^{\bullet T}(x)\boldsymbol{U}\boldsymbol{e}^{\bullet}(x)\mathrm{d}x ,\end{aligned} \tag{61}$$

$$\begin{aligned}\dot{v}_2(\boldsymbol{e}^{\bullet}(t)) &\leq \tfrac{h}{m}\boldsymbol{e}^{\bullet T}(t)\boldsymbol{U}\boldsymbol{e}^{\bullet}(t) - \tfrac{m}{h}\int_{t-\frac{h}{m}}^{0} \boldsymbol{e}^{\bullet T}(x)\mathrm{d}x\, \boldsymbol{U} \int_{t-\frac{h}{m}}^{0} \boldsymbol{e}^{\bullet}(x)\mathrm{d}x = \\ &= \tfrac{h}{m}\boldsymbol{e}^{\bullet T}(t)\boldsymbol{U}\boldsymbol{e}^{\bullet}(t) - \tfrac{m}{h}\boldsymbol{e}_{h1}^{\bullet T}(t)\boldsymbol{U}\boldsymbol{e}_{h1}^{\bullet}(t) .\end{aligned} \tag{62}$$

Under the same conditions

$$\begin{aligned}\dot{v}_3(\boldsymbol{e}^{\bullet}(t)) &= \int_{-\frac{h}{m}}^{0}\int_{\vartheta}^{0} \left(\dot{\boldsymbol{e}}^{\bullet T}(t)\boldsymbol{V}\dot{\boldsymbol{e}}^{\bullet}(t) - \dot{\boldsymbol{e}}^{\bullet T}(t+\lambda)\boldsymbol{V}\dot{\boldsymbol{e}}^{\bullet}(t+\lambda)\right)\mathrm{d}\lambda\mathrm{d}\vartheta = \\ &= \int_{-\frac{h}{m}}^{0} -\vartheta\dot{\boldsymbol{e}}^{\bullet T}(t)\boldsymbol{V}\dot{\boldsymbol{e}}^{\bullet}(t)\mathrm{d}\vartheta - \int_{-\frac{h}{m}}^{0}\int_{t+\vartheta}^{t} \dot{\boldsymbol{e}}^{\bullet T}(x)\boldsymbol{V}\dot{\boldsymbol{e}}^{\bullet}(x)\mathrm{d}x\mathrm{d}\vartheta = \\ &= \tfrac{1}{2}(\tfrac{h}{m})^2\dot{\boldsymbol{e}}^{\bullet T}(t)\boldsymbol{V}\dot{\boldsymbol{e}}^{\bullet}(t) - \int_{-\frac{h}{m}}^{0}\int_{t+\vartheta}^{t} \dot{\boldsymbol{e}}^{\bullet T}(x)\boldsymbol{V}\dot{\boldsymbol{e}}^{\bullet}(x)\mathrm{d}x\mathrm{d}\vartheta\end{aligned} \tag{63}$$

and, since (7), (56) implies,

$$\begin{aligned}&\int_{-\frac{h}{m}}^{0}\int_{t+\vartheta}^{t} \dot{\boldsymbol{e}}^{\bullet T}(x)\boldsymbol{V}\dot{\boldsymbol{e}}^{\bullet}(x)\mathrm{d}x\mathrm{d}\vartheta \geq \tfrac{2}{(\frac{h}{m})^2}\int_{-\frac{h}{m}}^{0}\int_{t+\vartheta}^{t} \dot{\boldsymbol{e}}^{\bullet T}(x)\mathrm{d}x\mathrm{d}\vartheta\, \boldsymbol{V}\int_{-\frac{h}{m}}^{0}\int_{t+\vartheta}^{t} \dot{\boldsymbol{e}}^{\bullet}(x)\mathrm{d}x\mathrm{d}\vartheta = \\ &\quad= \tfrac{2}{(\frac{h}{m})^2}\int_{-\frac{h}{m}}^{0} \left(\boldsymbol{e}^{\bullet}(t) - \boldsymbol{e}^{\bullet}(t+\vartheta)\right)^T\mathrm{d}\vartheta\, \boldsymbol{V}\int_{-\frac{h}{m}}^{0} \left(\boldsymbol{e}^{\bullet}(t) - \boldsymbol{e}^{\bullet}(t+\vartheta)\right)\mathrm{d}\vartheta = \\ &\quad= \tfrac{2}{(\frac{h}{m})^2}\left(\tfrac{h}{m}\boldsymbol{e}^{\bullet T}(t) - \int_{t-\frac{h}{m}}^{t} \boldsymbol{e}^{\bullet T}(x)\mathrm{d}x\right)\boldsymbol{V}\left(\tfrac{h}{m}\boldsymbol{e}^{\bullet}(t) - \int_{t-\frac{h}{m}}^{t} \boldsymbol{e}^{\bullet}(x)\mathrm{d}x\right) = \\ &\quad= \tfrac{2}{(\frac{h}{m})^2}\left(\tfrac{h}{m}\boldsymbol{e}^{\bullet T}(t) - \boldsymbol{e}_{h1}^{\bullet T}(t)\right)\boldsymbol{V}\left(\tfrac{h}{m}\boldsymbol{e}^{\bullet}(t) - \boldsymbol{e}_{h1}^{\bullet}(t)\right) ,\end{aligned} \tag{64}$$

then

$$\dot{v}_3(\mathbf{e}^\bullet(t)) \le \\ \le \tfrac{1}{2}\left(\tfrac{h}{m}\right)^2 \dot{\mathbf{e}}^{\bullet T}(t)\boldsymbol{V}\dot{\mathbf{e}}^\bullet(t) - \frac{2}{\left(\frac{h}{m}\right)^2}\left(\tfrac{h}{m}\mathbf{e}^{\bullet T}(t) - \mathbf{e}_{h1}^{\bullet T}(t)\right)\boldsymbol{V}\left(\tfrac{h}{m}\mathbf{e}^\bullet(t) - \mathbf{e}_{h1}^\bullet(t)\right) . \tag{65}$$

Constructing the composite vectors

$$\mathbf{e}^{\circ T}(t) = \left[\mathbf{e}^{\bullet T}(t)\ \mathbf{e}_{h1}^{\bullet T}(t)\ \mathbf{e}_{h2}^{\bullet T}(t)\ \mathbf{e}_{h3}^{\bullet T}(t)\right], \quad \mathbf{e}_g^{\circ T}(t) = \left[\mathbf{e}^{\circ T}(t)\ \boldsymbol{g}^{\bullet T}(t)\right], \tag{66}$$

introducing the notations

$$\boldsymbol{T}_A^\circ = \left[(\boldsymbol{A}^\bullet - \boldsymbol{J}^\bullet\boldsymbol{C}^\bullet)\ \boldsymbol{A}_h^\bullet\ \left[\boldsymbol{A}_h^\bullet \cdots \boldsymbol{A}_h^\bullet\right]\ \mathbf{0}\right], \ \boldsymbol{\Lambda}_1 = \begin{bmatrix} \mathbf{0} & \boldsymbol{T}_I^T\boldsymbol{P}^\bullet\boldsymbol{G}^\bullet \\ \boldsymbol{G}^{\bullet T}\boldsymbol{P}^\bullet\boldsymbol{T}_I & -\delta\,\boldsymbol{I}_{r_f+r_d} \end{bmatrix} \tag{67}$$

and using (44), then $\dot{v}_0(\mathbf{e}^\bullet(t))$ can be noted and rewritten as

$$\dot{v}_0(\mathbf{e}^\circ(t)) = \mathbf{e}_g^{\circ T}(t)\boldsymbol{\Lambda}_1\mathbf{e}_g^\circ(t) + \\ + \mathbf{e}^{\circ T}(t)\left(\boldsymbol{T}_I^T\boldsymbol{P}^\bullet\boldsymbol{T}_A^\circ + \boldsymbol{T}_A^{\circ T}\boldsymbol{P}^\bullet\boldsymbol{T}_I + \boldsymbol{T}_I^T(\boldsymbol{H}^\bullet - \boldsymbol{H}^\star)^T(\boldsymbol{H}^\bullet - \boldsymbol{H}^\star)\boldsymbol{T}_I\right)\mathbf{e}^\circ(t) . \tag{68}$$

Since, using (47), the matrix $\boldsymbol{T}_A^\circ$ can be modified as

$$\boldsymbol{T}_A^\circ = \boldsymbol{P}^{\bullet -1}\left[(\boldsymbol{P}^\bullet\boldsymbol{A} - \boldsymbol{P}^\bullet\boldsymbol{J}\boldsymbol{C})\ \boldsymbol{P}^\bullet\boldsymbol{A}_h\ \left[\boldsymbol{P}^\bullet\boldsymbol{A}_h \cdots \boldsymbol{P}^\bullet\boldsymbol{A}_h\right]\ \mathbf{0}\right] = \boldsymbol{P}^{\bullet -1}\boldsymbol{Y}^\circ\boldsymbol{T}_A , \tag{69}$$

then, considering (49) and introducing the notation

$$\boldsymbol{Y}^\bullet = \boldsymbol{P}^\bullet\boldsymbol{J}^\bullet = \operatorname{diag}\left[\boldsymbol{P}_1^\bullet\boldsymbol{J}\ \boldsymbol{P}_2^\bullet\right] = \operatorname{diag}\left[\boldsymbol{Y}\ \boldsymbol{P}_2^\bullet\right], \tag{70}$$

(68) takes the form

$$\dot{v}_0(\mathbf{e}^\circ(t)) = \mathbf{e}_g^{\circ T}(t)\boldsymbol{\Lambda}_1\mathbf{e}_g^\circ(t) + \\ + \mathbf{e}^{\circ T}(t)\left(\boldsymbol{T}_I^T\boldsymbol{Y}^\circ\boldsymbol{T}_A + \boldsymbol{T}_A^T\boldsymbol{Y}^{\circ T}\boldsymbol{T}_I + \boldsymbol{T}_I^T(\boldsymbol{H}^\bullet - \boldsymbol{H}^\star)^T(\boldsymbol{H}^\bullet - \boldsymbol{H}^\star)\boldsymbol{T}_I\right)\mathbf{e}^\circ(t) . \tag{71}$$

Same as above, using (44)-(46),

$$\dot{v}_1(\mathbf{e}^\circ(t)) \le \mathbf{e}^{\circ T}(t)\boldsymbol{T}_W^T \begin{bmatrix} \boldsymbol{W} & \mathbf{0} \\ \mathbf{0} & -\boldsymbol{W} \end{bmatrix} \boldsymbol{T}_W\mathbf{e}^\circ(t) = \mathbf{e}^{\circ T}(t)\boldsymbol{T}_W^T\boldsymbol{W}^\circ\boldsymbol{T}_W\mathbf{e}^\circ(t) , \tag{72}$$

$$\dot{v}_2(\mathbf{e}^\circ(t)) \le \mathbf{e}^{\circ T}(t)\boldsymbol{T}_U^T \begin{bmatrix} \boldsymbol{U} & \mathbf{0} \\ \mathbf{0} & -\boldsymbol{U} \end{bmatrix} \boldsymbol{T}_U\mathbf{e}^\circ(t) = \mathbf{e}^{\circ T}(t)\boldsymbol{T}_U^T\boldsymbol{U}^\circ\boldsymbol{T}_U\mathbf{e}^\circ(t) , \tag{73}$$

$$\dot{v}_{32}(\mathbf{e}^\circ(t)) = \frac{2}{\left(\frac{h}{m}\right)^2}\left(\frac{h}{m}\mathbf{e}^{\bullet T}(t) - \mathbf{e}_{h1}^{\bullet T}(t)\right)\boldsymbol{V}\left(\frac{h}{m}\mathbf{e}^\bullet(t) - \mathbf{e}_{h1}^\bullet(t)\right) = \mathbf{e}^{\circ T}(t)\boldsymbol{T}_V^T\boldsymbol{V}\boldsymbol{T}_V\mathbf{e}^\circ(t) . \tag{74}$$

Denoting

$$\mathbf{e}_c^\circ(t) = (\boldsymbol{A}^\bullet - \boldsymbol{J}^\bullet\boldsymbol{C}^\bullet)\mathbf{e}^\bullet(t) + \boldsymbol{A}_h^\bullet\mathbf{e}_h^\bullet(t) , \tag{75}$$

$$\boldsymbol{V}^\circ = \boldsymbol{P}^{\bullet -1}\boldsymbol{V}\boldsymbol{P}^{\bullet -1}, \quad \mathbf{e}_g^{\bullet T}(t) = \left[\mathbf{e}^{\bullet T}(t)\ \ \boldsymbol{g}^{\bullet T}(t)b^2\boldsymbol{G}^{\bullet T}\boldsymbol{V}\boldsymbol{P}^{\bullet -1}\right], \tag{76}$$

then it yields

$$\boldsymbol{P}^{\bullet}\boldsymbol{e}_c^{\circ}(t) = \boldsymbol{Y}^{\circ}\boldsymbol{T}_A\boldsymbol{e}^{\circ}(t)\,, \tag{77}$$

$$\begin{aligned}&\tfrac{1}{2}\left(\tfrac{h}{m}\right)^2\dot{\boldsymbol{e}}^{\bullet T}(t)\boldsymbol{V}\dot{\boldsymbol{e}}^{\bullet}(t) =\\ &= \left(\boldsymbol{P}^{\bullet}\boldsymbol{e}_c^{\circ}(t) + \boldsymbol{P}^{\bullet}\boldsymbol{G}^{\bullet}\boldsymbol{g}^{\bullet}(t)\right)^T b^2\boldsymbol{V}^{\circ}\left(\boldsymbol{P}^{\bullet}\boldsymbol{e}_c^{\circ}(t) + \boldsymbol{P}^{\bullet}\boldsymbol{G}^{\bullet}\boldsymbol{g}^{\bullet}(t)\right) =\\ &= \boldsymbol{e}^{\circ T}(t)\boldsymbol{T}_A^T\boldsymbol{Y}^{\circ T}b^2\boldsymbol{V}^{\circ}\boldsymbol{Y}^{\circ}\boldsymbol{T}_A\boldsymbol{e}^{\circ}(t) + \boldsymbol{e}_g^{\bullet T}(t)\begin{bmatrix}\boldsymbol{0} & \boldsymbol{T}_A^T\boldsymbol{Y}^{\circ T}\\ \boldsymbol{Y}^{\circ}\boldsymbol{T}_A & b^{-2}\boldsymbol{P}^{\bullet}\boldsymbol{V}^{-1}\boldsymbol{P}^{\bullet}\end{bmatrix}\boldsymbol{e}_g^{\bullet}(t)\end{aligned} \tag{78}$$

and the first term in (65) can be written as

$$\dot{v}_{31}(\boldsymbol{e}^{\circ}(t)) = \boldsymbol{e}^{\circ T}(t)\boldsymbol{T}_A^T\boldsymbol{Y}^{\circ T}b^2\boldsymbol{V}^{\circ}\boldsymbol{Y}^{\circ}\boldsymbol{T}_A\boldsymbol{e}^{\circ}(t) + \boldsymbol{e}_g^{\circ T}(t)\boldsymbol{\Lambda}_2\,\boldsymbol{e}_g^{\circ}(t)\,, \tag{79}$$

where

$$\boldsymbol{\Lambda}_2 = \begin{bmatrix}\boldsymbol{I}_{2n} & \boldsymbol{0}\\ \boldsymbol{0} & b^2\boldsymbol{G}^{\bullet T}\boldsymbol{V}\boldsymbol{P}^{\bullet -1}\end{bmatrix}\begin{bmatrix}\boldsymbol{0} & \boldsymbol{T}_A^T\boldsymbol{Y}^{\circ T}\\ \boldsymbol{Y}^{\circ}\boldsymbol{T}_A & b^{-2}\boldsymbol{P}^{\bullet}\boldsymbol{V}^{-1}\boldsymbol{P}^{\bullet}\end{bmatrix}\begin{bmatrix}\boldsymbol{I}_{2n} & \boldsymbol{0}\\ \boldsymbol{0} & b^2\boldsymbol{P}^{\bullet -1}\boldsymbol{V}\boldsymbol{G}^{\bullet}\end{bmatrix}. \tag{80}$$

Since, using (4), (46), it yields

$$\boldsymbol{P}^{\bullet}\boldsymbol{V}^{-1}\boldsymbol{P}^{\bullet} \geq 2\boldsymbol{P}^{\bullet} - \boldsymbol{V}, \qquad 2b^2 = a^4\,, \tag{81}$$

$$b^2\boldsymbol{T}_A^T\boldsymbol{Y}^{\circ T}\boldsymbol{P}^{\bullet -1}\boldsymbol{V}\boldsymbol{P}^{\bullet -1}\boldsymbol{P}^{\bullet}\boldsymbol{G}^{\bullet} \geq b^2\boldsymbol{T}_A^T\boldsymbol{Y}^{\circ T}(2\boldsymbol{P}^{\bullet -1} - \boldsymbol{V})\boldsymbol{P}^{\bullet}\boldsymbol{G}^{\bullet} \geq a^4\boldsymbol{T}_A^T\boldsymbol{Y}^{\circ T}\boldsymbol{G}^{\bullet}\,, \tag{82}$$

$$\boldsymbol{\Lambda}_2 \leq \begin{bmatrix}\boldsymbol{0} & a^4\boldsymbol{T}_A^T\boldsymbol{Y}^{\circ T}\boldsymbol{G}^{\bullet}\\ a^4\boldsymbol{G}^{\bullet T}\boldsymbol{Y}^{\circ}\boldsymbol{T}_A & b^2\boldsymbol{G}^{\bullet T}\boldsymbol{V}\boldsymbol{G}^{\bullet}\end{bmatrix}, \tag{83}$$

the derivative of the Lyapunov-Krasovskii functional can be written as

$$\dot{v}(\boldsymbol{e}^{\circ}(t), \boldsymbol{e}_g^{\circ}(t)) \leq \boldsymbol{e}^{\circ T}(t)\boldsymbol{P}^{\circ}\boldsymbol{e}^{\circ}(t) + \boldsymbol{e}_g^{\circ T}(t)\boldsymbol{P}_g^{\circ}\boldsymbol{e}_g^{\circ}(t) < 0\,, \tag{84}$$

where

$$\begin{aligned}\boldsymbol{P}^{\circ} &= \boldsymbol{T}_I^T\boldsymbol{Y}^{\circ}\boldsymbol{T}_A + \boldsymbol{T}_A^T\boldsymbol{Y}^{\circ}\boldsymbol{T}_I + \boldsymbol{T}_I^T(\boldsymbol{H}^{\bullet} - \boldsymbol{H}^{\star})^T(\boldsymbol{H}^{\bullet} - \boldsymbol{H}^{\star})\boldsymbol{T}_I -\\ &-\boldsymbol{T}_V^T\boldsymbol{V}\boldsymbol{T}_V + \boldsymbol{T}_U^T\boldsymbol{U}^{\circ}\boldsymbol{T}_U + \boldsymbol{T}_W^T\boldsymbol{W}^{\circ}\boldsymbol{T}_W + \boldsymbol{T}_A^T\boldsymbol{Y}^{\circ T}b^2\boldsymbol{V}^{\circ}\boldsymbol{Y}^{\circ}\boldsymbol{T}_A\,,\end{aligned} \tag{85}$$

$$\boldsymbol{P}_g^{\circ} = \begin{bmatrix}\boldsymbol{0} & \boldsymbol{T}_I^T\boldsymbol{P}^{\bullet}\boldsymbol{G}^{\bullet} + a^4\boldsymbol{T}_A^T\boldsymbol{Y}^{\circ T}\boldsymbol{G}^{\bullet}\\ \boldsymbol{G}^{\bullet T}\boldsymbol{P}^{\bullet}\boldsymbol{T}_I + a^4\boldsymbol{G}^{\bullet T}\boldsymbol{Y}^{\circ}\boldsymbol{T}_A & -\delta\boldsymbol{I}_{r_f+r_d} + b^2\boldsymbol{G}^{\bullet T}\boldsymbol{V}\boldsymbol{G}^{\bullet}\end{bmatrix}. \tag{86}$$

Therefore, using the Schur complement property, (85), (86) implies

$$\begin{bmatrix}\boldsymbol{\Delta}_{11} & * & * & *\\ \boldsymbol{Y}^{\circ}\boldsymbol{T}_A & -b^{-2}\boldsymbol{P}^{\bullet}\boldsymbol{V}^{-1}\boldsymbol{P}^{\bullet} & * & *\\ \boldsymbol{G}^{\bullet T}(\boldsymbol{P}^{\bullet}\boldsymbol{T}_I + a^4\boldsymbol{Y}^{\circ}\boldsymbol{T}_A) & \boldsymbol{0} & -\delta\boldsymbol{I}_{r_f+r_d} + b^2\boldsymbol{G}^{\bullet T}\boldsymbol{V}\boldsymbol{G}^{\bullet} & *\\ (\boldsymbol{H}^{\bullet} - \boldsymbol{H}^{\star})\boldsymbol{T}_I & \boldsymbol{0} & \boldsymbol{0} & -\boldsymbol{I}_{p_r}\end{bmatrix} < 0\,, \tag{87}$$

where $\boldsymbol{\Delta}_{11}$ is given in (43). It is noticed, (85) is a nonlinear matrix inequality and, using (81), then (87) has to be approximated as (42). ∎

7 Illustrative Example

To illustrate the proposed method, a system whose dynamics is described by equations (1), (2) is considered with

$$A = \begin{bmatrix} 2.6 & 0.0 & -0.8 \\ 1.2 & 0.2 & 0.0 \\ 0.0 & -0.5 & 3.0 \end{bmatrix}, \quad A_h = \begin{bmatrix} 0.00 & 0.02 & 0.00 \\ 0.00 & 0.00 & -1.00 \\ -0.02 & 0.00 & 0.00 \end{bmatrix}, \quad B = \begin{bmatrix} 1 & 3 \\ 2 & 1 \\ 1 & 1 \end{bmatrix}, \quad C = \begin{bmatrix} 1 & 2 & 1 \\ 1 & 1 & 0 \end{bmatrix},$$

$$B_f^T = \begin{bmatrix} 1 & 2 & 1 \end{bmatrix}, \quad B_d^T = \begin{bmatrix} 0.3 & 0.1 & 0.1 \end{bmatrix}, \quad h = 1.2\ s, \quad \operatorname{rank}(H) = p_r = 2\,.$$

Solving (23), (24) for $P^\diamond$, $Y^\diamond$, $H^\diamond$, δ and γ using SeDuMi package, the residual reference model parameter design problem was solved as feasible, where

$$J^\diamond = \begin{bmatrix} 2.8421 & 1.0896 \\ -2.1774 & 3.5106 \\ 9.5192 & -10.3716 \end{bmatrix}, \quad H^\diamond = 10^{-3} \begin{bmatrix} 0.1248 & 0.0673 & -0.1523 \\ -0.0014 & 0.0010 & 0.0005 \end{bmatrix},$$

$$\rho(A - J^\diamond C) = \{ -1.9127 \ -2.4471 \pm 2.7012\,\mathrm{i} \}, \quad \delta = 1.5808, \quad \gamma = 1.2488\,.$$

ensuring the stable eigenvalue spectrum of $A_e^\diamond = A - J^\diamond C$. Subsequently, solving (41), (42) with respect to the LMI matrix variables $P_1^\bullet$, $P_2^\bullet$, U, V, W, Y, H, δ, and with the references parameters $J^\diamond$, $H^\diamond$, the feasible solution gave the result

$$J = \begin{bmatrix} 6.3562 & -3.5187 \\ -3.6426 & 6.0964 \\ 15.4197 & -18.3528 \end{bmatrix}, \quad H = 10^{-3} \begin{bmatrix} 0.0109 & -0.0534 & -0.1270 \\ -0.0001 & -0.0003 & -0.0009 \end{bmatrix},$$

$$\rho(A - JC) = \{ -2.3157 \ -4.4764 \pm 1.1702\,\mathrm{i} \}, \quad \delta = 4.2766\,.$$

Figure 1 illustrates the residual signals concerning a fault on the input $f(t)$, where the fault was modeled as the short-circuit outage during $t \in \langle 30, 65 \rangle$ [s].

Note, a feasible solution was obtained with $m = 2$ for h up to $h_m = 1.37$ s.

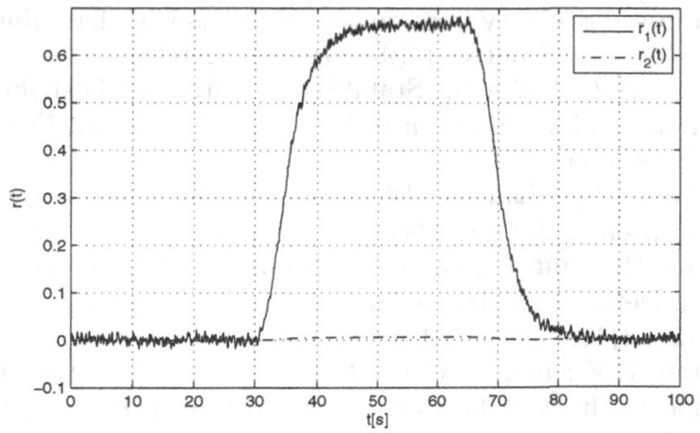

Fig. 1. The residual generator response to a fault

8 Concluding Remarks

Design conditions for robust FDFs, destined for systems with DsTD and unknown disturbances, are derived in the paper. The main idea goes with introducing a reference residual model, which design conditions are independent of a DsTD. Subsequently, the delay-dependent design conditions for FDF are established in terms of LMI as a feasible problem, accomplishing the manipulation in a manner giving guaranty of FDF asymptotic stability and respecting the limitations fought off by the existence of structured LMI variables. In presented form, decoupling of the the most important disturbances is not solved. The illustrative example confirms the effectiveness and applicability of the proposed method.

Acknowledgments. The work presented in the paper was supported by VEGA, the Grant Agency of the Ministry of Education and the Academy of Science of Slovak Republic, under Grant No. 1/0256/11. This support is very gratefully acknowledged.

References

1. Bai, L., Tian, Z., Shi, S.: Robust Fault Detection for a Class of Nonlinear Time-Delay Systems. J. Franklin Institute 344(6), 873–888 (2007)
2. Blanke, M., Kinnaert, M., Lunze, J., Staroswiecki, M.: Diagnosis and Fault-Tolerant Control. Springer, Berlin (2003)
3. Ding, S.: Model-Based Fault Diagnosis Techniques: Design Schemes, Algorithms, and Tools. Springer, Berlin (2008)
4. Feng, Z., Lam, J.: Integral Partitioning Approach to Robust Stabilization for Uncertain Distributed Time Delay Systems. Int. J. Robust and Nonlinear Control 22(6), 676–689 (2012)
5. Fiagbedzi, Y.A., Pearson, A.E.: A Multistage Reduction Technique for Feedback Stabilizing Distributed Time-Lag Systems. Automatica 23(3), 311–326 (1987)
6. Filasová, A., Gontkovič, D., Krokavec, D.: LMI Based Control Design for Linear Systems with Distributed Time Delays. Archives of Control Sciences 22(2), 125–139 (2012)
7. Filasová, A., Gontkovič, D., Krokavec, D.: Actuator Faults Estimation for a Class of Linear Distributed Time Delay Systems. In: 10th European Workshop on Advanced Control and Diagnosis, Copenhagen-Lyngby, Denmark, pp. 3.1–3.6 (2012)
8. Gu, K.: An Improved Stability Criterion for Systems with Distributed Delays. Int. J. Robust and Nonlinear Control 13(9), 819–831 (2003)
9. Suh, Y.S., Kang, H.J., Ro, Y.S.: Stability Condition of Distributed Delay Systems Based on an Analytic Solution to Lyapunov Functional Equations. Asian J. Control 8(1), 91–96 (2006)
10. Wang, H.B., Wang, J.L., Lam, J.: Worst-Case Fault Detection Observer Design: Optimization Approach. J. Opt. Theory and Applications 132(3), 475–491 (2007)
11. Zhang, J., Zhang, H., Yang, F., Wang, S.: Robust Fault Detection Filter Design for a Class of Time-Delay Systems via Equivalent Transformation. J. Contr. Theory and Applications 11(1), 54–60 (2013)
12. Zheng, F., Frank, P.M.: Robust Control of Uncertain Distributed Delay Systems with Application to the Stabilization of Combustion in Rocket Motor Chambers. Automatica 38(3), 487–497 (2002)
13. Zhong, M., Ye, H., Ding, S.X., Wang, G., Zhou, D.: Fault Detection Filter for Linear Time-Delay Systems. Nonlinear Dyn. Systems Theory 5(3), 273–284 (2005)

Fault Residual Functions Make Use of Linear Quadratic Control Performances

Dušan Krokavec, Anna Filasová, and Pavol Liščinský

Technical University of Košice, Faculty of Electrical Engineering and Informatics,
Department of Cybernetics and Artificial Intelligence,
Letná 9, 042 00 Košice, Slovakia
{dusan.krokavec,anna.filasova,pavol.liscinsky}@tuke.sk
http://web.tuke.sk/kkui

Abstract. An approach to fault detection systems design, adjusted for linear continuous-time systems, is proposed in the paper. Based on LMI approach to inverse infinite-time horizon LQ problem, the method blends state-space control principle with checking and evaluation of degradation in the equivalent performance index value due to faults in actuators, sensors or system dynamics. A simulation example, subject to different type of failures, demonstrates the effectiveness of the proposed form of the fault detection function.

Keywords: LQ control, state-space system description, performance degradation, fault detection schemes.

1 Introduction

Automated diagnosis is one of the more fruitful applications in sophisticated control systems. The real problem is usually to fix the system with faults so that it can continue its mission for some time with some limitations of functionality. Diagnosis is a part of larger problem known as fault detection, identification and reconfiguration (FDIR). The practical benefits of an integrated approach to FDIR seem to be considerable, especially when knowledge of available fault isolations and system reconfigurations is used to increase the reliability and utility of control. The classical approaches, mainly observer-based methods, parity space principle and parameter identification technique, have been eminently studied (see, e.g., [1], [3], [7], [8], [10], [16], and the references therein).

Following the linear quadratic (LQ) control performance [2], [9], the fault detection principle, based on LQ control properties for full state measurable linear discrete-time systems, was suggested in [11], [12], and generalized in [13]. If the LQ control problem is solvable, the method evaluates degradations in values of the LQ performance index in the fault detection scheme. In this paper is presented the substantial generalization of this method, as well as full adaptation of these results for linear continuous-time systems, being related with the extension to state-space control laws founded on other design methodologies than LQ control approach. Exploiting the inverse LQ problem [5] to obtain an equivalent LQ

J. Korbicz and M. Kowal (eds.), *Intelligent Systems in Technical and Medical Diagnostics*, Advances in Intelligent Systems and Computing 230,
DOI: 10.1007/978-3-642-39881-0_36, © Springer-Verlag Berlin Heidelberg 2014

performance index, the fault detection function, associated with the LQ control criterium, is devised and verified. The proposed principle, suitable chiefly for actuator and sensor faults detection, gives a relative simple algorithmic support potentially applicable in embedded diagnosis structures [4].

The paper is organized as follows. Starting with system model description in Section 2, in Section 3 the basic concepts is presented. The solution to the inverse LQ problem existence and the proposed fault detection function are given in Section 4. Section 5 shows an example to illustrate fault detection characteristics and Section 6 is devoted to a brief overview of the method properties.

Throughout the paper, the following notations are used: $\boldsymbol{x}^T$, $\boldsymbol{X}^T$ denotes the transpose of the vector $\boldsymbol{x}$ and matrix $\boldsymbol{X}$, respectively, for a square matrix $\boldsymbol{X} < 0$ means that $\boldsymbol{X}$ is a symmetric negative definite matrix, the symbol $\boldsymbol{I}_n$ indicates the n-th order unit matrix, $\mathbb{R}$ denotes the set of real numbers and $\mathbb{R}^{n \times r}$ refers to the set of $n \times r$ real matrices.

2 System Model

In the paper, there are taken into account the continuous-time linear dynamic systems, described as

$$\dot{\boldsymbol{q}}(t) = \boldsymbol{A}\boldsymbol{q}(t) + \boldsymbol{B}\boldsymbol{u}(t) + \boldsymbol{B}_f\boldsymbol{f}(t) \,, \tag{1}$$

$$\boldsymbol{y}_c(t) = \boldsymbol{I}_n\boldsymbol{q}(t) + \boldsymbol{f}_s(t), \quad \boldsymbol{y}(t) = \boldsymbol{C}\boldsymbol{q}(t) \,, \tag{2}$$

where $\boldsymbol{q}(t) \in \mathbb{R}^n$ stands up for the system state, $\boldsymbol{u}(t) \in \mathbb{R}^r$ denotes the control input, $\boldsymbol{y}(t) \in \mathbb{R}^m$ is the reference output, and matrices $\boldsymbol{A} \in \mathbb{R}^{n \times n}$, $\boldsymbol{B} \in \mathbb{R}^{n \times r}$, $\boldsymbol{B}_f \in \mathbb{R}^{n \times r_f}$, $\boldsymbol{C} \in \mathbb{R}^{m \times n}$ are finite valued. The system and sensor faults are modeled by unknown additive functions $\boldsymbol{f}(t) \in \mathbb{R}^{r_f}$, $\boldsymbol{f}_s(t) \in \mathbb{R}^n$, respectively. Considering all system state variables being measurable, with respect to infinite time horizon control task it is evidently necessary that $(\boldsymbol{A}, \boldsymbol{B})$ is controllable.

3 Basic Concept

Proposition 1. *The finite time horizon LQ control of the system (1), (2) is optimized with respect to the equivalent quadratic cost function*

$$J_T = \boldsymbol{q}^T(0)\boldsymbol{P}(0)\boldsymbol{q}(0) + \int_0^T p\big(\boldsymbol{q}(t), \boldsymbol{u}(t)\big)\, dt \,, \tag{3}$$

$$p\big(\boldsymbol{q}(t), \boldsymbol{u}(t)\big) = \begin{bmatrix} \boldsymbol{q}^T(t) \; \boldsymbol{u}^T(t) \end{bmatrix} \boldsymbol{J}(t) \begin{bmatrix} \boldsymbol{q}(t) \\ \boldsymbol{u}(t) \end{bmatrix}, \tag{4}$$

$$\boldsymbol{J}(t) = \begin{bmatrix} \boldsymbol{P}(t)\boldsymbol{A} + \boldsymbol{A}^T\boldsymbol{P}(t) + \dot{\boldsymbol{P}}(t) + \boldsymbol{Q} & \boldsymbol{P}(t)\boldsymbol{B} + \boldsymbol{S} \\ \boldsymbol{B}^T\boldsymbol{P}(t) + \boldsymbol{S}^T & \boldsymbol{R} \end{bmatrix}, \tag{5}$$

where $\boldsymbol{Q} \geq 0$, $\boldsymbol{Q} \in \mathbb{R}^{n \times n}$, $\boldsymbol{R} > 0$, $\boldsymbol{R} \in \mathbb{R}^{r \times r}$, $\boldsymbol{S} \in \mathbb{R}^{n \times r}$, $\boldsymbol{P}(t) > 0$, $\boldsymbol{P}(t) \in \mathbb{R}^{n \times n}$, and $\boldsymbol{J}(t) > 0$, $\boldsymbol{J}(t) \in \mathbb{R}^{(n+r) \times (n+r)}$.

Proof. (compare, e.g., [9]) The standard quadratic cost function in LQ control is defined as follows

$$J_T = \boldsymbol{q}^T(T)\boldsymbol{Q}^{\bullet}\boldsymbol{q}(T) + \int_0^T r\big(\boldsymbol{q}(t), \boldsymbol{u}(t)\big)\mathrm{d}t \,, \tag{6}$$

where

$$r\big(\boldsymbol{q}(t), \boldsymbol{u}(t)\big) = \boldsymbol{u}^T(t)\boldsymbol{R}\boldsymbol{u}(t) + \boldsymbol{q}^T(t)\boldsymbol{Q}\boldsymbol{q}(t) + \boldsymbol{q}^T(t)\boldsymbol{S}\boldsymbol{u}(t) + \boldsymbol{u}^T(t)\boldsymbol{S}^T\boldsymbol{q}(t) \,, \tag{7}$$

$$r\big(\boldsymbol{q}(t), \boldsymbol{u}(t)\big) = \big[\, \boldsymbol{q}^T(t) \; \boldsymbol{u}^T(t) \,\big] \, \boldsymbol{J}_J \begin{bmatrix} \boldsymbol{q}(t) \\ \boldsymbol{u}(t) \end{bmatrix} , \tag{8}$$

respectively, and $\boldsymbol{Q}^{\bullet} \geq 0$, $\boldsymbol{Q}^{\bullet} \in I\!R^{n \times n}$, $\boldsymbol{J}_J \in I\!R^{(n+r)\times(n+r)}$,

$$\boldsymbol{J}_J = \begin{bmatrix} \boldsymbol{Q} & \boldsymbol{S} \\ \boldsymbol{S}^T & \boldsymbol{R} \end{bmatrix} > 0 \Leftrightarrow \boldsymbol{Q} - \boldsymbol{S}\boldsymbol{R}^{-1}\boldsymbol{S}^T > 0, \; \boldsymbol{R} > 0 \,. \tag{9}$$

Defining Lyapunov function of the form

$$v(\boldsymbol{q}(t)) = \boldsymbol{q}^T(t)\boldsymbol{P}(t)\boldsymbol{q}(t) \,, \tag{10}$$

the derivative of the Lyapunov function along a solution of (1) is

$$\dot{v}(\boldsymbol{q}(t), \boldsymbol{u}(t)) = \dot{\boldsymbol{q}}^T(t)\boldsymbol{P}(t)\boldsymbol{q}(t) + \boldsymbol{q}^T(t)\boldsymbol{P}(t)\dot{\boldsymbol{q}}(t) + \boldsymbol{q}^T(t)\dot{\boldsymbol{P}}(t)\boldsymbol{q}(t) \,, \tag{11}$$

$$\dot{v}(\boldsymbol{q}(t), \boldsymbol{u}(t)) = \big[\, \boldsymbol{q}^T(t) \; \boldsymbol{u}^T(t) \,\big] \, \boldsymbol{J}_V(t) \begin{bmatrix} \boldsymbol{q}(t) \\ \boldsymbol{u}(t) \end{bmatrix} , \tag{12}$$

respectively, where

$$\boldsymbol{J}_V(t) = \begin{bmatrix} \boldsymbol{P}(t)\boldsymbol{A} + \boldsymbol{A}^T\boldsymbol{P}(t) + \dot{\boldsymbol{P}}(t) & \boldsymbol{P}(t)\boldsymbol{B} \\ \boldsymbol{B}^T\boldsymbol{P}(t) & \boldsymbol{0} \end{bmatrix} . \tag{13}$$

Defining, at the time instant T, the cumulative function V_T as

$$V_T = \int_0^T \dot{v}(\boldsymbol{q}(t), \boldsymbol{u}(t))\mathrm{d}t \,, \tag{14}$$

which is equivalent to

$$V_T = \boldsymbol{q}^T(T)\boldsymbol{P}(T)\boldsymbol{q}(T) - \boldsymbol{q}^T(0)\boldsymbol{P}(0)\boldsymbol{q}(0) \,, \tag{15}$$

then setting $\boldsymbol{P}(T) = \boldsymbol{Q}^{\bullet}$, adding (14) to (6) and subtracting (15) from (6), the performance index is brought to the form (3), where

$$p\big(\boldsymbol{q}(t), \boldsymbol{u}(t)\big) = r\big(\boldsymbol{q}(i), \boldsymbol{u}(t)\big) + \dot{v}\big(\boldsymbol{q}(t), \boldsymbol{u}(t)\big) \,. \tag{16}$$

It is evident that with $\boldsymbol{J}(t) = \boldsymbol{J}_J + \boldsymbol{J}_V(t)$ then (8), (9), and (12), (13) implies (4), (5). This concludes the proof. ∎

Proposition 2. *(infinite time horizon LQ control) The LQ control law, the gain has a constant value, is given by*

$$\boldsymbol{u}(t) = -\boldsymbol{K}\boldsymbol{q}(t) , \tag{17}$$

$$\boldsymbol{K} = \boldsymbol{R}^{-1}(\boldsymbol{P}\boldsymbol{B} + \boldsymbol{S})^T , \tag{18}$$

where $\boldsymbol{P} > 0$ is a solution of the continuous algebraic Riccati equation (CARE)

$$\boldsymbol{0} = \boldsymbol{A}^T\boldsymbol{P} + \boldsymbol{P}\boldsymbol{A} + \boldsymbol{Q} - (\boldsymbol{P}\boldsymbol{B} + \boldsymbol{S})\boldsymbol{R}^{-1}(\boldsymbol{P}\boldsymbol{B} + \boldsymbol{S})^T . \tag{19}$$

Proof. With $\dot{\boldsymbol{P}}(t) = \boldsymbol{0}$ then (4), (5) implies ([9])

$$\frac{\partial p(\boldsymbol{q}(t), \boldsymbol{u}(t))}{\partial \boldsymbol{u}^T(t)} = \left[\boldsymbol{0}\ \boldsymbol{I}_r\right] \boldsymbol{J} \begin{bmatrix} \boldsymbol{q}(t) \\ \boldsymbol{u}(t) \end{bmatrix} = \left[(\boldsymbol{P}\boldsymbol{B} + \boldsymbol{S})^T\ \boldsymbol{R}\right] \begin{bmatrix} \boldsymbol{q}(t) \\ \boldsymbol{u}(t) \end{bmatrix} = \boldsymbol{0} , \tag{20}$$

$$\frac{\partial p(\boldsymbol{q}(t), \boldsymbol{u}(t))}{\partial \boldsymbol{q}^T(t)} = \left[\boldsymbol{I}_n\ \boldsymbol{0}\right] \boldsymbol{J} \begin{bmatrix} \boldsymbol{q}(t) \\ \boldsymbol{u}(t) \end{bmatrix} = (\boldsymbol{A}^T\boldsymbol{P} + \boldsymbol{P}\boldsymbol{A} + \boldsymbol{Q})\boldsymbol{q}(t) + (\boldsymbol{P}\boldsymbol{B} + \boldsymbol{S})\boldsymbol{u}(t) = \boldsymbol{0} \tag{21}$$

and $\boldsymbol{I}_r \in I\!R^{r\times r}$, $\boldsymbol{I}_n \in I\!R^{n\times n}$ are identity matrices. It is obvious that (20) implies (18) and inserting (17), (18) in (21), then (19) is obtained. This concludes the proof. ■

Remark 1. The steady-state solution of (5) is

$$\boldsymbol{J} = \begin{bmatrix} \boldsymbol{P}\boldsymbol{A} + \boldsymbol{A}^T\boldsymbol{P} + \boldsymbol{Q} & \boldsymbol{P}\boldsymbol{B} + \boldsymbol{S} \\ (\boldsymbol{P}\boldsymbol{B} + \boldsymbol{S})^T & \boldsymbol{R} \end{bmatrix} > 0 . \tag{22}$$

Definition 1. *To the system (1), (2) without faults and under the LQ control (17), (18), $h\big(\boldsymbol{q}(t), \boldsymbol{u}(t)\big)$ is fault residual function if $h\big(\boldsymbol{q}(t), \boldsymbol{u}(t)\big) = 0$.*

Proposition 3. *With respect to a steady-state state solution of CARE, the function $h\big(\boldsymbol{q}(t), \boldsymbol{u}(t)\big)$ for the fault-free system (1), (2) under the LQ control (17), (18) in the unforced regime is*

$$h\big(\boldsymbol{q}(t), \boldsymbol{u}(t)\big) = p\big(\boldsymbol{q}(t), \boldsymbol{u}(t)\big) . \tag{23}$$

Proof. Using the steady-state values of (4) and (22) $p(\boldsymbol{q}(t), \boldsymbol{u}(t))$ can be rewritten as:

$$p(\boldsymbol{q}(t), \boldsymbol{u}(t)) = \boldsymbol{q}^T(t) \left[\boldsymbol{I}_n\ -\boldsymbol{K}^T\right] \boldsymbol{J} \begin{bmatrix} \boldsymbol{I}_n \\ -\boldsymbol{K} \end{bmatrix} \boldsymbol{q}(t) \tag{24}$$

and rearranging (24), it yields

$$p(\boldsymbol{q}(t), \boldsymbol{u}(t)) = \boldsymbol{q}^T(t) \left\{ \begin{matrix} \boldsymbol{Q} + \boldsymbol{A}^T\boldsymbol{P} + \boldsymbol{P}\boldsymbol{A} - (\boldsymbol{S} + \boldsymbol{P}\boldsymbol{B})\boldsymbol{K} - \\ -\boldsymbol{K}^T(\boldsymbol{S} + \boldsymbol{P}\boldsymbol{B})^T + \boldsymbol{K}^T\boldsymbol{R}\boldsymbol{K} \end{matrix} \right\} \boldsymbol{q}(t) . \tag{25}$$

Since (21) implies that the sum of items of the first row of (25) is equal zero, it is simple to verify that such selection of $\boldsymbol{K}$ as given in (18) leads to the next result

$$p(\boldsymbol{q}(t), \boldsymbol{u}(t)) = \boldsymbol{q}^T(t)\boldsymbol{K}^T(-(\boldsymbol{S} + \boldsymbol{P}\boldsymbol{B})^T + \boldsymbol{R}\boldsymbol{R}^{-1}(\boldsymbol{S} + \boldsymbol{P}\boldsymbol{B})^T)\boldsymbol{q}(t) = 0 \tag{26}$$

and (26) implies (23). Since $p(\boldsymbol{q}(t), \boldsymbol{u}(t)) = 0$ shows the case when there is no fault in the system that is why $p(\boldsymbol{q}(t), \boldsymbol{u}(t))$ can be applied as a fault residual function. This concludes the proof. ∎

Definition 2. *The forced regime for (1), (2) is given by the control policy*

$$\boldsymbol{u}(t) = -\boldsymbol{K}\boldsymbol{q}(t) + \boldsymbol{W}\boldsymbol{w}(t)\,, \tag{27}$$

where $r = m$, $\boldsymbol{w}(t) \in I\!R^m$ is desired output signal vector and $\boldsymbol{W} \in I\!R^{m\times m}$ is the signal gain matrix.

Remark 2. If the system (1), (2) is stabilizable, and

$$\operatorname{rank}\begin{bmatrix} \boldsymbol{A} & \boldsymbol{B} \\ \boldsymbol{C} & \boldsymbol{0} \end{bmatrix} = n + m\,, \tag{28}$$

then the matrix $\boldsymbol{W}$ in (27), designed by using the static decoupling principle, takes the form [13], [17]

$$\boldsymbol{W} = \big(\boldsymbol{C}\big(-(\boldsymbol{A} - \boldsymbol{B}\boldsymbol{K})\big)^{-1}\boldsymbol{B}\big)^{-1}\,. \tag{29}$$

Theorem 1. *With respect to a steady-state solution of CARE, the fault residual function $h\big(\boldsymbol{q}(t), \boldsymbol{u}(t)\big)$ for the fault-free system (1), (2) under the LQ control (17), (18) in the forced regime (27) is*

$$h\big(\boldsymbol{q}(t), \boldsymbol{u}(t)\big) = p\big(\boldsymbol{q}(t), \boldsymbol{u}(t)\big) - p_w\big(\boldsymbol{w}(t)\big)\,, \tag{30}$$

where

$$p_w(\boldsymbol{w}(t)) = \boldsymbol{w}^T(t)\boldsymbol{W}^T\boldsymbol{R}\boldsymbol{W}\boldsymbol{w}(t)\,. \tag{31}$$

Proof. Inserting (27) into the steady-state solution of (4), (22), then

$$\begin{aligned} &p_w(\boldsymbol{q}(t), \boldsymbol{w}(t)) = \\ &= \Big(\boldsymbol{w}^T(t)\,[\boldsymbol{0}\ \boldsymbol{W}^T] + \boldsymbol{q}^T(t)\,[\boldsymbol{I}_n\ -\boldsymbol{K}^T]\Big)\boldsymbol{J}\Big(\begin{bmatrix} \boldsymbol{I}_n \\ -\boldsymbol{K} \end{bmatrix} q(t) + \begin{bmatrix} \boldsymbol{0} \\ \boldsymbol{W} \end{bmatrix} \boldsymbol{w}(t)\Big)\,. \end{aligned} \tag{32}$$

Thus, writing the block partitioned structure of (22) as

$$\boldsymbol{J} = \begin{bmatrix} \boldsymbol{A}^T\boldsymbol{P} + \boldsymbol{P}\boldsymbol{A} + \boldsymbol{Q} & \boldsymbol{P}\boldsymbol{B} + \boldsymbol{S} \\ (\boldsymbol{P}\boldsymbol{B} + \boldsymbol{S})^T & \boldsymbol{R} \end{bmatrix} = \begin{bmatrix} \boldsymbol{J}_{11} & \boldsymbol{J}_{12} \\ \boldsymbol{J}_{12}^T & \boldsymbol{J}_{22} \end{bmatrix}, \tag{33}$$

after straightforward calculation it is obtained

$$\begin{aligned} &p_w(\boldsymbol{q}(t), \boldsymbol{u}(t)) = \\ &= \boldsymbol{w}^T(t)\boldsymbol{W}^T(\boldsymbol{J}_{12}^T - \boldsymbol{J}_{22}\boldsymbol{K})\boldsymbol{q}(t) + \boldsymbol{q}^T(t)(\boldsymbol{J}_{12} - \boldsymbol{K}^T\boldsymbol{J}_{22})\boldsymbol{W}\boldsymbol{w}(t) + \\ &\quad + \boldsymbol{w}^T(t)\boldsymbol{W}^T\boldsymbol{J}_{22}\boldsymbol{W}^T\boldsymbol{w}(t) + \\ &\quad + \boldsymbol{q}^T(t)(\boldsymbol{J}_{11} - \boldsymbol{J}_{12}\boldsymbol{K} - \boldsymbol{K}^T\boldsymbol{J}_{12}^T + \boldsymbol{K}^T\boldsymbol{J}_{22}\boldsymbol{K})\boldsymbol{q}(t)\,. \end{aligned} \tag{34}$$

Considering that (24), (33) gives

$$\boldsymbol{q}^T(t)(\boldsymbol{J}_{11} - \boldsymbol{J}_{12}\boldsymbol{K} - \boldsymbol{K}^T\boldsymbol{J}_{12}^T + \boldsymbol{K}^T\boldsymbol{J}_{22}\boldsymbol{K})\boldsymbol{q}(t) = 0 , \tag{35}$$

then (34) can be rewritten as

$$p_w(\boldsymbol{q}(t), \boldsymbol{u}(t)) = \begin{bmatrix} \boldsymbol{q}(t) \\ \boldsymbol{W}\boldsymbol{w}(t) \end{bmatrix}^T \begin{bmatrix} \mathbf{0} & \boldsymbol{J}_{12} - \boldsymbol{K}^T\boldsymbol{J}_{22} \\ \boldsymbol{J}_{12}^T - \boldsymbol{J}_{22}\boldsymbol{K} & \boldsymbol{J}_{22} \end{bmatrix} \begin{bmatrix} \boldsymbol{q}(t) \\ \boldsymbol{W}\boldsymbol{w}(t) \end{bmatrix} . \tag{36}$$

Since (18), (33) leads to

$$\boldsymbol{K} = \boldsymbol{R}^{-1}(\boldsymbol{P}\boldsymbol{B} + \boldsymbol{S})^T = \boldsymbol{J}_{22}^{-1}\boldsymbol{J}_{12}^T , \tag{37}$$

$$\boldsymbol{J}_{12}^T - \boldsymbol{J}_{22}\boldsymbol{K} = \boldsymbol{J}_{12}^T - \boldsymbol{J}_{22}\boldsymbol{J}_{22}^{-1}\boldsymbol{J}_{12}^T = \mathbf{0} , \tag{38}$$

respectively, then (36) implies

$$p_w(\boldsymbol{q}(t), \boldsymbol{w}(t)) = p_w(\boldsymbol{w}(t)) = \boldsymbol{w}^T(i)\boldsymbol{W}^T\boldsymbol{J}_{22}\boldsymbol{W}\boldsymbol{w}(t) . \tag{39}$$

Now, using the notation (33) then the equality (39) implies (31). This concludes the proof. ∎

Remark 3. It is clear that the best response dynamics of a residual function to faults can be obtained if this function is memory-free. One possibility is to use the property of the given relationship (4). Now, (39) is independent on the state-feedback and can be considered as an floating additive offset, and being attached into (30), the static residual function $h(\boldsymbol{q}(t), \boldsymbol{u}(t))$ in a regime without faults evidently takes the value

$$\begin{aligned} h(\boldsymbol{q}(t), \boldsymbol{u}(t)) = \boldsymbol{q}^T(t)\boldsymbol{J}_{11}\boldsymbol{q}(t) + \boldsymbol{q}^T(t)\boldsymbol{J}_{12}\boldsymbol{u}(t) + \boldsymbol{u}^T(t)\boldsymbol{J}_{21}\boldsymbol{q}(t) + \\ + \boldsymbol{u}^T(t)\boldsymbol{J}_{22}\boldsymbol{u}(t) - \boldsymbol{w}^T(t)\boldsymbol{W}^T\boldsymbol{J}_{22}\boldsymbol{W}\boldsymbol{w}(t) = 0 . \end{aligned} \tag{40}$$

This represents a constructive procedure for achieving given objective in the case of occurrence of sensor faults in the controlled systems.

Remark 4. Memory-free residual function (40) is indeed the best dynamics but is not suitable for detecting of such system parametric defects and actuator faults that do not cause system instability. Since (40) is derived using the steady-state solutions, availing the structure of (16), the adequate dynamic fault residual function can be obtained using (40) in the full dynamic form [13]

$$\begin{aligned} h(\boldsymbol{q}(t), \boldsymbol{u}(t)) = \dot{\boldsymbol{q}}^T(t)\boldsymbol{P}\boldsymbol{q}(t) + \boldsymbol{q}^T(t)\boldsymbol{P}\dot{\boldsymbol{q}}(t) + \boldsymbol{q}^T(t)\boldsymbol{Q}\boldsymbol{q}(t) + \\ + \boldsymbol{q}^T(t)\boldsymbol{S}\boldsymbol{u}(t) + \boldsymbol{u}^T(t)\boldsymbol{S}^T\boldsymbol{q}(t) + \boldsymbol{u}^T(t)\boldsymbol{R}\boldsymbol{q}(t) - \boldsymbol{w}^T(t)\boldsymbol{W}^T\boldsymbol{R}\boldsymbol{W}\boldsymbol{w}(t) . \end{aligned} \tag{41}$$

Note, this modification doesn't significantly change the computation procedure if the derivative of the system state vector is available.

4 General State Feedback Control

For most practical situations, the control policy takes the form (17), or (27), but the gain matrix $\boldsymbol{K}$ doesn't satisfy (18), (19) owing to that another design method (pole placement, bounded real lemma, Kalman-Yakubovitch-Popov lemma, etc.) was used in the control design step. This problem can be reasonably well circumvented by approximating the performance index using inverse LQ control principle.

Theorem 2. *(inverse LQ task) A solution of inverse LQ problem for given $\boldsymbol{A}$, $\boldsymbol{B}$ and $\boldsymbol{K}$ exists if exist symmetric positive definite matrices $\boldsymbol{P}_e \in I\!R^{n\times n}$, $\boldsymbol{R}_e \in I\!R^{r\times r}$ such that*

$$\boldsymbol{P}_e = \boldsymbol{P}_e^T > 0, \qquad \boldsymbol{R}_e = \boldsymbol{R}_e^T > 0\,, \tag{42}$$

$$\begin{bmatrix} \boldsymbol{A}^T\boldsymbol{P}_e + \boldsymbol{P}_e\boldsymbol{A} - \boldsymbol{K}^T\boldsymbol{R}_e\boldsymbol{K} & * \\ \boldsymbol{B}^T\boldsymbol{P}_e - \boldsymbol{R}_e\boldsymbol{K} & -\boldsymbol{R}_e \end{bmatrix} < 0\,, \tag{43}$$

$$\boldsymbol{A}^T\boldsymbol{P}_e + \boldsymbol{P}_e\boldsymbol{A} - \boldsymbol{K}^T\boldsymbol{R}_e\boldsymbol{K} < 0\,. \tag{44}$$

When the above conditions hold, the matrices $\boldsymbol{Q}_e \in I\!R^{n\times n}$, $\boldsymbol{S}_e^T \in I\!R^{r\times n}$ are given as

$$-\boldsymbol{Q}_e = \boldsymbol{A}^T\boldsymbol{P}_e + \boldsymbol{P}_e\boldsymbol{A} - \boldsymbol{K}^T\boldsymbol{R}_e\boldsymbol{K}\,, \tag{45}$$

$$-\boldsymbol{S}_e^T = \boldsymbol{B}^T\boldsymbol{P}_e - \boldsymbol{R}_e\boldsymbol{K} \tag{46}$$

and the structure of $\boldsymbol{J}_e$ is given in (33).

Here, and hereafter, $*$ denotes the symmetric item in a symmetric matrix.

Proof. If $\boldsymbol{A}$, $\boldsymbol{B}$ and $\boldsymbol{K}$ are given, following the structure of (19), it can be defined

$$\boldsymbol{0} = \boldsymbol{A}^T\boldsymbol{P}_e + \boldsymbol{P}_e\boldsymbol{A} + \boldsymbol{Q}_e - (\boldsymbol{P}_e\boldsymbol{B} + \boldsymbol{S}_e)\boldsymbol{R}_e^{-1}\boldsymbol{R}_e\boldsymbol{R}_e^{-1}(\boldsymbol{P}_e\boldsymbol{B} + \boldsymbol{S}_e)^T\,. \tag{47}$$

Then, with respect to the form of (18), the equation (47) takes the form

$$\boldsymbol{0} = \boldsymbol{A}^T\boldsymbol{P}_e + \boldsymbol{P}_e\boldsymbol{A} + \boldsymbol{Q}_e - \boldsymbol{K}^T\boldsymbol{R}_e\boldsymbol{K}\,, \tag{48}$$

where

$$\boldsymbol{K} = \boldsymbol{R}_e^{-1}(\boldsymbol{P}_e\boldsymbol{B} + \boldsymbol{S}_e)^T \tag{49}$$

and (48) implies (45). Analogously, (49) can be rewritten as

$$\boldsymbol{R}_e\boldsymbol{K} = (\boldsymbol{P}_e\boldsymbol{B} + \boldsymbol{S}_e)^T \tag{50}$$

and evidently (50) implies (46).

Using the inequality (9) in the equivalent form

$$\begin{bmatrix} -\boldsymbol{Q}_e & -\boldsymbol{S}_e \\ -\boldsymbol{S}_e^T & -\boldsymbol{R}_e \end{bmatrix} < 0\,, \tag{51}$$

substituting (44), (45) then (51) implies (43), (44) and equivalent block partitioned structure of (22) takes the form

$$\boldsymbol{J}_e = \begin{bmatrix} \boldsymbol{A}^T\boldsymbol{P}_e + \boldsymbol{P}_e\boldsymbol{A} + \boldsymbol{Q}_e & \boldsymbol{P}_e\boldsymbol{B} + \boldsymbol{S}_e \\ (\boldsymbol{P}_e\boldsymbol{B} + \boldsymbol{S}_e)^T & \boldsymbol{R}_e \end{bmatrix} = \begin{bmatrix} \boldsymbol{J}_{e11} & \boldsymbol{J}_{e12} \\ \boldsymbol{J}_{e12}^T & \boldsymbol{J}_{e22} \end{bmatrix} . \tag{52}$$

This concludes the proof. ■

Remark 5. Although there are only two direct parameters to be solved when the inverse LQ problem is implemented in the above proposed LMI form, for given linear control law gain this approximation results in (52) of the same structure as (41). This fault residual functions now may be achieved following the approach leading to (41) where, alternatively, the dynamic fault residual function is proceed using the form

$$\begin{aligned} h(\boldsymbol{q}(t), \boldsymbol{u}(t)) &= \dot{\boldsymbol{q}}^T(t)\boldsymbol{P}_e\boldsymbol{q}(t) + \boldsymbol{q}^T(t)\boldsymbol{P}_e\dot{\boldsymbol{q}}(t) + \boldsymbol{q}^T(t)\boldsymbol{Q}_e\boldsymbol{q}(t) + \\ &+ \boldsymbol{q}^T(t)\boldsymbol{S}_e\boldsymbol{u}(t) + \boldsymbol{u}^T(t)\boldsymbol{S}_e^T\boldsymbol{q}(t) + \boldsymbol{u}^T(t)\boldsymbol{R}_e\boldsymbol{q}(t) - \boldsymbol{w}^T(t)\boldsymbol{W}^T\boldsymbol{R}_e\boldsymbol{W}\boldsymbol{w}(t) . \end{aligned} \tag{53}$$

and the static fault residual function is

$$\begin{aligned} h(\boldsymbol{q}(t), \boldsymbol{u}(t)) &= \boldsymbol{q}^T(t)\boldsymbol{J}_{e11}\boldsymbol{q}(t) + \boldsymbol{q}^T(t)\boldsymbol{J}_{e12}\boldsymbol{u}(t) + \boldsymbol{u}^T(t)\boldsymbol{J}_{e21}\boldsymbol{q}(t) + \\ &+ \boldsymbol{u}^T(t)\boldsymbol{J}_{e22}\boldsymbol{u}(t) - \boldsymbol{w}^T(t)\boldsymbol{W}^T\boldsymbol{J}_{e22}\boldsymbol{W}\boldsymbol{w}(t) . \end{aligned} \tag{54}$$

Also in these cases the last term can be considered as an floating additive offset being fixed a prior for the known controller parameters.

5 Illustrative Examples

The feature of the proposed scheme, and the effectiveness of the proposed algorithm, are presented using an illustrative example.

The state space representation, describing the chemical reactor model [6], consists of the following matrices

$$\boldsymbol{A} = \begin{bmatrix} 1.380 & -0.208 & 6.715 & -5.676 \\ -0.581 & -4.290 & 0.000 & 0.675 \\ 1.067 & 4.273 & -6.654 & 5.893 \\ 0.048 & 4.273 & 1.343 & -2.104 \end{bmatrix}, \quad \boldsymbol{B} = \begin{bmatrix} 0.000 & 0.000 \\ 5.679 & 0.000 \\ 1.136 & -3.146 \\ 1.136 & 0.000 \end{bmatrix}, \quad \boldsymbol{C} = \begin{bmatrix} 1 & 0 \\ 0 & 0 \\ 0 & 1 \\ 0 & 0 \end{bmatrix}$$

and this system in closed loop structure under the state feedback (27) was used in the presented simulations.

Firstly, the gain matrix $\boldsymbol{K}_0$ was synthesized using the standard Matlab function `care(A,B,Q,R,S,E)` with respect to the LQ performance index parameters

$$\boldsymbol{Q} = \boldsymbol{E} = \boldsymbol{I}_4 , \quad \boldsymbol{R} = \boldsymbol{I}_2 , \quad \boldsymbol{S}^T = \begin{bmatrix} 0.3 & 0.6 & 0.3 & 0.0 \\ 0.0 & 0.3 & 0.3 & 0.3 \end{bmatrix} ,$$

where the following matrices were produced directly

$$\boldsymbol{P}_0 = \begin{bmatrix} 1.1259 & 0.0443 & 0.7022 & -0.4907 \\ 0.0443 & 0.0683 & 0.0504 & 0.0468 \\ 0.7022 & 0.0504 & 0.5302 & -0.2696 \\ -0.4907 & 0.0468 & -0.2696 & 0.4544 \end{bmatrix}, \quad \boldsymbol{K}_0^T = \begin{bmatrix} 0.7916 & -2.2093 \\ 1.0983 & 0.1414 \\ 0.8825 & -1.3681 \\ 0.4756 & 1.1481 \end{bmatrix}.$$

Secondly, the gain matrix $\boldsymbol{K}_1$ was computed using the standard Matlab function `place(A,B,r)` with respect to the closed loop system matrix eigenvalues vector $r = \begin{bmatrix} -0.2 & -6+2\,\mathrm{i} & -6-2\,\mathrm{i} \end{bmatrix}$. Using the obtained control law gain parameter

$$\boldsymbol{K}_1 = \begin{bmatrix} -0.2958 & 0.9665 & -0.1151 & 0.9324 \\ -2.9504 & -0.1404 & -1.7847 & 0.9614 \end{bmatrix},$$

the inverse LQ task (42)-(44) was solved with respect to the LMI matrix variables $\boldsymbol{P}_e$, $\boldsymbol{R}_e$ using Self–Dual–Minimization (SeDuMi) package for Matlab [14]. The inverse problem was solved as feasible with the results

$$\boldsymbol{P}_e = \begin{bmatrix} 0.2374 & 0.0321 & 0.0931 & 0.0283 \\ 0.0321 & 0.1410 & 0.0346 & -0.0057 \\ 0.0931 & 0.0346 & 0.1386 & 0.0094 \\ 0.0283 & -0.0057 & 0.0094 & 0.0796 \end{bmatrix} > 0, \quad \boldsymbol{R}_e = \begin{bmatrix} 0.6855 & -0.1091 \\ -0.1091 & 0.2094 \end{bmatrix},$$

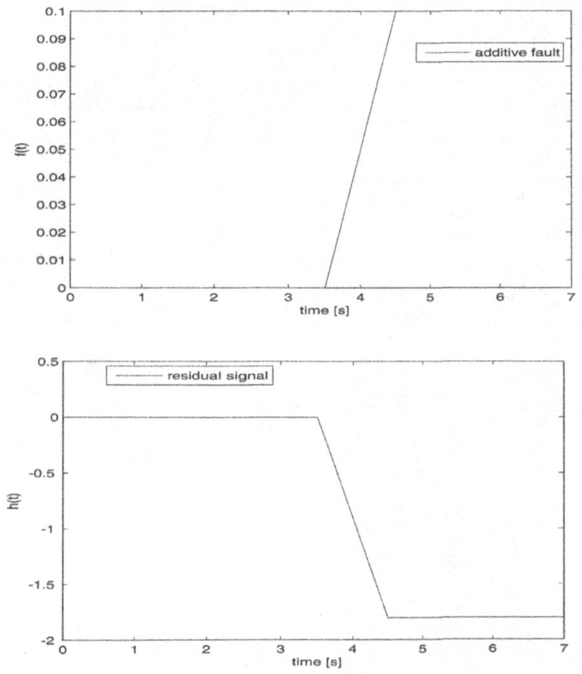

(a)

(b)

Fig. 1. Fault signal (a) and RF response to the fourth sensor additive fault (b)

$$\boldsymbol{Q}_e = \begin{bmatrix} 0.8730 & -0.1335 & -0.2382 & 0.3287 \\ -0.1335 & 1.6503 & -0.2794 & -0.0251 \\ -0.2382 & -0.2794 & 1.1998 & -0.7660 \\ 0.3287 & -0.0251 & -0.7660 & 1.1480 \end{bmatrix} > 0, \ \boldsymbol{S}_e = \begin{bmatrix} -0.2014 & -0.2926 \\ -0.1558 & -0.0259 \\ -0.2489 & 0.0748 \\ 0.4657 & 0.1290 \end{bmatrix}.$$

In all simulations the initial vector was $\boldsymbol{q}(0) = \boldsymbol{0}$ and the forced modes were established with the desired output vector $\boldsymbol{w}(t)$ and the signal gain matrices

$$\boldsymbol{w}(t) = \begin{bmatrix} 1.0 \\ 0.5 \end{bmatrix}, \quad \boldsymbol{W}_0 = \begin{bmatrix} 34.8562 & -5.9339 \\ -50.0413 & 11.0677 \end{bmatrix}, \ \boldsymbol{W}_1 = \begin{bmatrix} -6.1399 & 2.3932 \\ -80.6740 & 17.0028 \end{bmatrix}.$$

Figure 1 represents the additive fault signal (a) and the response of the residual function (RF), given in (54), to additive fault applied to the fourth sensor. Figure 2 illustrates the responses of the RF, given in (53), to a multiplicative fault starting and continuing from the time instant $t = 3.5s$ and being applied to the second actuator and reflecting the system without input noise (a) and with the input nose (b), respectively, where the fault and noise parameters are $m_a = 0.9$, $\mu_d = 0$, $\sigma_d^2 = 0.1$. Finally, Figure 3 gives the responses of the same RF to the $\boldsymbol{A}(2,2)$ parameter change from nominal to zero value, starting and continuing from the time instant $t = 3.5s$, while modeling the system with the input noise and the same noise parameters. Note, only if the dynamic RFs are used, the actuators and system parameter multiplicative faults can be detected.

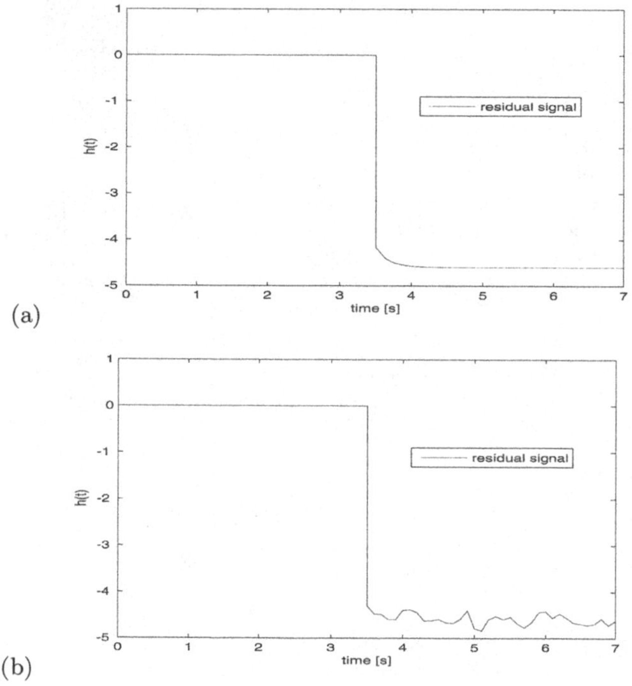

Fig. 2. RF responses to the first actuator multiplicative fault

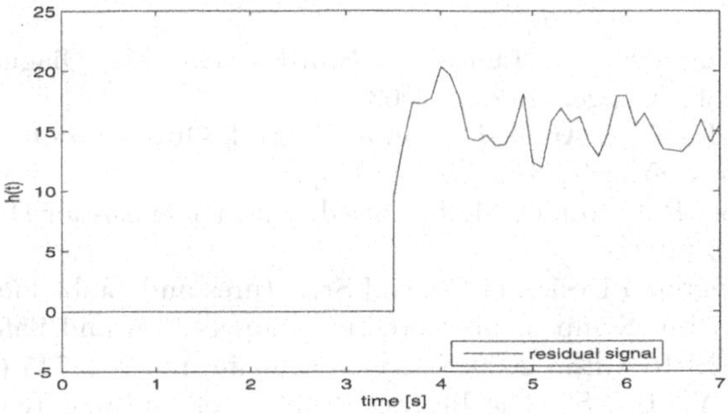

Fig. 3. RF response to a system parameter fault

6 Concluding Remarks

The fault detection scheme described in the paper is based on the LQ control performance index parameters evaluation, where the faults influences are analyzed with respect to the proposed fault residual functions under assumption of the full state feedback control. Based on the time-invariant continuous-time LQ control performance index parameters, obtained using the LMI approach to inverse LQ problem, a substantial generalization of the fault detection scheme is given in the paper.

The fault influences are analyzed according to equivalent quadratic criterion, and the fault detection design method for full states measurable case is achieved. Proposed method presents some new design features and modifications where it was emphasized that the advantage offered by such approach is a simple recursive algorithm with high sensitivity to additive sensor and actuator faults. Compared to observer-based fault detection methods, this method does not need extra dynamics so the computation consumption is lower. This constitutes a first step to develop a practical method based on a LQ control performance index representation. Extensions to linear-parameter varying systems, to static output control, as well as to all system states unmeasurable case will be the future work in this area.

Acknowledgment. The authors thank the anonymous reviewers, who have made several pertinent comments that have enhanced the quality of this paper.

The work presented in the paper was supported by VEGA, the Grant Agency of the Ministry of Education and the Academy of Science of Slovak Republic, under Grant No. 1/0256/11. This support is very gratefully acknowledged.

References

1. Blanke, M., Kinnaert, M., Lunze, J., Staroswiecki, M.: Diagnosis and Fault-Tolerant Control. Springer, Berlin (2003)
2. Bryson, A.E., Ho, Y.C.: Applied Optimal Control. Optimization, Estimation, and Control. Taylor & Francis, New York (1975)
3. Chen, J., Patton, R.J.: Robust Model-Based Fault Diagnosis for Dynamic Systems. Kluwer, Norwell (1999)
4. Ding, S.X.: Integrated Design of Control Structures and Embedded Diagnosis. In: Proc. 7th IFAC Int. Symp. Fault Detection, Supervision and Safety of Technical Processes, SAFEPROCESS 2009, Barcelona, Spain, pp. 734–745 (2009)
5. He, X., Chen, Y., Hu, S.: The Inverse Problem of Optimal Regulators and its Application. Geo-Spatial Information Science 3(9), 62–65 (2000)
6. Kautsky, J., Nichols, N.K., Van Dooren, P.: Robust Pole Assignment in Linear State Feedback. Int. J. Control 41(5), 1129–1155 (1985)
7. Khelassi, A., Theilliol, D., Weber, P.: Reconfigurability Analysis for Reliable Fault–Tolerant Control Design. Int. J. Applied Mathematics and Computer Science 21(3), 431–439 (2011)
8. Korbicz, J., Kościelny, J.M., Kowalczuk, Z., Cholewa, W.: Fault Diagnosis. Models, Artificial Intelligence, Applications. Springer, Berlin (2004)
9. Krokavec, D.: Convergence of Action Dependent Dual Heuristic Dynamic Programming Algorithms in LQ Control Tasks. In: Intelligent Technologies - Theory and Application. New Trends in Intelligent Technologies, pp. 72–80. IOS Press, Amsterdam (2002)
10. Krokavec, D., Filasová, A.: Dynamic System Diagnosis. Elfa, Košice (2007) (in Slovak)
11. Krokavec, D., Filasová, A.: Fault Detection Based on Linear Quadratic Control Performances. In: Proc. 10th Int. Science and Technology Conf. Diagnostics of Processes and Systems, DPS 2011, Zamošč, Poland, pp. 52–56 (2011)
12. Krokavec, D., Filasová, A.: Application of the Linear Quadratic Control Performance in Fault Detection. In: Proc. 9th European Workshop on Advanced Control and Diagnosis, ACD 2011, Budapest, Hungary, ID 14.pdf, p. 6 (2011)
13. Krokavec, D., Filasová, A.: Novel Fault Detection Criteria Based on Linear Quadratic Control Performances. Int. J. Applied Mathematics and Computer Science 22(4), 929–938 (2012)
14. Peaucelle, D., Henrion, D., Labit, Y., Taitz, K.: User's Guide for SeDuMi Interface 1.04. LAAS-CNRS, Toulouse (2002)
15. Philips, C.L., Nagle, H.T.: Digital Control System Analysis and Design. Prentice-Hall, Englewood Cliffs (1984)
16. Simani, S., Fantuzzi, C., Patton, R.J.: Model-Based Fault Diagnosis in Dynamic Systems Using Identification Techniques. Springer, London (2003)
17. Wang, Q.G.: Decoupling Control. Springer, Berlin (2003)

Multiple Fault Isolation Algorithm Based on Binary Diagnostic Matrix

Michał Bartyś

Institute of Automatic Control and Robotics, Warsaw University of Technology
św. A. Boboli 8, 02-525 Warsaw, Poland
bartys@mchtr.pw.edu.pl
http://www.iair@mchtr.pw.edu.pl

Abstract. The problem of the isolation of the multiple faults based on binary diagnostic matrix (BDM) is addressed in this paper. A novel multiple fault isolation approach (MFI) based on binary diagnostic matrix is presented and discussed here. MFI algorithm was optimized in respect to its effectiveness. The theoretical background of MFI was presented. A multiple fault generator based on a matrix of paths was introduced and discussed. The computational complexity of the algorithm was estimated and contrasted with the known MUFIA algorithm. An example of MFI application for on-line diagnostics of a steam-water line of the power boiler was presented.

Keywords: fault isolation, multiple faults, binary diagnostic matrix, on-line diagnostics, large scale systems.

1 Introduction

This paper is devoted to fault isolation based on binary diagnostic matrix (BDM) named by Gertler in [6] as a structure matrix of a residual set. Gertler defined basic features of structure matrices for single and multiple fault isolation. Particularly, he stated that the probability of effects of residual compensation that appear by multiple faults may be considered as negligible. Additionally, he assumed that the ideal signature in response to a combination of faults is the logical alternative combination of the component signatures. Both assumptions were adopted in this paper. Application of Reiter's theory [4,5,12,14] allow for the consideration of fault compensation effects. The diagnoses are generated as the minimal hitting sets of all minimal conflict sets. But fault isolation (FI) approaches based on Reiter theory found applications for the diagnostics of relatively simple systems only. Their applicability to the on-line diagnostics is strongly limited due to huge computational power demands and high development costs. The structural approach for the design of fault detection and isolation (FDI) in large scale industrial plants has been proposed in [2,3,15] where as in the applicable idea of FDI algorithm based on the concept of hierarchical system decomposition is given. A new approach to multiple fault diagnosis, based on a combination of diagnostic matrices, graphs, algebraic and rule-based models, is given in [12]. In

J. Korbicz and M. Kowal (eds.), *Intelligent Systems in Technical and Medical Diagnostics*, Advances in Intelligent Systems and Computing 230,
DOI: 10.1007/978-3-642-39881-0_37, © Springer-Verlag Berlin Heidelberg 2014

the case of diagnostics of industrial processes, making use of expert knowledge is the practical method of acquisition of the knowledge of the relations between faults and fault symptoms. The simplest method of notation of this knowledge has a form of commonly used binary diagnostic matrix or conditional if-then rules [1,6,13]. The method of creating Boolean signatures of multiple faults is given in [6,15]. The algorithms of on-line diagnostics for complex technological installations allowing the recognition of single and multiple faults are given in [7,8,9,10]. Staroswiecki [15] proposed a Boolean inference scheme and also a Hamming distance as a practical fault separability index in the case of single and multiple faults. Principally, model based diagnostics searches for, and afterwards makes profits from the relations between faults and residuals or their derivatives. Hence, the model based diagnostics is focused on: seeking for these relations, generating residuals by means of modelling, residual processing and inferring of faults. This paper deals with the process of inferring about faults. Modelling, and residual generation approaches are principally beyond the scope of this paper.

A novel, practical and easy to implement algorithm for the reasoning of multiple faults named MUFIA is presented in [11]. This algorithm consists of three parts:

Part A, where the necessary condition of applicability of the dynamic decomposition approach is checked. If this condition fails, then there is a mismatch between the observed symptoms and signatures of single faults.

Part B, where reduced binary diagnostic matrix is defined by determining a subset of possible faults and a subset of diagnostic signals useful for isolation of these faults. This matrix is based on knowledge of reference values of diagnostic signals.

Part C, where the matrix of system states with multiple faults and multiple faults are determined by searching this matrix.

The described above algorithm is highly effective particularly when searching for a relatively low number of system states with faults. This algorithm loses its effectiveness with higher fault multiplicities. This paper takes inspiration from this observation and proposes a new alternative solution named further as MFI algorithm.

The main objective of this paper is to present and prove efficiency of MFI algorithm. This algorithm is intended particularly for application in on-line diagnostics of large scale systems. Principally it is intended for applications where relatively high fault multiplicities are expected. In this case, MFI algorithm may be used instead or as supplementary or complementary to algorithm MUFIA.

The structure of the paper is as follows: In Sect. 2 a theoretical background of multiple fault isolation is given. Sufficient and necessary conditions of multiple faults existence are defined. Binary diagnostic submatrix of the first order and matrix of paths are introduced and further used for description of MFI algorithm. In Sect. 3 MFI algorithm is presented. In Sect. 4 its computational complexity is compared with complexity of MUFIA algorithm. Concluding remarks in Sect. 5 completes the paper.

2 Problem Formulation

Basically, this paper has been elaborated to tackle the problem of inferring of multiple faults based on binary diagnostic matrix (BDM). The challenge for this class of problems is to propose a practical inference scheme. Therefore, this paper focuses mainly on the elaboration of a fast and efficient algorithm of multiple fault isolation (MFI) particularly intended for use in on-line diagnostics in large scale systems.

2.1 Assumptions

The following main assumptions are considered:

1° bi-valued diagnostic matrix of the diagnosed system is known,
2° parallel reasoning scheme about faults is applied,
3° the fault compensation effect does not take place,
4° fault signatures are logically additive.

2.2 Theoretical Background

When reasoning about single faults, each column $V_i = [v_{1,i}, v_{2,i}, ..., v_{m,n}]^T$ of a binary diagnostic matrix V is associated with one and only one fault $f_i \in F, \forall i \in \{1..n\}$. When reasoning about multiple faults, each fault f_i can be allocated to a subset of faults $F_i^c \in F$ for which holds:

$$F_i^c = \{f_i \in F : V_i^s = \{V_k : V_i \cup V_k \cup S = S; \quad \forall k \in \{i..n\}\} \neq \varnothing\}\}. \quad (1)$$

where: $S = [s_1, s_2, ... \; s_m]^T$ is a bi-valued vector of diagnostic signal values. Further, F_i^c will be called as the set of multiple fault candidates.

Definition 1. *The non empty subset of all signatures V_i^s satisfying (1) will be further called as the subset of supplementary signatures to the fault f_i.*

Lemma 1. *Multiple faults do not exist if all supplementary subsets V_i^s of all faults f_i are empty.*

Proof. If all supplementary signatures $V_i^s = \varnothing$ of all faults f_i then from (1) all sets of multiple fault candidates are also empty :

$$F_i^c = \{f_i \in F : V_i^s = \varnothing\} = \varnothing. \quad (2)$$

It is obvious that if all sets of multiple fault candidates are empty then set of multiple faults is also empty.

Conclusion 1. If at least one supplementary subset V_i^s exists such that:

$$\exists V_i^s \neq \varnothing \quad ; \; \forall i \in \{1..n\} \quad (3)$$

then multiple faults may exist. This is the sufficient and necessary condition of multiple faults. Let cardinality of each set F_i be n_i and let N be a set of n_i elements: $n_i \in \{N\}$; $\forall i \in \{1..n\}$.

Lemma 2. *Fault multiplicity* k *by given binary diagnostic matrix and given set of diagnostic signals S is equal to a cardinal number of any nonempty candidate set.*

$$k = |F_i^c| : F_i^c \neq \{\varnothing\} \quad ; \ \forall i \in \{1..n\}. \tag{4}$$

Proof. If all candidate sets are empty and a set of diagnostic signals in not empty then a mismatch occurs between fault signatures and diagnostic signal values. In this case, a hypothesis regarding a unknown (undefined) fault or faults may be accepted. Let us suppose that it is not the case. All nonempty candidate sets F_i^c according to (1) contain all faults which signatures are considered as not to be contradictory with the diagnostic signals. If each candidate set contains all faults then the first of these candidates doubtlessly does too.

Lemma 2 has practical meaning, because it allows significant reduction of the number of searches for the candidate sets.

Conclusion 2. If all candidate sets are empty then binary diagnostic table is not sufficiently defined.

Conclusion 3. Candidate sets are either empty or contain k elements. Therefore it is sufficient to find the first and not empty candidate set to determine fault multiplicity k.

Lemma 3. *Maximal number of multiple faults K by given binary diagnostic matrix and given vector of diagnostic signals S is equal to a sum of all combinations without repetitions of all elements of candidate set.*

$$K = \sum_{l=1}^{k} \binom{k}{l} = 2^k - 1 \tag{5}$$

Proof. If a candidate set F_i^c consists of k elements then for the first element there is the possibility of $\binom{k}{1}$ single faults, for the second element $\binom{k}{2}$ double faults and for the k-th element $\binom{k}{k}$ one k-multiple fault. Hence, the total number of all combinations without repetitions equals (5).

Conclusion 4. There are a maximal k single faults possibilities and there is only one k-multiple fault possibility.

It is clear that in particular cases, the number of multiple faults may be lower than K. In fact, the number K determines the number of considered system states with faults. The higher k is the higher K is. Number K may be enormously high in large scale systems. Therefore, there is a need to find an effective algorithm of searching for multiple faults. Under some conditions, there is a possibility to reduce the size of the binary diagnostic matrix. This has further a crucial influence on the cardinality of the set of possible multiple faults and computational effort.

Definition 2. *Binary diagnostic submatrix $B_1(m - \sigma_1; k)$ of the first order of binary diagnostic matrix B, is a matrix formed by deleting σ_1 rows and $n - k$ columns of matrix B.*

Value of σ_1 is selected in such a manner that it lowers the cardinality of the set of candidates F_i^c without loss in the number of multiple faults. If either $\sigma_1 \neq 0$ or $k \neq 0$ then the number K of considered combinations of the system states with faults can be reduced. If we want to reduce cardinality of the set of candidates F_i then we should examine cases for which $f_i \notin F_i^c$. Now, let us assume that at least one value of diagnostic signal S exist such that:

$$\exists! s_j = 0 \quad ; \quad \forall j \in \{1..m\} \ . \tag{6}$$

If condition (6) holds, then set of supplementary signatures V_i^s does not contain signature V_i of the fault f_i iff $(v_{j,i} = 1)$ and in consequence $f_i \notin F_i^c$. Binary diagnostic submatrix will be further named briefly as BDS matrix.

Conclusion 5. All faults f_i for which $(s_j = 0) \wedge (v_{j,i} = 1)$ is true, may be deleted from the set of candidates F_i^c. Let us define sets I_1 and J_1 containing respectively the numbers of columns and rows that may be deleted in order to reduce dimension of the BDM.

$$I_1 = \{i \in \{1..n\} : (s_j = 0) \wedge (v_{j,i} = 1); \quad \forall j \in \{1..m\}\};\ card(I_1) \leqslant n \tag{7}$$

$$J_1 = \{j \in \{1..m\} : (s_j = 0); \quad \forall j \in \{1..m\};\ card(J_1) \leqslant m \tag{8}$$

The cardinality of the set J_1 is equal to the number of all diagnostic signal items for which $s_j = 0$.

$$|J_1| = \sigma_1 \tag{9}$$

Lemma 4. *Cardinality of the set I_1 is equal to the difference between number n of columns in binary diagnostic matrix and fault multiplicity k.*

$$|I_1| = n - k \tag{10}$$

Proof. Maximal cardinality of the set I_1 is not greater than maximal number of faults n because it is not possible to delete more columns than truly exist in BDM. Fault multiplicity k shows the maximum number of single faults that are constituents of multiple faults. Hence, the number of faults that may not be taken into consideration is equal to $n - k$. Deleting the redundant columns allows for reduction of the size of binary diagnostic matrix.

2.3 Matrix of Paths

Definition 3. *Let matrix of paths $P[m_P; n_P]$ be a structural transformation of any two-dimensional sparse matrix $M[m_M; n_M]$ that have the following features:*
a) number of rows of both matrices are equal i.e. $m_P = m_M$,
b) number of columns in matrix P is equal to the maximum number of all non-zero entries of all rows of matrix M. Hence,
$n_P = \bigvee_{j=1}^{m_M} (\sum_{i=1}^{n_M} 1 : m_{j,i} \neq 0)$
where $m_{j,i} \in M; \forall\, i \in \{1..n_M\}; \forall\, j \in \{1, m_M\}$.
c) P and M matrices are populated by the same number of non-zero elements.

Let us define the above transformation as one dimensional squeezing of matrix M and denote this transformation by symbol $\overrightarrow{sqz}$.

Definition 4. *Path P_i^z is any set of non zero elements such that each of its elements belongs to each separate row of matrix P.*

$$P_i^z = \{p_{j,i} \in P : \; p_{j,i} \neq \; 0; \; \forall j \in \{1, m_P\}\} \tag{11}$$

Lemma 5. *The number of all paths P_i^z of matrix P is equal to:*

$$z = \prod_{j=1}^{m_P} \sum_{i=1}^{n_P} b_{j,i} \quad ; \quad b_{j,i} = \begin{cases} 1 \Rightarrow p_{j,i} \neq 0 \\ 0 \Rightarrow p_{j,i} = 0 \end{cases} \tag{12}$$

Proof. Let σ_j denotes the number of non-zero elements in j-th row of matrix P. Each path of matrix P contains at most one element of each row. Hence, there is a possibility for each row σ_j combinations taken on at a time without repetitions. Hence, total number of paths for all rows is equal to the sum of all combinations σ_j for $j = 1..m_P$.

Conclusion 6. The maximal number z of paths $P_i^z$of non-sparse matrix P is equal to:

$$z = n_P^{m_P} \tag{13}$$

Proof. Let us assume that matrix P is not sparse. We get (13) directly from (12) because all $b_{j,i} = 1$.

Observation 1. Any two adjacent rows of the matrix of paths may be combined together. Combined row consists of n_P entries. Each entry of combined row is calculated from formula (14).

$$p_{j,i}^c = \bigcup_{i=1}^{n_P} \{p_{j,i} \cap \; p_{j+1,i}\}; \quad j \in \{1, ..., m_P - 1\}; \quad \forall i \in \{1, ..., n_P\} \tag{14}$$

Observation 1 has practical meaning. It allows for reduction of the size of the matrix P.

Lemma 6. *Matrix P may be reduced to one row by consequent combinations of its rows. One row matrix P contains all paths.*

Proof. Indeed one combination according to (14) extends each $p_{j,i}$ by additional one $p_{j+1,i}$ element from the row $(j+1)$. If (m_P-1) combinations for all remaining rows will be done then each element of combined matrix row will consist of all combinations of rows elements i.e. all possible paths.

3 MFI Algorithm

The multiple fault isolation algorithm based on binary diagnostic matrix basically points out all possible single and multiple faults. MFI generates multiple fault sequences based on matrix of paths of multiple faults. Path of multiple fault is equivalent to the system state with faults because it points out all faults of this particular state. In contrast to MUFIA algorithm, MFI does not need to check all hypothetic system states with multiple faults. This promises for the significant reduction of computational effort. MFI is performed in three main steps:

Step 1: First step relies on the formation of BDS matrix of the first order. This is the same step that is performed in MUFIA algorithm. Additionally, fault multiplicity k is determined.
Step 2: Second step of MFI algorithm rely on formation of the matrix of fault paths P. Step 2 consists of 3 stages. In the first stage all Boolean values equal 1 of each column of first order BDS submatrix are replaced by fault labels of this columns(f_i). In the second stage, the matrix of paths is created by means of squeezing transformation (Def. 3). It the third stage, the matrix is reduced in accordance with features of this class of matrices stated in Sect. 2.3. Additionally, in this step the number of multiple faults z is determined from (13).
Step 3: Third step rely on the generation of the multiple faults from the matrix of paths based on the combination operator defined in (14).

3.1 Example of Application of MFI Algorithm

Example 1. Let us recall binary diagnostic matrix (Tab. 1) obtained for steam-water line of power boiler for a given set of diagnostic signals S [11]. This matrix was used for exemplification of MUFIA algorithm.

Step 1: Construct BDS of order 1. Determine fault multiplicity k.

In Tab. 1 there is a mismatch between vector of diagnostic signals S and signatures of binary diagnostic matrix. So, we assume multiple faults. The subsets F_i^c of faults that are candidates to be constituents of multiple faults according to (1) are as follows: $F_1^c = \emptyset$; $F_2^c = \{f_5, f_6, f_{11}, f_{12}\}$; ... . Multiple faults may occur because at least one $F_i^c \neq \emptyset$. Fault multiplicity taken from (4) is equal $k = 4$ and maximal number of system states with faults in accordance with (5) is equal $K = 15$. While $(s_{j=4} = 0)$ and $(s_{j=5} = 0)$ then condition (6) is satisfied and rows

4 and 5 of the matrix shown in Tab. 1 will be deleted. Because set $I_1 = \{1, 6, 7\}$ is not empty then columns 1, 6 and 7 will be also deleted. Diagnostic submatrix of the first order of matrix from Tab. 1 is shown in Tab. 2.

Step 2: Construct matrix of paths. Determine the number of multiple faults z.

Matrix of paths P is formed in accordance to transformation defined in Def. 3.

This matrix does not contain any identical rows. Therefore reduction of the rows of this matrix is not possible. Total number of multiple faults in accordance with (13) is equal $z = 2 \cdot 1 \cdot 2 \cdot 1 = 4$.

Step 3: Generate multiple faults from the matrix of paths.

There are different approaches possible. For example we can generate multiple faults paths directly from Tab. 3 by tracking all paths. In this case we get: $\{f_5 f_{11}\}, \{f_6 f_5 f_{11}\}, \{f_5 f_{12} f_{11}\}, \{f_6 f_5 f_{12} f_{11}\}$.

Table 1. An example [11] of a binary diagnostic matrix

S/F	f_1	f_5	f_6	f_{11}	f_{12}	f_{14}	f_{16}		S
s_{12}	1	1	1						1
s_{13}		1							1
s_{15}				1	1				1
s_{16}							1		
s_{17}	1					1			
s_{18}				1		1	1		1

Table 2. Binary diagnostic submatrix of the first order of the matrix from Tab. 1

S/F	f_5	f_6	f_{11}	f_{12}
s_{12}	1	1		
s_{13}	1			
s_{15}			1	1
s_{18}			1	

Table 3. Matrix of paths of the BDS from Tab. 2

f_5	f_6
f_5	
f_{11}	f_{12}
f_{11}	

Table 4. Combined rows of matrix from Tab. 3

$\{f_5\}$	$\{f_5 f_6\}$
$\{f_{11}\}$	$\{f_{11} f_{12}\}$

Table 5. Combined rows of matrix from Tab. 4

$\{f_5 f_{11}\}, \{f_5 f_{11} f_{12}\}$	$\{f_5 f_6 f_{11}\}, \{f_5 f_6 f_{11} f_{12}\}$

It is also possible to minimize Tab 3. by applying three combination operations of matrix rows in accordance with (14). The result is shown in Tab. 4. Final diagnosis is shown in Tab. 5. $DGN = \{\{f_5 f_{11}\}, \{f_5 f_{11} f_{12}\}, \{f_5 f_6 f_{11}\}, \{f_5 f_6 f_{11} f_{12}\}\}$.

4 Computational Complexity

Computational complexity of any algorithm depends on quality of its implementation on a given machine. Therefore, there is not a possibility to uniquely assess this problem without any attempt of implementation. Despite this, computational complexity may be estimated roughly by counting at least elementary operations needed for implementation. In this section, there is assumed that all basic operations are characterised by equal computational load.

4.1 Computational Complexity of MUFIA Algorithm

In the first step of MUFIA algorithm [11] (part B) a binary diagnostic submatrix of the first order is formed. In this step, binary diagnostic matrix is reduced by deleting $(n-k)$ columns and $\{\sigma_1 \leqslant m\}$ rows. At most n columns and m rows may be deleted. Rows are deleted by appropriate checking of each element of the set of diagnostic signals S. Therefore, at most $(m-1)$ checking operations are performed. Each column is deleted by checking of each element from each row for which the equivalent value of diagnostic signal equals 0. Together, maximum $(n-1)$ columns and $(m-1)$ rows may be deleted. In worst case scenario, all elements of all rows must be checked. In this case the maximal number of checked operations is equal to $0,5(m-1)(n+2)$. Hence, maximal estimated number c_1 of checking and deleting operations in step 1 is equal:

$$c_1 = 2(m-1) + (n-1) + \frac{(m-1)(n+2)}{2} = 3m\frac{n(m+1)}{2} - 4 \tag{15}$$

It is beneficial to mention that MUFIA and MFI algorithms are using the same first step of algorithm. The second step of MUFIA algorithm (part C) forms a matrix of system states with possible multiple faults and then compare each state with the hypothetic state vector of size $[m, 1]$ in which all entries are set to 1. Because the size of the matrix of system states with faults may be high (5) then the number of considered fault multiplicity should be reasonably limited typically to single and double faults only. This may be considered as a disadvantage of this approach. Maximal number of single faults equals k. From (5) it is easy to determine the maximal number of considered system states with double faults.

$$s_2^2 = \binom{k}{2} = \frac{k!}{2!(k-2)!} = \frac{k(k-1)}{2} \tag{16}$$

Therefore, the maximal number of operations necessary to isolate all single and double faults is equal to the number of operations needed to form the matrix of system states with faults and number of comparisons of each column of this matrix with hypothetic state vector. The formation of each column of the system states matrix with double faults only need four together (read, adding, overwriting and comparison) operations.

$$c_{2MUFIA} = k + 2k(k-1) = 2k^2 - k \quad (17)$$

Hence, computational complexity of MUFIA in part C may be estimated as quadratic. If we assume isolation of all multiple faults then maximal number of considered system states with faults equals:

$$K_{MUFIA} = 2^k - (k+1) \quad (18)$$

Complexity of MUFIA algorithm in part C will be approximately equal to:

$$C_{MUFIA} = 2^{k+2} + 3(m-k) + \frac{n(m+1)}{2} - 8 \quad (19)$$

Computational complexity strongly depends on implementation and therefore in practice computational complexity is case dependent.

4.2 Computational Complexity of MFI Algorithm

The main savings in computational complexity are achieved in MFI by application of approach in which recognition of multiple faults does not need to check all hypothetic system states with multiple faults. Vice versa, MFI generates exclusively system states with multiple faults.

Algorithm. Complexity of the third step has the highest influence on computational complexity. Let us suppose that step 3 will be carried out by application of additional row squeezing algorithm working in row by row mode. Firstly, we prepare empty matrix of diagnosis $DGN[1:z]$. This needs at most z clearing operations. Next we copy all k entries of the first row of matrix P to matrix DGN. This needs at most k read and k copy operations. Next we start to analyse matrix of paths beginning from the first cell of the second row in matrix P. The entry of this cell is checked. If this entry equals zero, then next cell in the same row will be checked. If this entry is different than zero, then the first cell of the DGN will be checked. If both entries are equal then this entry is checked with the next cell of DGN matrix. If this entry is not equal then it is concatenated the with pointed out entry of DGN matrix. These operations will be repeated for all nonempty entries of DGN matrix. In worst case, for each DGN entry one update cycle takes place. Each update cycle in worst case needs to perform: 1 check operation of the cell in P matrix, 1 comparing operation with the actual cell of DGN matrix and 1 concatenation operation. Together, one update cycle needs at most 3 operations. Update cycles are performed for all cells of all remaining (m_1-1) rows of P matrix. Number of updates needed for one cell of matrix P needs therefore at most $3z$ operations. Total number of all updates for all (m_1-1) rows needs at most $3z \cdot n_P(m_1-1)$ operations, where n_P denotes number of columns of P matrix. Hence, overall complexity of MFI algorithm will be approximately equal to:

$$c_{MFI} \cong (2k-1)m_1 + z + 3k + 3z \cdot n_P \cdot (m_1-1) + 3m + \frac{n(m+1)}{2} - 4 \quad (20)$$

Table 6. Worst case estimated computational complexity of MUFIA and MFI algorithms

k	1	2	3	4	5	6	7	8	9	10
C_{MUFIA}/C_{MFI}	0.58	0.50	0.48	0.53	0.65	0.87	1.26	1.93	3.09	5.11

Table 7. Estimated computational complexity of MUFIA algorithm for steam-water line of power boiler diagnostics case

Operation	*row deletion*	*column deletion*	*checking*	*read*	*write*	*arithmetic*
Step B	2	3	19			
Step C			14	25	4	10

Table 8. Estimated computational complexity of MFI algorithm for steam-water line of power boiler diagnostics case

Operation	*row deletion*	*column deletion*	*checking*	*read*	*write*	*arithmetic*
Step 1	2	3	19			
Step 2			16		4	
Step 3			6	12	5	

Exemplary results of estimated computational complexity are put together in Tab. 6. Following parameters are adopted: $n = k + 5, m = n, n_P = 1, m_1 = n - 3; z = n_P \cdot m_1$. From Tab. 6 we see that MFI algorithm in worst case is much more effective for multiplicities of isolated faults greater then 6.

Example 2. Let us now apply MUFIA and MFI algorithm for case of diagnostics of steam-water line of power boiler from Example 1. The number of operations for both algorithms are presented in Tab. 7 and Tab. 8. The number of multiple faults equals 4. The total number of operations in case of application of MUFIA equals 77 and 67 for MFI algorithm 2. In this context, the MFI algorithm has better performance than its counterpart.

5 Summary

The theoretical analysis dealing with isolation of multiple faults based on binary diagnostic matrix has been presented in this paper. Among others, a necessary and sufficient conditions of the existence of multiple faults were formulated. A method of determining the set of candidates of multiple faults was presented. Novel algorithm of isolation of multiple faults was proposed. It does not need to search all system states with multiple faults. Therefore, it is highly effective. The algorithm called MFI is based on the concept of matrix of paths introduced in this paper. Principally MFI is intended for applications where relatively high fault multiplicity is expected, i.e. in on-line diagnostic systems of large scale systems. The example of application of MFI algorithm for diagnostics of a steam-water line of the power boiler illustrate its features.

Acknowledgments. This work was supported in part by the Polish National Science Centre in the frames of research project DEC-2011/01/B/ST7/06183.

References

1. Bartyś, M.: Generalised reasoning about faults based on diagnostic matrix. International Journal of Applied Mathematics and Computer Science 23(2) (accepted for printing, 2013)
2. Blanke, M., Staroswiecki, M.: Structural design of systems with safe behaviour under single and multiple faults. In: Proceedings of IFAC Symposium Safeprocess 2006, Beijing, pp. 511–515 (2006)
3. Blanke, M., Kinnaert, M., Lunze, J., Staroswiecki, M.: Diagnosis and Fault-Tolerant Control. Springer (2004)
4. de Kleer, J., Williams, B.C.: Diagnosing Mutliple Faults. Artificial Intelligence 32, 97–130 (1987)
5. de Kleer, J., Kurien, J.: Fundamentals of model-based diagnosis. In: Proceedings of 5th IFAC Symposium on Fault Detection, Supervision and Safety of Technical Processes SAFEPROCESS 2003, Washington, D.C., USA, pp. 25–36 (2003)
6. Gertler, J.: Fault Detection and Diagnosis in Engineering Systems. Marcel Dekker Inc., New York (1998)
7. Korbicz, J., Kościelny, J.M., Kowalczuk, Z., Cholewa, W. (eds.): Fault Diagnosis. Models, Artificial Intelligence, Applications. Springer, Heildelberg (2004)
8. Korbicz, J., Kościelny, J.M. (eds.): Modeling, Diagnostics and Process Control. Implementation in the DiaSter System. Springer (2010)
9. Kościelny, J.M., Bartyś, M.: Application of the method of dynamic decomposition for recognition of multiple faults in the large scale systems. In: Proceedings on 7th Workshop on Advanced Control and Diagnostics, ACD 2009, Zielona Góra, Poland (2009)
10. Kościelny, J.M., Bartyś, M., Syfert, M.: Method of Multiple Fault Isolation in Diagnostics of Large Scale Systems. In: Proceedings on 7th Workshop on Advanced Control and Diagnosis, ACD 2009,, Zielona Góra, Poland (2009)
11. Kościelny, J.M., Bartyś, M., Syfert, M.: Method of Multiple Fault Isolation in Large Scale Systems. IEEE-Trans. on Control Systems Technology 20(5), 1302–1310 (2012)
12. Ligęza, A., Kościelny, J.M.: A new approach to multiple fault diagnosis. Combination of diagnostic matrices, graphs, algebraic and rule-based models. The case of two-layer models. Int. Jour. of Applied Mathematics and Computer Science 18(4), 465–476 (2008)
13. Patton, R.J., Frank, P.M., Clark, R.N. (eds.): Issues of Fault Diagnosis for Dynamic Systems. Springer, London (2000)
14. Reiter, R.A.: Theory of Diagnosis from First Principles. Artificial Intelligence 32, 57–95 (1987)
15. Staroswiecki, M., Cassar, J.P., Declerck, P.: A structural framework for the design of FDI in large scale industrial plants. In: Patton, R., Frank, P., Clark, R. (eds.) Issues of Fault Diagnosis for Dynamic System. Springer (2000)

Distributed State Estimation Using a Network of Asynchronous Processing Nodes

Zdzisław Kowalczuk[1] and Mariusz Domżalski[2]

[1] Gdańsk University of Technology, ETI, Narutowicza 11/12, 80-233 Gdańsk, Poland
kova@pg.gda.pl
[2] Gdańsk University of Technology, ETI
mardo@eti.pg.gda.pl

Abstract. We consider the problem of distributed state estimation of continuous-time stochastic processes using a network of processing nodes. Each node performs measurement and estimation using the Kalman filtering technique, communicates its results to other nodes in the network, and utilizes similar results from the other nodes in its own computations. We assume that the connection graph of the network is not complete, i.e. not all nodes are directly connected, and that the nodes work asynchronously, i.e. they perform measurement and estimation in time moments independent of each other. We evaluate the impact of the way of propagation of information from most precise nodes over the network on the overall performance of distributed estimation.

Keywords: continuous-time stochastic processes, distributed systems, state estimation, Kalman filtering, sensor networks.

1 Introduction

Distributed (decentralized) estimation and identification [6,13,18] is an important and challenging issue in the field of control, robotics, and diagnostics. A typical distributed estimation system is composed of a set of processing nodes (local processors) connected through a network. Each of those nodes performs state estimation of the observed process and works on data obtained from two groups of sources. The first group are sensors integrated with a given node and supplying local measurements. The second group are the remaining nodes which are connected to the same network and provide (alternative) remote state estimates. Therefore, in a process is called data fusion, each node combines data supplied by its own sensors with the data from other processing nodes.

In this paper we assume that the nodes are connected via a network with an incomplete connection graph. It means that not all nodes are directly connected. Such a topology causes data propagation delays in the network, because it may take several estimation cycles for the estimates computed by one node to be incorporated by other distant nodes. An exemplary configuration of a distributed system is presented in Fig. 1.

The main advantages of such a completely distributed processing system lie in its robustness and flexibility. Any failure in any local node does not stop the

J. Korbicz and M. Kowal (eds.), *Intelligent Systems in Technical and Medical Diagnostics*, Advances in Intelligent Systems and Computing 230,
DOI: 10.1007/978-3-642-39881-0_38, © Springer-Verlag Berlin Heidelberg 2014

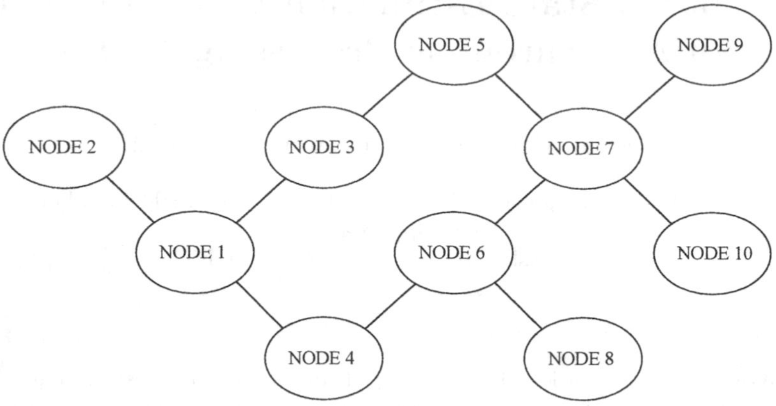

Fig. 1. Distributed estimation system with an incomplete connection graph

other processing nodes from functioning and communicating unless this node is an articulation point (a cut vertex) of the network graph (nodes 1, 6 and 7 in the system of Fig. 1). Moreover, mathematical models of the estimated process used in each processing node can be completely different. The models can, for example, be chosen specifically (optimally) to cope with data taken from the observed process by a given local sensor. What is more, as the data passed between the local systems are of the same type (state estimates), the existing infrastructure of the distributed estimation system can be easily extended to compose a new system structure, or to include new nodes in the existing system. Thus decentralization can effectively provide a new degree of scalability in estimation systems. Note that typical applications of such distributed estimation systems can be found both in air traffic control and in mobile robotics.

Recently, researchers have proposed a new group of distributed estimation algorithms for dynamic systems, called average consensus filters. Such solutions for distributed estimation of scalar random processes based on average consensus over a network of filters is presented in [16,15,21,20]. Distributed estimation of states of linear dynamic systems is discussed in [3,4]. In distributed systems based on the average consensus approach, at each time step, each node is taking a weighted average of differences between the estimates received from the other nodes and its own estimate. If this average is positive, the estimate for this particular node is increased proportionally to the average, and if this average is negative, the estimate is decreased. Both the amount of information exchanged between the nodes and the weights used in the average consensus algorithms, which can be chosen according to different strategies, affect the rate of convergence toward the asymptotic agreement between the nodes.

In contrast to the average consensus filters, the distributed method presented in this paper belongs to the estimation algorithms that are based on the set of state estimators exchanging full state information between themselves [17,2,7]. This information is usually comprised of current state estimates and their corresponding covariance matrices. In each of the computing nodes, weights used for data fusion of local and remote estimates are computed based on these covariance matrices and therefore are proportional to the quality of particular nodes.

There is an essential assumption in the consensus filter that all the nodes are using the same process model. Whereas in the presented system comprised of the local nodes, each node can utilize a different model for the processed, (assuming that the set of state variables is the same for each model, or the state variables can be transformed from one set to another; otherwise the fusion, or rather an aggregation has to be done in a completely different way).

Another important point is that processing nodes in the presented system can work asynchronously, as they can perform measurements and compute state estimates at different time moments. However, both the average consensus methods and the methods based on state estimators usually assume that the sensors take measurements and communicate with each other synchronously or that the sensors exchange data multiple times at a regular pace during each of measurement cycles.

In the paper we investigate two important issues in the distributed estimation system, namely the importance of position of high and low accuracy nodes within the network and the effects of information propagation in the network on the overall performance of distributed estimation. Both these issues are closely related to the important problem of convergence of the estimation algorithm.

The paper is organized as follows. In Section 2 a continuous-time stochastic process model and its discrete-time version are presented. The Kalman filter which is a basis for state estimation is recalled in 3. Next, data synchronization and data fusion are described in Section 4. A simulation example is presented in Section 5. Section 6 contains conclusions.

2 Process Model

To deal with asynchronous data a suitable linear continuous-time (c-t) process model [10,14] is to be considered, whose stochastic nature allows us to aptly describe data uncertainty. A corresponding sampled-data model will be obtained afterwards.

2.1 C-T Gauss-Markov Model

It is assumed that the dynamics of the process of interest can be described by the following n-dimensional linear stochastic differential equation

$$\begin{aligned} dX(t) &= [AX(t) + b]\,dt + \sigma dw(t), \quad 0 \le t < \infty \\ X(0) &= X_0 \end{aligned} \tag{1}$$

where w is the r-dimensional Brownian motion independent of the initial vector X_0, which has a given n-variate normal distribution. The $(n \times n)$, $(n \times 1)$ and $(n \times r)$ matrices A, b, and σ, respectively, are nonrandom and bounded.

It can be proved that the solution X of (1) is a Gaussian process with the strong Markov property, thus the finite-dimensional distributions of the process X are completely determined by the mean and the covariance functions [10,19].

2.2 A Sampled-Data Model

As the measurements and data fusion are performed at selected discrete-time moments, a sampled-data process model is necessary. Such a model can be obtained based on the above c-t model.

For any two particular time moments τ and t, $\tau \leq t$, the following transitional (reference) equation results from the c-t model described above:

$$x(t) = F(t,\tau)x(\tau) + u(t,\tau) + w(t,\tau) \tag{2}$$

with

$$F(t,\tau) = \Phi(t)\Phi^{-1}(\tau), \tag{3}$$

$$u(t,\tau) = \Phi(t)\int_{\tau}^{t}\Phi^{-1}(\eta)bd\eta, \tag{4}$$

$$w(t,\tau) = \Phi(t)\int_{\tau}^{t}\Phi^{-1}(\eta)\sigma dw(\eta). \tag{5}$$

where $\Phi(t)$ is a non singular matrix called the fundamental solution to the following homogeneous ordinary differential equation

$$\dot{\varsigma}(t) = A\varsigma(t). \tag{6}$$

The fundamental solution can be easily expressed, for example, as

$$\Phi(t) = e^{tA} \triangleq \sum_{i=0}^{\infty}\frac{t^i}{i!}A^i. \tag{7}$$

Equation (5) represents the Itô stochastic integral [10,19,14], which can be calculated by parts. The mean of (5) is $E\{w(t,\tau)\} = 0, \forall t,\tau$, and the covariance matrix of (5) is

$$\begin{aligned} Q(t,\tau) &\triangleq E\left\{w(t,\tau)\left[w(t,\tau)\right]^{\top}\right\} \\ &= \Phi(t)\left[\int_{\tau}^{t}\Phi^{-1}(\eta)\sigma\left[\Phi^{-1}(\eta)\sigma\right]^{\top}d\eta\right]\Phi^{\top}(t). \end{aligned} \tag{8}$$

An useful optimal solution for asynchronous sampling of two dimensional Gauss-Markov stochastic processes, based on explicit forms of the matrix exponential, is presented in [12].

A necessary discrete-time signal ($z_i \in \mathbb{R}^p$) of observations of the process $x(t)$, performed by any of the local sensors ($i = 1, \ldots, N$) can be described by the following equation:

$$z_i(t) = H_i(t)x(t) + \xi_i(t) \tag{9}$$

where $H_i(t)$ is a $p(i) \times n$ observation matrix, $p(i) \in \mathbb{N}_+$, and $\xi_i \in \mathbb{R}^{p(i)}$ represents zero-mean Gaussian discrete-time measurement noise with a known covariance matrix

$$R_i(t) \triangleq E\left\{\xi_i(t)\,[\xi_i(t)]^\top\right\}. \tag{10}$$

Note that the size $p(i)$ of the measurement vector z_i can be different for each local sensor.

3 Kalman Filter

State estimation performed in each processing node can be based on the standard Kalman filter paradigm [9,1], which results in a linear, unbiased, and minimum error-variance recursive procedure allowing to optimally estimate the state of an observed linear dynamic process. The basics of this theory and the resulting algorithm are presented below.

Applying the results from the previous time moment τ, the sampled-data model (2) leads to the following prior prediction of the process state in the next time moment t, performed by the i-th node considered:

$$\hat{x}_i(t|\tau) = F(t,\tau)\hat{x}_i(\tau|\tau) + u(t,\tau) \tag{11}$$

with the predictive covariance matrix

$$P_i(t|\tau) = F(t,\tau)P_i(\tau|\tau)\,[F(t,\tau)]^T + Q(t,\tau) \tag{12}$$

where $Q(t,\tau)$ is the process noise covariance matrix (8). The measurement prediction is simply

$$\hat{z}_i(t|\tau) = H_i(t)\hat{x}_i(t|\tau) \tag{13}$$

The covariance matrix of this (measurement) prediction is

$$S_i(t) = H_i(t)P_i(t|\tau)\,[H_i(t)]^T + R_i(t) \tag{14}$$

where $R_i(t)$ is the covariance matrix (10) of the measurement noise ξ_i.

The gain matrix of the Kalman filter is described as

$$K_i(t) = P_i(t|\tau)\,[H_i(t)]^T\,[S_i(t)]^{-1} \tag{15}$$

At each time instant t the following Kalman filter innovation is calculated:

$$v_i(t) = z_i(t) - \hat{z}_i(t|\tau) \tag{16}$$

where $z_i(t)$ is a measurement vector, gained from the corresponding i-th sensor, and associated with the time moment t.

The ultimate state estimate for time t is computed as

$$\hat{x}_i(t|t) = \hat{x}_i(t|\tau) + K_i(t)v_i(t) \tag{17}$$

with the effective covariance matrix expressed by the following Joseph formula:

$$P_i(t|t) = K_i(t)R_i(t)\,[K_i(t)]^T + [I - K_i(t)H_i(t)]\,P_i(t|\tau)\,[I - K_i(t)H_i(t)]^T \tag{18}$$

where I stands for an identity matrix.

4 Data Synchronization and Data Fusion

In the considered distributed estimation system each node ($i = 1, \ldots, N$) generates estimation data comprised of local state estimates and their corresponding covariance matrices. The processing nodes can circulate these data asynchronously among themselves.

To perform data fusion, in each data-fusion cycle the i-th node synchronizes all data obtained from the other processors to a common time moment. The best choice for such a synchronization moment is the time of the last local measurement. Then the respective measurement can be included in data fusion without any modifications. Such an approach will be exercised in this paper.

Let us consider an exemplary time scale, presented in Fig. 2, for the first processing node and nodes 2, 3 and 4 which form a subsystem of the distributed estimation network presented in Fig. 1.

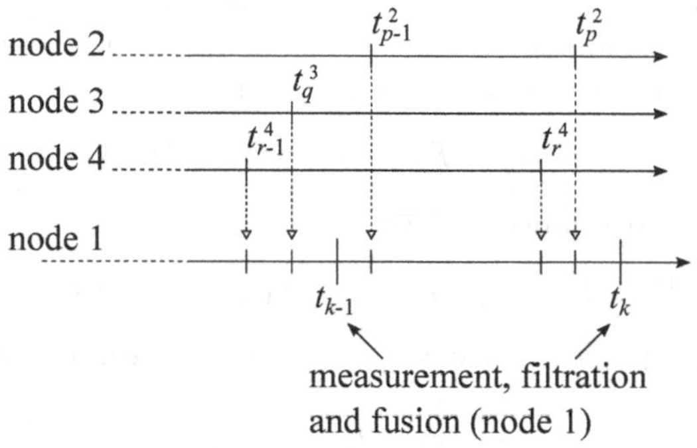

Fig. 2. An exemplary time scale

The time moments of both data fusion and sensor measurement for this processor are denoted as $t_k, k \in \mathbb{N}$. The time moments of data fusion in nodes 2, 3 and 4 are t_p^2, t_q^3 and t_r^4, respectively, with $p, q, r \in \mathbb{N}$. At time t_k the first local processor utilizes data from the second node and from the fourth one attributed to the time moments t_p^2 and t_r^4, respectively. Data from the second processor for time t_{p-1}^2 are not considered as newer data (from time t_p^2) are available. On the other hand, no new data from the third node are available (since the previous data fusion at time t_{k-1}).

In the first local processor, for synchronization purposes the estimate $\hat{x}_j(\tau|\tau)$ of the process state gained from the j-th local processor ($j = 2, 3, 4$) calculated using the data up to time τ can be predicted for time t_k according to the following:

$$\hat{x}_j(t_k|\tau) = F(t_k, \tau)\hat{x}_j(\tau|\tau) + u(t_k, \tau). \tag{19}$$

The above equation is based on the c-t model (2). The noise w in (2) is unknown, however the mean of this noise is 0. The corresponding covariance matrix is

$$P_j(t_k|\tau) = F(t_k, \tau)P_j(\tau|\tau)\left[F(t_k, \tau)\right]^\top + Q_j(t_k, \tau). \tag{20}$$

The state transition matrix $F(t_k, \tau)$, the input signal $u(t_k, \tau)$ and the covariance matrix $Q(t_k, \tau)$ of the process noise w, can be determined according to (3), (4) and (8), respectively. From (8) it is clear that the elements of the matrix Q are in proportion to the time difference between t_k and τ. In a sense, the matrix Q represents a loss of 'information' about the corresponding state estimate $\hat{x}_j$.

With the use of (19) and (20) all the state estimates considered in each data-fusion cycle in any local processor can by synchronized to a common time moment. Such the synchronization, executed by each of the processors, allows to perform data fusion. Note, however, that the exemplary time scale in Fig. 2 is for the first node. Other nodes perform filtration and fusion according to the same algorithm.

Next, based on the synchronized data, the first node performs data fusion, i.e. it combines its own estimate for time t_k with the synchronized results from other nodes. In the estimation algorithm presented in this paper we use the Covariance Intersection (CI) algorithm [8,5] which takes into account effects of common information between nodes.

CI is a useful method for data fusion in cases where cross correlation between estimation errors for different nodes is unknown or hard to calculate. Data fusion of n synchronized state estimates is performed according to the following equations (for better readability the time indices are omitted here)

$$(P_f)^{-1} = \omega_1 (P_1)^{-1} + \omega_2 (P_2)^{-1} + \ldots + \omega_n (P_n)^{-1} \tag{21}$$

$$\hat{x}_f = P_f \left[\omega_1 (P_1)^{-1} \hat{x}_1 + \omega_2 (P_2)^{-1} \hat{x}_2 + \ldots + \omega_n (P_n)^{-1} \hat{x}_n \right] \tag{22}$$

where $\hat{x}_j$ and P_j $(j = 1, \ldots, n)$ are the synchronized state estimates and their corresponding covariance matrices which are to be fused, $\hat{x}_f$ and P_f are the resulting (aggregate) state estimate and its corresponding covariance matrix for the considered node performing fusion, and $\omega_j \in [0, 1]$ $(\sum_{j=1}^{i=n} \omega_i = 1)$ are the weighting parameters. In practice, the parameters ω_j are chosen so as to minimize a given performance criteria, like the trace or the determinant of P_f. Such optimization must be performed to guarantee that the fusion algorithm is non-divergent. CI guarantees that the fused state estimate is consistent.

The distributed estimation algorithm using Covariance Intersection is described in full details in [11].

5 Simulation Example

Let us consider a simple problem of distributed trajectory estimation of a particle moving in a two-dimensional space.

For this purpose, the estimation network comprised of ten nodes in the configuration of Fig. 1 was chosen and the same kinematic model (1) was assumed for all processing nodes $(i = 1, \ldots, 10)$ with the following state:

$$X(t) \triangleq \left[\text{position}_x(t), \text{position}_y(t), \text{velocity}_x(t), \text{velocity}_y(t) \right]^T \tag{23}$$

and the system matrices

$$A = \begin{bmatrix} 0\,0\,1\,0 \\ 0\,0\,0\,1 \\ 0\,0\,0\,0 \\ 0\,0\,0\,0 \end{bmatrix}, \quad \sigma = \begin{bmatrix} 0\,0 \\ 0\,0 \\ 1\,0 \\ 0\,1 \end{bmatrix}, \quad b = \begin{bmatrix} 0 \\ 0 \\ 0 \\ 0 \end{bmatrix} \tag{24}$$

The observation matrix was the same for all sensors and equal to

$$H_i = H = \begin{bmatrix} 1\,0\,0\,0 \\ 0\,1\,0\,0 \end{bmatrix} \tag{25}$$

We considered two classes ($\mathcal{C}$ and $\widetilde{\mathcal{C}}$) of sensors distinguished by the measurement noise covariance matrix. Sensors in the first class $\mathcal{C}$ were more accurate and the measurement noise covariance matrix was R and the covariance matrix for the less accurate sensors of the second class $\widetilde{\mathcal{C}}$ was $\widetilde{R}$, where

$$R = \begin{bmatrix} 1.0 & 0 \\ 0 & 1.0 \end{bmatrix} \quad \text{and} \quad \widetilde{R} = \begin{bmatrix} 4.0 & 0 \\ 0 & 4.0 \end{bmatrix} \tag{26}$$

In each node one sensor from the class $\mathcal{C}$ or the class $\widetilde{\mathcal{C}}$ was used according to one of the four configurations described in Table 1. Thus, for example, in configuration II only the node 2 had a more accurate sensor from the class $\mathcal{C}$.

Table 1. Sensor distribution in the nodes of the estimation network

	Configuration			
class	I	II	III	IV
$\mathcal{C}$	—	2	1	1,7
$\widetilde{\mathcal{C}}$		remaining		

All the sensors performed measurements independently of each other in random times with the exponential distribution with the rate parameter equal to 1 (it means that each sensor performed on average one measurement per second).

The average estimation errors of position_x (the results for position_y are essentially the same) averaged over 100 simulation runs each performed for time t in the interval $[0, 100]$ are presented in Table 2. For comparison, the results for the system in which there are no edges in the connection graph (nodes do not exchange data) are added in Table 2.

The first conclusion is that the results for the system in configuration I are more accurate than those for the system whose nodes do not exchange information. Therefore, the overall estimation of the network is better, if nodes can exchange information, even if all the nodes have a sensor from the same class $\widetilde{\mathcal{C}}$.

The second conclusion is that the node with a more accurate sensor can significantly improve the results only in those surrounding nodes, which are directly connected to it. It is caused by the relatively large process-noise covariance represented by the matrix σ (24). Such large process uncertainty causes the past state estimates to become obsolete very soon after their calculation.

Table 2. Estimation errors for all sensors in all configurations

		Configuration			
sensor	no edges	I	II	III	IV
1	5.65	4.33	2.70	1.88	1.85
2	5.85	4.70	1.79	2.86	2.60
3	5.55	4.51	3.86	2.73	2.39
4	5.92	4.42	3.85	2.58	2.50
5	5.72	4.40	4.35	3.89	2.70
6	5.78	4.35	4.26	3.53	2.58
7	5.83	4.30	4.55	4.08	1.85
8	5.82	4.62	4.62	4.47	4.32
9	5.59	4.78	4.76	4.57	2.72
10	5.69	4.67	4.79	4.51	2.85

Finally, from configurations III and IV it can be concluded that using a more accurate sensor from the class $\mathcal{C}$ in a node that is coupled to the plurality of nodes, improves the overall quality of the network estimation.

6 Conclusions

In the paper we presented a framework for distributed asynchronous estimation of continuous-time stochastic processes using a network of estimators. We considered the case of networked estimation systems with connection graphs being not complete. In such systems, the lack of connections between certain nodes induces a certain transmission delay. This delay leads to degradation in estimation performance, because the delay invalidates the exchanged estimates of the state in case of stochastic dynamic processes. Due to the stochastic nature of the observed process, the passage of time reduces the importance of the knowledge about the process state that was obtained at a specific time in the past. Such effect of data degradation is even more evident for systems comprised of a large number of nodes. Therefore, a great care should be taken in selecting the system configuration with incomplete connections. In such cases we recommend placing more accurate sensors in nodes having 'effective' positions. Moreover, a general conclusion can also be drawn that an acceptable delay introduced by a networked system should be comparable with the time constants (dynamics) and uncertainty of the observed process.

References

1. Bar-Shalom, Y., Li, X.R.: Estimation and Tracking: Principles, Techniques, and Software. Artech House, Boston (1993)
2. Bar-Shalom, Y., Li, X.R.: Multitarget-Multisensor Tracking: Principles and Techniques. YBS Publishing, Storrs (1995)

3. Carli, R., Chiuso, A., Schenato, L., Zampieri, S.: Distributed kalman filtering based on consensus strategies. IEEE Journal on Selected Areas in Communications 26(4), 622–633 (2008)
4. Cattivelli, F., Sayed, A.: Diffusion strategies for distributed Kalman filtering and smoothing. IEEE Transactions on Automatic Control 55(9), 2069–2084 (2010)
5. Chen, L., Arambel, P.O., Mehra, R.K.: Estimation under unknown correlation: covariance intersection revisited. IEEE Transactions on Automatic Control, AC-47(11), 1879–1882 (2002)
6. Hall, D.L., Llinas, J.: An introduction to multisensor data fusion. Proceedings of the IEEE 85(1), 6–23 (1997)
7. Hall, D.L., Llinas, J.: Handbook of Multisensor Data Fusion. CRC, Boca Raton (2001)
8. Julier, S., Uhlmann, J.: A non-divergent estimation algorithm in the presence of unknown correlations. In: Proc. of the American Control Conference, pp. 2369–2373 (1997)
9. Kalman, R.: A new approach to linear filtering and prediction problems. Transactions of the ASME, Journal of Basic Engineering 82, 34–45 (1960)
10. Karatzas, I., Shreve, S.E.: Brownian Motion and Stochastic Calculus. Springer, NY (1991)
11. Kowalczuk, Z., Domżalski, M.: Asynchronous distributed state estimation based on a continuous-time stochastic model. International Journal of Adaptive Control and Signal Processing 26(5), 384–399 (2012)
12. Kowalczuk, Z., Domżalski, M.: Optimal asynchronous estimation of 2D Gaussian Markov processes. Int. Journal of Systems Science 43(8), 1431–1440 (2012)
13. Liggins, M., Chong, C., Kadar, I., Alford, M., Vinnicola, V., Thomopoulos, S.: Distributed fusion architectures and algorithms for target tracking. Proceedings of the IEEE 85(1), 95–107 (1997)
14. Oksendal, B.: Stochastic Differential Equations: An Introduction with Applications. Springer, Berlin (2003)
15. Olfati-Saber, R.: Distributed Kalman filtering for sensor networks. In: Proc. of the 46th IEEE Conference on Decision and Control, New Orleans, USA, pp. 5492–5498 (2007)
16. Olfati-Saber, R., Shamma, J.: Consensus filters for sensor networks and distributed sensor fusion. In: Proc. of the 44th IEEE Conference on Decision and Control, and 2005 European Control Conference (CDC-ECC), pp. 6698–6703 (2005)
17. Rao, B., Durrant-Whyte, H., Sheen, J.: A fully decentralized multi-sensor system for tracking and surveillance. The International Journal of Robotics Research 12(1), 20–44 (1993)
18. Ribeiro, A., Giannakis, G.B., Roumeliotis, S.: SOI-KF: Distributed Kalman filtering with low-cost communications using the sign of innovations. IEEE Transactions on Signal Processing 54(12), 4782–4795 (2006)
19. Rogers, L., Williams, D.: Diffusion, Markov Processes and Martingales: Itô Calculus, vol. 2. Cambridge University Press, UK (2000)
20. Speranzon, A., Fischione, C., Johansson, K.: Distributed and collaborative estimation over wireless sensor networks. In: Proc. of the 45th IEEE Conference on Decision and Control, San Diego, USA, pp. 1025–1030 (2006)
21. Xiao, L., Boyd, S., Lall, S.: A scheme for robust distributed sensor fusion based on average consensus. In: Proc. of the 4th International Symposium on Information Processing in Sensor Networks, Los Angeles, USA, pp. 63–70 (2005)

Transformer Windings Diagnostics with Combined Frequency Response Analysis and Vibroacoustic Methods

Szymon Banaszak and Eugeniusz Kornatowski

West Pomeranian University of Technology, Szczecin, Poland
{szymon.banaszak,eugeniusz.kornatowski}@zut.edu.pl

Abstract. The paper presents the idea of performing combined analysis of measurements results conducted with FRA and V-SSM methods. Both of them give some information on the mechanical condition of transformer's active part. However none of these methods allows for direct and straightforward interpretation of measured data. The application of FRA and V-SSM measurements together may give detailed information of the mechanical condition of windings. The idea has been experimentally verified on the real transformer, during controlled deformations.

Keywords: transformer, winding, Frequency Response Analysis, Vibroacoustic Method, Spectral Subtraction Method.

1 Introduction

Reliable diagnostics of power transformers is necessary in asset management of distribution companies. The complete information on technical condition of transformers population, together with information on each unit's importance in the system, leads to economically and technically reasonable management [1]. Nowadays there are many diagnostic methods, some are still under development. One of important fields in transformer diagnostics is assessment of the mechanical condition of an active part: windings, core and leads. Various high current events, overvoltages and insulation deterioration may lead to loosening original winding clamping and subsequently to deformations of windings. Also a core may lose its mechanical integrity over time. There are different diagnostic methods which may be used for such assessment, such as: measurements of transformer ratio, exciting currents, DC winding resistance or magnetic balance test. In addition some information on mechanical condition of active part can be indirectly obtained from oil analysis or vibrations monitoring [2]. However mentioned traditional methods are not giving clear results, enough for modern maintenance strategies, they give very limited information on the actual condition of active part the kind, scale or location of the problem. In most cases it is not clear whether the unit can be operated normally or should be sent to internal inspection or repairs, having the risk of catastrophic failure. There are also two methods which give more detailed information on the condition of the active part: Frequency Response Analysis (FRA) and Vibroacoustic Method (VM).

J. Korbicz and M. Kowal (eds.), *Intelligent Systems in Technical and Medical Diagnostics*, Advances in Intelligent Systems and Computing 230,
DOI: 10.1007/978-3-642-39881-0_39,

2 Frequency Response Analysis – FRA

The transformer winding can be described by a set of local capacitances, self and mutual inductances and resistances. Every change in winding geometry leads to change of these parameters, which influences the shape of the transfer function. The analysis of Frequency Response measurements results is based on comparison of data presented usually as sine signal damping along frequency spectrum in logarithmic scale. Such curve can be compared to results recorded for a transformer in time intervals, between phases, between twin or sister units or with help of computer models. Time based comparison is the most reliable, but for most of old transformers there is no fingerprint data available. The next two approaches are usually applied in industrial practice, however they are quite uncertain and may lead to misinterpretations. Each transformer can have differences in FRA curve compared between phases or, if compared to other units, due to constructional differences [3]. Helpful results can be obtained from controlled deformations, but this method cannot be applied in mass scale and generalized [4]. Applicability of FRA method and various techniques of test systems and connection setups have been widely discussed [4,5,6]. At the current stage of the FRA method it is possible to perform repetitive results thanks to the use of suitable connection techniques, which is summarized in CIGR document [7] in which recommendations are provided and at the present the IEEE (PC57.149/D8) and the IEC PT 60076-18 are working on the elaboration of standards under which test procedures and FRA instrument requirements are established.

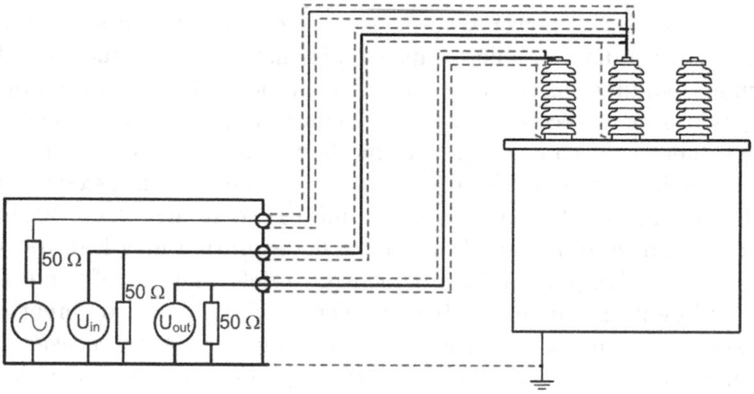

Fig. 1. FRA measurement principle and configuration

All FRA tests were conducted in three leads setup: Source, Reference and Measurement. All signals are given against earth potential (joined together tank, leads screens, measuring device zero potential). Measurements results presented in the paper were taken with the Source and Reference leads connected together to the bushing at the beginning of tested winding, while Measurement lead to the bushing at its end, which is called 'end-to-end' measurement. The Source lead gives sine signal with frequency from 20 Hz to 2 MHz or higher, Reference is used for measurement of this signal at the 'input' bushing, while Measurement at the 'output' bushing. Such three leads setup allows reliable measurements, not depending on the route of the leads.

The still pending problem of the FRA method is the interpretation of test results. In the most of cases the interpretation relies on human experts who analyze the results on the base of their experience and by accessing to large database of FRA measurements of transformers. However in many cases it is not clear whether differences observed e.g. between phases of sister units are the result of windings deformations or are just the effect of constructional details. There are obvious differences, like between side and middle phase, due to different flux distribution, but also being characteristic for given type of transformer, e.g. between phase A and C. Therefore sometimes it is hard to determine whether observed deviations in test results come from deformations or are typical.In such cases application of complementary method would help to clarify the interpretation. Authors propose to use VM for this purpose.

3 Vibroacoustic Method – VM

The vibroacoustic diagnostics is used at present in many fields of engineering and science. Its advantage is no need to interfere in a structure of tested object. Also in the case of mechanical condition assessment of an active part of the transformer this method is used more and more often. The application of the Vibroacoustic Method (VM) in Poland was introduced in eighties of the last century to assessment of technical condition of transformers' cores [8]. Recommendations used for vibroacoustic measurements of transformers are over 30 years and do not include modern metrological and computational possibilities and may lead to serious misinterpretations [9].

The idea of transformer's construction vibrations is based on the analysis of a signal recorded with an accelerometer attached to the tank surface. Inside a transformer electrodynamical and magnetostriction forces influence windings and a core, causing their vibrations. These vibrations, transferred by an insulating liquid, cause vibrations of tank walls. The reason of a core vibrations is magnetostriction, leading to changes of geometrical dimensions of magnetic mater ial located in a magnetic field. The intensity of magnetic field in the core placed inside of a winding depends on a voltage applied to the winding. On the base of Faraday law it was showed in [10] that the change of core length placed in the coil is directly proportional to the square of a momentary value of applied voltage $\Delta L \propto [U_{max}\cos(\omega t)]^2$, where ω is pulsation of the voltage. A momentary value of vibrations acceleration caused by magnetostriction equals $a_c(t) = d^2\Delta L/dt^2$ and its basic harmonic frequency is two times bigger than applied voltage frequency.

Vibrations of transformer construction recorded with an accelerometer attached to transformer's tank are a superposition of mentioned above vibrations of a core and windings. The latter are influenced by electrodynamical forces proportional to the square value of a current [11]. Because the force is directly proportional to the acceleration, it can be assumed that vibrations acceleration of windings is directly proportional to the square value of the current: $a_w \propto I^2$. With assumption that $I = I_{max}\cos(\omega t)$ and knowing that $cos^2\omega t = 0.5(\cos 2\omega t + 1)$, also in this case a frequency of basic vibrations harmonic is twice as big as supplied power frequency, in other words the same as in the case of a core.

Loosening and deformations of windings and core sheets are the direct causes of significant changes in frequency spectrum of recorded vibrations signal. Apart from harmonics with basic frequency 100 Hz (for 50 Hz power supply), the frequency spectrum will contain also other harmonics having various amplitudes. The shape of amplitude spectrum is strictly correlated to the character of mechanical defects in a core and windings.

Most of vibroacoustic methods used for transformers diagnostics is bases of the analysis of frequency spectrum of vibrations acceleration in steady state. Analyzed amplitude spectrum is characterized with high redundancy of information, which makes efficient concluding very difficult. Harmonic frequencies distribution is a representation of a core and windings vibrations and depends from load current and applied voltage. Large amount of information resulting from spectral analysis makes it impossible to find it unambiguously which of these factors has dominant influence on recorded frequency spectrum.

The assessment of vibrations in a transient state, in the first 10-20 seconds from powering a transformer without any load, may reduce the influence of some factors causing vibrations. The method of tank vibrations assessment in transient state is based on following assumptions:

1. The influence of magnetostriction phenomena on recorded vibrations is identical for transient and steady states of transformer operation, with assumption that voltage amplitude is constant.
2. The amplitude of magnetizing current in period shorter that 0.5 s from powering the transformer reaches the value several or over ten times higher than nominal currents [12], leading to vibrations of windings caused by electrodynamical forces.
3. In steady state without load the main source of transformer tank vibrations core are vibrations caused by magnetostriction phenomena Windings vibrations are negligible low due to the lack of a load, in windings there is a magnetizing current having the amplitude considerably lower than nominal current.

On the Fig.2 there is presented example of recorded signal of tank vibrations of transformer without load in period approx. 20 s from powering.On the graph showing momentary values of the acceleration (Fig. 2a) and vibrations spectrogram (Fig. 2b), there are visible two ranges of vibrations: the first one from start to approx. $6 \div 8$ s – vibrations stabilization, and the second one – vibrations with stabilized amplitude.

Because of 'interfering' influence of magnetostriction phenomena the analysis of transient state may be difficult. For example, problematic might be accurate

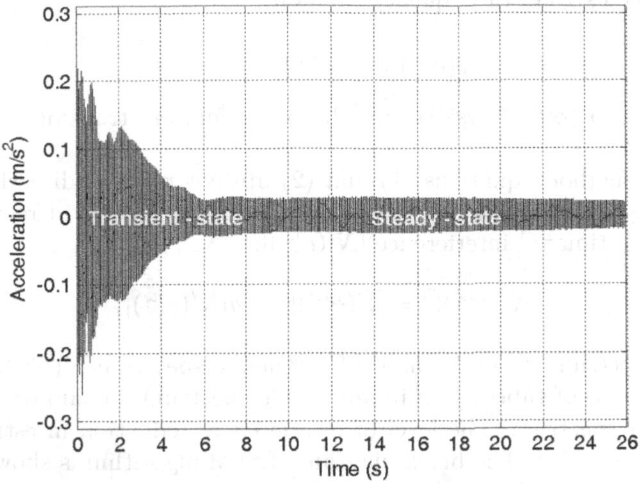

(a)

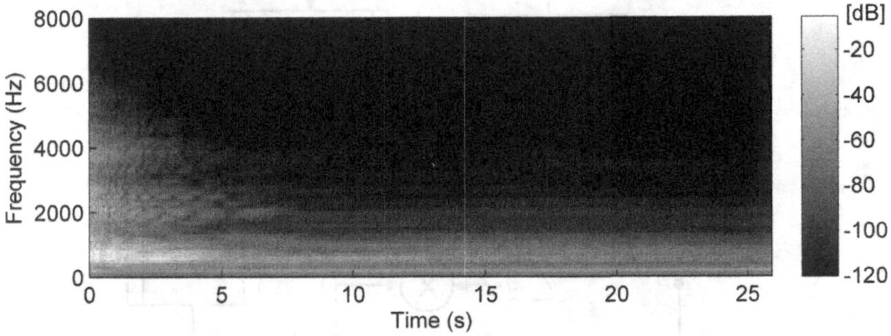

(b)

Fig. 2. a) Momentary value of transformer tank vibrations, b) vibrations spectrogram

assessment of vibrations time, caused by transient state. Vibroacoustic comparative diagnostics of windings condition of similar type transformers also is not reliable, because core's technical condition of tested transformers might be different which influences amplitude of vibrations acceleration and amplitude spectrum in steady state operation without load.

After considering assumption 1 of the presented method it is possible to reduce the influence of magnetostriction phenomena on recorded vibrations signal. It can be obtained by using a Spectral Subtraction Method – SSM [13]. SSM is one of basic methods used for reduction of interferences having additive character from sound recording and is used in sound engineering. The idea of this method can be described in the simplest realization as following: if recorded discrete signal is present with additive interference $n(k)$, then:

$$x(k) = y(k) - n(k) \tag{1}$$

is noninterfered signal. In frequency domain:

$$X(e^{j\omega}) = Y(e^{j\omega}) - N(e^{j\omega}) \tag{2}$$

where $X(e^{j\omega}), Y(e^{j\omega}), N(e^{j\omega})$ are discrete Fourier transforms of signals $x(k), y(k), n(k)$.

In a SSM method equations (1) and (2) are not realized directly, but there are performed calculations based only on amplitude spectra of recorded signal $|Y(e^{j\omega})|$ and estimated interference $|N'(e^{j\omega})|$:

$$|X'(e^{j\omega})|^{\beta} = |Y(e^{j\omega})|^{\beta} - a|N'(e^{j\omega})|^{\beta} \tag{3}$$

leaving in spectrum $X'(e^{j\omega})$ phase of frequency spectrum $Y(e^{j\omega})$. Coefficient $\beta = 1$ in the case of subtraction in amplitude spectrum domain or $\beta = 2$ in the case of power spectrum. The level of interferences reduction in estimated with coefficient $a \in <0, 1>$ The block diagram of SSM algorithm is shown at Fig. 3.

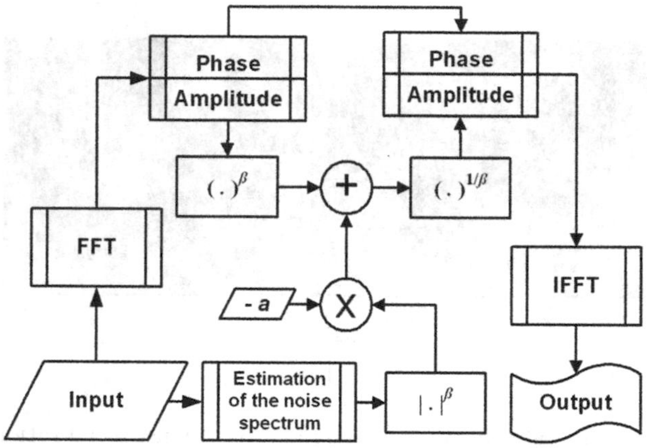

Fig. 3. The algorithm of Spectral Subtraction Method – SSM

In the sound engineering the reduction of interferences is achieved by using an estimate of amplitude spectrum from 'silence' periods in recorded signal. The reduction of the magnetostriction based vibrations influence is realized by treating steady state vibrations signal as the additive interference. If in equation (3) $\beta = 1, a = 0.1$ and using the signal presented on Fig. 2a the final result was obtained, shown on Fig. 4: momentary value of transformer tank vibrations acceleration (Fig. 4a) and vibrations spectrogram (Fig. 4b). The value of coefficient a in equation (3) has been determined on the base of assumption that vibrations amplitude reduction caused by magnetostriction is -40 dB, while value of $\beta = 1$ because the algorithm of spectral subtraction had been realized in signal amplitudes domain.

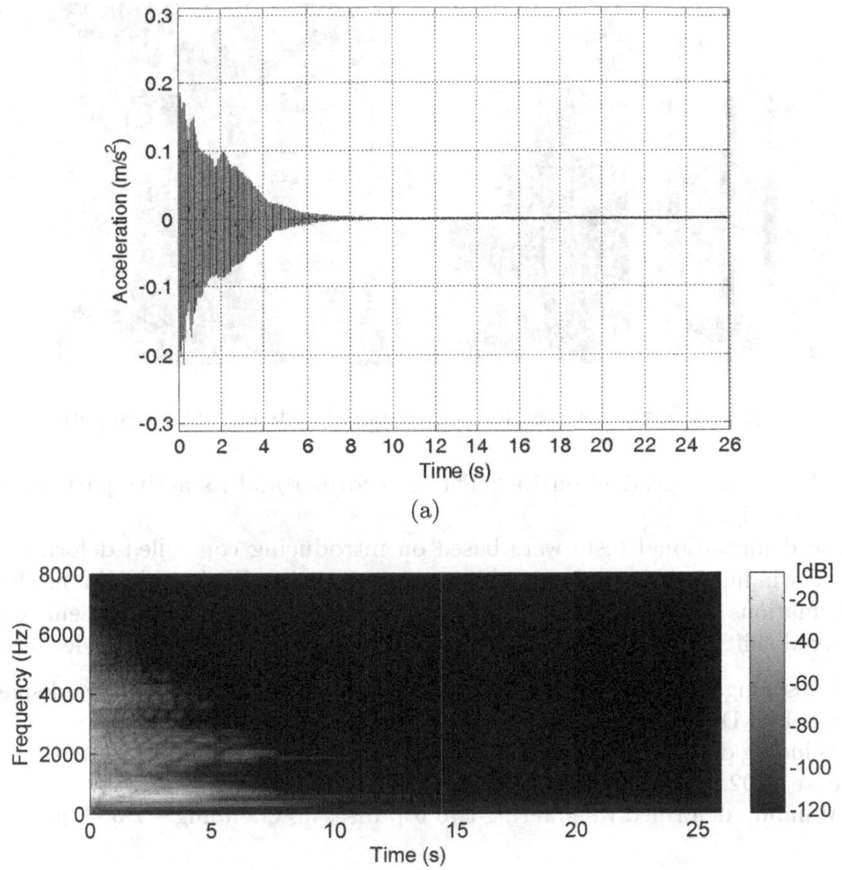

Fig. 4. a)Momentary value of transformer tank vibrations acceleration after reducing magnetostriction phenomena, b) vibrations spectrogram

The application of SSM leads to reduction of magnetostriction based components in recorded signal (Fig. 4b).The analysis of such signal allows assessment of windings and core vibrationsinfluence, resulting from electrodynamical forces, on the transformer's tank vibrations.

4 Combined Methods Diagnostics of Winding's Controlled Deformations

To find out complementary possibilities of both diagnostic methods there was prepared deformational experiment on a real transformer. It was a unit replaced in distribution company by a new one because of age, however it was in a very good technical condition. Its parameters were: type TONa 800/15, 15/0.4 kV,

Fig. 5. Transformer used for deformational tests and its active part

800 kVA, year of production 1962.The transformer and its active part are presented on Fig. 5.

The deformational tests were based on introducing controlled deformations to the windings and conducting measurements with FRA and VM methods. Deformations introduced into winding were based on clamping loosening and the axial shift of whole discs. There were three levels of controlled defects:

1. Loosening the winding clamping, without any additional discs displacements – D1.
2. Winding deformed by lowering the single top disc, no clamping as in previous case – D2.
3. Winding deformed by lowering two top disc, no clamping – D3 (Fig. 6).

Fig. 6. Transformer winding with deformation D2 – top two discs lowered to minimum

To perform necessary operation in the windings, it was removed from the transformer with crane and placed in the tank afterwards. The transformer for measurements was completely mounted, with all screws tightened, original oil in the tank and original bushings. Therefore measurements results were not influenced by any factors other than controlled defects.

4.1 FRA Measurements

The FRA method is capable of detecting defects resulting in the change of geometry or electric configuration of the active part. Such changes influence local capacitances, inductances and couplings giving changes in Frequency Response (FR) curve. Therefore only such defects are to be considered in FR results analysis.

Frequency Response in the experiment has been measured with a commercial device FRAnalyzer from Omicron. High voltage winding, used for deformations, was delta connected, therefore the signal was also transferred through other phases,so the Frequency Response had its specific shape.

Frequency Response of the winding is presented on Fig. 7. It can be seen that changes in FR curve for baseline measurement and after introducing the deformation are not significant and are visible in higher frequencies. The frequency range has been divided into three parts: Low Frequency gives information on magnetic circuit of a transformer, Medium Frequency on significant deformations and High Frequency on smaller deformations, influence of leads etc. [14]. Detailed ranges limits depend on the construction of the transformer and its geometrical size the smaller unit is the higher frequencies show its Frequency Response.

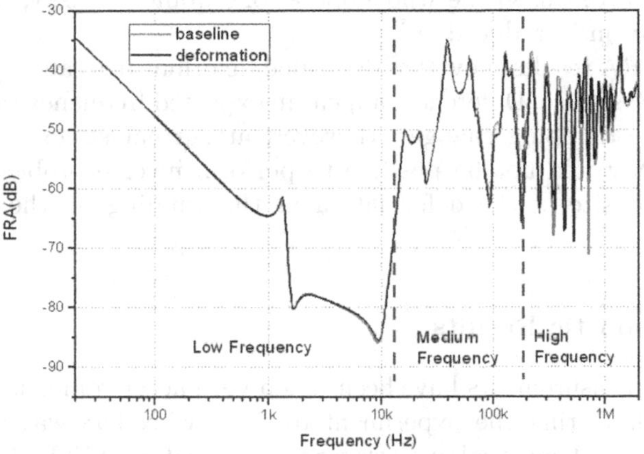

Fig. 7. Frequency Response of tested transformer the whole frequency range

In this case differences are visible in Medium and High Frequency ranges. It is clear that they come from deformations in the winding. Fig. 8. presents zoomed frequency range with visible differences between curves for 'healthy' winding and after two kinds of deformations. Axial shift of discs influenced mainly damping of FR. The deformation was not significant, as the shift of top two discs reduced the oil gap less than 1 cm (Fig. 3). Nevertheless some resonances have shifted vertically even by 10 dB (350 kHz, 700 kHz). The conclusion is that such deformation can be easily detected by FRA method. No further deformation were

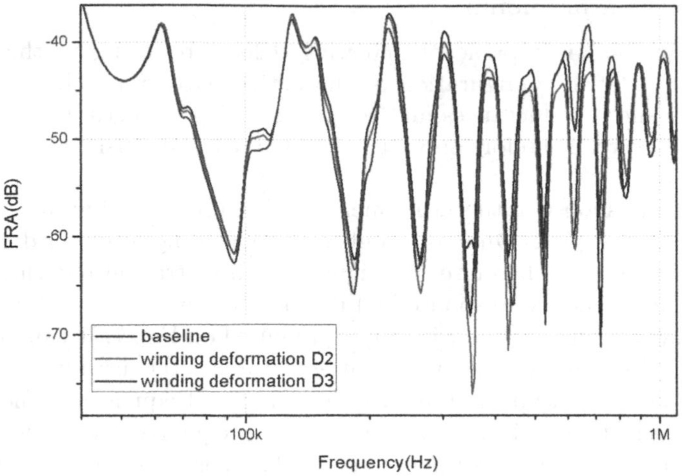

Fig. 8. Zoomed frequency range with visible differences of Frequency Response for two deformations

introduced because the transformer had to be powered in order to perform vibroacoustic recordings, so the whole operations would be risky, especially with radial deformation introduced.

Presented FRA results show that described deformations cause typical changes in Frequency Response. Differences appear in expected frequency range, and have high amplitude damping changes. However, at current stage of FRA method development it would not be possible to perform inverse problem to identify the location or size of the deformation in the winding on the base of FRA measurements.

4.2 Vibroacoustic Results

Vibroacoustic measurements have been taken with accelerometric vibration sensor SVAN 958. During the experiment device SVAN 958 was used as signal recorder, while mathematical analysis was conducted in MATLAB environment. The accelerometer was mounted on the right side of the tank (on column A), in the middle of its height. For vibroacoustic analysis with SSM algorithm (V-SSM) there was taken the first ten seconds of recorded signal, starting from powering transformer without any load. Fig. 9 presents stabilization process of vibrations acceleration (envelope) with reduced influence of the magnetostriction.

Developing simulated degradation of windings mechanical condition unambiguously influences the envelope shape of vibrations acceleration signal: by loosening and deforming the winding increases the amplitude of vibrations and the time of transient state vibrations suppression is changing. From Fig. 9 it can be concluded that e.g. after time approx. 1 s from powering the transformer the acceleration is 0.022 m/s2, 0.032 m/s2and 0.040 m/s2, respectively for windings

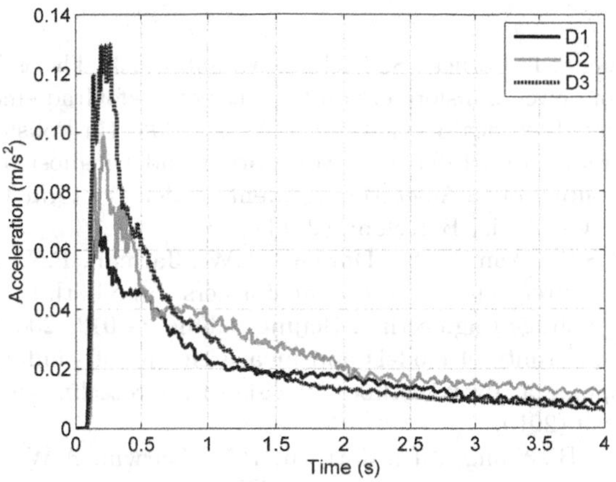

Fig. 9. Stabilization process of tank vibrations for three windings deformations

defects D1, D2 and D3. However it shall be noted that even though magnetostriction phenomena has been reduced, in transient state vibrate both windings and the core. Therefore the measurement with V-SSM in real conditions cannot unambiguously determine the mechanical condition of windings only. Obtained results shall be verified with FRA method.

5 Summary

The idea of combined FRA and VM measurements for the assessment of mechanical condition of transformer windings has been investigated and introduced into experimental deformational tests. A new approach for VM method has been presented, based not on the traditional analysis of an amplitude spectrum, but coming from analysis of stabilization of tank vibrations' acceleration in the transient state. The vibroacoustic signal reflected vibrations of construction caused by electromagnetic forces affecting the core and windings. The influence of magnetostriction had been reduced by application of Spectral Substraction Method (V-SSM). The experiment performed on the real transformer, by the means of controlled deformations and FRA and VM measurements showed that it was possible to use both methods for the complementary analysis. The application of FRA and V-SSM measurements may give detailed information of the mechanical condition of transformer's active part.

Acknowledgements. The research was financed by Polish National Science Center, grant nr. N N510 698240 titled "The algorithm for identification of transformer winding deformations on the base of frequency response measurements".

References

1. Szrot, M., Płowucha, J., Borucki, S., Cichoń, A., Subocz, J.: The assessment of technical condition of power transformers on the base of TrafoGrade method. Przeglad Elektrotechniczny (Electrical Review) (10), 8–11 (2008) (in Polish)
2. Velasquez Contreras, J.L.: Intelligent Monitoring and Diagnosis of Power Transformers in the Context of an Asset Management Model. Doctoral Thesis, Polytechnic University of Catalonia, Barcelona (2011)
3. Jayasinghe, J.A.S.B., Wang, Z.D., Darwin, A.W., Jarman, P.N.: Practical issues in making fra measurements on power transformers. In: XIVth International Symposium on High Voltage Engineering, Beijing, China, G-013 (2005)
4. Banaszak, S.: Conformity of models and measurements of windings deformations in frequency response analysis method. Przeglad Elektrotechniczny (Electrical Review) (7), 278–280 (2010)
5. Jayasinghe, J.A.S.B., Wang, Z.D., Jarman, P.N., Darwin, A.W.: Winding movement in power transformers: a comparison of FRA measurement connection methods. IEEE Transactions on Dielectrics and Electrical Insulation 13(6), 1342–1349 (2006)
6. Christian, J., Feser, K.: Procedures for detecting winding displacements in power transformers by the transfer function method. IEEE Transactions on Power Delivery 19(1), 214–220 (2004)
7. Mechanical condition assessment of transformer windings using frequency response analysis (FRA), Report of CIGRE Working Group A2.26 (2008)
8. Diagnostics of transformers' technical condition. Principles of tests and assessment criteria. Report of Energopomiar Company, Gliwice, Poland (1983) (in Polish)
9. Borucki, S.: Diagnostic of transformer cores by use of vibroacoustic methods, Opole, Poland (2012)
10. Shenghang, J., Yongfen, L., Yanming, L.: Research on extraction technique of transformer core fundamental frequency vibration based on OLCM. IEEE Transactions on Power Delivery 21(4), 1981–1988 (2006)
11. Kornatowski, E.: Time-frequency vibration analysis of the transformer construction. Przeglad Elektrotechniczny (Electrical Review) (11b), 268–271 (2012)
12. Halinka, A., Szewczyk, M., Rzepka, P., Szablicki, M.: Overvoltage protection operation in transient states during switching on MV transformers without load. Elektroinfo (3), 24–27 (2010) (in Polish)
13. Czyżewski, A.: Digital sound. Chosen theoretical problems, technology, applications. EXIT, Warszawa, Poland (1998) (in Polish)
14. Banaszak, S.: Sensitivity of FRA measurements to various failure modes. Przeglad Elektrotechniczny (Electrical Review) (3b), 270–272 (2013) PL ISSN 0033-2097

Structural Diagnostics of Composite Beams Using Optimally Selected Fractional B-spline Wavelets

Andrzej Katunin and Piotr Przystałka

Silesian University of Technology, Institute of Fundamentals of Machinery Design, Konarskiego 18A, 44-100 Gliwice, Poland
{andrzej.katunin,piotr.przystalka}@polsl.pl

Abstract. The method of structural diagnostics of composite structures presented in this paper is based on discrete wavelet transform of displacements of modal shapes with usage of fractional B-spline wavelets. An application of such wavelets makes possible to improve the sensitivity of the method, which allows for detection and identification of even small damages occurred in the structure. In order to select the most sensitive wavelet bases for the analysis the optimization study was carried out, where the highest ratio of peak, which indicated the damage, to other values of detail coefficients and possibly short support of a peak in the location of the damage were chosen as the optimization criteria. Obtained results allow for the evaluation of the possible best wavelet bases for the damage identification of one-dimensional composite structures.

Keywords: structural diagnostics, fractional wavelet transform, composite structures, multicriteria optimization, damage identification.

1 Introduction

Polymer-based laminated composite structures are characterized by light weight and high ratios of strength and stiffness to weight in comparison with metals and metal alloys, therefore they have been widely applied in lightweight constructions, primarily in aircraft and aeronautical industries, automotive industry and others. Obviously, such structures are subjected to the damages occurred during their manufacturing and operation. The most common damages are matrix and fiber cracks and delaminations, which occur due to the degradation processes, overloading or shock exposure. Therefore, one of the most important problems, which should occur during their operation is the development of proper methods in order to detect, localize and identify the damage on possibly early stage of its propagation.

There are a lot of methods used for testing of composite structures, however their application is often limited to the laboratory studies. A group of methods, which allow to apply them in industrial conditions and characterizes by comparatively simple and effective testing are the vibration-based methods. In several methods based on the modal analysis the modal shapes of vibration are considered during the testing procedure. However, in order to detect and identify small damages the modal shapes should be processed using advanced signal

J. Korbicz and M. Kowal (eds.), *Intelligent Systems in Technical and Medical Diagnostics*, Advances in Intelligent Systems and Computing 230,
DOI: 10.1007/978-3-642-39881-0_40, © Springer-Verlag Berlin Heidelberg 2014

processing tools. One of the promising tools applied for the damage identification is the wavelet transform, which gain a popularity for such problems in the last decade. The problem of cracks identification was discovered by the Greek scientific group in [5,6]. The authors used continuous wavelet transform (CWT) for detection and localization of cracks in beams based on numerical and experimental data. The authors of [2,12] used CWT with Gabor wavelets in order to detect cracks in beams; the methods were tested based on numerical data. An experimental study of damage identification in beam structures using CWT-based method with Gabor wavelets was presented in [16]. Based on above-cited references it could be proved that Gabor wavelets are the most suitable for the damage identification studies.

Another approach based on discrete wavelet transform (DWT) was proposed by the first author of the paper in [8,9] with further generalization to 2D structures [10]. The main advantage of an application of DWT in damage identification problems is the reducing of computational time. Following to the statements of DWT only the orthogonal compactly-supported wavelet bases could be applied for the analysis, which exclude Gabor wavelets reported in [2,12,16] as the most suitable. However, the proposed method was based on B-spline wavelets, which fulfill the orthogonality property and have compact support and converge asymptotically to the Gabor wavelets [15]. Results obtained using DWT-based method in numerical and experimental studies proved its effectiveness.

In order to improve the sensitivity of the DWT-based method the authors applied fractional B-spline wavelets proposed in [3]. Moreover, for the selection of the most suitable wavelets the multicriteria optimization was applied. The optimization algorithms together with wavelet transform were used in several studies. The authors of [7] used a multi-objectives evolutionary algorithm for parameter optimization in order to improve a Wavelet-Transform-based ECG delineation method. Rafiee et al. [13] used artificial neural networks together with genetic algorithm in order to select the most suitable wavelet for diagnosing the gear faults. A two-step CWT-based damage detection method was proposed in [17]: in the first step the wavelet transform of the modal shapes was performed, while in the second step the finite element (FE) models were constructed in order to fill the database of damaged cases and based on particle swarm optimization the damages were detected. As it could be noticed, such a method besides its accuracy requires a lot of computational time and could not be used in industrial conditions. The method proposed in this paper allows for selecting the most suitable wavelet base, which could detect, precisely localize (with minimal distortion of the damage indicator) and identify the occurred damage. The authors used two approaches for selecting the optimal wavelet: standard and multicriteria evolutionary algorithms. Obtained results confirmed the improvement of sensitivity of the proposed method for the damage identification, which implies that more tiny damages could be recognized. The study was carried out based on FE numerical data and the same data with additional noise in order to simulate measurement conditions.

2 Description of Damage Identification Method

In this section we will present fundamental definitions of fractional B-spline wavelets and scaling functions. Then the proposed damage identification method will be described and analyzed. Finally, we will describe the optimization procedure in order to select the most suitable wavelets for the analysis.

2.1 Fractional B-spline Wavelets and Scaling Functions

The fundamental approach of DWT is the dyadic multiresolution analysis, introduced by Mallat, where the B-spline scaling functions $\beta(x)$ constitute the space of square integrable functions $L^2(\mathbb{R})$ and generate a sequence of the functional spaces:

$$\{0\} \subset \ldots \subset V_{-1} \subset V_0 \subset V_1 \subset \ldots \subset L^2(\mathbb{R}) . \tag{1}$$

The general form of the fractional B-spline scaling function is given by the following equation [1]:

$$\beta_\tau^\alpha(x) = \sum_{k=0}^{\infty} (-1)^k \begin{vmatrix} \alpha+1 \\ k-\tau \end{vmatrix} \rho_\tau^\alpha(x-k) , \tag{2}$$

where α is the fractional order of B-spline scaling function, τ is its shift parameter and ρ_τ^α is the function given by

$$\rho_\tau^\alpha(x) = -\frac{\cos \pi\tau}{2\Gamma(\alpha+1)\sin\frac{\pi}{2}\alpha}|x|^\alpha - \frac{\sin \pi\tau}{2\Gamma(\alpha+1)\cos\frac{\pi}{2}\alpha}|x|^\alpha \operatorname{sgn}(x) , \tag{3}$$

$\Gamma(\alpha+1)$ is the Euler's gamma function, which allows for partial factorization. Note, that for $\alpha \in \mathbb{Z}$ and $\tau = (\alpha+1)/2$ the classic B-splines could be obtained.

In terms of Mallat's multiresolution algorithm the scaling function fulfill the two-scale relation:

$$\beta_\tau^\alpha(x) = 2^{-\alpha} \sum_{k\in\mathbb{Z}} \begin{vmatrix} \alpha+1 \\ k-\tau \end{vmatrix} \beta_\tau^\alpha(2x-k) . \tag{4}$$

The explicit form of fractional B-spline wavelets was defined by Unser and Blu in [14]:

$$\psi_\tau^\alpha(x/2) = \sum_{k\in\mathbb{Z}} \frac{(-1)^k}{2^\alpha} \sum_{l\in\mathbb{Z}} \binom{\alpha+1}{l} \beta_0^{2\alpha+1}(l+k-1)\, \beta_\tau^\alpha(x-k) . \tag{5}$$

The fractional B-spline scaling and wavelet functions satisfy the most properties of classic B-spline analogues. The only property, which is not satisfied is an existence of non-compact support for $\alpha \notin \mathbb{Z}$. However, the fractional discrete wavelet transform (FrDWT) could be constructed basing on Fourier series (see [14] for details).

2.2 Damage Identification Method

It is well known, that the wavelet transform could be interpreted as filtering using the sequence of the high- and low-pass filters. The proposed method is based on single-level filtering procedure of displacements of modal shapes using the set of filters of fractional order. Considering the two-scale relation of the fractional B-spline scaling function (4) the high-pass (scaling) filter could be defined by its impulse response [1]:

$$H_{\tau}^{\alpha}\left(e^{j\omega}\right)=2^{-\alpha}\left(1+e^{j\omega}\right)^{\frac{1}{2}(\alpha+1)-\tau}\left(1+e^{-j\omega}\right)^{\frac{1}{2}(\alpha+1)+\tau} \tag{6}$$

and the low-pass (wavelet) filter could be presented in the same way as:

$$G_{\tau}^{\alpha}\left(e^{j\omega}\right)=-e^{-j\omega}H_{\tau}^{\alpha}\left(-e^{-j\omega}\right)A^{\alpha}\left(-e^{j\omega}\right), \tag{7}$$

where

$$A^{\alpha}\left(e^{j\omega}\right)=\sum_{k\in\mathbb{Z}}\beta_{\tau}^{2\alpha+1}\left(k\right)e^{-j\omega k}. \tag{8}$$

The proposed algorithm consider a single-level decomposition of displacements of the modal shapes s_n using (6) and (7) and than the downsampling procedure. As a result of these operations the approximation a_n and detail d_n coefficients are obtained. If the singularity exists in the signal it could be visible as the local increase of detail coefficients in its position. However, it was observed in previous studies [8,10], that the magnitudes of detail coefficients are strongly dependent on the magnitudes of displacements, i.e. if the location of the damage coincides with the node of a normal node the damage will be undetectable in the set of detail coefficients. In order to avoid this disadvantage more modal shapes should be considered during the analysis. As it was observed [11], for small α the detail coefficients consist a trend, which should be removed before further operations. It could be achieved by the polynomial approximation of each set of detail coefficients. Additionally, for emphasizing the damage presence the sum of sets of detail coefficients of considered M modes could be determined. The addition operation could compensate the magnitudes in the location of a damage due to the changes of sign of the detail coefficients, thus this operation should be carried out on absolute values of detail coefficients. The schematic representation of the damage identification method was presented in Fig. 1.

2.3 Optimal Selection of the Wavelet

The great influence on the damage identification procedure has the scaling and wavelet functions applied for the analysis. In order to determine the most suitable functions, which allow for minimizing the blurring of the peak, which indicate the damage and simultaneously find the cases, when the peak will have the greatest magnitude in comparison with other detail coefficients in the set D.

As to be expected, it could be achieved by optimizing parameters as the fractional order α, the shift factor τ and the degree of the polynomial

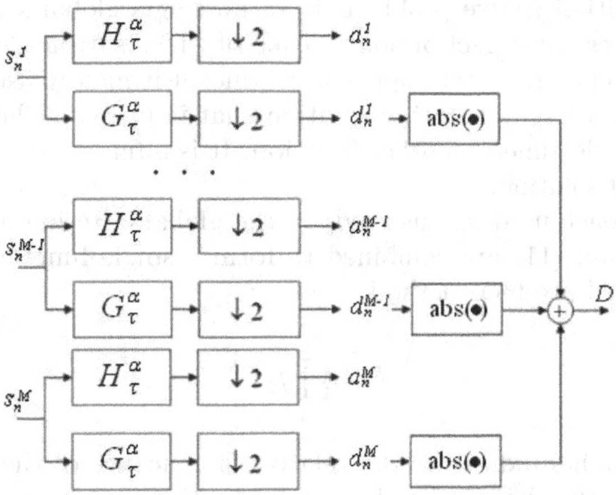

Fig. 1. The scheme of damage identification

approximation r. The aim of the optimization process is to adjust these parameters in order to minimize a multiple objective function $\mathbf{F}$ which could be formulated due to the following two criteria. Assuming that both objectives are not conflicted the optimization task is posed as follows:

$$\min_{\alpha,\tau,r} \mathbf{F}(\alpha,\tau,r) = \left[f_1(\alpha,\tau,r),\ f_2(\alpha,\tau,r)\right]^T, \tag{9}$$

where $\alpha,\ \tau \in \mathbb{R}, r \in \mathbb{Z}$ and $0 < \alpha < \alpha_c$, $1 < \tau < \tau_c$, $1 < r < r_c$. It should be noted that α could be gretaer than -0.5 following the definition presented in [14], however initial tests show that the scaling and wavelet function with $-0.5 > \alpha > 0$ are not suitable for the investigated problem (for instance see [11]).

Taking into account this consideration, the first payoff function is inversely proportional to all the magnitudes of the peaks:

$$f_1 = \left[1 + \sum_{i=1}^{H} \max(D_i)\right]^{-1}, \tag{10}$$

where H is the number of notches, detail coefficients in the set $D_1 = D$ and in $D_i = D_{i-1} \setminus d^*_{i-1}$, $i \neq 1$ where d^*_{i-1} is the point in the set D_{i-1} for which the maximum magnitude of the peak is obtained. Taking it a step further, the second objective is connected with the blurring of the peak and other detail coefficients in the set D, therefore it can be written as:

$$f_2 = \sum_{i=1}^{|D^*|} d_i, \tag{11}$$

where $D^* = D \setminus d^*_1 \setminus d^*_2 \cdots \setminus d^*_H$.

Typically, multi-objective problems have no single global solution, and it is reasonable to investigate a set of points, each of which satisfies the objectives. In this paper, predominant Pareto optimality concept is mainly used. A solution is Pareto optimal if there is no other solution that improves at last one objective function without detriment another function. It is often considered the same as a non-dominated solution.

Another approach used in this study is the global criterion method in which objectives (10) and (11) are combined to form a single function. It is realized using the weighted product method:

$$U = \prod_{i=1}^{2} f_i^{w_i}, \tag{12}$$

where w_i are weights indicating the relative significance of the objective functions. In this way, the above-defined objectives with different orders of magnitude may have similar meaning and their transformation is not required.

Stochastic search methods, such as evolutionary algorithms, are particularly applied herein to find the optimal solution. Standard optimization methods (e.g. gradient-based methods) cannot be adopted in this context, mainly due to the form of the objective function (10) and the domain of the variable r that leads to discontinuous cost surfaces. The multi-objective evolutionary algorithm proposed in [4] is applied to solve the task defined as (9). Well-known genetic operators for multi-objective optimization are used to guarantee convergence to a solution. As a result, the problem is solved by identifying the Pareto front, hence the set of evenly distributed non-dominated optimal solutions are achieved. On the other hand, in the case of the scalar function (12) the single-objective evolutionary optimization is realized.

3 Damage Identification Procedure and Results

The proposed damage identification method was tested on numerical data prepared based on FE models. Three cases were considered: the beam with a single notch, the beam with two notches and the delaminated beam. Obtained results describe the optimal wavelets selected for the analysis and its influence on the sensitivity of the damage identification also with the presence of noise.

3.1 Data Preparation

The numerical model of a cantilever composite laminate beam was prepared using MSC Marc/Mentat commercial software. The beam was modelled as the 3D solid in order to simulate the damages properly. The dimensions of a beam were as follows: length of 0.2 m, width of 0.01 m and thickness of 0.0024 m. The laminate consists of 12 orthotropic layers made of epoxy resin with carbon fiber reinforcement with the orientation and material properties as in [11].

The damages were modelled as follows. In the first case the crack was modelled at the distance of 0.15 m from the clamped side with a gap size of 0.0005 m and the depth of 0.0004 m (the thickness of two layers). In the second case an additional crack was modelled at the distance of 0.1 m from the clamped side with the same properties. The last considered model consists a delamination between two most upper layers at the distance of $0.1 \div 0.15$ m, which was modelled by deactivation of contact in this area.

During the numerical analysis the normal modes of vibration of the considered models were computed. The displacements of selected modal shapes were collected. Only first five bending modes were considered for further analysis.

3.2 Analysis of Results

A scalar and multi-objective evolutionary optimization was carried out applying MATLAB® with the Genetic Algorithm and Direct Search Toolbox. The trial and error procedure as well as heuristic rules were employed to get the best results from the evolutionary algorithm. The fitness function was declared following to (9) or (12) while the upper bounds were set as follows: $\alpha_c = 12$, $\tau_c = 12$, $r_c = 30$ basing on results of previous studies [10]. It was decided that individuals in the population are composed of genes representing real numeric values of parameters α, τ and r (the integer part of the last parameter was used during the computations). The feasible population method was adapted to create a random well-dispersed initial population that satisfies all bounds in (9). Fitness scaling was realized using the rank method, whereas the selection of the parents to the next generation was achieved by applying the stochastic uniform method. Additionally, two reproduction options – the elite count and crossover fraction p_c were chosen. The first one specifies the number of individuals that are guaranteed to survive to the next generation (it was equal 2). The second deals with the fraction of the next generation, other than elite children, that are produced by crossover. It was decided to use a simple heuristic crossover operator that returns a child that lies on the line containing the two parents in such a way that the distance between the child and the better parent is determined by the user-defined parameter (it was set to 1.2). The remaining individuals (other than elite and crossover children) are mutation children. They were obtained using the adaptive feasible method.

In this study, the convergence of the evolutionary algorithm was verified specially according to the population size N and crossover fraction p_c for the case of the beam with a single notch and the fitness function in a scalar form (12). These parameters are critical to ensure, as far as possible, to find the optimal solution of the problem. Following this point of view, several testing trails with multiple realizations have been carried out to examine the convergence of the optimization algorithm for different numbers of individuals ($N = \{5, 10, 20, 30\}$) and different values of the crossover fraction ($p_c = \{0.4, 0.5, \cdots, 1\}$). For each pair of these numbers the optimization was run ten times. Afterwards, the averaging procedure was affected with the use of the second-order exponential approximation. Obtained results for this step are demonstrated in Fig. 2(a). This plot presents the evolution of the fitness function approximation of the best individuals over the number of the evaluations (the size of populations $\times$ the iteration number).

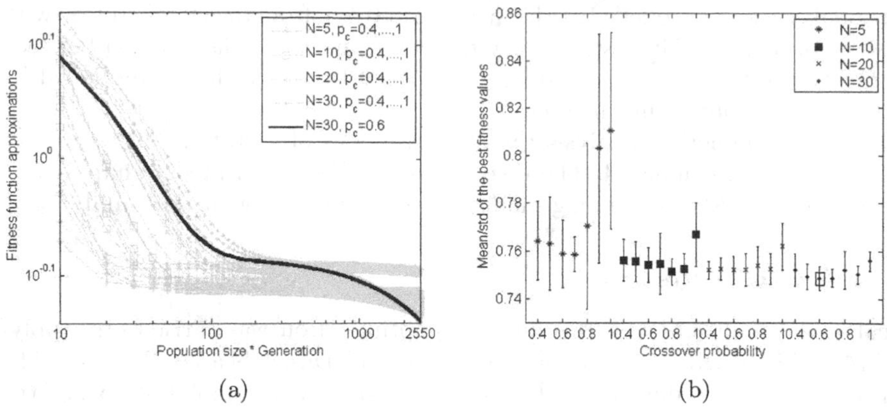

Fig. 2. The influence of the population size N and crossover fraction p_c on the performance of an evolutionary algorithm

Besides, Fig. 2(b) shows the means and standard deviations of the best fitness results for different values of N and p_c. It could be observed, that there is a global convergence around a value of 0.75 and it is obtained, in the average sense, for $p_c = 0.6$, $N = 30$ with one of the smallest standard deviation value. It could be also seen in Fig. 2(a) that for such values of these parameters the total number of generations equals 85 is enough to get the solution of the problem. From this experiment it was concluded that these values of parameters were used for further investigations.

The first example of the damage identification problem was concentrated on the case of the beam with a single notch. As it was mentioned before, the profound effect on the damage identification procedure has the scaling and wavelet functions as well as the polynomial approximation. Therefore, two different ways were taken into account in order to find the set of optimal parameters for the damage identification purpose.

In the first case, the single-objective evolutionary optimization was employed in the same manner as it was described above in the trial and error procedure. Fig. 3(a) presents the evolution of the fitness function (12) of the best and mean individuals over the number of generations. It can be observed that the algorithm was able, as far as possible, to reach the minimum value of the fitness function. The best individual in the last generation has a fitness of 0.7464 and the mean fitness of the individuals is 0.7465. It was achieved for the fractional order $\alpha = 1.0896$, the shift factor $\tau = 7.9825$ and the degree of the polynomial approximation $r = 20$. Fig. 3(b) presents that the obtained values of parameters guaranteed the detection and localization of a single notch damage the with sufficient accuracy. It should be noted here that the peak is narrow and it is located near the point of the damage position. It can be also observed, that the level of the noise is small relative to the magnitude of the peak.

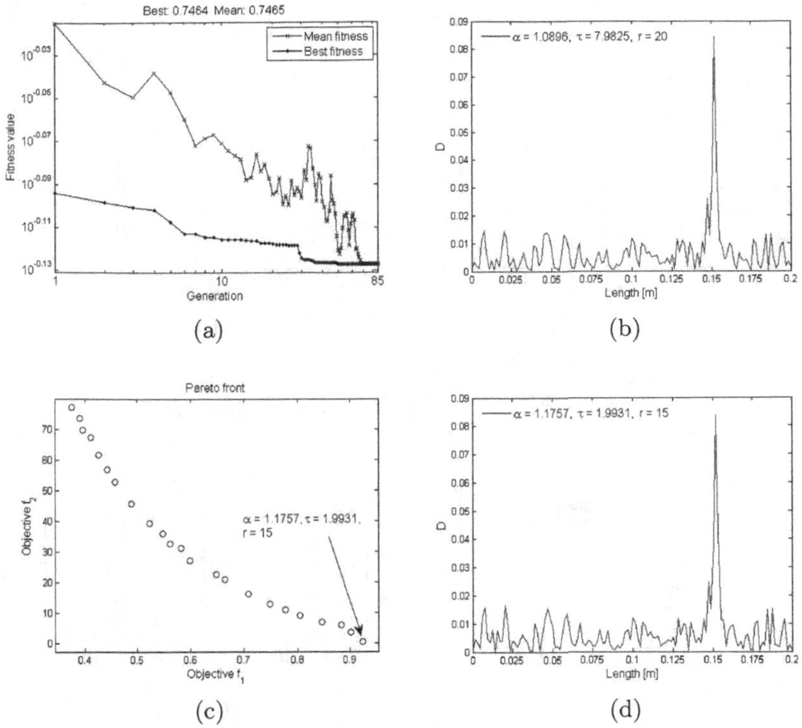

Fig. 3. Results of single notch identification for a scalar optimization (a,b) and a multi-objective optimization (c,d)

In the second case, the multi-objective evolutionary optimization was applied for searching the solution of the problem defined in the vector form (9). In this approach, basic features of the algorithm and values of its parameters were set as in the previous case. Besides that, a multi-objective evolutionary algorithm needs also to determine the Pareto fraction parameter for specifying the fraction of individuals to keep on the first Pareto front while the solver selects individuals from higher fronts. Hence, the Pareto fraction was set to 0.9. The key results from the optimization are shown in Fig. 3(c). This plot presents the Pareto front that means the set of non-dominated solutions. The optimal solution ($\alpha = 1.1757$, $\tau = 1.9931$, $r = 15$) was selected in such a way that the minimum level of the noise ($f_2 = 0.6207$) was obtained with the highest possible magnitude of the peak ($f_1 = 0.9228$). The resulted plot of D-coefficients was presented in Fig. 3(d). It is not hard to see, that these results are very similar to the previous example. The significant difference was observable only for the values of τ and r, which means that there exist more than one global optima.

The next study considered a double-notched beam. A single- and multi-objective optimization were carried out in order to find the solution to this case. However, the results achieved with the second approach are only described

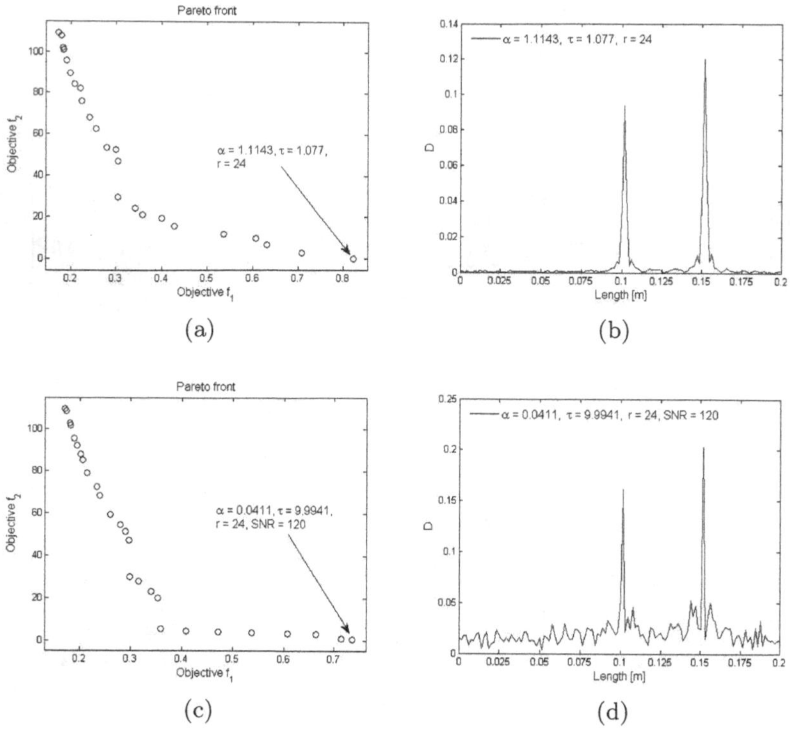

Fig. 4. Results of double notch identification with the use of a multi-objective optimization for the case without noise (a,b) and the case with noise (c,d)

here. Fig. 4(a) presents the Pareto optimal solutions, whereas Fig. 4(b) shows the result of damage identification, which was computed using the best scaling and wavelet functions. As is shown, the damage detection and localization are also possible for multiple damages of this type. In this study, there was also considered the influence of the measurement noise on the sensitivity of the damage identification method.

Results presented in Fig. 4(c,d) show that the proposed method of the damage identification could be successfully applied despite of the presence of noise. The data were noised by the normally distributed random noise on various levels. The presented cases were noised on the level of the signal-to-noise ratio of 120 dB, which is typical for the high-precision SHM measurement systems.

The last considered damage case was the delamination between two most upper layers at the distance of $0.1 \div 0.15$ m from the clamp. There are two plots presented in Fig. 5(a,b) which show the results obtained for this kind of damage. It could be observed that only the borders of the dalamination could be located, which could be explained by the specificity of the algorithm, which recognizes only the singularities (in this case the non-monotonicity of displacements). However, this plot is similar to the results obtained for the case of the beam with two

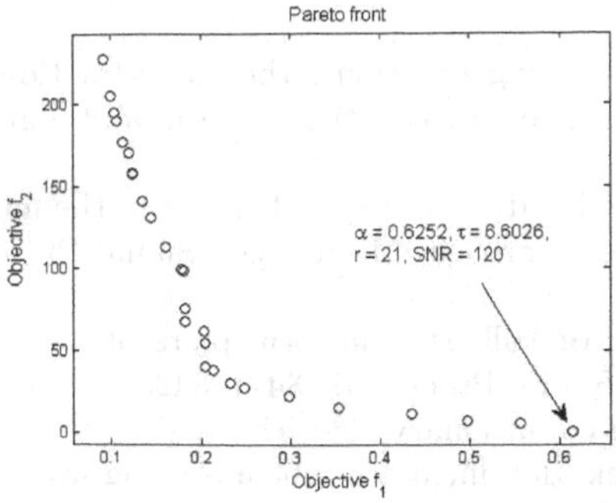

(a)

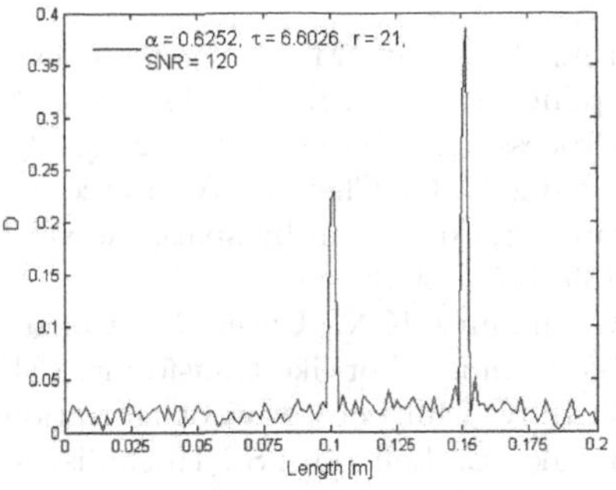

(b)

Fig. 5. Results of damage identification for the delaminated beam

notches. The damage isolation is possible by the deeper analysis of the width of the obtained peaks of D-coeficients.

4 Conclusions

The investigated problem of the damage detection, localization and isolation in polymeric composites was based on the novel approach, which used the fractional wavelet transform applied to the displacements data of the modal shapes of a beam. The application of optimally selected fractional B-spline wavelets allows to improve the sensitivity of the method, which results in a possibility of detection and localization of the damages in the early phase of their evolution. It can be summarized, that either a single- or multi-objective evolutionary optimization may be successfully employed to tune relevant parameters of the proposed method. The given examples of damage identification with the presence of noise have confirmed that the proposed method could be applied in the industrial conditions as well. The further studies will be concentrated on the improvement of the distinguishing of the damages and experimental verification of the method. A special attention should be given to the influence of measurement noise, which may dominate the artifacts in the signal caused by damages. In order to detect and localize the damages properly the high-precise measurement systems should be applied.

Acknowledgements. The research project was financed by the National Science Centre (Poland) granted according the decision No. DEC-2011/03/N/ST8/06205. The authors acknowledge a possibility of carrying out computations on the IBM BladeCenter HS21 in the Academic Computer Centre CYFRONET AGH under the computational grant No. MNiSW/IBM_BC_HS21/PŚląska/012/2013.

References

1. Blu, T., Unser, M.: A complete family of scaling functions: the (α,τ)-fractional splines. In: Proc. IEEE International Conference on Acoustics, Speech, and Signal Processing, vol. 6, pp. 421–424 (2003)
2. Chang, C.-C., Chen, L.-W.: Detection of the location and size of cracks in the multiple cracked beam by spatial wavelet based approach. Mech. Syst. Signal Pr. 19, 139–155 (2005)
3. Chaudhury, K.N., Unser, M.: Construction of Hilbert transform pairs of wavelet bases and Gabor-like transforms. IEEE T. Signal Proces. 57, 3411–3425 (2009)
4. Deb, K.: Multi-objective optimization using evolutionary algorithms. Wiley (2009)
5. Douka, E., Loutridis, S., Trochidis, A.: Crack identification in beams using wavelet analysis. Int. J. Solids Struct. 40, 3557–3569 (2003)
6. Douka, E., Loutridis, S., Trochidis, A.: Crack identification in double-cracked beams using wavelet analysis. J. Sound Vib. 277, 1025–1039 (2004)
7. Dumont, J., Hernandez, A., Carrault, G.: Improving ECG Beats Delineation With an Evolutionary Optimization Process. IEEE Trans. Biomed. Engineering 57(3), 607–615 (2010)
8. Katunin, A.: Identification of multiple cracks in composite beams using discrete wavelet transform. Sci. Probl. Mach. Oper. Maint. 45, 41–52 (2010)
9. Katunin, A.: The construction of high-order B-spline wavelets and their decomposition relations for fault detection and localisation in composite beams. Sci. Probl. Mach. Oper. Maint. 46, 43–59 (2011)
10. Katunin, A.: Damage identification in composite plates using two-dimensional B-spline wavelets. Mech. Syst. Signal Pr. 25, 3153–3167 (2011)
11. Katunin, A.: Crack identification in composite beam using causal B-spline wavelets of fractional order. Model. Eng. (in press)
12. Li, Z., Xia, S., Wang, J., Su, X.: Damage detection of cracked beams based on wavelet transform. Int. J. Impact Eng. 32, 1190–1200 (2006)
13. Rafiee, J., Tse, P.W., Harifi, A., Sadeghi, M.H.: A novel technique for selecting mother wavelet function using an intelligent fault diagnosis system. Expert Syst. Appl. 36, 4862–4875 (2009)
14. Unser, M., Blu, T.: Fractional splines and wavelets. SIAM Rev. 42, 43–67 (2000)
15. Unser, M., Aldroubi, A., Eden, M.: On the asymptotic convergence of B-spline wavelets to Gabor functions. IEEE T. Inform. Theory 38, 864–872 (1992)
16. Wu, N., Wang, Q.: Experimental studies on damage detection of beam structures with wavelet transform. Int. J. Eng. Sci. 49, 253–261 (2011)
17. Xiang, J., Liang, M.: A two-step approach to multi-damage detection for plate structures. Eng. Fract. Mech. 91, 73–86 (2012)

Control Allocation Using Reliability Measures for Over-Actuated System

Frederique Bicking, Philippe Weber, Didier Theilliol, and Christophe Aubrun

Université de Lorraine, CRAN - CNRS UMR 7039
Faculte des Sciences et Technologies - B.P. 70239, 54506 Vandoeuvre-les-Nancy, France
{Frederique.Bicking,Philippe.Weber,Didier.Theilliol, Christophe.Aubrun}@univ-lorraine.fr

Abstract. An optimal control law synthesis conditioned by the reliability importance measures and degradation of components in the presence of failures is presented. The aim is to preserve the health of the actuators and the availability of over-actuated system both in the nominal situation and in the presence of some actuator failures. The reliability importance measures are computed by mean of a Bayesian Network (BN) and degradation is modeled by a Dynamic Bayesian Network (DBN) since BN and DBN are well suited to compute actuators reliability through the bayesian inference. It is applied on a system to estimate its reliability and those of its components and provide the parameters to synthesize the control laws. The performance of the developed method is illustrated on an hydraulic system.

Keywords: Reliability Importance Factors, Control allocation, Bayesian Network.

1 Introduction

In order to satisfy the increasing economic demand for high plant availability and system safety, dependability is an essential need for industrial automation. In this context Fault-Tolerant Control (FTC) is introduced. A FTC is a system which have the capability of detecting the occurrence of faults and retaining satisfactory system performance in the presence of faults. In certain cases, it may be necessary to modify the objectives of the system in order to maintain some critical performance level [1]. Among the various methods of FTC allocation [15,16], few methods only consider the dependability of the components. Authors in [17,18] introduce some degradation model and reliability measures as indicators to elaborate the control allocation.

Dependability analysis of a system with several components allows to determine the relationship between the performance level of the system and the performance levels of its components. The reliability importance measures are one way to characterize this relationship. Reliability importance factors allow to rank the components according to their contribution to the functioning of the system. System reliability is usually computed through the Structure Function (SF). The limit encountered using SF is the system modeling complexity. Bayesian Networks (BN), used as a modeling method, are able to compute the reliability taking into account observations (evidences) about the state of the components [13] and are useful to compute the reliability importance measures.

J. Korbicz and M. Kowal (eds.), *Intelligent Systems in Technical and Medical Diagnostics*, Advances in Intelligent Systems and Computing 230,
DOI: 10.1007/978-3-642-39881-0_41, © Springer-Verlag Berlin Heidelberg 2014

Section 2 of this paper presents the control allocation problem statement. Section 3 introduces the reliability importance factors definitions and their computation with a BN. In section 4, health indicators and their computation with a Dynamic Bayesian Network (DBN) are described. Section 5 and 6 are dedicated to the application example. Finally conclusions are given in section 7.

2 Control Allocation Problem Statement

From a general point of view, let us consider a LTI system described by:

$$\begin{cases} \dot{x}(t) = Ax(t) + B_u u(t) \\ y(t) = Cx(t) \end{cases} \tag{1}$$

where $A \in \mathbb{R}^{n \times n}$, $B \in \mathbb{R}^{n \times m}$ and $C \in \mathbb{R}^{p \times n}$ are respectively, the state, the control and the output matrices. $x \in \mathbb{R}^n$ is the system state, $u \in \mathbb{R}^m$ is the control input, $y \in \mathbb{R}^p$ is the system output and (A, B_u) is stabilizable. Control allocation is generally used for over-actuated systems, where the number of operable control is greater than the controlled variables. Let us assume that $rank(B_u) = k < m$. This implies that B_u can be factorized as:

$$B_u = B_v B$$

where $B_v \in \mathbb{R}^{n \times k}$ and $B \in \mathbb{R}^{k \times m}$. A virtual actuator is computed through a classical controller synthesis and classical control allocation optimization methods can be considered in order to distribute the load to each actuator (see [2] or [3]).

As proposed recently by [4] and implemented on a test bed [18] , another state space representation of (1) can be considered such as:

$$\begin{cases} \dot{x}(t) = Ax(t) + B_v v(t) \\ v(t) = Bu(t) \\ y(t) = Cx(t) \end{cases} \tag{2}$$

where $v \in \mathbb{R}^k$ is the virtual control input, called as the total control effort produced by the actuators and defined by the controller.

The control allocation problem can be expressed as a constrained linear mapping problem based on the relationship:

$$v(t) = Bu(t) \tag{3}$$

$$u_{\min} \leq u \leq u_{\max} \tag{4}$$

where (4) represents the physical limits of the actuator controls. Optimized based control allocation methods aim to find an optimal solution. If there is no exact solution, the optimal control is the feasible one such that $Bu(t)$ approximates $v(t)$ as close as possible. The optimal control input can be obtained by minimizing:

$$\Psi = \underset{u_{\min} \leq u \leq u_{\max}}{\arg\min} \frac{1}{2} \|Bu - v\|_2 \tag{5}$$

$$u = \arg\min_{u \in \Psi} \|W_u(u - u_d)\|_2 \tag{6}$$

where Ψ is the set of feasible solutions subject to the objective criterion (6). u_d is the desired control input. The weighting matrix $W_u \in \mathbb{R}^{m \times m} \succ 0$ is used to give a specific priority level to the actuators.

In order to preserve the safety of the system and to keep the set of the actuators available as long as possible, a specific choice of the weighting matrix W has been proposed based on the actuators reliability [5]. The weighing matrix W_u is considered as a key to manage the over-actuators and contributes the system reliability improvement. However, in order to perform simultaneously the control law synthesis and to solve the control allocation problem, the desired efforts have to be distributed.

The optimal control input $u^* = (u_1^*, u_2^*, \ldots, u_m^*)$ solution to the control allocation problem (5) and (6) is defined according to the values of the weighting matrix as follows:

$$W_u(t) = \begin{pmatrix} w_1(t) & & & 0 \\ & w_2(t) & & \\ & & \ddots & \\ 0 & & & w_m(t) \end{pmatrix} \succ 0 \tag{7}$$

where $w_i(t)$ depends on reliability measures and health indicators of system and actuators states.

This paper is devoted to the study of $W_u(t)$ behavior according to the variation of control input considering the systems input/output constant.

3 Reliability Importance Measures for Control Allocation

Reliability importance measures are used to evaluate the impact of the components on the system reliability. The evaluation of these importance factors is related to the probability of functioning (or malfunctioning) of the system according to the state of the components. In control design systems, it will be interesting to pay attention to the reliability importance factors of the components that indicates which component (actuator) is the most suitable to be used according to the system failure "immunity" regarding to this component [1].

Let us consider the actuator $i \in \{1 \ldots m\}$ for which the state is defined as a random variable E_i with two states $\{Up, Down\}$ and S is the system state with also two states $\{Up, Down\}$. We define $\mathbf{e} = (e_1, \ldots, e_i, \ldots, e_m)$ as the state vector of the m components with $e_i = 1$ also noted 1_i, component i is functioning else $e_i = 0$ or 0_i the component i is failed. The probability $p_i = Pr\{E_i = 1\}$ is the reliability of the component i and $R(\mathbf{p}) = Pr\{S = Up\}$ the system reliability with $\mathbf{p} = (p_1, \ldots, p_i, \ldots, p_m)$. The probabilities vector $\mathbf{p}$ can be fixed a priori or computed according to the structure function of the system reliability or with an analytical expression of the system reliability.

In the next subsection, the reliability measures used to compute the matrix $W_u(t)$ (equation 7) are defined and their computation is explained.

3.1 Reliability Importance Factors Definition

Authors in [22] have reviewed importance measures for coherent binary systems. In this paper, we focus on reliability importance factors in the context of systems with components stochastically independent.

The Marginal Importance Factor (MIF), introduced by Birnbaum [6], is defined as a measure of the variation of the system reliability according to the reliability of a component. The marginal important factor of the component i can be written as:

$$MIF(i;\mathbf{p}) = \frac{\partial R(\mathbf{p})}{\partial p_i} = R(1_i;\mathbf{p}) - R(0_i;\mathbf{p}) \tag{8}$$

Eq. 8 evaluates how fast the system reliability increases according to the increase of the component reliability. So, to increase the system reliability, the reliability of the component having the greatest MIF value should be increased.

Taking into account the MIF of each component in the control allocation strategy means that the greatest MIF of a component is the smallest load level the component will have. This will be defined by the load level indicator, noted ll_i:

$$ll_i = (1 - MIF(i;\mathbf{p})) \tag{9}$$

The load of component i depends on its load level indicator. However, the component load has to depend also on the component state and/or its dependability or reliability at time t. This dependability measure makes sense only if the system has not failed.

The Diagnostic Importance Factor (DIF), defined by Fussel-Vesely [7,8] is a time dependant reliability importance of component i. DIF is defined as the probability that the functioning of component i contributes to the functioning of the system given that the system is not failed. Assuming that a mission time is implicit and fixed, the DIF can be defined as follows:

$$DIF(i;\mathbf{p}) = Pr\{E_i = 1/S = Up\} \tag{10}$$

The more the $DIF(i;\mathbf{p})$ is closed to 1, the stronger the contribution of the component i to the system functioning is. So this component should be less loaded.

MIF and DIF are classically computed using the structure function of the system. In this paper, a Bayesian Network (BN) is used. The variable handled by BN are not limited to Boolean random variables contrary to the standard reliability analysis as Fault Tree or Reliability Block Diagram [12]. The principles of this modeling tool are explained in [9].

3.2 MIF and DIF Computation with BN

Bayesian networks (BN) [9] are directed acyclic graphs representing a joint probability density function (Pdf). Nodes represent variables and directed arcs indicate causal relations between the connected nodes. The existence of arc between two variables indicates a direct probability dependency between them. For more details, readers can refer to [10].

Node that has no parents is described by a marginal distribution: unconditional probability. So it is characterized by a probability table (PT) composed of the class belonging

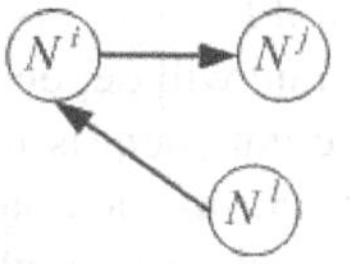

Fig. 1. Example of BN

probabilities $p(N^i)$ over all the possible states of N_i. Nodes having at least one parent (children nodes) are characterized by a conditional probability distribution. So, Conditional Probabilities Tables (CPT) are associated with these types of nodes. PT and CPT together with the independence assumption defined by the graph present a unique joint distribution over all the variables. Thus, it is possible to quantify probabilities for each variable conditioned by all the possible values of the others variables in the graph.

The Marginal Important Factor of the component i can be written in Bayesian formalism as:

$$MIF(i;\mathbf{p}) = Pr\{S = Up/E_i = 1\} - Pr\{S = Up/E_i = 0\} \tag{11}$$

Eq. 11 sets that the MIF factor of the component i in a coherent system can be defined as the probability that the failure of i induces the failure of the system S.

The Diagnostic Importance Factor is computed by setting the observation that the system is not failed ($S = Up$) and collecting the probability defined by (10).

4 Health Indicators for Control Allocation

As said previously, in control design systems, only few authors pay attention to the degradation of the components (actuators) of the system and their impact to the system reliability. In our case, the actuators are used in the control strategy proportionally to their ability and necessity of functioning.

4.1 Component Degradation Model with Load Variation

The operating conditions of a particular component depend on the applied load [20]. Several mathematical models have been developed to estimate the component aging in the nominal functional mode [21]. In this work, the 3-parameters Weibull distribution function is considered and the failure rate of a component is defined as follows:

$$\lambda(t) = \frac{\beta(t-\gamma)^{\beta-1}}{\eta^{\beta}} \tag{12}$$

where η is the scale parameter, β is the shape parameter and γ the location parameter. In many cases, the location parameter may be assumed to be zero, but we will take a closer look at this location parameter and show how it can be used in FTC allocation context. As the name implies, the location parameter, γ, locates the distribution along the time. Changing the value of γ has the effect of "sliding" the distribution and its associated function. As the life period t to $t+\gamma$ is a failure free operating period, we consider that the component failure rate will depend on the process load with an effect on the location parameter γ. If the component is not loaded, its aging is stopped and its failure rate becomes equal to zero during the non operating period $\gamma_i(t+\Delta t)-\gamma_i(t)$. When the component is loaded, the failure rate evolves and depends on the location parameter variation (eq. 13).

$$\lambda_{i,u(k)}(k+1)=\frac{\beta_i(k-\gamma_{i,u(k)}(k+1))^{\beta_i-1}}{\eta_i^{\beta_i}} \tag{13}$$

The load impact on the failure rate is managed by means of the location parameter. The subscript u denotes that the failure rate depends on the input control u^*.

4.2 Dynamic Bayesian Network for the Component Degradation Computation

DBN includes a temporal dimension managed by time-indexed random variables. The process is represented at time step kT, denoted k thereafter, by a node N_k^i with a finite number of states (2 states in our case) and arcs represent dependencies across time points. Let $p(N_k^i)$ denotes the probability distribution over the variables states at time step k. Nodes corresponding to the state variables can be partitioned into two sets: those corresponding to the state of the variable at the current time step k and those corresponding to system state at the following time step $k+1$. In this case, the variable is represented at successive times. In the DBN, shows in (Fig. 2), the temporal evolution is represented in two slices of time k and $k+1$. Defining the transition-probabilities between the states of the variable at time step k and time step $k+1$ leads to define CPT relatively to inter-time slices. So, given any $p(N_k^i)$ at time step k, the network induces a unique distribution $p(N_k^i)$ at time $k+1$ which is calculated by inference. Starting at time $k=0$ with distribution $p(N_0^i)$, the computation of $p(N_T^i)$ can be realized by iterative inferences.

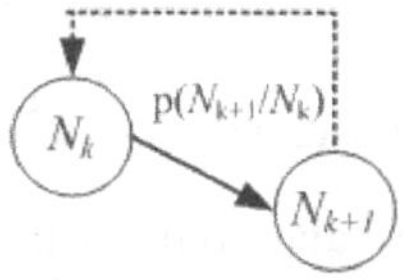

Fig. 2. A DBN modelling MC or semi MC with node N_k

The knowledge of the distribution probabilities and the CPT allows the calculation of the distribution probabilities at time $k+1$ by making inference. So, with this model, the future $(k+1)$ is conditionally independent of the past given the present (k), which means that the CPT or $p(N_{k+1}/N_k)$ respects the Markov property [11]. In the classical MC representation, parameters are calculated for all the possibilities of states variables leading to the state space Ω, and the notion of independence between variables is unexploited. Contrary to this, DBN allows to take into account this consideration to represent joint Pdf in a factorized manner. In real world application, the number of system states is very important, so a matrix representation is untractable and usually unfeasible. The use of DBN modeling to analyze the deterioration system is more suitable to solve this problem, it permits to minimize considerably the number of parameters of the joint Pdf [14].

4.3 DBN Model Computation of Weighting Matrix for Control Allocation

The system is modeled by a DBN. Probabilities of each component is set a priori at the beginning. At each step time, the system reliability is computed by the inference algorithm of the DBN [9]. The MIF and DIF factors of each component are computed using equations 11 and 10 as explained in [19]. The load level of each component is deduced and the load of the component i, (w_i), is calculated. w_i is given by the product of its load level indicator ll_i and its DIF:

$$w_i(t) = ll_i(t) \times DIF(i;\mathbf{p}) \tag{14}$$

Remember that w_i integrates the degradation of the component because its failure rate $\lambda(t)$ is also re-evaluated on line.

Next, $W_u(t)$, defined in eq. 7, is re-estimated and changed on-line according to the reliability importance measures and health indicators of each component/actuator. A new distribution of the desired control efforts is obtained. Therefore, if an actuator i is unavailable (actuator out of order) $w_i(t) = 0$, then it assumes that the system is still working because it is an over-actuated system.

In failure case, the component failure is introduced in the DBN by setting the observation $Pr\{E_i = 0\}$. Thanks to the inference algorithm, the probabilities of other components will be modified accordingly. The failure rates of the remaining components are also re-evaluated.

The weighting matrix W_u is re-estimated resulting in a new a control allocation.

5 Application

5.1 System Description

Let us consider a system with four actuators. The corresponding Reliability Block Diagram is depicted on figure 3. The system is considered coherent, non repairable, with binary state $(\{Up, Down\})$.

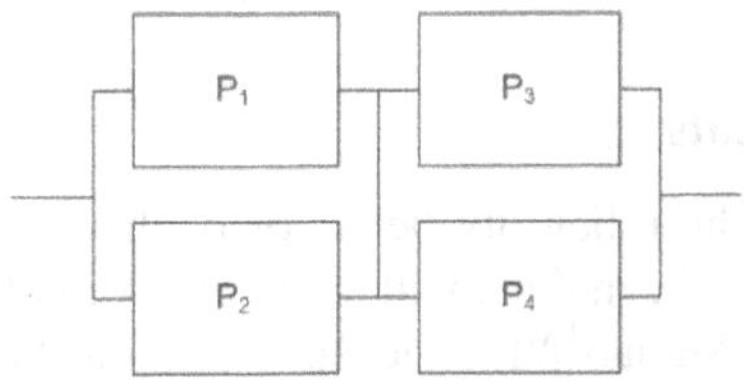

Fig. 3. System Reliability Block Diagram

The main objective is to determine the input control allocation in order to assume the operating mission using a dependability criteria. Reliability importance measures (MIF and DIF) and health indicator (aging, load) are used as a help to the input control allocation method.

5.2 BN Synthesis and Synoptic

The BN reliability model is given in Fig. 4. The nodes on the left part of the figure model the reliability of each active component as a DBN and the availability of the actuators. The node ANDi and ORi models the combination operator between the components to compute the system reliability $R(\mathbf{p})$ knowing the success paths $\mathscr{P}_a = \{P1, P4\}; \mathscr{P}_b = \{P1, P3\}; \mathscr{P}_c = \{P2, P4\}$ et $\mathscr{P}_d = \{P2, P3\}$.

The temporal aspect in the BN is represented, on the right part by the link $Pi(k)$ to $Pi(k+1)$ that is a semi markov chain with a transition rate equal to the failure rate at time t (eq. 13). The conditional probabilities table (CPT) for each component i is:

$$\begin{bmatrix} 1-\lambda_{i,u}(k) & \lambda_{i,u}(k) \\ 0 & 1 \end{bmatrix} \tag{15}$$

The BN part is developed thanks to Bayesialab^{TM}some Bayesian Engines dedicated software in association with an Application Program Interface (API) (`http://www.bayesia.com`).

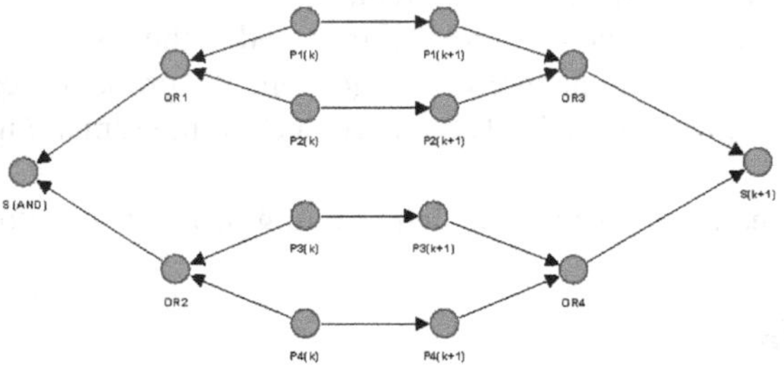

Fig. 4. Dynamic Bayesian Network (DBN) of the system

Each component reliability is set for each node P$i(k)$ at initial time step $k = 0$. At each step time, MIF, DIF and loads levels are computed using the DBN, the matrix $W_u(t)$ is evaluated and the input control allocation is determined. The resulting reliabilities are then computed with the inference in the DBN.

5.3 Results and Comments

In this part, we focus on the reliability behavior of the components and the system. So, several simulations have been done with different Weibull parameters for the components. It is done under Matlab^{TM}for the optimization, the control and the system simulation. At the beginning, a failure-free case is considered and next a failure is introduced.

Failure-free Case

Actuators P1 and P3 (resp. P2 and P4) have the same Weibull parameters:
$\beta_1 = \beta_3 = 2.1; \eta_1 = \eta_3 = 1000$; and $\beta_2 = \beta_4 = 3; \eta_2 = \eta_4 = 400$.
On fig. 5a, the load levels are different for P1 and P3 (resp. P2 and P4) according to the Weibull model. The loads, w_i on fig. 5c, are the same for P1 and P3 (resp. P2 and P4) until $t = 700$h. After $t = 700$h, the loads of P2 and P4 decrease faster than the loads of P1 and P3 consequently to the variation of their DIF (fig. 5b). System and components reliabilities are also depicted on fig. 5d).

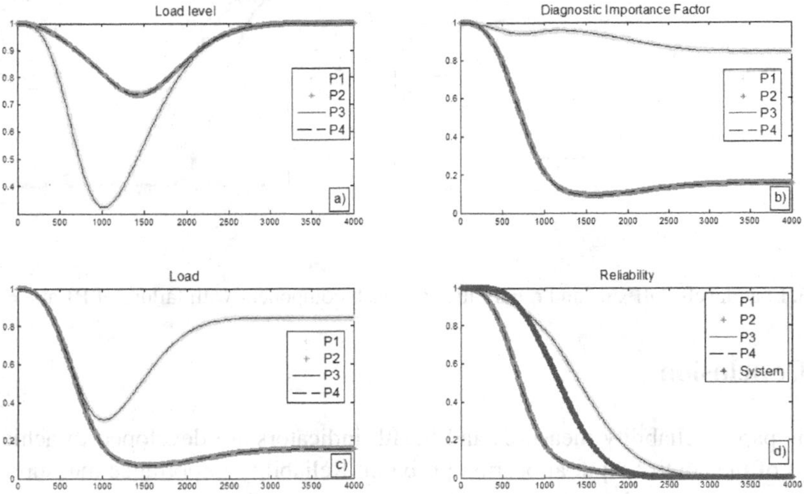

Fig. 5. Load levels, DIF , w_i and reliabilities for each component in failure-free case

Failure Case

Actuators P1 and P3 (resp. P2 and P4) have the same Weibull parameters:
$\beta_1 = \beta_3 = 2.1; \eta_1 = \eta_3 = 1000$; and $\beta_2 = \beta_4 = 3; \eta_2 = \eta_4 = 400$. A failure is assumed to occur on P1 at $t = 1000$h.

Before $t = 1000$h, the load levels (fig. 6a) of actuators P1 and P3 are the same as those of actuators P2 and P4 which is consistent since MIF of P1 and MIF of P3 (resp. MIF of P2 and MIF of P4) are the same. Unfortunately after $t = 1000$h, actuator $P1$ is out of order, consequently the load level becomes equal to zero as well as the load (fig. 6c) . The load level of P2 evolves according to its MIF which only depends on the reliabilities of P3 and P4. For these last actuators, the load levels of P3 (resp. P4), depending only on reliability of actuators P2 and P4 (and P3) rapidly increase.

The DIF of each component (fig. 6b), computed by the BN at each time step shows that P2 becomes a critical component ($DIF_2 = 1$). The system is reconfigured using the success paths with components P2, P3, P4. Figure 6d shows clearly that the system reliability drop toward zero due to the failure of P1. The reliability of P4 is immediately impacted due to its low DIF.

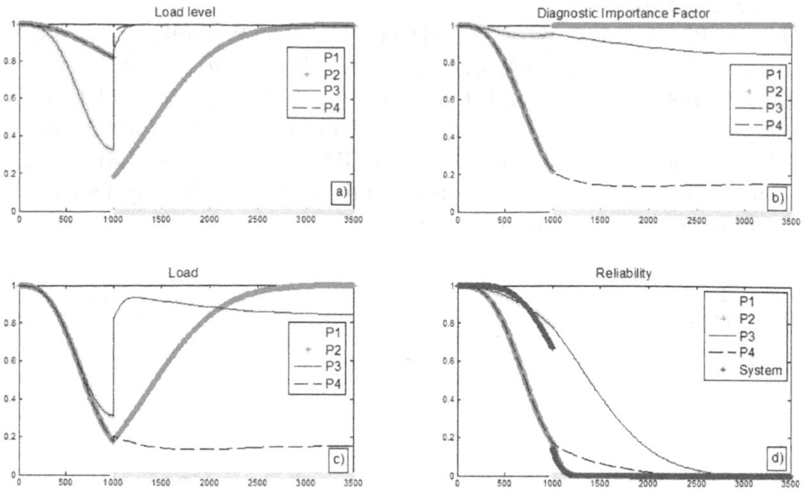

Fig. 6. Load levels, DIF , w_i and reliabilities for each component with failure of P1 at $t = 1000$h

6 Conclusion

In this paper, reliability measures and health indicators are developed to achieve the design of the control input allocation problem. Reliability importance measures allow to take into account the importance of a component in the functioning of the system. Health indicator assumes to be representative of the degradation of the components when loads are applied or not. Therefore, a generic BN used to easily model the reliability of over-actuated system is also used to compute importance measures and the functioning probabilities of each actuator is computed with a DBN inference. This method provides a control re-allocation solution that is based on the on-line actuators reliability estimation. Given that it exists nowadays smart / intelligent actuators with self diagnostic, this technology could be used in our scheme. The delivered information from self diagnostic actuator could be integrated as evidence in the DBN model of the system and so will contribute to the control re-allocation.

References

1. Blanke, M., Kinnaert, M., Lunze, J., Staroswiecki, M.: Diagnosis and Fault Tolerant Control. Control Systems Series. Springer, London (2006)
2. Durham, W.: Constrained control allocation. Journal of Guidance, Control, and Dynamics 16, 717–725 (1993)
3. Harkegard, O., Glad, S.: Resolving actuator redundancy-optimal control vs. control allocation. Automatica 41, 137–144 (2004)
4. Picasso, B., Vito, D.D., Scattolini, R., Colaneri, P.: An MPC approach to the design of two-layer hierarchical control system. Automatica 46, 823–831 (2010)

5. Khelassi, A., Weber, P., Theilliol, D.: Reconfigurable Control Design for Over-actuated Systems based on Reliability Indicators Conference on Control and Fault-Tolerant Systems, SysTol (2010)
6. Birnbaum, Z., Krishnaiah, P.R.: On the importance of different components in a multicomponent system Multivariate Ananlysis II. Academic Press (1969)
7. Fussell, J.B.: How to hand-calculate system reliability charateristics. IEEE Transactions on Reliability 24, 169–174 (1975)
8. Vesely, W.: A time dependent methodology for fault tree evaluation. Nuclear Engineering and Design 13, 337–360 (1970)
9. Jensen, F.V.: An introduction to Bayesian Networks. Springer, New York (1996)
10. Jensen, F.V.: Bayesian Networks and Decision Graphs. Springer, New York (2001)
11. Kjaerulff, U.: dHugin: a computational system for dynamic time-sliced Bayesian networks. Internationnal Journal of Forecasting 11, 89–111 (1995)
12. Weber, P., Medina-Olivia, G., Simon, C., Iung, B.: Overview on Bayesian networks Applications for Dependability, Risk Analysis and Maintenance areas. Engineering Applications of Artificial Intelligence 25(4), 671–682 (2012)
13. Weber, P., Jouffe, L.: Reliability modelling with Dynamic Bayesian Networks. In: 5th IFAC Symposium on Fault Detection, Supervision and Safety of Technical Processes, Washington, D.C., USA, June 9-11 (2003)
14. Weber, P., Munteanu, P., Jouffe, L.: Dynamic Bayesian Networks modelling the dependability of systems with degradations and exogenous constraints. In: 11th IFAC Symposium on Information Control Problems in Manufacturing, INCOM 2004, Salvador-Bahia, Brazil, April 5-7 (2004)
15. Cassavola, A., Garone, E.: Adaptive fault tolerant actuator allocation for overactuated plants. In: American Control Conference, New York city, pp. 3985–3989 (2007)
16. Staroswiecki, M., Berdjag, D.: A general fault tolerant linear quadratic control strategy under actuator outages. Internal Journal of Systems Science 41(8), 971–985 (2010)
17. Khelassi, A., Weber, P., Theilliol, D.: Reconfigurable Control Design for Over-actuated Systems based on Reliability Indicators, Conference on Control and Fault-Tolerant Systems (SysTol 2010), Nice, France (2010)
18. Chamseddine, A., Sadeghzadeh, I., Zhang, Y., Theilliol, D., Khelassi, A.: Control Allocation for a Modified Quadrotor Helicopter Based on Reliability Analysis. In: AIAA Intelligent Autonomy for Space and Unmanned Systems Grove, CA, USA, June 19-21 (2012)
19. Bicking, F., Simon, C., Mechri, W.: Mesures d'importance par reseaux bayesiens. In: Qualita 2013, Compiegne, France (2013)
20. Toscano, R., Lyonnet, P.: On-line reliability prediction via dynamic failure rate model. IEEE Trans. on Reliability 57, 452–457 (2008)
21. Gertsbakh, I.: Reliability theory with applications to preventive maintenance. Springer (2000)
22. Kuo, W.K., Zhu, X.: Importance measures in reliability, risk and optimization: Principles and applications. Wiley and Sons Ltd. (2012)

Human Factors and Functional Safety Analysis in Designing the Control Rooms of Industrial Hazardous Plants

Kazimierz T. Kosmowski

Gdansk University of Technology, Faculty of Electrical and Control Eng., Poland
k.kosmowski@ely.pg.gda.pl

Abstract. In this work some aspects of human factors and functional safety analysis in designing the control rooms of industrial hazardous plants are presented. The "defence in depth" (D-in-D) concept is outlined. Some designing issues of the alarm system (AS) with regard to human factors to be supported using the human reliability analysis (HRA) are discussed. The AS and decision support system (DSS), when properly designed, would contribute to decreasing the human error probability (HEP) in various abnormal situations of the plant. The aim of research undertaken is to develop the methods and software tools consisting of the risk models and probabilistic models with relevant data and knowledge bases for computer aided assessment of human factors in performing HRA as well as optimizing the architectures of safety-related systems.

Keywords: Hazardous plants, defence in depths, instrumentation and control, functional safety, human factors, procedures, tasks, human reliability analysis.

1 Introduction

The research on the causes of industrial accidents indicate that broadly understood human errors, resulting often from organizational inadequacies, are the main determining factors in 70-90% of cases [15], depending on industrial sector and the plant category. Because several defences against potential accidents are used in hazardous plants to protect people and environment, it is clear that multiple faults have contributed to most of industrial accidents.

It has been emphasized that accidents usually arose from a combination of latent and active human errors. They are to be committed during the design, operation and maintenance [18], [19]. The characteristic of latent errors is that they do not immediately degrade the safety functions, but in combination with other events, such as random equipment failures, internal or external disturbances and active human errors, may contribute to major accident with serious consequences. Some categorizations of human actions and related errors have been proposed in references [11], [18], [20].

J. Korbicz and M. Kowal (eds.), *Intelligent Systems in Technical and Medical Diagnostics*, Advances in Intelligent Systems and Computing 230,
DOI: 10.1007/978-3-642-39881-0_42, © Springer-Verlag Berlin Heidelberg 2014

Traditionally, potential human and organisational influences in industrial plants are to be incorporated into the probabilistic models through the failure events with relevant probabilities evaluated using a method of human reliability analysis (HRA) [18]. Careful analysis of expected human behaviour (including context oriented diagnosis, decision making and intentional actions) and potential errors is an essential prerequisite of correct risk assessment and rational safety-related decision making, particularly in dynamic situations. The probabilities of the failure events depend significantly on various human, organisational, environmental and technical factors being categorised as a set of performance shaping factors (PSFs) relevant to the situation under consideration [21].

Nowadays the operators supervise the process and make decisions based on information from the alarm system (AS) [2] and the decision support system (DSS) being in fact a safety-related advisory software [3]. They should be designed especially carefully for abnormal situations and potential accidents, also for cases of faults and failures within the electric/electronic/programmable electronic (E/E/PE) systems [7] or the safety instrumented systems (SISs) [8]. The AS and DSS, when properly designed, would contribute to decreasing of the HEP in various plant states and thus reducing the risk of potential accidents with serious consequences.

The aim of this article is to present the current issues of human factors and functional safety analysis in the context of the control rooms designing for industrial hazardous plants. There are still many challenges to deal with [13], [16], also those related to introducing the programmable safety-related systems in the process industry [8] and digital or hybrid technology of the instrumentation and control (I&C) systems [9] in the nuclear power plants [5], [6].

2 Defence in Depth and Defining the Main Safety Functions

In Fig. 1 a concept of defence in depth of hazardous plants is presented. The primary means of preventing and mitigating the consequences of accidents is "defence in depth" (D-in-D) that is implemented primarily through combination of a number of consecutive and preferably fully independent levels of protections that would have to fail before harmful effect could be caused to individual, people or to environment [6], [17]. Five layers of D-in-D are illustrated in this figure that:

- Prevent disturbances, system failures and deviations from normal operations, and keep installation integrity;
- Detect and intercept deviations from normal operating states to prevent anticipated operational occurrences from escalating to accident conditions;
- Control the consequences of accident conditions;
- Confine toxic or radioactive material in the event of severe accidents;
- Mitigate the consequences of toxic or radioactive release.

The Instrumentation and Control (I&C) systems support each of the above D-in-D levels of relevant layers/barriers designed. In traditional I&C designs, different systems support usually one of the defence lines. Strong independence should be provided between safety-related (S-R) systems implementing safety functions. The actuation systems of the engineering safety features (ESFs) and the reactor trip (RT) system use different actuation logics. In addition, the signal and functional diversity are to be provided in new plants so that shared data and environment would not jeopardize multiple lines of defence [5].

The design of computer-based I&C systems face now new issues which, if not properly dealt with, may jeopardize independence between lines of defence or independence between redundant elements within a line of defence [7], [9]. The architecture of most computer-based I&C systems is fundamentally different from that of traditional I&C [6].

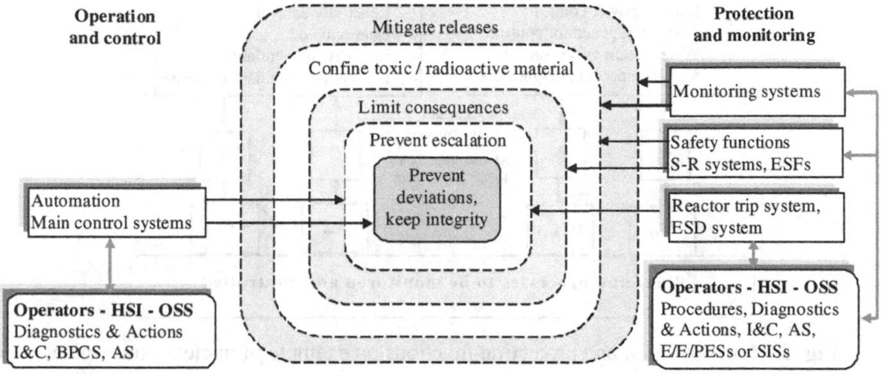

Fig. 1. A concept of defence in depth in hazardous plants

The human system interface (HSI) and the operator support system (OSS) must be designed with regard to relevant methods of human factors engineering (HFE) to be functionally effective, reliable and safe [18]. In the process industry sector the safety functions are implemented using the safety instrumented systems (SISs) [8] The SIS may perform in particular a function of emergency shut-down (ESD). The main control systems are named the basic process control systems (BPCS) [7], [8]. The alarm system (AS) can be design within BPCS or as a separated AS having own sensor subsystem, logical subsystem and indicators [2, 15].

In a computer-based systems a failure of an individual component affects not one, but many functions and may degrade operation of the I&C supporting two or more lines of defence. The scope of failures in computer-based systems may therefore be greater than in traditional systems, unless the computer-based system is carefully designed to avoid such situations and analysed to identify potential vulnerabilities with confirmation that they have been appropriately addressed [6].

Fig. 2 shows a simplified functional overview of the I&C in a NPP. To ensure a safe and reliable plant operation under all plant conditions, the I&C systems have to monitor and control hundreds or thousands of plant parameters. Thus, nuclear power plant I&C systems are very complex. Subdividing the plant I&C according to its functions facilitates understanding and designing of the entire system. An important role plays the human system interface (HIS) that makes the plant state supervision and control more effective and reliable by human operators [6]. In case of an abnormal or accident situation the operators have to follow relevant procedure developed before by a group of experienced engineers and operators. Nowadays the procedures are computerised to be used in the context of abnormal event or accident situation.

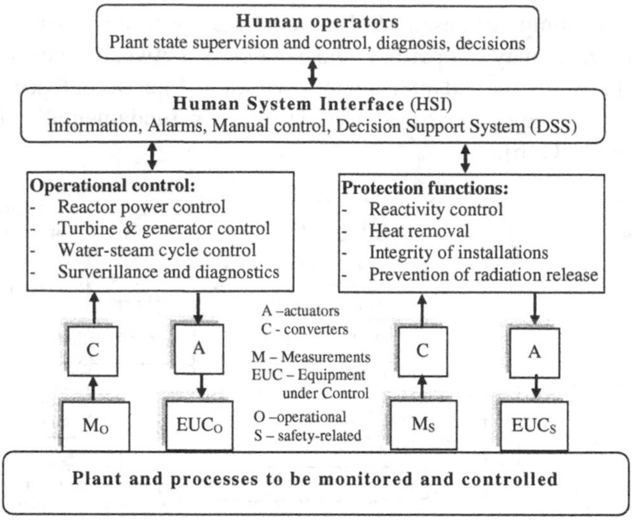

Fig. 2. Main operation and protection functions on example of nuclear power plant

3 Human-Centered Design of the Control Room

Fig. 3 shows the human operators of hazardous industrial plant and their activity context. Several goals have been distinguished in industrial practice, such as the organisation strategy and goals, current plant performance goals, operational and safety goals. The operators must take into account the current operation, diagnostic and preventive maintenance (PM) plans. They control the plant using the human system interface (HIS) and the operator support system (OSS) that can be divided into operation-related (O-R) and safety-related (S-R). A safety function to achieve given safety-related goal can be subdivided into tasks and activities of two kinds (Fig. 4): reading displays for detecting abnormalities and diagnosis the plant state based on a set of symptoms, and undertaking current decision and action to control a plant state.

Depending on the complexity of the tasks or function, there can be several levels of activities distinguished (Fig. 4). The high level function is broken into sub-functions. The sub-functions can be broken into tasks, the tasks into task steps. The steps can be further broken into activities. Activities are at the lowest level of analysis and describe behaviours such as monitor reactor control system temperature [5]. Tasks in a sequence tend to cycle through these categories, although well-designed and skilfully performed tasks do not necessarily show distinct categories. The benefit of this framework is that it directs the analyst's attention to the necessary components of deliberate, rule based (i.e. procedural) behaviour [4], [11]. The task steps level of such analysis specifies critical details that may be associated with each task activities.

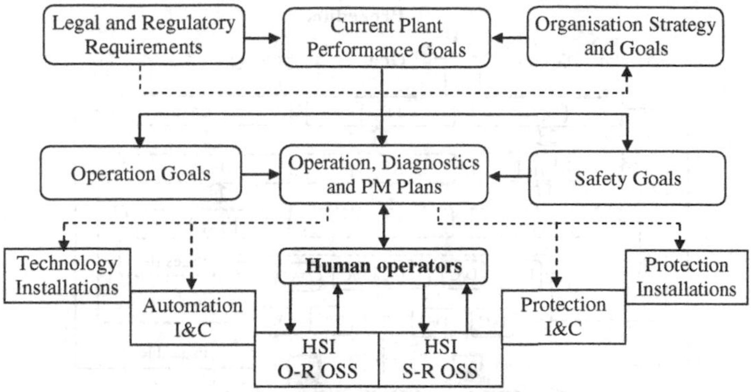

Fig. 3. Human operators and their activity context

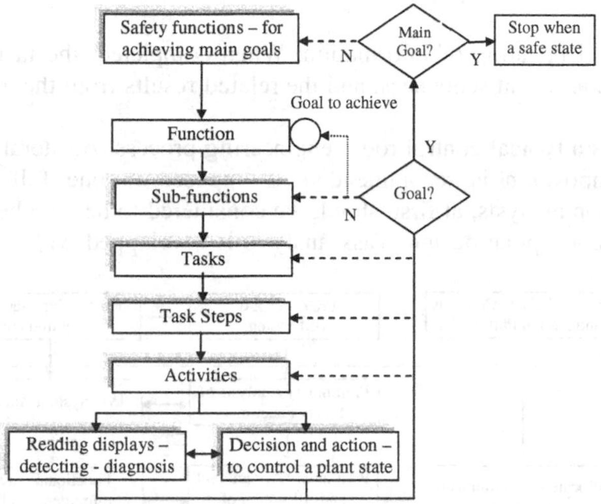

Fig. 4. Hierarchy of goals, functions, tasks and human operator activities

To achieve consecutive goal the operators use a procedure from a set of predefined procedures developed for some categories of transients, abnormalities and emergency situations. The structure of a function based displays using the results of task analysis and function decomposition is shown in Fig. 5. A few goals can be extracted from the procedure, and these goals can be broken into more detail functions. These functions can be then decomposed into tasks. Thus this figure shows the display design model of a function based display distinguishing three levels of pages: I - for function (a page with concise information), II - for sub-functions and III – for tasks consisting of more detailed information.

The task requirements and the sequence information may be significant inputs in procedure development. In fact, draft procedures can be written directly from the task analysis, especially when new tasks are issued from the function allocation. The documentation should be produced to verify human factors involvement in the control room design. The task related data should be stored on a database system to allow

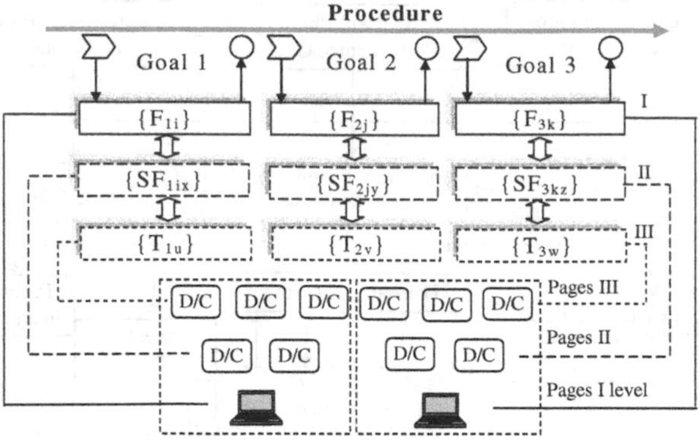

Fig. 5. Functions (F), sub-functions (SF) and tasks (T) based HSI design model with three levels of display/control (D/C) pages

manipulation and updating of information. When completed, the task area database will incorporate all event sequences, and the related results from the analysis of those sequences [5].

Fig. 6 shows a typical control room engineering process. An iterative approach to function decomposition is recommended as opposed to one full comprehensive analysis. Function analysis, at first, should be considered to be a higher level analysis supporting the conceptual design. Task analysis is overlapped with function analysis

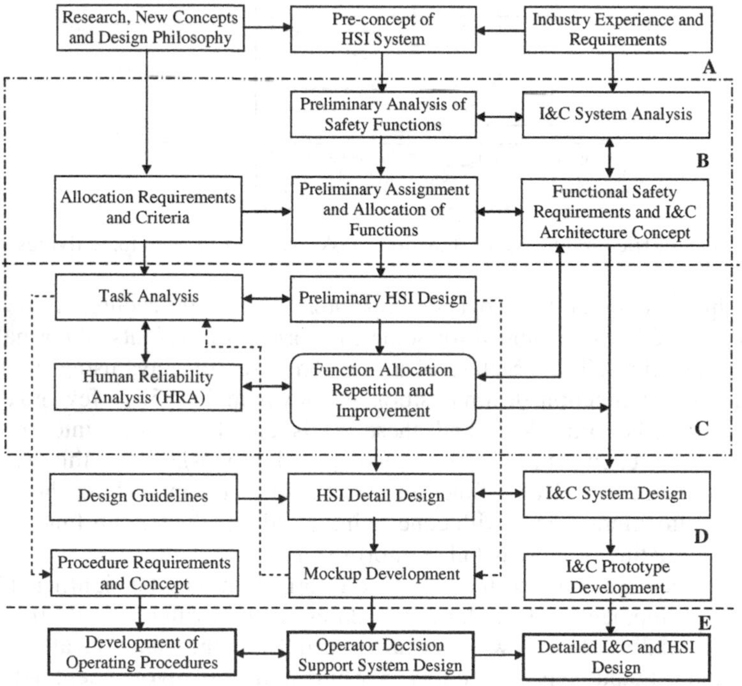

Fig. 6. Designing process of I&C systems and HSI for hazardous plant (based on [5], p.30)

and is performed for more detailed level of operational analysis. Thus, the video displays consist of: specific task oriented display, procedure based display, physical process flow based display and function based display [5].

The task analysis methods are described in a guide [10] and the human reliability analysis (HRA) methods in publications [4], [11], [21]. A framework for addressing human factors in functional safety design is proposed in a report [1].

4 Layer of Protection Analysis Including Human Factors

In Fig. 7 a protection system is presented that consists of three layers [15]:

- PL1 – the Basic Process Control System (BPCS);
- PL2 – the Human-Operator (OPERATOR), who supervises the process and intervene in cases of abnormal situations and during emergencies that are indicated by the alarm system (AS) or the Decision Support System (DSS);
- PL3 – the Safety Instrumented System (SIS), which can perform an Emergency ShutDown (ESD) function.

These systems should be physically and functionally independent; however, it is not always achievable in practice.

These layers should be independent what requires appropriate technical and organizational solutions. In case of PL1 and PL3 it can be achieved using separate measurement lines (input elements), modules for information processing, e.g. in safety PLCs, and actuators (final elements). Required safety integrity level (SIL) of the BPCS and SIS for given safety function can be achieved by designing appropriate architectures of their subsystems [14] taking into account the probabilistic criteria given in standards [7], [8] for verifying SIL of the SIS.

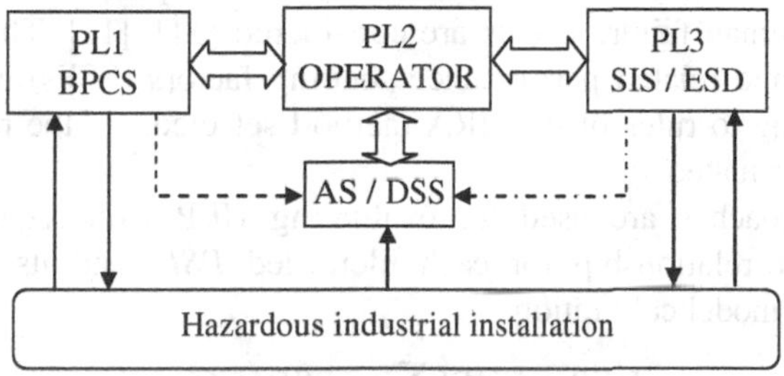

Fig. 7. BPCS, OPERATOR and Alarm System (AS) in Protection System

If the risk reduction requirement concerns all protection layers should be properly distributed between the BPCS, OPERATOR and SIS, e.g. if 10^{-4} is for all layers then it can be is distributed as follows: 10^{-1} (SIL1), 10^{-1} (HEP at the level of SIL1) and 10^{-2} (SIL2), which are values achievable without difficulty for the hardware architectures used in industrial practice [15].

There is, however, a considerable problem in some cases concerning the layer PL2, i.e. OPERATOR, who needs relevant information through the HIS from the

BPCS, AS and/or DSS. Only in case of independence of these layers the frequency of i-th accident scenario F_i may be calculated from the formula [15]

$$F_i = F_i^I \cdot PFD_{i;PL1} \cdot PFD_{i;PL2} \cdot PFD_{i;PL3} = F_i^I \cdot PFD_i \quad (1)$$

where F_i^I is the frequency of i–th initiating event I [a^{-1}] and $PFD_{i;PLj}$ are probabilities of failure on demand of j-th protection layer shown in Fig. 7. In case of the second layer $PFD_{i;PL2} = HEP_{i;PL2}$ and relevant HEP is to be evaluated using appropriate HRA method.

Generally, the frequency of accident scenarios for layers considered should be evaluated using a formula consisting of conditional probabilities [15]

$$F_i^Z = F_i^I \cdot P(X_{i;PL1} \mid I) \cdot P(X_{i;PL2} \mid I \cap X_{i;PL1}) \cdot P(X_{i;PL3} \mid I \cap X_{i;PL1} \cap X_{i;PL2}) = F_i^I \cdot PFD_i^Z \quad (2)$$

where: $X_{i;PLj}$ denote the events that represent failure in performing safety-related functions on demand by consecutive protection layers (j = 1, 2, 3) that should be considered for i-th initiating event.

The results of analyses have shown that assuming dependencies between the layers in probabilistic modelling significantly increases the failure probability on demand even an order of magnitude or more, thus $PFD_i^Z >> PFD_i$ - see formulas (1) and (2). Significant meaning in reducing dependencies of mentioned layers may have appropriate designing of the alarm system (AS) and the decision support system (DSS) as well as the quality of HSI characterized by relevant factors that are assessed when performing the HRA.

The HEP is evaluated when potential human failure event is placed into the probabilistic model structure of the system. In the HRA performed within PSA only more important human failure events are considered [11], [12]. Then, the abnormal situation context and related performance shaping factors (PSFs) are identified and evaluated according to rules of the HRA method selected. As the result the specific value of HEP is evaluated.

Different approaches are used for evaluating HEP with regard to PSFs, e.g. assuming a linear relationship for each identified PSF_k and its weight w_k, with constant C for the model calibration

$$HEP = NHEP \sum_k w_k PSF_k + C \quad (3)$$

Various approaches are used for evaluating human error probability (*HEP*) with regard to a set of *performance shaping factors* (PSFs). For instance, in the SPAR-H method [21] it is proposed to distinguish two cases of human error probability: for a diagnosis HEP_D and HEP_A for action that follow diagnosis. The following formulas are used for calculation of HEP_{Dj} for diagnosis of j-th abnormal situation

$$HEP_{Dj} = NHEP_D \cdot S_{Dj} \quad (4)$$

and HEP_{Aj} for action that follows diagnosis of j-th abnormal situation

$$HEP_{Aj} = NHEP_{A} \cdot S_{Aj} \tag{5}$$

where: $NHEP_D$ is the nominal HEP for diagnosis, in SPAR-H method evaluated to be equal 0.01; $NHEP_A$ is the nominal HEP for action, in SPAR-H method equal 0.001; S_{Dj} is the composite PSF of j-th situation, and S_{Aij} is the composite PSF for action that follows diagnosis of j-th situation.

These composite PSFs are evaluated from following formulas:

$$S_{Dj} = \prod_{i} S_{Dij} \tag{6}$$

$$S_{Aj} = \prod_{i} S_{Aij} \tag{7}$$

where: S_{Dij} is the value of i-th factor for diagnosis of j-th situation taken from a set of values; S_{Aij} is the value of i-th factor for action that follows diagnosis of j-th situation taken from a set of values.

When three or more values of S_{Dij} or S_{Aij} are greater than 1 (some of them can reach value up to 50), to keep the probability values of HEP_{Dj} and HEP_{Aj} below or equal 1, other formulas have to be used [21]:

$$HEP_{Dj} = \frac{NHEP_{D} \cdot S_{Dj}}{NHEP_{D}(S_{Dj} - 1) + 1} \tag{8}$$

$$HEP_{Aj} = \frac{NHEP_{A} \cdot S_{Aj}}{NHEP_{A}(S_{Aj} - 1) + 1} \tag{9}$$

In a similar way are evaluated HEPs in the HEART method [11], although in that method the human error includes both diagnosis and action resulting in one failure event with relevant HEP.

The resulting human error probability for j-th situation is proposed to be calculated from the following formula:

$$HEP_{j} \cong HEP_{Dj} + HEP_{Aj} \tag{10}$$

when the values of HEP_{Dj} and/or HEP_{Aj} are small, i.e. below 0.1, and from the formula:

$$HEP_{j} = HEP_{Dj} + HEP_{Aj} - HEP_{Dj} \cdot HEP_{Aj} \tag{11}$$

when the values of HEP_{Dj} and/or HEP_{Aj} are equal or greater than 0.1.

In the method SPAR-H eight factors (consecutive i-th factor in numbered from 1 to 8) are to be evaluated by HRA analysts/experts:

1. Available time;
2. Stress/stressors;
3. Complexity;

4. Experience/training;
5. Procedures;
6. Ergonomics/HSI;
7. Fitness for duty, and
8. Work processes.

In case of diagnosis, if all PSF ratings are nominal ($S_{Dij} = 1$, for i from 1 to 8), then the $HEP_{Dj} = NHEP_D = 0.01$. In remaining cases the calculations of HEP_{Dj} are made according to the formulas (4) and (6) or (8).

In case of action, if all PSF ratings are nominal ($S_{Aij} = 1$, for i from 1 to 8), then the $HEP_{Aj} = NHEP_A = 0.001$. In remaining cases the calculations of HEP_{Aj} are made according to the formulas (5) and (7) or (9).

The values of $NHEP_D$ and $NHEP_A$ were assessed in the context of SPAR-H probabilistic models by experts and confirmed in the verification process [21]. Nevertheless, further research should be carried out taking into other categories of hazardous plants, not only from the nuclear energy sector.

Also the set of values for consecutive factors S_{Dij} and S_{Aij} have been evaluated by experts taking into account own experience and comparative assessment of results obtained from other HRA methods. It is worth to mentioned that the highest values of 50 can be assigned to factors 5 and 6: S_{D5j} (Procedures *not available*) and S_{D6j} (Ergonomics/HSI *missing/misleading*).

The evaluations these factors (5. Procedures and 6. Ergonomics/HSI) should be based on careful analysis of solutions considered with regard to hierarchy of goals, functions, tasks and human operator activities (see Fig. 4) as well as functions (F), sub-functions (SF) and tasks (T) in HSI design model with relevant levels of display/control (D/C) pages (see Fig. 5). The HRA should be performed in designing process of *instrumentation and control* (I&C) systems and HSI for hazardous plant (see Fig. 6).

Due to importance of the problem in assessing the HEP for emergency situations considered and the frequency of accident scenarios to evaluate correctly the risk levels the research works have been carried out to include issues of human factors in the context of functional safety analysis [15]. They are aimed at the development of a methodology and software tools consisting of the probabilistic models and relevant data-knowledge bases (D-KBs) with regard to computer supported assessments of human factors in performing the HRA.

The functional safety oriented framework offers additional possibilities for more comprehensive human reliability analysis (HRA) with emphasis on contextual human-operator behaviour in abnormal situations, also those related to danger failures of the control and protection systems. The analyses outlined provide better understanding how to design the safety features and functions to be implemented by means of the programmable control and protection systems. Without a doubt the integrated design of the control and protection systems, and advanced control room should be human-centred.

5 Conclusions

In the paper an integrated approach of the human factors and functional safety analysis in designing of the control room of industrial hazardous plants is outlined. The concept of "defence in depth" (D-in-D) in hazardous plants was presented. The primary means of preventing and mitigating the consequences of accidents using D-in-D relay on implementing a combination of consecutive and independent levels of protection.

The instrumentation and control (I&C) systems should support D-in-D and relevant safety barriers identified in the risk analysis. In traditional I&C designs, different systems often supported each of the lines of defence. Strong independence should be provided between the safety-related systems. The engineering safety features (ESFs) actuation systems and the reactor trip systems obviously should use different actuation logics. In addition, the signal and functional diversity are to be considered and provided when justified so that shared data and environment would not jeopardize multiple lines of defence.

Subdividing the plant I&C systems according to their functions facilitates understanding of the entire system. An important role plays the human system interface (HSI) to make the supervision and control of the plant states more effective and reliable by human operators. Each safety function is to be considered to achieve given safety-related goal. It is subdivided into tasks and activities of two kinds: reading displays for detecting abnormalities and diagnosis of the plant state, and undertaking decisions and actions to control the plant to reach a safe state.

The functional safety oriented framework offers additional possibilities for more comprehensive human reliability analysis (HRA) with regard to contextual human-operator behaviour in abnormal situations, also those related to danger failures of the control and protection systems. The analyses outlined provide better understanding how to design the safety functions to be implemented by means of the programmable control and protection systems. The integrated design of the control and protection systems, and advanced control room should be human-centred.

Additional research is needed to obtain more comprehensive insights into the reliability and safety aspects useful for designing advanced human-centred interactive solutions, especially for more dynamic hazardous events in industrial plants.

Acknowledgements. The research outlined in this paper has been carried out as a part of works aimed at developing methods and prototype software tools for supporting the functional safety management. These works are supported by the Ministry for Science and Higher Education – Center for Research in Warsaw: the research project VI.B.10 (2011-13) that concerns the safety and reliability of programmable control and protection systems in hazardous industrial plants.

References

1. Carey, M.: Proposed Framework for Addressing Human Factors in IEC 61508. A Study prepared by Amey VECTRA Ltd. for Health and Safety Executive (HSE), U.K., Research Report 373 (2001)

2. EEMUA Publication 191: Alarm Systems, A Guide to Design, Management and Procurement. The Engineering Equipment and Materials Users' Association, London (2007)
3. Froome, P., Jones, C.: Developing Advisory Software to comply with IEC 61508. Contract Research Report 419. HSE Books (2002)
4. Gertman, I.D., Blackman, H.S.: Human Reliability and Safety Analysis Data Handbook. A Wiley-Interscience Publication, New York (1994)
5. IAEA Nuclear Energy Series No. NP-T-3.10: Integration of Analog and Digital Instrumentation and Control Systems in Hybrid Control Rooms, Vienna (2010)
6. IAEA Nuclear Energy Series No. NP-T-3.12: Core Knowledge on Instrumentation and Control Systems in Nuclear Power Plants, Vienna (2011)
7. IEC 61508: Functional Safety of Electrical/ Electronic/ Programmable Electronic Safety-Related Systems, Parts 1-7. International Electrotechnical Commission, Geneva (2010)
8. IEC 61511: Functional safety: Safety Instrumented Systems for the Process Industry Sector. Parts 1-3. International Electrotechnical Commission, Geneva (2003)
9. IEC 61513: Nuclear power plants – Instrumentation and control for systems important to safety – General requirements for systems, International Electrotechnical Commission, Geneva (2011)
10. Kirwan, B.: A Guide to Practical Human Reliability Assessment. CRC Press, London (1994)
11. Kosmowski, K.T., Degen, G., Mertens, J., Reer, B.: Development of Advanced Methods and Related Software for Human Reliability Evaluation within Probabilistic Safety Analyses. Berichte des Forschungszentrum 2928, Jülich (1994)
12. Kosmowski, K.T.: Incorporation of human and organizational factors into qualitative and quantitative risk analyses. In: Proceedings of International Conference on Probabilistic Safety Assessment and Management (PSAM 7- ESREL 2004), vol. 3, pp. 2048–2053. Springer (2004)
13. Kosmowski, K.T.: Functional Safety Concept for Hazardous System and New Challenges. Journal of Loss Prevention in the Process Industries 19(1), 298–305 (2006)
14. Kosmowski, K.T. (ed.): Functional Safety Management in Critical Systems. Gdansk University of Technology. Publishing House: Fundacja Rozwoju Uniwersytetu Gdańskiego. Gdansk (2007)
15. Kosmowski, K.T.: Functional Safety Analysis including Human Factors. International Journal of Performability Engineering 7(1), 61–76 (2011)
16. Kosmowski, K.T.: Current challenges and methodological issues of functional safety and security management in hazardous technical systems. Journal of Polish Safety and Reliability Association 3(1), 39–51 (2012)
17. LOPA: Layer of Protection Analysis, Simplified Process Risk Assessment. Center for Chemical Process Safety. American Institute of Chemical Engineers, New York (2001)
18. OECD Report: Critical Operator Actions – Human Reliability Modeling and Data Issues. Nuclear Safety, NEA/CSNI/R; OECD Nuclear Energy Agency (1998)
19. Rasmussen, J., Svedung, I.: Proactive Risk Management in a Dynamic Society. Swedish Rescue Services Agency, Karlstad (2000)
20. Reason, J.: Human Error. Cambridge University Press (1990)
21. SPAR-H: Human Reliability Analysis Method, NUREG/CR-6883, INL/EXT-05-00509, US NRC (2005)

Modeling Discrete-Event Systems with Hard Synchronization Constraints

Lothar Seybold[1], Piotr Witczak[2], Paweł Majdzik[2], and Ralf Stetter[3]

[1] RAFI GmbH, Berg, Germany
lothar.seybold@rafi.de
[2] Institute of Control and Computation Engineering,
University of Zielona Gora, Poland
p.witczak@weit.uz.zgora.pl, p.majdzk@issi.uz.zgora.pl
[3] Faculty of Mechanical Engineering,
Ravensburg-Weingarten University of Applied Sciences, Weingarten, Germany
stetter@hs-weingarten.de

Abstract. An issue of prototyping Systems of Concurrent Cyclic Processes (SCCP), in which a number of concurrently running processes compete to access to a set of resources is considered. The main outcome of this paper is an automated procedure of performance evaluation for such systems with desired values of a wide range of system functional characteristics, such as a schedule of processes flows and the system period. This stage is realised by a procedure of automatic building of an analytical model of SCCP, which is based on the (max, +) algebra formalism. In most cases the (max, +) algebra is used to model of discrete-event systems, while the synchronization is based on rendezvous protocol. However, in this paper - modeling of the systems is based on mutual exclusion protocol, moreover buffers are included. As an illustrative example example RAFI Battery Assembly System was given in this paper.

Keywords: repetitive processes, analytical models, (max, +) algebra, scheduling, Petri networks.

1 Introduction

There are two main approaches to performance evaluation of SCCP. The first approach is based on a computer simulation of system state models [3,8]. It can be used for any class of SCCP, however it requires tedious simulation runs and cannot give an understanding of the dependence of parameters changes on important system properties, e.g. cycle time. The second approach allows obtaining the system period and processes flow using an algebraic model. The analytical methods aimed at design and performance analysis of SCCP with required parameters [2] as well as deadlock/starvation avoidance is still under development [4]. Those methods may use different modeling techniques, e.g. Petri nets and event-graphs [5], stochastic Markov chains, Nonlinear programming [7],

J. Korbicz and M. Kowal (eds.), *Intelligent Systems in Technical and Medical Diagnostics*, Advances in Intelligent Systems and Computing 230,
DOI: 10.1007/978-3-642-39881-0_43,

(max, +) algebra [1,6], and other models [2]. An algebraic approach to SCCP modeling has been used in a few works so far [6].

Energy storage and energy buffering are key technologies for upcoming generations of new smarter environments. Efficient battery systems are one of the major research issues in recent years and will be for the future. High performance batteries are characterised by a high energy density of more than $140Wh/kg$, a low volume footprint, long service of life due to passive and active thermo-management, high structural performance to deal with shocks of e.g. $50G$ or pulses of e.g. $15ms$ in all 6 directions. The battery chemicals are field of continuous improvements and varieties of different galvanic materials like lithium-ion materials. Since a lot these combinations are hazardous materials or explosive they have to be treated with care and attention. These properties have also influence on the production process, once in terms of flexibility and second in terms of employment protection. The flexibility of the production process owing to a high momentum in development of battery technology and applications requests to produce a lot of different battery types with different specifications. At the same time, the production process has to be on a very high automation level in order to protect production stuff as best as possible. Therefore a flexible battery assembly system with autonomous robots is planned for high volume serial production. Designed system falls into SCCP class of systems.

2 Systems of Concurrent Cyclic Processes

An SCCP consists of a set of concurrently-executing cyclic processes compete to access to a set of resources. There are two subsets of resources: local ones which are used by only one process, and other subset: shared resources which are utilised by two or more processes. In the SCCP processes are synchronised by means of a mutual-exclusion protocol (i.e. only one process may be allocated on a shared resource at a given time), which is realised through dispatching rules assigned to shared resources.
The SCCPs meet the following assumptions:

- the subsequent operation of a process starts immediately when the current operation has been finished, providing that the next resource may be used by the process;
- a resource being currently utilised by a process it not released by this process, unless the next resource to be used by the process is not being utilised by another process;
- a process is not pre-emptive, i.e. a process cannot be released from a resource, until it finishes its current operation on this resource.

From the above assumptions it is in fact modeling of the class of discrete event systems that contain a finite number of resources (such as machines, processors, robots, buffers) that are shared by a finite number of processes (such as assembly processes, tasks of parallel computation). Thus typical examples are serial production lines, production systems with a fixed routing schedule, railway networks and some kind of Real-Time Synchronous Systems.

2.1 Specification of SCCP

An SCCP is described by two sets: $P = \{P_1, P_2, \dots, P_n\}$ and $T = \{T_1, T_2, \dots, T_n\}$. Therefore, the pair (P, T) specifies the system structure and the parameters of the individual component processes, e.g. execution times and the order of each process operations. A System of Concurrent Cyclic Processes will be denoted by $S(P, T)$.

A set of resources is defined as follows $R = \{R_1, R_2, \dots, R_q\}$, where R_i is the i-th resource. The cyclic process is a vector $P_i = (p_{i,1}, p_{i,2}, \dots, p_{i,l})$ whose elements specify operations used by the process in the course of its execution, where $p_{i,j}$ is the j-th operation executed by the i-th process. A component process utilizes a set of resources, e.g. the process defined as $P_1 = (1, 5, 4)$ uses resources R_1, R_5 and R_4. A time representation $T_i = (t_{i,1}, t_{i,2}, \dots, t_{i,l})$ of a cyclic process is a vector whose elements specify the execution times of individual operations realised by the i-th process, where $t_{i,j}$ is the execution time of the j-th operation performed by the i-th process.

An initial state S_0 of the system is a vector, whose elements specify the allocations of individual processes on given resources: $S_0 = (P_j, P_k, \dots, \lambda, P_l)$, where: $||S_0||$ is a number of system resources, $crd_i S_0 = P_j$ if the j-th process is allocated on the i-th resource in the initial state S_0, $crd_i S_0 = \lambda$ if no process is allocated on the i-th resource in the initial state.

The conflict of processes occurs in the case of when at the same time a set of processes waits for an access to the same shared resource. The hard synchronization constraint can be defined as a fixed dispatching rule i.e. is such a rule, that for the same resource conflict allocates always the same process. In this paper, a dispatching rule for the i-th resource is represented by the following vector $\sigma_i = (P_j, P_k, \dots, P_l)$, whose elements specify the order in which processes will access the i-th resource.

3 Battery Assembly as SCCP

To show an active, real example of implementation above approach, system planned by RAFI company is considered.

This production system is based on transport and manipulation robots, with additional hand assembly stations. In the first stage only the transport robots are implemented. There are two different products (Box based stationary battery system with 24V in two different colours and Rack based stationary battery system with 100V) to be produced. The production system contains two assembly cycles. The first cycle is operated with two transportation robots (transportation robot 1, 2) and covers the assembly of the battery frames for the two different products. The second cycle is also operated with two transportation robots (transportation robot 3, 4) and covers the final assembly of the products. The sequence of the first production cycle starts with the robots in a starting setting. The robot moves in the next step to either the corpus storage of Box or Frame Product to pick up the empty battery frames. In the next step the robot moves with the battery frame aboard to the cell storage. The next step is the assembly of basic Lithium-ion cell packages into the battery frames.

The number of basic cell assembled depends of rated voltage of the battery. 11 cells for a 24V battery and 45 cells for the 100V battery. The ready assembled battery is moved to the relevant battery stores in the next step. Finally the robot returns to the starting position.

The sequence of the second production cycle starts also with the robots in a starting position. Depending on the final product the robot moves to the storages of the coloured box housings (but this step is omitted to fit system into this paper). Alternatively it moves to a hand assembly station to pick up additional parts and wiring for the 100V rack version and then to the rack housing storage.

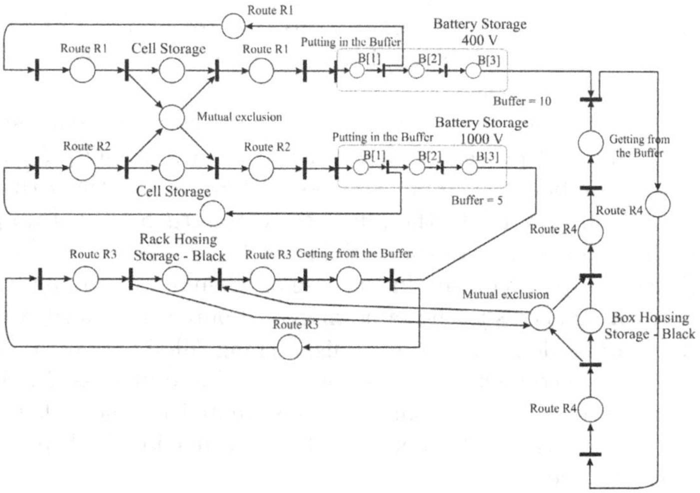

Fig. 1. Simplified System's Petri network

From these storages the robot moves to either the 24V or 100V battery storage with the Rack or Box housing aboard. Next step is the assembly of the batteries, controllers into the housings. Finally the ready assembled product is moved to either the final rack or box storage. For there the robot returns to the starting position. Above system is drawn on (Fig. 1), considered to be an example of System of Concurent Cyclic Processes.

4 (Max, +) Algebra Model

The (max,+) algebraic structure (R_{max},$\oplus$,$\otimes$) is defined as follows:

- $R_{max} = R \cup \{-\infty\}$, where R is the field of real numbers,
- $\forall a, b \in R_{max}, a \oplus b = max(a, b)$,
- $\forall a, b \in R_{max}, a \otimes b = a + b$.

Its main properties are:
$\forall a \in R_{max} : a \oplus \epsilon = a$ and $a \otimes \epsilon = \epsilon$
$\forall a \in R_{max} : a \otimes e = a$
For matrices $A, B \in R_{max}^{m \times n}$ and $C \in R_{max}^{n \times p}$ we have

$$(A \oplus B)_{ij} = (A)_{ij} \oplus (B)_{ij} = \max((A)_{ij}, (B)_{ij})$$

$$(A \otimes C)_{ij} = \left(\bigoplus_{k=1}^{n} (A)_{ik} \otimes (C)_{kj}\right) = \max_{k=1,\dots,n} ((A)_{ik} + (C)_{kj})$$

Further definitions and theorem related to the (max,+) algebra formalism can be found in [1].

The input data to the analytical model building procedure are as follows:

- the system structure,
- execution time of operations performed by component processes,
- an initial state of the system,
- set of priority dispatching rules - sequences of utilized shared resources,

The procedure consists of the following two main steps:

Step 1. Specification of the state vector and the operations vector.

The size of the operations vector is equal to the number of operations carried out in the systems and it is defined as:

$$p = (p_{1,1}, p_{1,2}, \dots, p_{1,m1}, p_{2,1}, p_{2,2}, \dots, p_{2,m2}, \dots, p_{n,1}, p_{n,2}, \dots, p_{n,mn}),$$

where $p_{i,j}$ is the j-th operation of the i-th process. The execution times vector is defined as follows:

$$t = (t_{1,1}, t_{1,2}, \dots, t_{1,m1}, t_{2,1}, t_{2,2}, \dots, t_{2,m2}, \dots, t_{n,1}, t_{n,2}, \dots, t_{n,mn}),$$

where $t_{i,j}$ is the execution time of the j-th operation by the i-th process, m_i is the number of operations carried out by i-th process during one cycle. The state vector is defined as:

$$x(k) = (x_1(k), x_2(k), x_3(k), \dots, x_r(k)),$$

where $x_i(k)$ represents operations crd_ip, and the value $x_i(k)$ denotes the moment of starting the operation crd_ip in the k-th iteration. Thus, the size of vector $x(k)$ is equal to size of vector p.

The realization of this step can be perceived as determination of the number of operations performed by the component processes of the system. Clearly, the operations may be ordered in some fashion, but the same order must be used in all consecutive steps of the procedure.

Step 2. Determining the components of the state equation: SCCP can be described by a model of the form:

$$\begin{aligned} x(k) &= A_0 x(k-1) \oplus A_1 x(k), \\ x(0) &= x_0 \end{aligned} \tag{1}$$

where: A_0 and A_1 are matrices size of $n \times n$ (n is the size of the state vector), x_0 is the state vector containing the start moments of operations in the 0-th iteration. The sufficient conditions allowing to create matrices C, A_0 and A_1 are presented in the next section.

The state equation (1) is equivalent to:

$$x(k) = Ax(k-1), x(0) = x_0 \tag{2}$$

where

$$\begin{aligned} A &= A^* \otimes A_0, \\ A^* &= \left(\bigoplus_{i=0}^{\infty} A_1^i\right) \end{aligned} \tag{3}$$

If the matrix A is correspondent to the strongly connected graph, than it possesses an unique eigenvalue in the (max,+) plus algebraic sense [1].

$$Av = \lambda v \tag{4}$$

where: λ is eigenvalue of the matrix A (i.e. the maximal mean cycle weight taken over all cycles in the graph corresponding to A) and v is an eigenvector of matrix A. Thus, if $x(k)$ is an eigenvector of A, then

$$x(k) = Tx(k-1) \tag{5}$$

where T is the system period. Following the solution of the spectral equation (4), and assigning $x_0 = v$ the cyclic steady state execution can be expressed as follows:

$$\begin{aligned} x(k) &= Tx(k-1), \\ x(0) &= x_0 \end{aligned} \tag{6}$$

Also, the vector x_0 must meet the state equation (1), what leads to the spectral equation of the matrix A:

$$x(k) = Ax(k-1) = Tx(k-1) \tag{7}$$

The system period and the time-stamps of the operations are the main outcome of this procedure. The next step takes advantage of the results of the synthesis procedure. Hence, it is possible to determine the states of individual processes in any arbitrary iteration. This entails that it is possible to determine processes flow by means ofGanttś chart.

4.1 Sufficient Conditions for Designing of (Max, +) Algebra Model

In this section the conditions allowing to create matrices C, A_0 and A_1 are presented. The matrix C is defined by systems structure $S(P,T)$, and a set of the hard synchronization constraint represented by the vectors: $\sigma_1, \sigma_2, \ldots, \sigma_k$. The size of the matrix C is equal to the number of operations performed by component processes. The structure of the matrix C specifies dependencies between the operations performed by the processes, i.e. the matrix represents the conditions that must be satisfied so that individual operations might be executed. The matrix C is created additionally, because matrices A_0 and A_1 are created as a result of splitting the matrix C. Moreover, the result of splitting depends on the assumed initial state of the system. Thus, the matrix C allows to obtain the elements of state equation (matrices A_0, A_1) for a different initial states. The conditions of building the matrix C are as follows:

$$C_{i,j} = \begin{cases} t_{i,k} & \text{if i)} \\ e & \text{if ii)} \\ \varepsilon & \text{otherwise} \end{cases}$$

i) $\exists p_{l,k} \in p, l \in \{1, 2, \ldots, n\}, k \in N$
$\forall j \in \{1, 2, \ldots, L(x_k(k))\} : \Sigma_{a=1}^{l} L(p_{a,k}) \le j \le \Sigma_{a=1}^{l-1} L(p_{a,k})$
$i = j + 1$ if $(j \bmod \Sigma_{a=1}^{l} L(p_{a,k})) \neq 0$ or
$i = \Sigma_{a=1}^{l-1} + 1$ if $(j \bmod \Sigma_{a=1}^{l} L(p_{a,k})) = 0$

ii) $\forall r \in \{1, 2, \ldots, L(\theta)\}, q = \text{idx}(\text{crd}_r \theta), m = L(\sigma_q)$,
$\forall l \in \{1, 2, \ldots, m), d = l \bmod m$
if $\text{idx}\ (\text{crd}_{d+1}\sigma_q) \neq \text{idx}\ (\text{crd}_l \sigma_q)$ then
$i = \Sigma_{a=1}^{b-1} L(p_{a,k})) + c, b = \text{idx}(\text{crd}_{d+1}\sigma_q), \text{crd}_c(\text{crd}_{d+1}\sigma_q) = q$
$j = \Sigma_{a=1}^{b-1} L(p_{a,k})) + c + 1, b = \text{idx}(\text{crd}_{d+1}\sigma_q), \text{crd}_c(\text{crd}_l \sigma_q) = q$

where: idx denotes an index of the element, $L(h)$ is the size of vector h.

The above conditions are satisfied for the assumed order of operations in the state vector i.e. first elements of state vector are related to the operations performed by process P_1 and next operations are adequately related to process P_2, P_3, etc. From the these conditions it follows, that the process of the matrix C building depends on the systems structure and assumed rules controlling access to the shared resources. The condition i) is related to the order of operations carried out by a given process. From this condition it follows that $C_{i,j} = t_{l,k}$ if the i-th operation precedes the j-th operation, where $t_{l,k}$ is the execution time of the j-th operation. In other words, a given process can start executing the i-th operation if the j-th operation has finished, which lasts $t_{l,k}$. The conditions ii) is connected with the representation of the mutual-exclusion protocol within the structure of the matrix C, i.e. the assumed sequence of shared resources utilisation. A given process can start its operation on a shared resource, if the process, which precedes the given process in that sequence, starts execution of its operation on the next resource (according to that process route). In the case when in the sequence of shared resources utilisation a given process occurs two or more times (e.g. $P_1, P_1, P_3, \ldots$), only condition, which has to be meet, is related to the order of operations carried out by a given process (the condition i)). Thus, if the condition $idx(crd_{d+1}\sigma_q) \neq idx(crd_l \sigma_q)$ does not meet then the

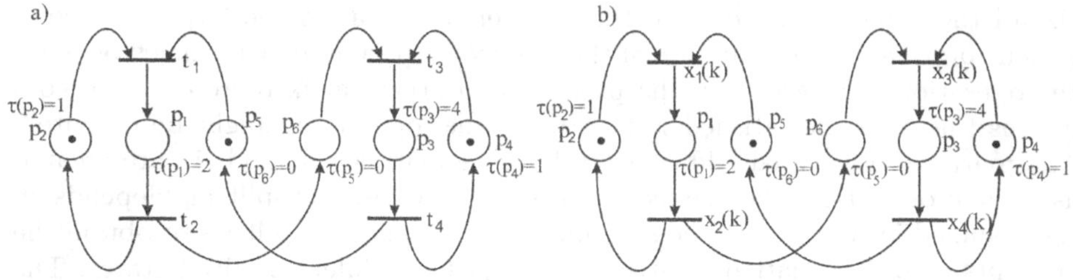

Fig. 2. Models of Petri Net for the following sequence P1, P2

arc of weight e from node i to node j do not exist. From this condition, it follows that $C_{i,j} = e$ if the start of the i-th operation is a required condition to start the j-th operation.

To explain better the above conditions, the connection between (max, +) algebra and the time Petri net of type P will be shown. To this end let us consider the simple system $S(P_2, T_2)$, where: $P_2 = \{P_1 = (1,2), P_2 = (1,3)\}$, $T_2 = \{T_1 = (2,1), T_2 = (4,1)\}$ in which two processes compete to access to shared resource R_1 (Fig. 2). Assuming that the moment of starting operation of process P_1 on shared resource R_1 is represented by $x_1(k)$ in the direct graph and by the transition t_1 in the model of Petri net, the moment of starting operation on local resource R_2 is represented by $x_2(k)$ in the direct graph and by the transition t_2 in the model of Petri net, moment of starting the operation of process P_2 on shared resource R_1 is represented by $x_3(k)$ and by the transition t_3 and the moment of starting operation on local resource R_3 is represented by $x_4(k)$ and by the transition p_4. Of course, the place p_i in the Petri net represents performing of i-th operation in the system (Fig. 2a). The i-th transition corresponds to i-th elements ($x_i(k)$) of state vector $x(k)$. The model of Petri Net gives fundamentals to the components of the (max,+) algebra state equation. The formula of construction of matrix C is as follows: $C_{i,j}$ is equal to $\tau(p_l)$ if the place p_l is the input place of transition t_i (or $x_i(k)$ in the Fig. 2a) and the output place of the transition t_j (or $x_j(k)$ in the Fig. 2b) and $C_{i,j} = \varepsilon$ in the other cases. For the above assumptions, the matrix C created bases on the presented conditions and the graph corresponding with the matrix C is presented in (Fig. 3). Thus, the elements of matrix C can be determined by using the above conditions as well as by creating the model of Petri Net of type P.

$$C = \begin{array}{c} \begin{array}{cccc} x_1 & x_2 & x_3 & x_4 \end{array} \\ \begin{bmatrix} \epsilon & 1 & \epsilon & e \\ 2 & \epsilon & \epsilon & \epsilon \\ \epsilon & e & \epsilon & 1 \\ \epsilon & \epsilon & 4 & \epsilon \end{bmatrix} \end{array} \begin{array}{c} \\ x_1 \\ x_2 \\ x_3 \\ x_4 \end{array}$$

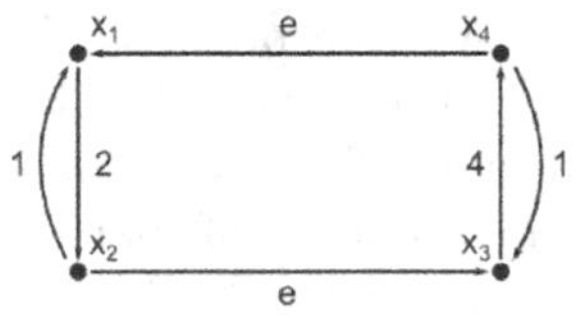

Fig. 3. The matrix C and the corresponding graph

The matrices A_0 and A_1 are created as a result of splitting the matrix C. The result depends on the assumed initial state of the system. This means that the matrices A_0 and A_1 are determined on the matrix C and the assumed initial state.

Let the matrix C be decomposed as follows $C = A_0 \oplus A_1$ where A_0 and A_1 are the elements of the state equation (1). If the following conditions hold, then the system modelled by (6) meets the assumptions provided in the section 2 (considered SCCP classes). The conditions to build the matrices A_0 and A_1 are:

$$A_{0\ i,j} = \begin{cases} C_{i,j} \text{ if the } j\text{-th operation belongs to initial state} \\ \varepsilon \qquad\qquad\qquad\qquad\qquad\qquad\qquad \text{otherwise} \end{cases}$$

$$A_{1\ i,j} = \begin{cases} e \quad \text{if } A_{0\ i,j} = C_{i,j} \\ C_{i,j} \qquad \text{otherwise} \end{cases}$$

The matrix A_1 contains elements related to all operations carried out by processes, except operations performed in the initial state. The graph associated with the matrix A_1 is non-cyclic. This states that sum of consecutive powers of matrix A_1 contains the maximal execution time of operations sequences except operations carried out in the initial state (given in the matrix A_0).

Instead of applying the above conditions the model of Petri Net can be utilized. The algorithm of creating of matrices A_0 and A_1 (Fig. 4) consists of the following steps:

Step 1. Create the model time Petri Net of type P for a given structure of system and for a assumed access sequences to shared resource.
Step 2. Find all cycles occurring in the Petri Net.
Step 3. Mark the Petri Net in order to the only one mark occur in each cycle.
Step 4. Create the matrices A_0 and A_1 according to the following conditions:

i) $A_{0i,j}$ is equal to $\tau(p_l)$ if the place p_l is empty and if p_l the input place of the transition t_i (or $x_i(k)$ in the Fig. 2b) and the output place of the transition t_j (or $x_j(k)$ in the Fig. 2b) and $C_{i,j} = \varepsilon$ in the other cases.
ii) $A_{1i,j}$ is equal to $\tau(p_l)$ if the place p_k has a mark and if p_l the input place of the transition t_l (or $x_i(k)$ in the Fig. 2b) and the output place of the transition t_j (or $x_j(k)$ in the Fig. 2b) and $C_{i,j} = \varepsilon$ in the other cases.

$$A_1 = \begin{bmatrix} \epsilon & 1 & \epsilon & e \\ \epsilon & \epsilon & \epsilon & \epsilon \\ \epsilon & \epsilon & \epsilon & 1 \\ \epsilon & \epsilon & \epsilon & \epsilon \end{bmatrix} \qquad\qquad A_0 = \begin{bmatrix} \epsilon & \epsilon & \epsilon & \epsilon \\ 2 & \epsilon & \epsilon & \epsilon \\ \epsilon & e & \epsilon & \epsilon \\ \epsilon & \epsilon & 4 & \epsilon \end{bmatrix}$$

Fig. 4. Matrices A_1 and A_0 given form a matrix C

5 An Example of SCCP

In this section the procedure of building (max,+) algebra model for the Battery Assembly System is applied on real example. By the reason of a large size of real Batter Assembly System (BAS) only some part (Fig. 1) of real system has been considered.

The BAS can be represented by SCCP, in which the robots are represented by cyclic processes, routes by local resources, cell storages and battery storages by shared resources.

For the partial system from (Fig. 1) the specification of SCCP is as follows: $S(P,T)$, $P = \{P_1, P_2, \ldots, P_4\}$ and $T = \{T_1, T_2, \ldots, T_4\}$, where: $P_1 = (p_{1,1}, p_{1,2}, \ldots, p_{1,5})$, $P_2 = (p_{2,1}, p_{2,2}, \ldots, p_{2,5})$, $P_3 = (p_{3,1}, p_{3,2}, \ldots, p_{1,5})$, $P_4 = (p_{4,1}, p_{4,2}, \ldots, p_{4,5})$, whose elements specify, respectively, operations performed by the robot1, robot2, robot3 and robot4.

A time representations for each process: $T_1 = (2,2,2,1,3)$, $T_2 = (2,2,2,1,3)$, $T_3 = (2,2,2,1,3)$, $T_4 = (2,2,2,1,3)$, whose elements specify the execution times of individual operations realised by the i-th robot (i-th process), where $t_{i,j}$ is the execution time of the j-th operation performed by the i-th robot (i-th process). The dispatching rules for the two shared resources $\sigma_1 = (P_1, P_2)$ and $\sigma_2 = (P_3, P_4)$, which represented in the Battery Assembly System respectively by Cell Storage and Box Housing Storage have been assumed (Fig. 1). The state vector is defined as $x(k) = (x_1(k), x_2(k), x_3(k), \ldots, x_{22}(k))$. The size of vector $x(k)$ is equal to the number of all operation performed in the system. The first twenty elements represent operations carried out by individual processes (robots) and four additional elements are related with operations carried out in the buffers. The matrix C, A_0 and A_1 are created according to the conditions presented in section 4.1. Due to matrix size it has been limited for illustration (Fig. 5). Matrix A is computed according to (3). The eigenvalue and eigenvector of the matrix A are respectively equal to:

- $\lambda = 11$ - the value of a system period T,
- $x_k(0) = (7,9,0,2,4,4,4,9,0,2,4,6,6,6,7,9,0,2,9,0,4,2)$ - vector elements representing the operation starting time.

$$C = \begin{bmatrix} \epsilon & \epsilon & \epsilon & \epsilon & 3 & \epsilon & \epsilon & \epsilon & \epsilon & \ldots \\ 2 & \epsilon & \epsilon & \epsilon & \epsilon & \epsilon & \epsilon & \epsilon & \epsilon & \\ \epsilon & 2 & \epsilon & \epsilon & \epsilon & \epsilon & \epsilon & \epsilon & \epsilon & \\ \epsilon & \epsilon & 2 & \epsilon & e & \epsilon & \epsilon & \epsilon & \epsilon & \\ \epsilon & \epsilon & \epsilon & 1 & \epsilon & e & \epsilon & \epsilon & \epsilon & \\ \epsilon & \epsilon & \epsilon & \epsilon & e & \epsilon & e & \epsilon & \epsilon & \\ \epsilon & \epsilon & \epsilon & \epsilon & \epsilon & e & \epsilon & \epsilon & \epsilon & \\ \epsilon & \epsilon & \epsilon & \epsilon & \epsilon & \epsilon & \epsilon & \epsilon & \epsilon & \\ \epsilon & \epsilon & e & \epsilon & \epsilon & \epsilon & \epsilon & 2 & \epsilon & \\ \vdots & & & & & & & & & \ddots \end{bmatrix}$$

Fig. 5. System matrix C

6 Concluding Remarks

The main result of this paper is related to determining the sufficient conditions allowing the conversion of structure of the system (composed of a set of cyclic processes) into its matrix representation for an assumed initial state and a set of dispatching rules. Moreover, two equivalent way of determining of the components (max, +) algebraic state equation have been presented. But it should be note that presented conditions allow to determine the structure of matrix C automatically, without the necessity of creating the Petri Net. This property is very important in the construction of a computer-aided system prototyping tool. The algorithm and conditions has been applied in the tool supporting automated prototyping of the SCCP with desired values of performance indices.

References

1. Baccelli, F.L.: Synchronization and linearity. An algebra for discrete event systems. John Wiley & Sons, Toronto (1992)
2. Cassandras, C.G.: Discrete state systems. Modelling and performance analysis. Boston, MA, Aksen (1993)
3. Abrams, et al.: Visual analysis of parallel and distributed programs in the time, event, and frequency domains. IEEE Trans. on Parallel and Distrib. Syst. 3, 672–685 (1992)
4. Polak, et al.: The performance evaluation tool for automated prototyping of concurrent cyclic processes. Fundamenta Informaticae 60, 269–289 (2004)
5. Proth, J.M., Hillion, H.P.: Performance evaluation of job-shop systems using timed event-graphs. IEEE Trans. Aut. Contr. 34, 3–9 (1989)
6. Banaszak, Z.A., Majdzik, P., Wójcik, R.: An automatic synthesis of performance models for systems of concurrent cyclic processes. In: Proc. of the 15th Int. Conference on System Science, Wroclaw, Poland, pp. 281–292 (2004)
7. Blaszczyk, J., Szynkiewicz, W.: Optimization-based approach to path planning for closed chain robot systems. Int. J. Appl. Math. Comput. Sci. 21, 659–670 (2011)
8. El Moudni, A., Yan, F., Dridi, M.: An autonomous vehicle sequencing problem at intersections. Int. J. Appl. Math. Comput. Sci. 23, 183–200 (2013)

Selection Pressure in the Evolutionary Path Planning Problem

Łukasz Kuczkowski and Roman Śmierzchalski

Gdańsk University of Technology, Faculty of Electrical and Control Engineering,
G. Narutowicza 11/12 80-233 Gdańsk
lkuczkowski@ely.pg.gda.pl, romsmier@pg.gda.pl

Abstract. This article compares an impact of using various post-selection methods on the selection pressure and the quality of the solution for the problem of planning the path for a moving object using the evolutionary method. The concept of selection pressure and different methods of post-selection are presented. Article analyses behaviour of post-selection for four options of evolutionary algorithms. Based on the results achieved, waveform diagrams were presented showing best and average fitness score depending on the generation. Those allowed to determine the selection pressure for each of the variants. The results presented allow to choose a post-selection method that maintains the population's variety and the compromise between the exploration and exploitation phases. In study case problem of avoiding collisions at sea is analysed. The modelled environment includes static constraints (lands, canals, etc.) and dynamic objects (moving ships).

Keywords: evolutionary algorithms, path planning, selection pressure, post-selection.

1 Introduction

One of the tasks for controlling a moveable object (i.e. mobile robot, autonomous marine vehicle) is to get an object from a specified starting point to its destination or to the task (mission) area. To achieve this a path has to be plotted again a specific criterion i.e. the shortest time to reach the destination. The path has to avoid obstacles which are treated as static and dynamic limitations. Usually the dynamic obstacles are treated as moveable objects, which move along certain trajectory with specific speed. The problem presented was reduced to an optimization task with static and dynamic obstacles [1]. The problem was solved using an evolutionary method [2], [3]. Based on the Evolutionary Planner/Navigator (EP/N) planning concept [4], modified version (υEP/N++) of the system [5], [6] was developed. υEP/N++ takes into account specific nature of the process of avoiding collisions. The present version of the system uses different types of static and moving constraints to model the real environment of moving targets and their dynamic characteristics.

The main goal of the paper was to analyze the impact of the post-selection on the selection pressure in the evolutionary path planning method. Different variants of post-selection was used to control of the selection pressure. This allowed to maintain a balance between the exploration and exploitation phases and control the algorithm's ability to search the solution space. Undertaken research allow to find the best solution for presented problem and decided which method of post-selection ensures that the balance between population's variety and selection pressure is maintained.

This article is organized in the following way: in chapter two the path planning system is presented. Chapter 3 describes the selection pressure and the post-selection methods, while the following chapter presents the simulations. The final part of the paper summarizes the research results.

2 Path Planning

The problem of path planning occurs in numerous technical applications, such as, for instance, motion planning for mobile robots [4], [7], ship weather routing in ocean sailing [8] or safety path planning for a ship in a collision situation at sea [6], [9]. The problem is defined in the following way: having given a moving object and the description of the environment, plan the path for an object motion between the beginning and end location which avoids all constraints and satisfies certain optimization criteria. The problem can be divided into two basic tasks: an off-line task, in which we look for the path of the object in a steady environment, and an on-line task, in which the object moves in the environment that meets the variability and uncertainty restrictions. The on-line mode of the path planning relates to the control of the moving object in the non-stationary environment, in which parts of some obstacles reveal certain dynamics.

One of the examples of path planning is the problem of avoiding collisions at sea. To solve this problem, the safe trajectory for the so-called own ship has to be determined. We look for a trajectory that compromises the cost of a necessary deviation from a given route, or from the optimum route leading to a destination point, and the safety of passing all static and dynamic obstacles, here referred to as strange ships (or targets). In this paper the following terminology is used: the term own ship means the ship, for which the trajectory will be generated, and strange ship or target mean other ships in the environment, i.e. the objects which will be avoided. All trajectories which meet the safety conditions reducing the risk of collision to a satisfactory level constitute a set of feasible trajectories. The safety conditions are, as a rule, defined by the operator based on the speed ratio between the ships involved in the passing manoeuvre, the actual visibility, weather conditions, navigation area, manoeuvrability of the ship, etc.

Other constraints resulting from formal regulations (e.g., traffic restricted zones, fairways, etc) are assumed stationary and are defined by polygons – in a similar manner to that used in creating the electronic maps. When sailing in the stationary environment, the own ship meets other sailing strange ships/targets (some of which constitute a collision threat).

It is assumed that a dangerous target is a target that has appeared in the area of observation and can cross the estimated course of the own ship at a dangerous distance.

The actual values of this distance depend on the assumed time horizon. Usually, the distances of 5-8 nautical miles in front of the bow, and 2-4 nautical miles behind the stern of the ship are assumed as the limits for safe passing. In the evolutionary task, the targets threatening with a collision are interpreted as the moving dangerous areas having shapes and speeds corresponding to the targets determined by the ARPA system.

The problem of collision avoidance consists of plotting a path ***P***, as a part of a route the ships travels from current position (starting point p_0 (x_0, y_0)) to the destination point p_e (x_e, y_e). The path is composed of linear segment sequences p_i $(i = 1, ..., n)$, connected with turning nodes (x_i, y_i). The start and destination points are chosen by the operator. Considering this, a path ***P*** is feasible (is a part of the save route set) if each of its segments p_i $(i = 1, \dots, n)$ remains within the boundaries of the environment and does not cross with neither dynamic nor static obstacles. The paths which cross the restricted areas generated by the static and dynamic constrains are considered unsafe or dangerous paths (target 1, point *PPK* (x, y), Fig. 1) [1].

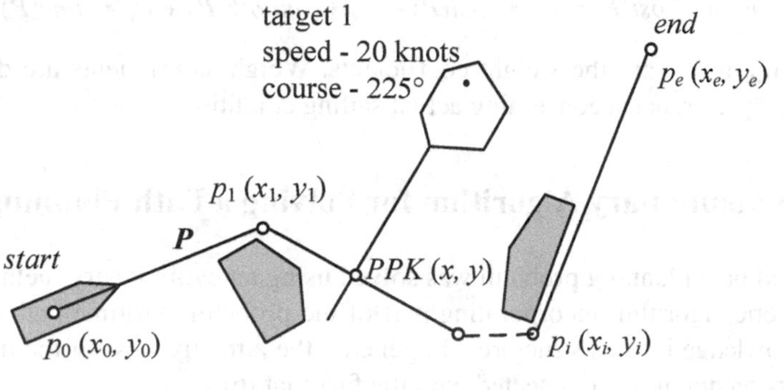

Fig. 1. Potential collision scenario

The safety conditions are met when the trajectory does not cross the fixed navigational constraints nor the moving areas of danger. The actual value of the safety cost function is evaluated as the maximum value defining the quality of the turning points with respect to their distance from the constraints.

The level of adaptation of the trajectory to the environment determines the total cost of the trajectory, which includes both the safety cost and that connected with the economy of the ship motion along the trajectory of concern. One can set the algorithm to weighed the particular element of the path cost so that i.e. the safest paths will be evaluated as the ones with the highest fitness even though they will at the same time represent the longest route. The level of adaptation of the trajectory to the environment determines the total cost of the trajectory, which includes both the safety cost *Safe_Cost*(***P***) and that connected with the economy *Econ_Cost*(***P***) of the ship motion along the trajectory of concern. The total cost of the trajectory (fitness function) is defined as [5]:

$$Total\_Cost(\boldsymbol{P}) = Safe\_Cost(\boldsymbol{P}) + Econ\_Cost(\boldsymbol{P}) \quad (1)$$

The safety conditions are met when the trajectory does not cross the fixed navigational constraints, nor the moving areas of danger. The actual value of the safety cost function *Safe_Cost*(***P***) is evaluated as the maximum value defining the quality of the turning points p_i with respect to their distance from the constraints:

$$Safe\_Cost(\boldsymbol{P}) = w_c * clear(\boldsymbol{P}) \tag{2}$$

where: $clear(\boldsymbol{P}) = max_{i=1}^{n} c_i$, w_c is the weight coefficient, c_i is the length difference between the distance to the constraint (the closest turning point p_i) and the safe distance d.

The trajectory cost connected with the economic conditions *Econ_Cost*(***P***) includes: the total length of the trajectory ***P*** consisting of n line sections p_i, the function of the maximum turning angle between particular trajectory sections at turning points p_i, the time needed for covering the trajectory ***P***. The total cost of the trajectory adaptation to the environment, resulting from the economic conditions, is equal to:

$$Econ\_Cost(\boldsymbol{P}) = w_d * dist(\boldsymbol{P}) + w_s * smooth(\boldsymbol{P}) + w_t * time(\boldsymbol{P}) \tag{3}$$

where: w_d, w_s, w_t are the weight coefficients. Weight coefficients are determined by navigator preferences, concerning actual sailing conditions.

3 Evolutionary Algorithm for Solving a Path Planning Problem

Presented path planning problem was solved using an evolutionary method. The EP/N is a genetic algorithm incorporating part of the problem maritime path planning specific knowledge into its structures. In general, the attractiveness of the use of the evolutionary techniques is connected with the fact that [6]:

- random search is believed to be the most effective in dealing with NP-hard problems and in escaping from local minima;
- parallel search actions not only secure high speed but also provide opportunities for interactions between search actions, all this acting in favour of better efficiency of the optimization;
- intelligent behaviour can be treated as a composition of simple reactions to a complex world;
- a planner can be much more simplified, and still much more efficient and flexible, and increase the quality of the search if it is not confined to the action within a specific map structure;
- it is better to equip the planner with the flexibility to change the optimization goals than the ability to find the absolute optimum solution for a single particular goal.

The EP/N realizes all the above ideas by incorporating part of the problem specific knowledge into the evolutionary algorithm. What is evenly important and not quite obvious, due to the unique design of the chromosome structure and genetic operators the EP/N does not need a discrete map for search, which is usually required by other planners. Instead, the EP/N "searches" the original and continuous environment by generating paths with the aid of various evolutionary operators. The objects in the environment can be defined as collections of straight-line "walls". This representation refers both to the known objects as well as to partial information of the unknown objects obtained from sensing. As a result, there is little difference for the EP/N between the off-line planning and the on-line navigation. In fact, the EP/N realizes the off-line planning and the on-line navigation using the same evolutionary algorithm and chromosome structure.

A crucial step in the development of the evolutionary trajectory planning systems was made by introducing the dynamic parameters: time and moving constraints (υEP/N++). Chromosome consists of path nodes (turning points), that are described by own vessel course, speed and coordinates of actual position. In the evolutionary algorithm used for trajectory planning eight genetic operators were used, which were: soft mutation, mutation, adding a gene, swapping gene locations, crossing, smoothing, deleting a gene, and individual repair [6].

Further work on the program was focused on introducing the library of genetic algorithm components: GALib [10]. Application of the GALib library allowed to easily change the evolutionary algorithm parameters such as fitness function scaling [11], [12], [13] or selection schemes [14]. Based on the built-in library mechanisms: the research on the dedicated to the path planning problem niching mechanism have been undertaken [15], a new specialized genetic operators have been introduced [16], [17], different variants of evolutionary algorithms have been used [18] and a concept of evolutionary hierarchical agent decision support system for marine traffic coordination has been developed [19], [20].

4 Selection Pressure

4.1 Selection Pressure Definition

Evolutionary algorithm operates based on processing population members. The algorithm consists of the following phases: pre-selection (reproduction), genetic operations and post-selection (succession) [21]. Pre-selection is responsible for choosing population members for the temporary population $\boldsymbol{T}^t$. The members with highest fitness score are favoured. The temporary population created in pre-selection process is processed using genetic operations. Child population $\boldsymbol{O}^t$ is created by using genetic operations on the temporary population $\boldsymbol{T}^t$. The post-selection phase creates a new base population $\boldsymbol{P}^{t+1}$. This process chooses individuals both from child population $\boldsymbol{O}^t$ and previous base population $\boldsymbol{P}^t$.

Evolutionary algorithm's attribute to improve the average fitness score of the base population is called the selection pressure (4). Increasing the selection pressure increases the amount of the copies of the best individual in the new base population. Post-selection defines the way of creating a new base population.

$$Selection\ pressure = \frac{\frac{\sum_{i=1}^{popSize} Total\_Cost(\boldsymbol{P}_i)}{popSize}}{Total\_Cost(\boldsymbol{P}_{best})} \tag{4}$$

where: $\boldsymbol{P}_i$ - i path in the population; $\boldsymbol{P}_{best}$ - best path in generation; $Total\_Cost(\boldsymbol{P}_i)$ - fitness of path $\boldsymbol{P}_i$; *popSize* - size of the population.

4.2 Post-Selection Methods

Post-selection with complete replacement. Post-selection with complete replacement (Fig. 2a) removes all the members of the previous generation. The new base population consists of all the individuals of the child generation. This post-selection does not introduce selection pressure. Simple Genetic Algorithm (SGA) [2], [10] is an example here.

Post-Selection with Partial Replacement. The process of creating a new base population in partial replacement post-selection replaces some members of the previous generation with individuals from the child population (Fig. 2b). The amount of the replaced individuals is chosen by the user and is contained between 0 – 100%. Removal of the individuals from the previous generation can be processed using the following schemas:

- removal of the worst adapted individuals;
- removal of the best adapted individuals;
- removal of random individuals;
- removal of the parent individuals.

Incremental Genetic Algorithm (IGA) [10] is an example here.

Elite Post-selection. Elite post-selection (Fig. 2b) is a variant of the partial replacement post-selection. n best individuals from the child population are being selected ($n \in < 0, m)$, m - population size). Afterwards the chosen individuals are added to the new base population where they are sorted. To maintain fixed size of the base population n worst individuals are being removed. The process ensures the survival of at least one best individual. Steady-State Genetic Algorithm (SSGA) [10] is an example here.

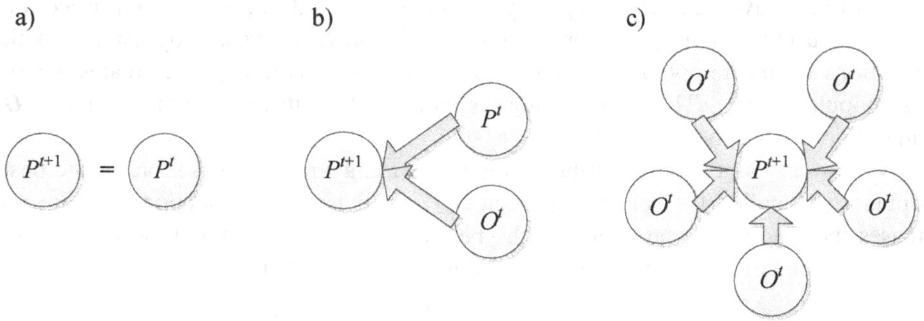

Fig. 2. Post-selection schemas in the evolutionary algorithms

Elite Post-selection in a Multi-Population Algorithm. This post-selection is based on the elite post-selection. The base population role is played out by the elite population (Fig. 2c). In each evolutionary independent population n best individuals are chosen. Those individuals are introduced to the base population. In the final step the worst individuals are removed to keep fixed population size. Distributed Genetic Algorithm (DGA) [22], [23] is an example here.

5 Simulations

To analyze the post-selection variants vEPN++ was used. A collision favouring environment used in this algorithm is presented on Fig. 3a. Target ships were set with the following parameters:

- target 1 – speed 10 knots, course 90°;
- target 2 – speed 11 knots, course 270°.

Simulations were performed for 4 different types of the post-selection variants for four options of evolutionary algorithms:

- Simple GA (SGA);
- Steady-State GA (SSGA);
- Distributed GA (DGA);
- Incremental GA (IGA).

For IGA the following parameters were chosen for simulation:

- removal of the worst adapted individuals;
- removal of the best adapted individuals;
- removal of random individuals;
- removal of the parent individuals.

For SSGA and DGA simulations were performed with replacing 6 and 15 individuals. Regardless of test case, the following parameters of the evolutionary algorithm were chosen:

- population size – 30 individuals;
- number of populations (DGA) - 3;
- crossover probability – 0.8;
- mutation probability – 0.15;
- ranked selector;
- generation number – 400.

10 random initial populations were chosen for the simulations. They were processed with each of the post-selection variants. Example solution is presented on Fig. 3b. The achieved mean results present the dependence between fitness score (F) and

generation number (*nGen*). Solid line marks change with the value of the best individual. The dotted line mark changes in the value of the average fitness score. Due to the form of the fitness function (1), (2), (3) and because of the weight values which has been empirically chosen, if the path violated the boundaries and cross with either dynamic or static obstacles the fitness score will exceed 1000.

Based on the definition shown in chapter 4.1, increasing the selection pressure increases the number of copies of the best adopted individual in the new base population. In accordance with the above, the ratio between the fitness of the best adapted individual and the average fitness in the population is considered the measure of the selection pressure.

Fig. 4 presents the diagram of dependence between best fitness score, average fitness score and generation number for SGA. The Simple Genetic Algorithm removes all the members of the previous generation and replace them with individuals from new temporary population. For this reason, there is no selection pressure. The waveforms of the best and average fitness score are independent from one another.

SSGA waveforms are presented on Fig. 5a (6 individual replacement) and Fig. 5b (15 individual replacement). In both cases it can be seen that high selection pressure leads to loosing population variety after 10 populations. Post-selection method which has been used provide that the algorithm is convergent but it causes the loss of the ability to search the solution space and settlement solutions within the local optimum.

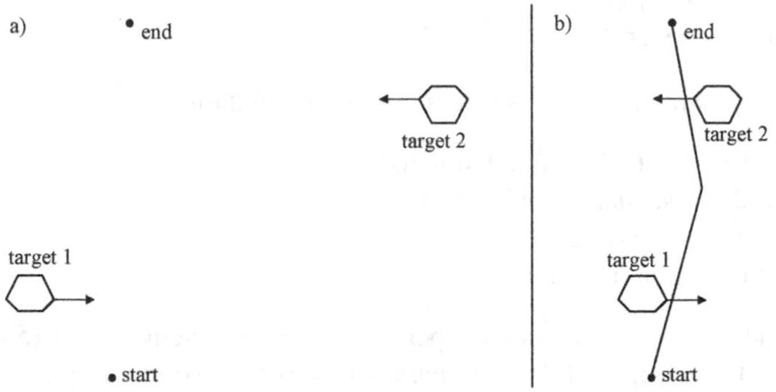

Fig. 3. a) Research environment b) Example solution after 400 generations

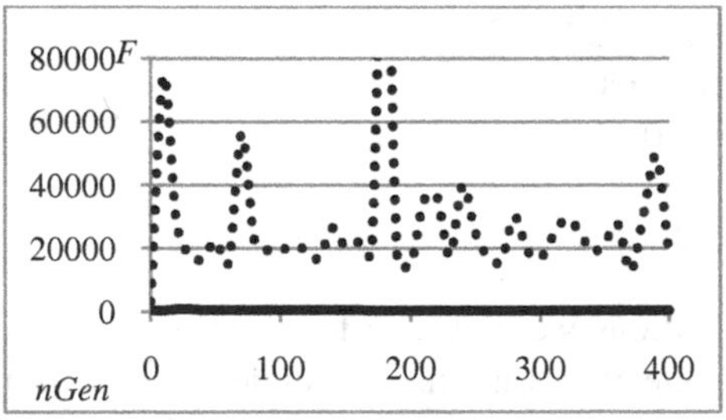

Fig. 4. SGA results

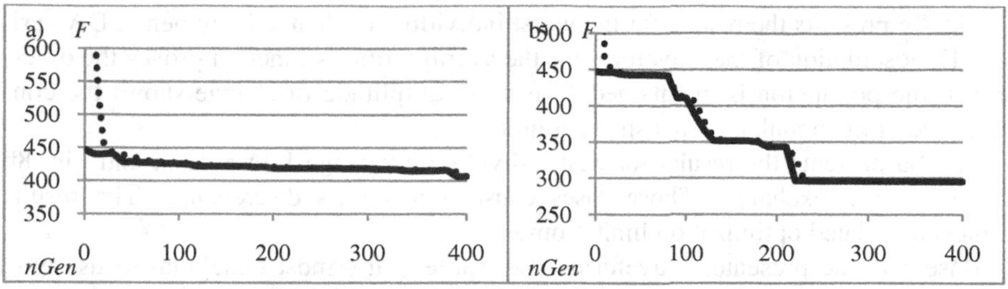

Fig. 5. a) SSGA with 6 individuals replacement results, b)SSGA with 15 individuals replacement results

DGA waveforms are presented on Fig. 6a (6 individual replacement) and Fig. 6b (15 individual replacement). Overlapping waveforms of the average and best fitness score are a result of high volume of the selection pressure. The same kind of post-selection as for Steady-State GA is applied in this case but extended for the multi population. Using multitude populations allows to maintain a variety of the solutions. In contrast to the Steady-State GA this results in a compromise between the exploration and exploitation phase. As a result, the algorithm maintains the ability to search the solution space for whole simulation allowing for continuous improvement of solution.

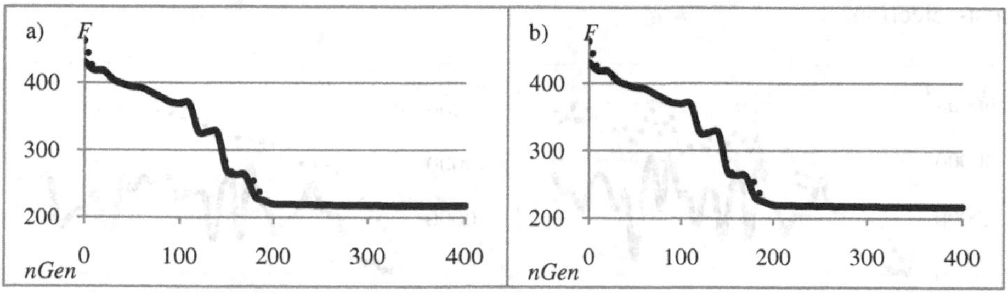

Fig. 6. a) DGA with 6 individuals replacement results, b) DGA with 15 individuals replacement results

Following waveforms consider Incremental GA. Fig. 7a shows the results for the random individual exchange variant. Smaller variation of the average fitness score were observed though.

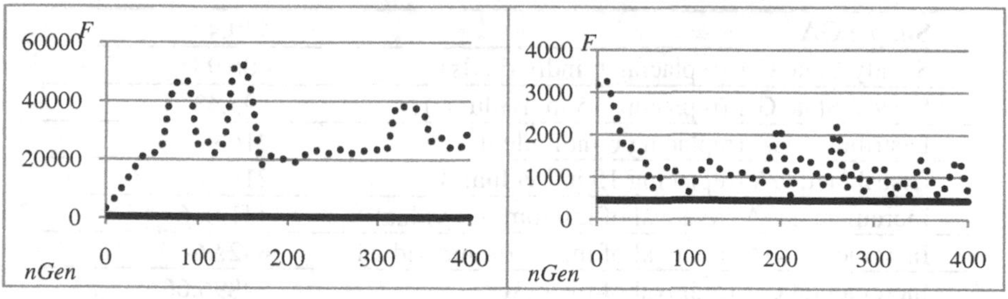

Fig. 7. a) IDA with random individual exchange, b) IDA with worst individual exchange

Fig. 7b presents the results for the worst individual exchange Incremental DA variant. The oscillation of the waveform for the average fitness function proves the diversity of the population is maintained. The small amplitude of change shows the convergence of the population to a single solution.

Fig. 8a presents the results for best individual exchange IDA variant and Fig. 8b for the parent exchange. Those cases cause algorithm's discrepancy. The results achieved violated optimization limitations.

Based on the presented waveforms and Table 1 it's most beneficial to use elite post-selection (Steady State GA) with replacing 15 individuals or its multi-population variation (Distributed GA). Both cases allow the algorithm to converge to a feasible solution and introduce strong selection pressure.

Using the multi-population algorithm additionally allows to balance the proportions between exploration and exploitation phases, eliminating the negative effect of the selection pressure (quick expiry of the possibility of search on the wide spectrum of solutions) and most importantly gives the best available solution (best fitness value = 216.96). Using post-selection with partial individuals exchange (Incremental GA), variants with the best individual exchange and the parent exchange, lead to algorithm's discrepancy. The generated paths cross with dynamic or static obstacles. As a consequence of that the fitness value of best individual in this case significantly exceeded 1000. Other post-selections gave results worse than those given by the elite post-selection.

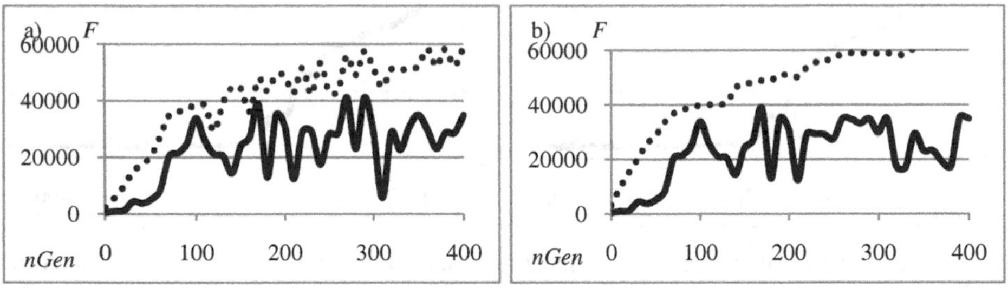

Fig. 8. a) IDA with best individual exchange, b) IDA with the parent exchange

Table 1. Fitness score comparison of all the post-selection variants

Post-selection variant	Best fitness value
Simple GA	449,4
Steady-State GA (replacing 6 individuals)	406,93
Steady-State GA (replacing 15 individuals)	295,43
Distributed GA (replacing 6 individuals)	216,96
Distributed GA (replacing 15 individuals)	216,96
Incremental GA (removal of random individual)	441,066
Incremental GA (removal of the worst individual)	442,966
Incremental GA (removal of the best)	34896,66
Incremental GA (removal of the parent)	34936,66

6 Conclusions

This paper presents the impact of the post-selection variants in the evolutionary path planning. Using vEPN++, four types of evolutionary algorithms were examined. Each algorithm used different post-selection method. In the undertaken research the amount of exchanged individuals and variants of replacement means were taken into account. Based on the results achieved waveform diagrams were presented showing best and average fitness score depending on the generation. Those allowed to determine the selection pressure for each of the variants.

The results presented allow to choose a post-selection method that maintains the population's variety. This is responsible for the exploration phase, which increases the probability of finding a global solution. The control of the selection pressure allows to maintain the compromise between the exploration and exploitation phases. This supports continues search of the solution space and directing the algorithm towards finding a better solution.

References

1. Śmierzchalski, R.: Trajectory planning for ship in collision situations at sea by evolutionary computation. In: Proc. of the IFAC MCMC 1997, Brijuni, Croatia (1997)
2. Goldberg, D.E.: Genetic Algorithms in Search, Optimization, and Machine Learning. Addison-Wesley Longman Publishing Co., Inc., Boston (1989)
3. Michalewicz, Z.: Genetic Algorithms + Data Structures = Evolution Programs. Spriger - Verlang. Spriger (1996)
4. Xiao, J., Michalewicz, Z.: An Evolutionary Computation Approach to Planning and Navigation. In: Soft-Computing and Mechatronics. Physica-Verlag (1999)
5. Śmierzchalski, R., Michalewicz, Z.: Modeling of a Ship Trajectory in Collision Situations at Sea by Evolutionary Algorithm. IEEE Transaction on Evolutionary Computation 4(3) (2000)
6. Śmierzchalski, R., Michalewicz, Z.: Path Planning in Dynamic Environments. In: Innovations in Machine Intelligence and Robot Perception, pp. 135–154. Springer (2005)
7. Yap, C.K.: Algorithmic Motion Planning. In: Schwartz, J.T., Yap, C.K. (eds.) Advances in Robotics. Algorithmic and Geometric Aspects of Robotics, vol. 1, pp. 95–143. Lawrence Erlbaum Associates (1987)
8. Medyna, P., Wiśniewski, B.: Sources and the Structure of the Weather Data Used for the Calculations of the Ship's Route. In: Proc. of the ExploShip, Świnoujście, Poland (2012)
9. Śmierzchalski, R.: Ships' domains as collision risk at sea in the evolutionary method of trajectory planning. In: Computer Information and Applications, vol. II, pp. 117–125 (2004)
10. Wall, M.: GAlib: A C++ Library of Genetic Algorithm Components. MIT (1996)
11. Śmierzchalski, R., Kolendo, P., Jaworski, B.: Liniowe skalowanie funkcji przystosowania w ewolucyjnej metodzie planowania ścieżek przejść. In: Postępy automatyki i robotyki. Part 1, pp. 66–78 (2011)
12. Kolendo, P., Smierzchalski, R., Jaworski, B.: Scaling Fitness Function in Evolutionary Path Planning Method. In: 20th IEEE International Symposium on Industrial Electronics Proceedings, IEEE-ISIE (2011)

13. Kolendo, P., Smierzchalski, R., Jaworski, B.: Experimental research on evolutionary path planning algorithm with fitness function scaling for collision scenarios. In: Methods and Algorithms in Navigation Marine Navigation and Safety of Sea Transportation, Gdynia, Poland, pp. 85–91 (2011) ISBN 978-0-415-69114-7
14. Kolendo, P., Jaworski, B., Śmierzchalski, R.: Comparison of Selection Schemes in Evolutionary Method of Path Planning. In: Jędrzejowicz, P., Nguyen, N.T., Hoang, K. (eds.) ICCCI 2011, Part II. LNCS (LNAI), vol. 6923, pp. 241–250. Springer, Heidelberg (2011)
15. Śmierzchalski, R., Kolendo, P., Kuczkowski, L., Jaworski, B., Witkowska, A.: The niching mechanism in the evolutionary method of path planning. In: Rutkowski, L., Korytkowski, M., Scherer, R., Tadeusiewicz, R., Zadeh, L.A., Zurada, J.M. (eds.) ICAISC 2013, Part II. LNCS (LNAI), vol. 7895, pp. 101–112. Springer, Heidelberg (2013)
16. Kuczkowski, Ł., Kolendo, P., Jaworski, B., Śmierzchalski, R.: Mean Crossover in evolutionary path planning method for maritime collision avoidance. Scientific Journals Martime University of Szczecin 102, 70–77 (2012)
17. Kuczkowski, Ł., Kolendo, P., Jaworski, B., Śmierzchalski, R.: Zastosowanie krzyżowania uśredniającego do ewolucyjnej metody wyznaczania ścieżki przejścia na morzu. Elektronika - konstrukcje, technologie, zastosowania 12, 67–71 (2012)
18. Śmierzchalski, R., Kuczkowski, Ł., Kolendo, P., Jaworski, B.: Distributed Evolutionary Algorithm for Path Planning in Navigation Situation. In: Proc. TransNav, Gdynia (2013)
19. Jaworski, B., Kolendo, P., Kuczkowski, Ł., Śmierzchalski, R.: Evolutionary Hierarchical Agent Decision Support System For Marine Traffic Coordination. In: Proc of 9th IFAC Conference on Manoeuvring and Control of Marine Craft, Arenzano, Italy (2012)
20. Jaworski, B., Kuczkowski, Ł., Śmierzchalski, R.: Extinction Event Concepts for the Evolutionary Algorithms. Przeglad Elektrotechniczny 10b, 252–255 (2012)
21. Arabas, J.: Lectures on Evolutionary Algorithms. Technical and Scientific Publishing, Warsaw, Poland (2004) ISBN 978-8-320-42970-1
22. Tanese, R.: Distributed Genetic Algorithms.. In: Proc. of 3rd Int. Conf. Genetic Algorithms, pp. 432–439 (1989)
23. Belding, T.C.: The Distributed Genetic Algorithms Revised. In: Proc. of 6th Int. Conf. Genetic Algorithms, pp. 114–121 (1995)

Author Index